Studies in Ecological Economics

Volume 6

Series editor
R. Kerry Turner, University of East Anglia School of Environmental Sciences, Norwich, UK

The series "Ecological Economics" aims to present monographs and edited volumes that offer fresh perspectives and overviews on a range of topics in Ecological Economics. Special attention will be given to integrative research methods and techniques and to pluralistic approaches. Fundamental and critical discussions are invited of: theoretical assumptions, ethical starting points, behavioural models, (co) evolutionary change, integration concepts, the ecosystem approach and ecosystem services environmental macroeconomics, industrial ecology, spatial dimensions, thermodynamics and production functions, policy goals and instruments, international policy dimensions, alternative welfare measures, valuation and benefits transfer. Particular emphasis will be placed on the interaction between valuation, modelling and evaluation in a multidisciplinary setting; the link between ecology, biodiversity, ecosystem services, economics politics and environmental management; the incorporation of physical flows in economic models; and the interface between development, poverty, technology and sustainability. In addition, applied and policy oriented research is welcomed, addressing specific resources, substances, materials, regions, sectors, countries or environmental problems. International comparative studies are also encouraged. The ultimate aim of this series is to present a rigorous but broad perspective on contemporary and future environmental policy questions.

More information about this series at http://www.springer.com/series/6526

Stanislav Shmelev
Editor

Green Economy Reader

Lectures in Ecological Economics and Sustainability

Springer

Editor
Stanislav Shmelev
Environment Europe Ltd
Oxford, Oxfordshire, UK

ISSN 1389-6954
Studies in Ecological Economics
ISBN 978-3-319-38917-2 ISBN 978-3-319-38919-6 (eBook)
DOI 10.1007/978-3-319-38919-6

Library of Congress Control Number: 2016953624

Printed on acid-free paper

This Springer imprint is published by Springer Nature
The registered company is Springer International Publishing AG Switzerland

Foreword

Free Thinking Pursuit of Sustainable Development

Since the concept of a 'green economy' first came into widespread use at the Rio+20 Conference in 2012, there has been growing interest in the relationship between economics, the environment and our common future. By 2015, that fledgling concept had matured into growing public and private engagement in securing the transition to an 'inclusive green economy'. Today, it plays a critical role in delivering global commitments to the 2030 Agenda for Sustainable Development and the Paris Agreement on Climate Change.*The Green Economy Reader* emphasizes this evolution by bringing together different perspectives on ecological economics and reflecting on the need to substantially rethink the way we organize and execute economic policy in this age of the Anthropocene, when human development makes an indelible mark on the earth and its atmosphere as never before.

This dynamic dialogue on economics and the environment is largely thanks to the work of the thought leaders who have contributed to this publication, including Herman Daly, Robert Ayres and Tim Jackson. It has been complemented by the expansion of the work of UNEP and its partners in areas such as equitable consumption, critical natural capital and the central role of institutions and governance in shaping market outcomes. And it will continue to evolve as we strengthen our understanding of key drivers and obstacles through initiatives like the UNEP Inquiry into the Design of Sustainable Financial Systems, which looks at the role of capital markets in determining sustainability pathways.

I recommend *The Green Economy Reader* to all free thinkers in pursuit of alternative approaches to delivering an inclusive green economy for a healthy planet and healthy people. By questioning the status quo, the authors depart from 'business as usual' and compel us to consider how social and economic norms can be more deeply aligned with the ecological resilience required to achieve a more equitable and prosperous future for all. I would like to thank them and all of our partners for their determination in the search for sustainable solutions.

Nairobi, Kenya

Achim Steiner

Contents

Contributors

Gar Alperovitz The Democracy Collaborative and Department of Government and Politics, University of Maryland, College park, MD, USA

Robert U. Ayres Economics and Political Science and Technology Management, Management and the Environment, Emeritus, INSEAD, Fontainebleau, France

Terry Barker Department of Land Economy, University of Cambridge, Cambridge, UK

School of Environmental Sciences, University of East Anglia, Norwich, UK

Martin Bruckner Institute for Ecological Economics, Vienna University of Economics and Business (WU), Vienna, Austria

Robert Costanza Crawford School of Public Policy, Australian National University, Canberra, Australia

Herman Daly Emeritus, School of Public Policy, University of Maryland, College Park, MD, USA

Federico Demaria ICTA, Institute of Environmental Science and Technology, Autonomous University of Barcelona, Barcelona, Spain

David Elliott The Open University, Milton Keynes, UK

Joshua Farley Department of Community Development and Applied Economics, and Gund Institute for Ecological Economics, University of Vermont, Burlington, VT, USA

Carol Franco Woods Hole Research Centre, Falmouth, MA, USA

Anthony Friend OIKOS, Ottowa, Canada

Stefan Giljum Institute for Ecological Economics, Vienna University of Economics and Business (WU), Vienna, Austria

Herbert Girardet The Club of Rome, Winterthur, Switzerland

University of the West of England, Bristol, UK

Tim Jackson Centre for the Understanding of Sustainable Prosperity (CUSP), University of Surrey, Guildford, UK

Ida Kubiszewski Crawford School of Public Policy, Australian National University, Canberra, Australia

Stephan Lutter Institute for Ecological Economics, Vienna University of Economics and Business (WU), Vienna, Austria

Joan Martinez-Alier ICTA, Institute of Environmental Science and Technology, Autonomous University of Barcelona, Barcelona, Spain

David W. Orr Oberlin College, Oberlin, OH, USA

Juliet Schor Department of Sociology, Boston College, Chestnut Hill, MA, USA

Stanislav Shmelev Environment Europe Ltd, Oxford, Oxfordshire, UK

Irina A. Shmeleva Institute of Sustainable Development Strategy, St Petersburg, Russia

ITMO University, St Petersburg, Russia

Peter Söderbaum School of Economics, Society and Engineering, Mälardalen University, Västerås, Sweden

Stefan Speck European Environment Agency, Copenhagen, Denmark

Achim Steiner United Nations Environment Programme, Nairobi, Kenya

Leah Temper ICTA, Institute of Environmental Science and Technology, Autonomous University of Barcelona, Barcelona, Spain

Jeroen C.J.M. van den Bergh ICREA, Barcelona, Spain

Institute for Environmental Science and Technology, Universitat Autònoma de Barcelona, Bellaterra, Spain

Faculty of Economics and Business Administration and Institute for Environmental Studies, VU University Amsterdam, Amsterdam, The Netherlands

Peter Victor Faculty of Environmental Studies, York University, Toronto, ON, Canada

Mariana Walter ICTA, Institute of Environmental Science and Technology, Autonomous University of Barcelona, Barcelona, Spain

Roberto Zoboli IRCrES Institute of Research on Sustainable Economic Growth, Catholic University, Milan, Italy

Introduction

The *Green Economy Reader: Lectures in Ecological Economics and Sustainability* is aimed to fill the gap in understanding ecological economics as one of the important foundations for green economy trend in the new economic thinking. The green economy is defined by the United Nations Environment Programme (UNEP 2011) as the economy that results in 'improved human well-being and social equity, while significantly reducing environmental risks and ecological scarcities'. The original term 'green economy' has been introduced at the end of the 1980s in the famous report by the UK economists led by David Pearce, 'Blueprint for a Green Economy' (Pearce et al. 1989). At the same time, green economy ideas have been much influenced by the works of ecological economists, collected in the present volume.

Ecological economics is an interdisciplinary field focused on issues of sustainable development that have emerged in response to the difficulties in solving the global and local environmental problems. Sustainable development is understood as a harmonious process of development, where ecological, social and economic aspects are taken into account. Ecological economics and sustainability science both draw on the multidisciplinary expertise—economics, ecology, physics, environmental sciences, sociology, psychology, complex systems theory, etc.—to address the current challenges the world is facing, including climate change, the loss of biodiversity, water crisis and achieving sustainability.

The concepts, methods and theories of ecological economics have influenced the green economy policy proposal focused on such possible solutions as renewable energy, sustainable waste management, regenerative and sustainable cities and so on. The need for a Keynesian push for a green, smart and creative economy after the recent crisis was identified by the United Nations as early as 2009. A range of alternative paradigms and methods, however, were introduced much earlier: multi-criteria decision aid in the 1960s, material flow analysis in the early 1970s, non-linear dynamics in the 1970s, econometric input-output modelling in the 1980s, the theory of basic human values in the 1980s and alternative measures of progress in the 1990s.

Researchers working in ecological economics propose a range of analytical approaches, largely neglected by the mainstream: analysis of biophysical measures of economic performance complementing the GDP, systems dynamics modelling of sustainability pathways or potentially chaotic events, environmentally extended and econometric input-output analysis and multidimensional sustainability assessment. Some of these approaches offer a more realistic and accurate perspective of economic development than more traditional approaches that assume equilibrium and rational economic behaviour. Although ecological economists were not involved in modelling financial markets, the 'green economy' solution to the financial crisis put forward by the United Nations has been inspired at least to some extent by ecological economics.

Ecological economics has been adopting a critical perspective right from the early years of its existence. At the time when the global CO_2 concentrations in the atmosphere reached an unprecedented level of 400 ppm, half of world's biodiversity has been lost in the past 40 years and over 260 million tonnes of plastics are currently circulating in the world's oceans; it is both justified and necessary. Many ecological economists have embraced recent developments in mainstream economics (among others, by Nobel Prize winners Elinor Ostrom, Daniel Kahneman and Amartia Sen). There is in fact a significant process going on within the traditional economics, aiming at a wide-ranging reform, which has recently manifested itself in the Institute for New Economic Thinking, supported by such prominent mainstream economists as Joseph Stiglitz, David Hendry, Axel Leijonhufvud and Ian Goldin.

This book was inspired by a series of lectures, which became known as the ECI Ecological Economics Lectures.[1] It seemed very important at the time to allow the students and staff of Oxford University to get acquainted with the ideas, concepts and methods of ecological economics; a new conceptual framework emerged at the end of the 1980s in response to the inability of traditional economic approaches to tackle the environmental crisis. The speakers we invited included Professor Robert Ayres, the author of the concepts of material flows analysis, life cycle analysis and industrial ecology, who introduced the concept of 'environmental externality' in 1968 (Ayres and Kneese 1969); Professor Peter Söderbaum, a supporter of an institutional perspective of economy-environment interactions and the author of a concept of a political-economic person (Söderbaum 2000); Professor Jeroen van den Bergh, an author of a seminal paper on the differences between ecological and environmental economics (van den Bergh 2001), who explored issues of ecological-economic modelling, climate change, resource, energy and biodiversity issues and evolutionary economics approaches; Professor Terry Barker, who worked with Nobel Prize-winning economist Richard Stone on the Cambridge Multisectoral Dynamic Model of the British economy and later became the leader in developing EU and global macroeconometric energy-economy-environment input-output models (Barker and Petersen 1988); Professor Martin O'Connor, affiliated with the University of Versailles Saint Quentin-en-Yvelines who explored multi-stakeholder and multi-criteria deliberation techniques in the context of sustainable development; and Professor Beat Burgenmeier, who focused on international interactions

[1] http://www.eci.ox.ac.uk/news/events/ecol-economics08.php

and promoted ecological economic ideas in the French-speaking world. This lecture series has been the main impetus to prepare this volume.

In 2012, Environment Europe started an independent educational initiative, which took form of Oxford Summer and Winter Schools in Ecological Economics, which were held at St Hugh's College, St Hilda's College and Balliol College in Oxford and brought together the leading ecological economics thinkers and participants from over 35 different countries on all 6 continents. The participants represented various United Nations organizations, including UNEP, UNDP and ILO; government ministries, including DEFRA; businesses; NGOs; and numerous universities. The countries taking part in the Oxford Summer and Winter Schools in Ecological Economics included Canada, the USA, Mexico, Costa Rica, Ecuador, Peru, Colombia, Brazil, Argentina, Morocco, Ghana, Nigeria, Iceland, Portugal, Spain, the UK, France, Sweden, Norway, Italy, Switzerland, Austria, the Netherlands, Belgium, Denmark, Germany, Czech Republic, Poland, Latvia, Bosnia, Greece, Lebanon, India, China, Thailand, Taiwan, and Australia. The Summer and Winter Schools are focused on new interdisciplinary science of ecological economics and cover issues of macroeconomic sustainability, sustainable urban development, climate change mitigation, renewable energy, ecosystems and the economy and analytical methods and tools that help to make better decisions taking environmental, social and economic aspects into account.

Professor David Elliott, the most prominent UK advocate of renewable energy (Elliott 2013); Dr Stefan Speck, focusing on the green economy issues at the European Environment Agency; and Professor Herbert Girardet, the leading proponent of regenerative cities and sustainable urban development (Girardet 2014), all taught at the Oxford Summer and Winter Schools in Ecological Economics organized by Environment Europe. Professor David Orr, one of the key environmentalists in the USA has been invited to speak within the Oxford University Environmental Change Institute energy lectures in 2009. Dr Stefan Giljum presented the global database on material flows at the Environmental Change Institute in 2009. Professors Joan Martinez Alier and Professor Robert Costanza who have been both Presidents of the International Society for Ecological Economics have spoken at the Environmental Change Institute on other occasions. We are particularly privileged to welcome the contribution by a group of authors led by Robert Costanza focused on the sustainable pathway recommendations for the Kingdom of Bhutan, which was co-authored by Professor Herman Daly, the author of the Index of Sustainable Economic Welfare, which has now been incorporated in the Sustainable Development Goals (Daly and Cobb (1989); Professor Tim Jackson, the author of *Prosperity without Growth* (Jackson 2011); Professor Peter Victor, the author of *Managing Without Growth: Slower by Design, Not Disaster* (Victor 2008); and other colleagues.

The present volume brings together the authors of high international calibre and excellent reputation. Their past works included Orr (1992), Söderbaum (2000), Söderbaum (2008), Ayres (1997), Ayres and Warr (2010), Shmelev (2012), Jäger et al. (2009), Barker et al. (1994); van den Bergh and Bruinsma (2008), Girardet (2008), Girardet (2014), Elliott (2009), Martinez-Alier (2003), Healy et al. (2012),

Shmeleva (2006), Costanza et al. (1997), and R. Costanza, L.J. Graumlich and W. Steffen (2006). We sincerely hope that scholarly work listed above will provide a good foundation for the present volume.

The book is organized in three parts: Part I. Ecological Economics – Alternative Perspectives (Orr, Söderbaum, Ayres, Daly, Friend, Shmelev); Part II. Ecological Economics of Natural Resources, Climate Change and Renewable Energy (Speck, Giljum, Girardet, Shmelev, Barker, van den Bergh, Elliott); Part III. Ecological Economics of Social Change (Martinez-Alier, Shmeleva, Costanza et al.). Such a structure illustrate a wide interdisciplinary scope of the contributions, which can be explored as additional reading with graduate courses in Ecological Economics, Green Economy, Sustainability and new economic paradigms.

In Chapter 1, David Orr sets the scene by presenting the current state of affairs in the development of world economy viewing it through a prism of 'long emergency', caused by combustion of fossil fuels, which powered an economic expansion for the last two centuries. He questions the 'progress' in such a context, highlighting the significant costs and risks, associated with such a development pattern, in particular, the rapid accumulation of CO_2 in the atmosphere. Professor Orr points to the potential cause of the situation that emerged in the mismatch between the economic theory, pioneered by Adam Smith in 1776 and the 'rules that guide earth systems', which are 3.8 billion years old. He attributes the causes of the long emergency to a massive political failure associated with the overarching power of the fossil fuel vendors in the global political economy and argues for a rapid transition to renewable energy and energy efficiency.

In Chapter 2, Peter Söderbaum introduces the Scandinavian tradition of institutional economics as an alternative, non-orthodox branch of economics, well suited to address sustainable development issues. He questions the role of science and economics in dealing with the existing development trends, which are unsustainable in more ways than one, for example, concerning climate change, loss of biological diversity, loss of fish stocks and risks of nuclear accidents. Professor Söderbaum asserts that economics claims to supply a conceptual framework and theory for efficient resource allocation at various levels; at the level of individuals, business corporations and society. At the same time, he points out the fact that university departments of economics educate students in one way, nationally and globally, the so-called neoclassical economics. This theory can offer some ideas about how to deal with sustainability issues. But neoclassical theory has been dominant in a period when serious problems related to sustainability have emerged. It is therefore probably wise to also consider alternatives to neoclassical theory, such as institutional economics. In this chapter, Peter Söderbaum suggests a way of opening up economics to make the field more compatible with democracy and argues that economics has to move from the present monism to pluralism and become more sensitive to value or ideological issues in present society.

In Chapter 3, Professor Robert Ayres explores the 'failure' of neoclassical economics to deal with energy. He argues that useful energy (exergy) is an essential ingredient for economic activity, especially growth despite the widely held convention of the neoclassical theory that the only two 'factors of production' are capital and human labour and that raw materials are 'produced' by some combination of capital and labour. Professor Ayres emphasizes that energy must be a third factor of production. He argues that the process of decarbonization will employ a lot of capital and labour, which will promote growth, thereby stimulating the development of green economy through green growth.

In Chapter 4, Prof. Herman Daly presents extensive and detailed arguments for going beyond GDP in measuring progress of societies towards sustainable development. His work has been highly influential since the publication of the books *Toward a Steady State Economy* (1973), *For the Common Good* (1989) and *Beyond Growth: Economics of Sustainable Development* (1997). GDP has been misused and should not be employed as a measure of welfare, argues Herman Daly, much in the same spirit as Simon Kuznetz, the author of GDP concept, has been warning earlier. In this chapter, Herman Daly argues against desirability and possibility of infinite growth in the absence of physical limits and draws our attention to the limits of substitution between the factors of production.

In Chapter 5, Dr Stanislav Shmelev offers an alternative to the macro-sustainability assessment methodologies based on a single integrating index approach in the form of a new macro-sustainability assessment methodology applicable at the national scale. Such a methodology based on multi-criteria decision aid is applied to a range of world economies: resource-rich large countries such as Brazil, China and the USA and dynamic developed economies of France, the UK and Germany. The use of multi-criteria decision aid tools for sustainability assessment allows a focused consideration of trade-offs among economic, social and environmental indicators. The results presented in this chapter reveal the new taxonomy of countries in relation to sustainability issues: the countries that reached a relative degree of harmony between their economic, social and environmental policies and the countries where economic development happens at the expense of the environmental or social degradation and a so-called turning point is observed.

In Chapter 6, Anthony Friend proposes a system of entropy accounting, which could be a uniting element in joining the description of economic and physical processes in economic systems. Exploring the writings of early economists William Petty, Adam Smith, David Ricardo and Alfred Marshall; more contemporary writings of Joseph Schumpeter, Pierro Sraffa and Philip Mirowski; and the works of ecological economists Kenneth Boulding and Nicholas Georghescu-Roegen, he discusses the Flow-Fund model and discusses the importance of entropy in economic processes.

Chapter 7, written by Dr Stefan Speck and colleagues from the European Environment Agency, outlines the green economy success stories in the EU. Doctor Speck discusses the enabling policies and factors, such as eco-innovation, the open circulation of green knowledge and the availability of financial resources for investing in the long-term transition as well as fiscal reforms and carbon pricing schemes in the context of EU policies. The specific policy focus and the perspective of the European Environment Agency make this chapter particularly interesting for policymakers.

In Chapter 8, Dr Stefan Giljum outlines a new set of resource use indicators, which could contribute towards increasing resource efficiency at different levels: products, companies, regions or countries. Based on a brief review of the current state of the art of resource use indicators, this paper suggests a new set of resource use indicators, covering the cores resource input categories of materials and water and land area, and includes the output category of greenhouse gas emissions. It can be regarded as a general indicator framework, based on which more specific indicators can be calculated.

Chapter 9 by Professor Herbert Girardet proposes a concept of 'regenerative cities', seeking to address the relationship between cities and their hinterland and beyond that with the more distant territories that supply them with water, food, timber and other vital resources. He argues that we need to re-enrich the landscapes on which cities depend, and this includes measures to increase their capacity to absorb carbon emissions. Creating a restorative relationship between cities, their local hinterland and the world beyond means harnessing new opportunities in financial, technology, policy and business practice. Professor Girardet argues that the established horizon of urban ecology should be expanded to include all the territories involved in sustaining urban systems. Urban regeneration thus takes on the meaning of eco-regeneration.

Chapter 10 by Dr Stanislav Shmelev explores the world's most dynamic megacities from the point of view of sustainability. It considers London, New York, Hong Kong, Los Angeles, Sao Paolo, Rio de Janeiro, Paris, Berlin, Singapore, Shanghai, Sydney and Tokyo and applies multi-criteria decision aid tools to assess urban sustainability performance and to compare these cities on the range of dimensions. The tools chosen for this assessment are ELECTRE III, NAIADE and APIS. The results have shown that Singapore dominates the sustainability rankings in most multi-criteria applications, showing particular strength in economic and environmental dimensions and a slightly less strong performance in the social dimension according to the APIS results. The chapter explores innovative sustainability strategy and new governance structures in Singapore and discusses the reasons for such success.

In Chapter 11, Professor Terry Barker offers an unorthodox analysis of climate change economics and policy. He argues that the mainstream thinking about climate change has shifted with the Stern Review from a single-discipline focus on cost-benefit analysis to a new interdisciplinary risk analysis. Professor Barker asserts that the traditional equilibrium approach fails to provide an adequate understanding of observed behaviour either at the micro or macro scale. This chapter sets out four issues of critical importance to the new conclusions about avoiding dangerous climate change, each of which has been either ignored by the traditional literature or treated in a misleading way that discounts the insights from other disciplines: the complexity of the global energy-economy system (including the poverty and sustainability aspects of development), the ethics of intergenerational equity, the understanding from engineering and history about path dependence and induced technological change and finally the politics of climate policy.

Chapter 12, written by Professor Jeroen van den Bergh criticizes existing cost-benefit studies justifying the necessity of climate change mitigation actions and offers 12 conceptual reasons to justify the beneficial nature of climate mitigation investment. This chapter will argue instead that safe climate policy is not excessively expensive and is indeed cheaper than suggested by most current studies. To this end, climate cost-benefit analysis and policy cost assessments are critically evaluated, and as a replacement, 12 complementary perspectives on the cost of climate policy are offered.

In Chapter 13, Professor David Elliott looks at the way renewable energy technology has been developed in the UK in the context of its overall response to climate change. In particular, it highlights the impacts that differing views on the role of market competition have played. It argues that the market-oriented approach to the support of renewable energy adopted by the UK has been a key reason why it has, with some exceptions, been relatively slow to develop its very large renewable energy resource compared to most other EU countries. It suggests that, under present policies, this relatively poor showing may not improve, especially given the UK's strong commitment to expanding nuclear power.

Chapter 14, written by Professor Juan Martinez-Alier, draws on results of the project entitled EJOLT (Environmental Justice Organizations, Liabilities and Trade) focused on the analysis of ecological distribution conflicts across the world. The chapter offers comparative data on India and Latin America, and also for some variables on Africa and Europe, exploring the links between increases in the social metabolism and the appearance of ecological distribution conflicts. Professor Martinez-Alier also analyses successful resistance movements led by environmental justice organizations and the 'valuation languages' deployed by them.

In Chapter 15, Professor Irina Shmeleva explores the changes of values in modern society. Analysis of theoretical insights and empirical research has shown that human values determine the formation of ecological consciousness, pro-environmental behaviour, the behavioural change and finally the implementation of sustainable development policy. During the last decades, interest in the study of environmental values, applying S. Schwartz methodology, has grown considerably. Universalism has been emphasized as a highly significant component of the value system being in opposition to and conflict with the power values. Universalism corresponds to understanding, tolerance and protection of the well-being of all peoples and nature and historically evolved from the need of individuals and groups to survive. Professor Shmeleva presents the empirical results focused on the fundamental differences in value structures of various stakeholder groups.

Chapter 16, written by Robert Costanza with Gar Alperovitz, Herman Daly, Carol Franco, Tim Jackson, Ida Kubiszewski, Juliet Schor and Peter Victor, presents a synthesis of ideas about what the new sustainable economy could look like and how we might get there. In particular, parts of this chapter were crafted during discussions and consultations with the Royal Government of Bhutan in preparation of its draft outcome statement for its high-level meeting at the United Nations on 2 April 2012. The purpose of this chapter is to lay out a new model of the economy based on the worldview and principles of 'ecological economics', which uses better measures of progress that go well beyond GDP and begin to measure human well-being and its sustainability more directly.

I would like to dedicate this volume to my teachers, many of whom have kindly contributed a chapter.

Oxford, 2016

References

Ayres RU (1997) Turning point: end of the growth paradigm. Earthscan, London, 192pp.
Ayres RU, Kneese AV (1969) Production, consumption and externalities. Am Econ Rev 59(3):282–97
Ayres RU, Warr B (2010) The economic growth engine: how energy and work drive material prosperity, Edward Elgar, 448pp.
Barker T, Petersen W (1988) The cambridge multisectoral dynamic model. Cambridge University Press, 532pp.
Barker T, Ekins, P Johnstone N (1994) Global warming and energy demand. Routledge, 356 pp.
Costanza R, Cumberland J, Daly H, Goodland R, Norgaard R (1997) Introduction to ecological economics, CRC Press, 288 pp.
Costanza R, Graumlich LJ, Steffen W (2006) Sustainabiligy or collapse?: an integrated history and future of people on earth, MIT Press, 518pp.
Daly H, Cobb J, Cobb C. (1989) For the common good: redirecting the economy toward community, the environment, and a sustainable future. Beacon Press, Boston, 482 pp.
Elliott DA (2009) Sustainable Energy: Opportunities and Limitations (Energy, Climate and the Environment), Palgrave, 285pp.
Elliott DA (2013) Renewables: a review of sustainable energy supply options. IOP Publishing, Bristol, 168pp.
Girardet H (2008) Cities, people, planet, John Wiley and Sons, 304 pp.
Girardet H (2014) Creating regenerative cities, Routledge, 216pp.
Healy H, Martinez-Alier J, Temper L, Walter M, Gerber JF (2012) Ecological economics from the ground up. Routledge
Jackson T (2011) Prosperity without growth: economics for a finite planet. Routledge, 288pp.
Jäger J, Bohunovsky L, Giljum S, Hinterberger F, Omann I, Schnepf D (2009) One planet: how much more can Earth take?, Haus Publishing, 224pp.
Martinez-Alier J (2003) The environmentalism of the poor: a study of ecological, Edward Elgar, 328 pp.
Orr D (1992) Ecological literacy: education and the transition to a postmodern world, 210 pp.
Pearce DW, Markandya A, Barbier EB (1989) Blueprint for a green economy. Routledge, 192 pp.
Shmelev SE (2012) Ecological economics: sustainability in practice. Springer, 248 pp.
Shmeleva IA (2006) Psychology of ecological consciousness, St Petersburg University Press, 208 pp.
Söderbaum P (2000) Ecological economics: a political economics approach to environment and development. Earthscan, 224 pp.
UNEP (2011) Towards a green economy: pathways to sustainable development and poverty eradication, www.unep.org/greeneconomy
Van den Bergh JCJM (2001) Ecological economics: themes, approaches and differences with environmental economics. Reg Environ Chang 2(1):13–23
Van den Bergh JCJM, Bruinsma FR (2008) Managing the transition to renewable energy: theory and practice from local, regional and macro perspectives, Edward Elgar, 400 pp.
Victor P (2008) Managing without growth: slower by design, not disaster. Edward Elgar, 288 pp.

Part I
Ecological Economics: Alternative Perspectives

Chapter 1
The Life Required. Political Economy in the Long Emergency

David W. Orr

The cost of a thing is the amount of what I will call life which is required to be exchanged for it, immediately or in the long run. (Henry David Thoreau)

We cannot go on living like this. (Tony Judt)

Abstract The neo-classical economy has assumed the dominant role in global affairs, riding roughshod over political systems and society alike. The present economy is predicated on continual growth, externalizing its full costs, and inequality, and is powered by fossil fuels that are rapidly changing the climate. The ratio of true wealth to 'ilth' has shifted to the latter. The prospects for a durable and fair economy will depend greatly on strengthening democratic controls. It will begin from the bottom up and is evident in the gathering momentum of social movements worldwide.

Keywords Ecology • Ecosphere • Capitalism • Circular economy • Climate change • Democracy • Efficiency • Energy return on investment • Environment • National accounts • Neo-liberalism • Nonviolence • Renewable energy • Resources • Sufficiency • Technology • Utility

It feeds, clothes, houses, entertains, transports, employs, invests, disinvests, and showers all manner of things on those with money. It also poisons and pollutes while making some rich and many others poor and some fat while others starve. It flattens mountains, destroys ecologies, acidifies oceans, destabilizes the climate, creates continental-sized gyres of trash in the oceans, and corrupts democracy. It is creating amazing new ways to displace humans in favor of robots and render our carbon based minds obsolete in favor of those made of silicon (Kurzweil 2005). It provides a multitude of ways to communicate 24/7 while making it less likely that

D.W. Orr (✉)
Oberlin College, Oberlin, OH, USA
e-mail: dorr@oberlin.edu

S. Shmelev (ed.), *Green Economy Reader*, Studies in Ecological Economics 6,
DOI 10.1007/978-3-319-38919-6_1

we can talk sensibly with our neighbors. It entertains and infantilizes perhaps while "consuming itself" (Barber 2007: pp. 3–37; Bell 1976). Both bane and blessing, it is the global, capitalist economy. Three centuries in the making, it has grown into a world-straddling colossus that levels cultural differences from Shanghai to Madrid. It dominates our politics and news. A slight blip up or down in the market can cause mass euphoria or gloom. It is said to have begun in the ancient urge to "truck and barter" and in the turbulent winds of greed, envy, and fear. Sometimes, however, it unleashes more benign forces of creativity, innovation, and philanthropy. It is a vast and incomprehensible machinery of banks, financiers, investors, entrepreneurs, arbitragers, corporations, tax-dodgers, foundations, workers, child laborers, capital flows, government agencies, legislative committees, lobbyists, business schools, professional economists, TV savants, peddling advertisers, compulsive consumers, hidden networks of influence, black marketers, organized crime, cyber-thieves, drug lords, and trillions of dollars of investment capital washing around the world every day in search of a tenth of a percent higher rate of return. Those left behind are a growing and ominous reservoir of discontent. Anyone who purports to understand it, at best, does so only in part and mostly in hindsight. Their various opinions and theories seldom penetrate below the surface of things. Deeper explanations must reckon with the ecosphere of which the economy is a subsystem, with the structure of class and privilege, and the roots of human behavior past our "animal spirits" all the way down to the ancient reptilian brain stem where goblins and ghouls lurk in the shadows. The result is a jerry-built system that lurches from boom to bust exhibiting the extraordinary madness of crowds that seizes on tulips in one time and smart phone apps in another.

The theoretical foundations of modern economics were first described by Adam Smith in "The Wealth of Nations" (Smith 1776). But Smith had previously written a major work about the bonds of sympathy that hold societies together and was working on a new edition of it when he died (The Theory of Moral Sentiments) (Smith 1759). In one he is said to have argued the importance of self-interest; in the other the importance of empathy and sympathy. Whatever one's opinion about what Smith really thought, subsequent generations of economists built the shambling edifice of economic theory on the foundations of self-interest not sympathy, individual not public interest, private wealth not commonwealth, the present not the future, and assumptions of insatiable wants, endless growth, infinite substitution for resource scarcity, know-how over know why, and the chimerical creature called "economic man" who maximizes the foggiest of foggy concepts called "utility," a particle in the social/psychological universe that has never been seen nor its tracks ever detected. One's utility is, therefore, whatever one believes it to be (including, I suppose, having utility in banishing the concept of utility forever). The mainstream neo-classical version of economic theory ignores the laws of thermodynamics, presumes away limits imposed by ecology, devalues nature as mere resources, assumes a model of human nature that would not pass muster among mentally healthy psychologists, advocates behavior repugnant to any ethical ethicist, defies the requirements for adequate safety margins that elsewhere inform good engineering, mixes its description of economic behavior with its prescription for proper behavior, and confuses rationality with our bottomless capacity to rationalize almost anything including the most abhorrent, depraved, ridiculous, idiotic, improbable, and hair-brained things in the catalog of human behavior. With some notable exceptions (such as Arthur Pigou, John Kenneth Galbraith, Robert Heilbroner, and Herman Daly), economic theory from Smith to Milton Friedman works best if the questions are small, the accounting narrow, time horizons short, and its practitioners, true believers (Hoffer 1951). Yet it towers virtually unassailable against legions of critics, including some of the most distinguished economists.[1] Its mathematical models, comprehensible only to the suitably inducted, are otherwise virtually bullet proof against logic, data, bio/physical reality, actual experience, and the consequences flowing from the profession's own modest predictive performance which, some believe to be on a par with readers of chicken entrails, palm readers, and TV weatherpersons. Yet more than any other body of thought economics has come to define us as self-maximizing economic automatons independent of society, not as thoughtful, attentive citizens, or as dutiful community members, caring parents, spiritual creatures, or as the beneficiary of the labors of earlier generations and ancestors of those yet to be born.[2] It has taken shallow to a whole new depth. By the circular logic of self-interest it purports to explain the saintly behavior of Mother Teresa and the bizarre megalomania of, say, Donald Trump thereby conflating self-interest (unavoidable) with selfishness (a choice). It is said to explain everything. Nobel prize winning economist Gary Becker, for example, announced that:

> the economic approach is a comprehensive one that is applicable to all human behavior, be it behavior involving money prices or imputed shadow prices, repeated or infrequent decisions, large or minor decision, emotional or mechanical ends, rich or poor persons, men or women, adults or children, brilliant or stupid persons, patients or therapists, businessmen or politicians, teachers, or students. (Becker 1976, p. 8.)

Professor Becker's revelation comes as a considerable relief to those still vainly laboring in now obsolete fields such as history, philosophy, psychology, politics, sociology, linguistics, theology, and literature. They may now stand down, thereby saving financially stressed institutions large amounts of money and thousands of students the aggravation of becoming "well rounded" when flatness is more efficient for career success. Perhaps the newly unemployed professors will have time to go bowling together. But I digress.

It comes, then, as no surprise that all problems are presumed to be economic and so can be solved only by economic solutions that mostly have to do with selling more of something unneeded to people who can't afford it in order to increase the wealth of those already over-burdened with too much and further accelerate the

---

[1] In Robert Skidelsky and Edward Skidelsky's words Economics is not just any academic discipline. It is the theology of our age, the language that all interests high and low, must speak if they are to win a respectful hearing in the courts of power." They attribute this to "failure of other disciplines to impress their stamp on political debate." (Skidelsky and Skidelsky, 2012, p. 92).

[2] Naomi Klein writes that the belief "that we are nothing but selfish, greedy, self-gratification machines … is neoliberalism's single most damaging legacy." (Klein 2014, p.62).

speed of the treadmill. This is called "neo-liberalism" which is merely turbocharged capitalism. It is not an innocent theory. As David Harvey writes:

> Neoliberalism has, in short become hegemonic as a mode of discourse. It has pervasive effects on ways of thought to the point where it has become incorporated into the common-sense way many of us interpret, live in, and understand the world.

Its main accomplishment has been "to redistribute, rather than to generate, wealth and income" (Harvey 2005, pp 2–3, 159). This, one might suspect, was the intention all along. As Maggie Thatcher once said "Economics is the method, the object is to change the heart and soul" (Klein 2014, p.60). And in contrast to all who believed that the study of economics was merely to elucidate a complicated field of human behavior, it did change hearts, minds, and souls creating a sizeable cadre for whom the abstraction of the market has become scriptural.

More progressive economists admit many of the flaws of economics in theory and practice but believe nonetheless that the Capitalist economy can be remodeled as "green" Capitalism but without the exertion of examining and re-forming its foundational assumptions.[3] A smarter, circular, solar-powered economy would offset the negative aspects of institutionalized greed, perpetual growth, and consumerism. A few smart adjustments at the margin, a policy shift here, better technology plus a change in taxation and viola …sustainability! Most of the proposed changes would be an improvement, of the necessary but insufficient sort; perhaps the economic version of what theologian Dietrich Bonhoeffer once called "cheap grace." Alas, a sterner and less forgiving reality is rapidly forcing us to rethink the basic premises of economics and change the economy to accord more closely with: (1) the way the world works as a physical system, (2) the basic rights of present and future generations, and (3) the obligations that go with being a "plain member and citizen" of the community of life, and do so in large measure for reasons of our own self-preservation if we can find no other. The considerable power of greed without guardrails can drive GDP into the stratosphere and generate technological miracles for a time, but it can also drive civilization predicated on our worst traits over a cliff. The economy organized to promote the seven deadly sins, mass consumption, novelty, and expansion without end is autistic to human needs as distinguished from manufactured wants. It knows nothing of the possibility, however slight of moral improvement, conviviality without consumption, basic fairness within and between generations, and limits whether rooted in the conception of "the good life," or in biophysical reality. On the contrary, economic doctrine, In Robert and Edward

[3] For a thoughtful and useful analysis of a better kind of Capitalism, Lovins and Cohen (2011), is indispensable; Helm (2015) is excellent as well; Better methods of accounting would certainly help as Jane Gleeson-White writes in Gleeson-White (2015), but there is she writes "a logical inconsistency at the heart of the six capitals model which will prevent it alone from saving the planet: it seeks to account for non-financial value but can see it only in terms of financial value. This is because the entity it seeks to govern, the corporation as we know it, is legally bound to make decisions in favour of financial capital." (Gleeson-White 2015: p. 282). Robert Reich finds little evidence that corporations will be "socially responsible, at least not to any significant extent." Further, what often passes for social responsible actions are nothing more than gussied up efforts to reduce costs (Reich 2007: pp 170–171).

Skidelskys' words, "has allowed the profession to maintain an attitude of cheerful indifference to the facts of human psychology" (Skidelsky and Skidelsky 2012, p. 101).

Economic theory, however, did not develop in a vacuum. Rather it co-evolved with business practice and particularly in interaction with increasingly powerful concentrations of capital called corporations. Both theory and practice were thereby shaped by political and judicial decisions that placed the corporation at the center of our burgeoning economic life. From the fossil record we can trace its origins back to the Dutch East India Company and it British cousin, the East India Company. Both existed as licensed monopolies acting in lieu of the State. Subsequent opinions vary widely on the evolution of this hybrid but now dominant creature. On the one hand corporations "allowed society to use markets much more effectively," while on the other they "locked in a societal focus on the market providing material, not social goods" (Colander and Kupers 2014). They can be credited greatly for the material abundance Americans enjoy and equally blamed, among other things, for overflowing landfills. They can be credited for our mobility and thermal comfort and blamed for climate change. They are a source of mixed blessings.

****

In the wake of the economic collapse of 2008, the inestimable Alan Greenspan, Chairman of the Federal Reserve, famously found a flaw in his economic thinking.[4] He suffered "shocked disbelief" and was "very distressed" to discover that foxes make indifferent guardians of financial chicken coops. Greenspan's aha moment was rather like a commercial pilot flying at 35,000' suddenly discovering something interesting about the Law of Gravity that he'd previously overlooked. Neither the passengers nor the chickens would be greatly amused. For Greenspan a stroll through the streets of Detroit or Youngstown a few years earlier might have illuminated theoretical flaws not otherwise visible from the commanding heights atop the Federal Reserve. Of economic theory generally, economist Paul Krugman says that the previous three decades of macroeconomics was "spectacularly useless at best, and positively harmful at worst." (Orrell 2010: p. 106). Even so it is easy to lose sight of how strange and recent the market fetish is in fact. In the late Tony Judt's words:

> much of what appears 'natural' today dates from the 1980s: the obsession with wealth creation, the cult of privatization and the private sector, the growing disparities of rich and poor. And above all, the rhetoric which accompanies these: uncritical admiration for unfettered markets, disdain for the public sector, the delusion of endless growth (Judd 2010: p.2).

It being assumed that this is just the way things are in the best of all possible economies, many became oblivious to the increasingly threadbare public estate[5] and their

[4] *New York Times*. October 23, 2008. p. 1.

[5] That includes bridges, roads, water systems, schools that educate our children, public transportation, and might have also included clean, efficient, high-speed rail systems found to be efficacious for many of the countries that we whupped up on in bygone wars.

increasingly precarious private circumstances and in that state of mind happily embarked on the political equivalent of a lengthy Australian walkabout. In the meantime others busied themselves over "the last three decades methodically unravelling and destabilizing [public institutions]…the dikes laboriously set in place by our predecessors …are we so sure [Judt asks] that there are no floods to come? (Ibid, p. 224). The obvious answer is that floods will come, and we have good reasons to believe that they will be larger than those before, but the public capacity to foresee, forestall, or at least to repair the damage may be only a distant memory of a competent civic culture that once existed at least in part. How did this happen?

One answer has to do with the diminished expectations and performance of those who profess to lead. On the ramparts of the economy one may still behold the captains of finance and business strangely oblivious, in this age of information and science, to the basic truths about how the earth works as a physical system, why such knowledge, whether deployed or not, bears importantly on their management of the commerce of the nation, and why the quaint notion of civic obligation should still rouse their curiosity about possible connections between the two. In the academy, where they were once diligent and perhaps idealistic young scholars, we witness a Byzantine system of training and certification that could only anesthetize and deaden their otherwise burgeoning idealism. That instructional mechanism is further embedded in a vast system of 'knowledge production' that may indeed range between useless and harmful. Perhaps it is a small consolation to laborers in the vineyard of economics that the medical profession itself may not have crossed the breakeven point where by one estimate it made a positive contribution to the health of those it purported to serve until sometime early in the twentieth century (Illich 1976). Before that time a visit to the Doctor actually lowered one's chances of survival. The word is "iatrogenic," meaning "physician induced illness." To its many critics, it is not clear whether or not economics has graduated from of its iatrogenic phase. Leaving that question aside, however, it is a good time to examine the fundamentals of the discipline and what the great economist Joseph Schumpeter once called its "pre-analytic assumptions," or what is now more casually and colorfully called the unseen "elephant in the room," a creature so taken for granted as to remain unnoticed, unremarked, and therefore unstudied, and unaccounted for.

Actually, there is a herd of elephants in the seminar room. Many of the factors that gave rise to the industrial economy, as noted, have changed or no longer exist. Things once presumed true are known to be less true than once believed or altogether false. Most important of these is the belief that fossil fuels are either inexhaustible or can be replaced by something even better. Imaginatively deployed faith, stubborn naiveté, and garden variety corruption have caused those in power to sit on their hands. Consequently, in the half century after the first Oil embargo we still do not have a coherent energy policy which means at this late hour we still have no *de jure* climate policy. So, the question of what will safely power the next economy remains unanswered. Whatever the source(s) they will not likely have the same return on investment or energy density that we got with oil early in the twentieth century. They will also come with costs and consequences yet to be revealed. Other issues of policy have also been deferred, including how we remove from the ledger books—assuming that we do—the portion of fossil fuels that we cannot burn without laying waste to the human prospect on one hand or causing undue hardship on the other (McKibben 2012).

Our energy choices will affect others, including the food system. Americans, for not altogether good reasons, pay comparatively less for food than everyone else on earth. But the true cost is hidden beneath multiple subsidies for energy, land, and capital that keep the costs of food artificially low. As a consequence one calorie of food on the plate requires between 11 and 70 calories of fossil energy to grow, transport, process, refrigerate, and cook.[6] Further, climate instability will cause increasing havoc on farms from drought, heat, flooding, and novel ecological conditions. In Ohio, for example, we should not presume a perpetual flow of foods from elsewhere at prices we can afford and at a volume that we need. Neither should anyone else. Similarly, the reality of growing water scarcity looms darkly over the future of the Southwest and in the Mid-west, where agribusiness is earnestly draining the last half of the Ogallala aquifer dry (Cook et al 2015). But the prospects are even more dismal in many parts of the Middle East, Africa, and South Asia affected by both permanent desiccation and rising sea levels (Kelley et al 2015).

Other, more technical, arcana in the arsenal of economics must also be recalibrated to different and more constraining realities. When the future was confidently thought to be provisioned by an unfailing cornucopia of plenty, investors could discount the likelihood of catastrophes thought to have more or less the same probability as an asteroid falling on Wall Street during its few hours of work. In less beneficent and less predictable times, however, the rate at which such events are "discounted" back to "net present value" requires much rethinking (Price 1993). Economists will long debate how the prospect of ill-tidings should be considered in making long-term financial decisions. Truth be told, no amount of academic quibbling over the appropriate discount rate relative to scarcely imaginable events at some time in the indistinct future can change the reality that the economic implications of climate destabilization are beyond mere human reckoning (Stern 2007, 2015; Nordhaus 2013). But they are not less real or less consequential for being difficult to fathom with tolerable precision in advance of their occurrence. We are entering a period of extreme climate uncertainty that will stress economic theories and business practices honed [fashioned] for less demanding times.

In short, Adam Smith and those dutifully following in his tracks innocently presumed broad and continuing progress, measured by material conditions in tons, acre feet, cubic feet, square footage, sales, and above all, profits. They could confidently expect that such progress would be vouchsafed far into the future by inexhaustible supplies: of fossil fuels, non-fuel minerals, abundant forests, deep soils, unlimited fisheries, sustaining and stable ecologies that offered a full menu of natural services, the expanding technological ingenuity to weave these into growing economies, and, not the least, they could assume a stable climate. These conditions were simply taken for granted as well they should have been given the state of knowledge in their

[6] Beef is the worst …

time and the much smaller scale of the population and the economy.[7] Similarly, they were disposed to assume the superiority of British and Western culture over all others. The proof was said to be in the pudding. They were empiricists, who having tasted the pudding, became optimists of the imperializing sort. Their ideas, culture, technologies, and economy were considered to be timeless, at least for a while.

Economic theory followed in due course. Virtually every assumption of classical and neo-classical economics, pre-analytic or otherwise, is an outgrowth of conditions existing at the dawn of the industrial world. Theory followed facts presumed to be permanent. Reality has a persistently rude habit of making fools of those who presume too much.

****

A great deal has changed. Against what appears to be an increasingly bleak horizon questions arise. We might pause to ask, for example, what useful economic theory might better fit our different circumstances and offer practical guidance for the perplexed? The issue, of course, is not whether we have an economy or not or even economic theory and a discipline called "economics," but rather what kind of economy, operating by what rules, and who qualifies by virtue of what training and experience as a helpful guide on matters pertaining to the economy and its appropriate niche in the wider field of human ecology. It is worth noting as well that previous societies somehow got along tolerably well without a specialized caste of economists. And some that we would deem primitive in all respects provided decent lives for its members entirely without a priesthood of economists and much that would be recognized as economic theory in any respectable economics department (Gowdy 1997; Sahlins 1972). In our far more complex world, however, economists can play a useful if perhaps more muted role than they presently do with, as John Maynard Keynes once proposed, the social standing roughly equivalent to that of dentists.[8]

It would be presumptuous, however, to say what exactly what flag economists should rally around or the level of status they should rightly aspire to. Accordingly, I will eschew overreach and confine myself to what I think is obvious and noncontroversial and so risk violating the other pole of possible critical outrage, that of being tedious and boring. But in truth, the facts, ecological and economic, are well known making it all the more odd that they have not been applied with dispatch as remedies toward making a better and more durable economy. With those misgivings, I suggest an economics and resulting business practices oriented around four well-known principles.

1. The economy is a subsystem of the ecosphere (Daly 2007, 2014a, b) and is thereby bound by its limits and is subordinate to the bio-geo-chemical cycles,

[7] David Ricardo for example once described "the original and indestructible powers of the land alongside other gifts of nature which exist in boundless quantity." Quoted in (Rist 2011: p. 171).

[8] Colander and Kupers similarly believe "it is useful for society to keep some standard economists around to remind society of fundamental rules. But it hardly justifies and entire scientific discipline which should aim at a more comprehensive understanding." (Colander and Kupers 2014; p. 73).

energy flows, and ecological functions that govern the earth and the health of its constituent parts. But the relationship is entirely asymmetrical. The ecosphere has no need or affection for an unruly and ungrateful tenant and could proceed quite well through the subsequent billion years or so without the spindly-legged, big-brained, narcissistic, and perpetually delinquent upstart, that proudly calls itself *Homo sapiens*. The economy, in other words, must conform to the rules set by the larger system in which it is embedded or sooner or later cause its own destruction. The demands of the economy for resources and energy and for absorbing its waste products, including the ~100,000+ chemicals and all of their various combinations, must not exceed what the larger system or its component ecosystems can provide in perpetuity. Further, the larger system is known to be "non-linear," unpredictable, and prone to sudden changes (Barnosky et al 2012). To the alert and prudent, the possibility of nasty surprises would suggest the kind of precaution that keeps wide margins. For an accident-prone, juvenile species it would further suggest sobriety in our interventions in natural systems that we understand only imperfectly. Yet we seem to live in the faith, as biologist Robert Sinsheimer once put it, that nature lays no booby traps for unwary species (Sinsheimer 1978).

The point is that no subsystem can grow carelessly and indefinitely within its larger system without destroying itself and its host. "Perpetual growth," someone once said, "is the ideology of the cancer cell." Nonetheless, the faith in endless economic growth and the continual material expansion on a finite and "full" planet, persists as the keystone myth of our time. The slightest mention of a "steady-state" economy, once proposed by John Stuart Mill in 1848, typically triggers an avalanche of ridicule and the superior disdain of the pedigreed (Mill 1848: pp. 746–751). "The Limits to Growth" published in 1972 is still widely dismissed as irretrievably errant which it is not (Meadows et al 1972; Meadows et al 2004; Turner 2008; Heinberg 2011; Higgs 2014). Beneath the surface of credentialed incredulity I suspect that there is something more at work that critics are loath to confess. As long as economies expand we can defer difficult and contentious issues such as the fair distribution of wealth, the effects of employment, and the content of the things we make and sell to each other. When growth, measured as rising quantity, ends—either because it collides with the finiteness of the earth or because we can no longer manage the rising complexity resulting from the massive scale of the growth economy—we will have to reckon with the many problems associated with the highly skewed distribution of wealth and its relation to domestic tranquility (Tainter 1988; Diamond 2004; Homer-Dixon 2000).

We will also have to reckon with the fact that we are not nearly as rich as we presume ourselves to be. The prices we have been paying seldom reflect the full costs of things purchased. Instead we have been offloading "external" costs onto others in some other place or at some other time (Srinivasan et al. 2008). In Juliet Schor's words: "When we finally and fully tally up the costs of fishery collapse, soil erosion, desertification, wildfires, loss of tropical forests, toxic releases, and a mass extinction of species, the price tag will loom large" (Schor 2010, p. 94). Some costs such as climate destabilization, soil erosion, biological extinctions, and human exploitation beyond some point are simply beyond reckoning. Had we been on a pay as you go plan would we have industrialized differently? Or not at all? Or might we have "developed" in more modest ways? Such is the value of hindsight.

In the meantime all of the economic gauges, dials, and indicators record indiscriminate expansion as if it could go on forever. But it won't and the reasons are well known. The four cubic miles/year of primeval goo dug up and burned each year to miraculously power the industrial economy is also its Achilles' heel. In contrast to all previous civilizations ours is powered by the exploitation of a one-time endowment of fossil fuels—ancient sunlight heated, compressed, and rendered dense and portable by millions of years of geology. The vast scaffolding of the modern growth economy was erected, accordingly, on the flimsy faith that the endowment from the carboniferous age was inexhaustible and further that it could be burned with impunity, that is to say without cost to the health of people, land, wildlife, and waters.[9] Americans accordingly increased their energy consumption 150-fold between 1850 and 1970 (Johnson 2014: p.5). Virtually everything we make, use, eat, wear, build, refrigerate, light, and transport depends on burning fossil fuels. But fossil fuels did much more; they also changed our experience of the world. Night became day. Distance shrunk. Time was compressed. Fossil fuels are lodged in our muscle memory and psyche as the thrill of power and in our sense of space and time as the exhilaration of speed. Combustion changed how we think and what we think about. In some ways it made us a dumber people unable to think clearly about limits and the work of repair, among other things. As historian Bob Johnson writes: "Having inadvertently skipped around the energetic limits to the solar economy, Americans became subsequently vaccinated against talk of ecological constraints." (Johnson 2014: p.12)

It was, however, a Faustian bargain and the devil will have his due. The carbon moved from where geology had safely stored it to the atmosphere where it will cause global havoc for a long time to come. But the deal was coming undone for other reasons as well. Energy analyst Richard Heinberg, for example, has tracked the energy it takes to find, extract, process, and transport energy otherwise called the energy return on investment (EROI). The evidence roughly parallels the declining rate of fossil fuel reserves revealing a marked decline in the energy return on the energy invested (Heinberg 2005). A century ago a hundred units of energy could be extracted for the expenditure of one unit for exploration, drilling, mining, refining, and transport. The EROI, once estimated to be 100:1 now ranges between 5:1 and 25:1 and will continue to fall as energy deposits are discovered farther out, deeper down, and often in places where people don't much like us. Without an outbreak of intelligence we will be stuck to a tarbaby of unsolvable, expensive, and interminable conflicts for decades to come with people who refuse to abide by the Marquis of Queensbury's rules of fair play. In other words, we blew through the easy stuff with great exuberance and now must spend more and more to reach, process, and fight for our access to what's left. Will we run out of oil, gas, and coal? Not likely, but we

[9] A National Research Council panel estimated these to be in excess of $120 billion in 2005, a number that by their admission "substantially underestimates the damages." (National Research Council 2010).

have already exhausted the cheap and easily accessible reserves and what remains will be bitterly fought over (Klare 2012). As historian Bob Johnson writes: "In coming into our energy inheritance, we behaved badly …the modern self is in crisis, and so getting sober and waking up to where we have been during that spree might not be the worst place to start" (Johnson 2014: pp. 173–174).

Could we power our present civilization by hyper-efficiency and sunlight in its various forms? Optimists like Amory Lovins believe it is possible if we are smart enough to operate with consistently flawless competence at the frontier of smarter technological possibilities (Lovins 2011). I want very much to believe this to be true. Ozzie Zehner, on the contrary, writes: "little convincing evidence supports the fantasy that alternative-energy technologies could equitably fulfill our current energy consumption, let alone an even larger human population living at higher standards of living" (Zehner 2012: p. 169).[10] But even if we could, why would we want to? Why would we choose to maintain a standard of living based on waste, frantic overconsumption, ecological ruin, fantasy, multiple addictions, exploitation of the powerless, growing inequality, and most assuredly, perpetual conflicts? In other words, why would we choose to do efficiently and with renewable energy many things that should not be done in the first place?[11] But it may be possible for renewable energy to power a less frenetic society reconfigured around efficiency, social justice, frugality, and the long-term, one as Gandhi once said that met everyone's needs but not their greeds (Trainer 2007). Whether that society would be capitalist or something else altogether remains to be seen.

2. A means, not an end. The purpose of a good economy is to provide and fairly distribute sustenance for living, a means of reliably earning a decent livelihood, provide the infrastructure necessary for the public welfare, increase the competence and capabilities of people, and improve their lives now and in the future. But it would not grow merely for its own sake. A good economy provides basic human needs such as food, shelter, education, security, healthcare, and opportunities for good work as well as things beyond necessities such as art, beauty, kindness, solace, and conviviality for everyone not just for the privileged. It would not be driven by the massive deflection of human consciousness and attention otherwise called advertising that preys on children and ruthlessly exploits our needs for status, affection, and connection (Hamilton 2004, pp. 91, 219). Robert Kennedy once made the same point saying that:

> Gross National Product counts air pollution and cigarette advertising, and ambulances to clear our highways of carnage. It counts special locks for our doors and the jails for the people who break them. It counts the destruction of the redwood and the loss of our natural wonder in chaotic sprawl. It counts napalm and counts nuclear warheads and armored cars for the police to fight the riots in our cities. It counts Whitman's rifle and Speck's knife, and

[10] In either case, "getting off fossil fuels … will be one of the most difficult challenges modern civilization has ever faced, and it will require the most sustained, well-managed, globally cooperative effort the human species has ever mounted." (Wagner and Weitzman 2015).

[11] Wendell Berry writes "We must understand that fossil fuel energy must be replaced not just by 'clean' energy, but also by *less* energy. The unlimited use of *any* energy would be as destructive as unlimited economic growth." (Berry 2015: p. 71).

> the television programs which glorify violence in order to sell toys to our children. Yet the gross national product does not allow for the health of our children, the quality of their education or the joy of their play. It does not include the beauty of our poetry or the strength of our marriages, the intelligence of our public debate or the integrity of our public officials. It measures neither our wit nor our courage, neither our wisdom nor our learning, neither our compassion nor our devotion to our country, it measures everything in short, except that which makes life worthwhile.[12]

Simon Kuznets, the author of our present system of national accounts by which we measure economic performance, noted that "the welfare of a nation can, therefore, scarcely be inferred from a measure of national income" (Rowe 2008). Others, of course, reply that growth directly or indirectly improves the lives of people and up to a point and in some ways it does. But economic growth conceals all manner of contradictory things as well as a large number of goods and services that are deleterious to human well-being and to the larger human prospect. Capitalists, it is said, will sell anything for a profit, but some things such as climate stability, human health and dignity, sacred groves, children's well-being, and grandmothers should not be for sale at any price (Sandel 2012).

A good economy would also facilitate that vague but important indicator of social health called happiness which is notoriously difficult to define and measure. But the U.S. economy, on the contrary, generates high rates of depression, autism, loneliness, violence, and addictions of various kinds which are easier to quantify. By reputable measures happiness in America peaked in the 1950s when people had less stuff and there was far less of it to buy.[13] The rising gap between things possessed and flat-lined happiness is an otherwise inexplicable embarrassment to the tribe of growthers who believe assuredly that these march in lock step. If that were true, along about 9 am on August 7th 1992 when the Mall of America first opened its doors we would have crossed over into the state of perpetual Nirvanic bliss. Alas, that didn't happen and that widening gap says a great deal about the chasm between the needs of the economy and those of flesh and blood human beings.

3. Nonviolence. If the economy is to exist harmoniously within the ecosphere, it cannot do violence to the larger host system without fatal consequences to itself. Neither can a durable and decent economy work by violence in its various forms without harming the people it purports to serve and to the social and ecological conditions that underwrite their well-being. A bit of perspective may help. Our economic ideas originated in a culture with a long history of violence from religious crusades, inquisitions, imperialism, militarization, the making and permanent threat to use nuclear weapons, and continual wars over one thing or another. As a nation we grew rich from the proceeds of violence done to people brought here as slaves and to displaced Native Americans. Some benefited by doing great violence to landscapes, soils, forests, and wetlands. Violence was implicit in the extractive economy that prospered by wrenching wealth from earth and people alike. Violence was also

[12] Speech at Kansas University, March 18, 1968.

[13] Architect Lance Hosey reports that in 1994 "there were half a million different consumer goods for sale in the United States, and now Amazon alone offers 24 million.(Hosey 2012: p. 96).

present at the founding of science. Frances Bacon, the founder of the Royal Academy of Sciences in London, once described the method of science as putting "nature on the rack and torturing her secrets out of her." Ever since our motto has been, as William McDonough puts it, "if brute force doesn't work you're not using enough of it." The modern economy, in particular, depends a great deal on the research, manufacture, and selling of more efficient ways to kill. The defense budget including the cost of wars, "security," surveillance, and 800 or more military bases worldwide exceeds a trillion dollars. Much of the economy depends on that constant, unquestioned, and well-distributed flow of largesse. We glorify violence in our movies, advertisements, politics, and sports, and all too easily overlook the violence happening in our name in places like Guantanamo, Abu Ghraib, and "black ops" sites where unnamed and unrepresented detainees are tortured for reasons that torture morality, logic, law, reason, bodies, and souls alike. That we have become a violent, gun-toting society, however, is no new insight. But violence also permeates the larger culture in less obvious ways. Our meat is mostly raised in animal gulags made efficient and profitable in direct proportion to the suffering inflicted on the confined animals. The application of machine intensive, industrial methods to agriculture lands has cost us upwards of half of our topsoil in the United States. Farms, forests, and private lawns are managed mechanically and chemically which is to say, violently. Modern medicine applies industrial methods to the human body. After a century of promiscuous chemistry every baby born in the U.S. arrives "prepolluted" with several hundred chemicals delivered through the mother's umbilical cord. *(pre cancer panel) We have indeed as Bacon advised in The Great Instauration constrained and vexed nature …"forced [her] out of her natural state, and squeezed and moulded [her] …[in order] to effect all things possible" (Bacon 1620: p. 25).

All the while other possibilities both economic and scientific existed. British economist, E. F. Schumacher writes: "Wisdom demands a new orientation of science and technology towards the organic, the gentle, the non-violent, the elegant and beautiful." His proposal for "Buddhist economics" is premised on "simplicity and non-violence" (Schumacher 1973: pp. 31, 54). Non-violent methods of natural systems farming and forestry are well-known and known to be both practical and profitable. Similarly, biomimicry, the study of how nature makes virtually everything without combustion, toxic chemicals, pollution, and ecological ruin can transform methods of manufacturing. Some of the most promising methods of avoiding disease, healing, and treatment of dissidents represent a merger of Eastern and Western practice and philosophy.

4. Primacy of Politics. The scale, purposes, and content of the economy are not first and foremost economic but political issues about "who gets what, when, and how." The "economy," accordingly, reflects things that occur in the political realm including laws, regulations, and customs by which wealth is created and apportioned. Karl Marx thought otherwise arguing at considerable length and density that the political system is an outgrowth of the economy. With no small irony, many, if not most, neo-classical economists would agree that economics is more fundamental than politics and mostly determines political reality.

The relationship between the economic and political realms is certainly intimate and reciprocal, but if democracy really matters there can be no useful debate about their priority. The rules that govern the economy should be made publicly by properly elected representatives serving an informed electorate, not by an oligarchy meeting behind closed doors. The public should participate in decisions having to do with the distribution of wealth, risk, reward, and the sustainability of the entire human enterprise. They should be made aware of the connection between otherwise technical economic and fiscal decisions, taxation, purchasing, and investment and the social, political, and ecological consequences. To abdicate that responsibility and assign such decisions to the abstraction of the market, as economist Karl Polanyi once said "would result in the demolition of society" (Polanyi 1944: p. 73). It is not only the decisions given over to abstract powers, but a deeper crisis of people caught in an inhuman system. In Vaclav Havel's words:

> A person who has been seduced by the consumer value system, whose identity is dissolved in an amalgam of the accoutrements of mass civilization, and who has no roots in the order of being, no sense of responsibility for anything higher than his or her own personal survival, is a *demoralized* person. The system depends on this demoralization, deepens it, is in fact a projection of it into society (Havel 1990: p. 62).

The upshot is that most economic crises, recessions, depressions, and other perturbations are not accidents or anomalies, but rather the normal working out of the logic built into the rules of the economic system. The logic of unrestrained and minimally regulated capitalism has lead relentlessly to greater concentration of wealth, economic monopoly, and ever larger ecological, social, and political crises. But the system rules that inform the design, structure, and workings of the economy are seldom discussed publicly by economists in ways comprehensible to the lay public. In fact, "proficiency in obscure and difficult language," as John Kenneth Galbraith once noted, "may even enhance a man's professional standing" (Galbraith 1971: p.32). This may explain, in part, the public somnolence in the face of the widening chasm of wealth distribution and the lack of public accountability. By one estimate, for instance, 80 persons worldwide have more net wealth than the bottom 3.6 billion.[14] In the United States it is said that 400 of the wealthiest have more wealth than the bottom 185 million.

Such facts would not have surprised Karl Marx who, with remarkable thoroughness, analyzed the dynamics of Capitalism and its tendency to concentrate wealth in fewer and fewer hands. The result was the enrichment of a diminishing few and the immiseration of the many, a pattern sometimes said to be the precursor of revolution. French economist Thomas Piketty, in similarly exhausting detail, has updated the trends of income and wealth showing that "inequality reached its lowest ebb in the United States between 1950 and 1980" largely because of policies aimed to deal with the Depression and World War II. After 1980, however, inequality "exploded" mostly because of shifts in taxation and finance (Picketty 2014: p.294). Piketty is not a determinist as was Marx. Rather he writes that "if we are to regain control of

[14] Oxfam Issue Briefing, January, 2015, p. 3.

capitalism, we must bet everything on democracy" (Picketty 2014: p. 573). The reason is that inequality eats away at the fabric of society. As Richard Wilkinson and Kate Pickett show in The Spirit Level, virtually every bad social trend from crime to obesity is strongly linked to the unequal distribution of income and opportunity, risk and reward (Wilkinson and Pickett 2009; Lepore 2015). The fact is that the global capitalist economy is trending toward greater concentration of wealth and is therefore increasingly prone to lurch from crisis to crisis compounded by growing public disaffection. It is widely believed that the collapse of the Soviet Union owed to flaws that do not afflict Capitalism. Were that reality be so simple. In fact, Communism and Capitalism share many similarities including their dependence on economic growth and an industrial paradigm rooted in a rationalistic philosophy that regarded the world as so much dead material waiting to be transformed to human use and then discarded without consequence. Both ideologies worship at the same alter, while in dispute about who owns the church. In the larger sweep of history this is a Lilliputian-scale dispute. Certainly different in important respects, the flaws in both systems bear a family resemblance, causing thinkers as different as Vaclav Havel and Jane Jacobs to conjecture that the collapse of Capitalism cannot be far behind.[15]

Others are not so sure. David Harvey, for one, writes that "Capitalism will never fall on its own. It will have to be pushed. The accumulation of capital will never cease. It will have to be stopped. The capitalist class will never willingly surrender its power. It will have to be dispossessed" (Harvey 2010). Maybe so but others like ecologist, Howard Odum argue that "general systems principles of energy, matter, and information are operating to force society into a different stage in a long-range cycle" (Odum and Odum 2001). Either way, the large numbers that drive historical trends—growing human discontent, declining net energy ratios, and ecological ruin—are converging toward a systems change.

****

The problem is not that we do not know about better alternatives to present economic arrangements, but that we don't yet feel them as real possibilities, and we don't feel them as anything more than idle curiosities because we still live in thrall to speed, accumulation, and convenience and believe that with still more clever technology we can get away with it—even as the biophysical and moral underpinnings of the old economy are disintegrating before our eyes. Nonetheless, it is still easier, as someone recently remarked, for us to envision the end of the world than the end of capitalism. The making of a better economy, capitalist or otherwise, hinges not only on our imagination to see what is untried yet possible but more important, our ability to recollect older and sometimes saner ways of doing things that were not so

[15] Jane Jacobs for example writes that "Today the Soviet Union and the United States each predicts and anticipates the economic decline of the other. Neither will be disappointed." (Jackobs 1984, p. 200).

much improved upon as they were elbowed out of existence.[16] But the tried and true is often discounted to yard sale prices and talk of anything before, say, 1995 or any form of communications that works slightly slower than the speed of light is to risk committing an unspeakable heresy. It is, in such times, clarifying to note, as the Welsh anthropologist Alwyn Rees once observed, "when you have reached the edge of an abyss, the only thing that makes sense is to step back" (Kohr 1957: p.221).

At this late hour standing on the edge of an abyss, is it possible to step back and still build a resilient, fair, prosperous and durable economy? The only useful answer is a contingent 'yes.' Contingent because the laws, regulations, tax system, politics, media, advertisers, habits of the herd, presumptions of doctrine, failures of imagination, and the sheer power of money madness stand athwart better possibilities and urgent matters of human survival. Contingent, also, because we Americans, full of self-congratulation as God's anointed, have not yet finally decided whether we wish one day to be known as history's most zealous and prodigious consumers and the earth's most earnest and ingenious wreckers or whether we wish to be celebrated for our art, literature, music, compassion, good humor, good schools, fairness, wise governance, great cities, the loveliness of our countryside, the health of our children, resilient prosperity, foresight, the life in our means of livelihood, and our fidelity to preserve the hard-won gains for human dignity?[17] Whichever we choose our economy, for better or for worse, will be the clearest portrait of who we are as a people and what we might—for better or worse—become. On such matters we remain divided.

Our cultural default is to be "optimistic" and fervently believe that "the future is better than you think" (Diamandis and Kotler 2012). Presumably it will be better because it requires no change in our aspirations and behavior, which is to say no deep improvement in us. The party, in other words, goes on, but with our sins absolved by virtue of "breakthrough" technologies. Such is the optimism on the edge or maybe just whistling past the graveyard near midnight. Pure optimists as distinguished from the merely hopeful, have not a quiver of doubt about our capacity to get out of self-induced, technologically amplified, and increasingly complicated messes. They exist happily in a wonderland of gee-whiz gadgetry that purifies, generates, grows, and makes, and makes money which is, one may presume is the driver in the process. So, "considering the gravity" of our plight," they are loath to "reign in our imaginations," which mostly excludes political, ethical, moral, or even economic imagination. They are prone to celebrate, for example, the ingenuity of Coca-Cola's recent breakthroughs in water conservation even as the Company prospers by depleting groundwater worldwide in order to sell more caffeinated sugared water in aluminum cans to people who need fewer junk calories, much less sugar, but more hydration, nutrition, and local, not corporate, control of their lands, water, and lives. True to form, such optimists categorically dismiss dissenters as "luddites"

[16] Highland clearances, etc.

[17] As the great economist Nicholas Georgescu-Roegen once put it "man's nature being what it is, the destiny of the human species is to choose a truly great but brief, not a long and dull, career (Georgescu-Roegen 1974, p. 304).

who might cause us to backtrack to the Amish level and thereby miss the "vast" benefits "afforded by all this technology" (Diamandis and Kotler 2012). One suspects, the actual history of the Luddite movement remains unread along with the companion volumes on Fukishima, Bhopal, endocrine disrupters, nuclear weapons, as well as those yet to be written about the cozy and sinister relationship between the National Security Agency and its kin, the CIA, and the giant information technology companies. They long for "breakthroughs," and with what must be either world class naiveté or elephantine ignorance, assume that if and when these occur they will have no hidden costs or impose no unforeseen collateral damages. They are technological fundamentalists whose manner of thinking, reminiscent of other fundamentalists, is to offer one solution—more of the same—to all problems including those that are first and foremost deeply human, moral, and political. They do not deal in dilemmas which are not solvable.

Sobriety on such portentous issues is more difficult where things are new and still shiny, and there are few physical ruins and festering psychological wounds that remind the observant of human fallibility, stupidity, and, yes, evil. White America is still too young and too self-assured in the manner of adolescence to have been sufficiently tempered by the memory of its sins, screw-ups, malice, and the thousands of ways things fall apart or technology bites back. Perhaps we are learning in spite of our predilections.

If real improvements economic and other – do come, and I think they will, they will begin without the hallucinations and magical thinking that wishes away our history and overlooks the limits by carrying capacity, complexity, and our own ignorance (Kunstler 2012). If and when improvements happen they will, as they always have, begin at the periphery of power and wealth, in the places and situations that are too small to be noticed or too insignificant to attract organized opposition, rather like the mouse-scale mammals that once scurried between the feet of dinosaurs. In fact, the transition began decades ago in such places: neighborhoods and communities both rural and urban, here and elsewhere. It continues to gather force.[18] But there is no overall strategy across a spectrum that includes people engaged in sustainable agriculture, slow foods, slow money, urban farmers, green builders, wind farmers, solar installers, bikers, inner-city businesses, environmental educators, pioneers, bioneers of all kinds, public servants, communicators, organizers, and those trying to level the economic playing field. In due course, this congregation might transform the larger culture and someday, perhaps, they might cause tectonic political change as well.

I leave it to others better equipped and wiser than I to describe the critical macro changes that must eventually be made at the state and national level in such things as ownership, taxation, investment, finance, and public policy (see e.g. Alperovitz 2004; Alperovitz and Daly 2008; Daly and Cobb 1994; Korten 2010; Schor 2010; Schweickart 2002; Shuman 2015; Speth 2012). But amidst promising shoots and hints of change we will need a compass to clarify our central convictions and serve as a reminder of what really matters. I know of none better than that once proposed

[18] The transition town movement led by Rob Hopkins is a notable example.

by John Ruskin: "the great, palpable, inevitable fact—the rule and root of all economy—that what one person has, another cannot have; and that every atom of substance, of whatever kind, used or consumed, is so much human life spent …for every piece of wise work done, so much life is granted; for every piece of foolish work, nothing; for every piece of wicked work, so much death" (Ruskin 1860). Accordingly a good economy would:

- exist within the limits of human scale (Sale 1980);
- place upper and lower boundaries on wealth;
- distinguish sufficiency from efficiency (Sachs 1999)[19]
- designate times when nothing is bought or sold except to save lives;
- help people grow, not them render hooked, dependent, and gullible;
- include debt forgiveness as a regular economic and human practice;
- provide good work that ennobles and dignifies (Schumacher 1979; Gill 1983);
- tax advertisements and ban those aimed to exploit children[20];
- confiscate all profits from the making and selling of weapons;
- distribute costs, risks, benefits fairly within and between generations;
- require that prices include the full life-cycle costs;
- help us transition from a having culture to a being culture (Fromm 1976)
- protect common heritage of oceans, climate, wildlife, lands, waters (Bollier 2014).

These changes, however, will not and cannot begin in the market place alone or in debates about how best to further self-interest at any scale. They, by which I mean the conditions necessary to the survival of civilization, will require a larger scaffolding of laws, regulations, and understandings that mesh biophysical realities, intergenerational and inter-species morality, cultures, histories, and public engagement—something Herman Daly once described as macro control with micro variability. In other words, a set of ordering is principles necessary, as Blaise Pascal once wrote, "in order to make [human] madness as little harmful as possible."[21]

References

Alperovitz G (2004) America beyond capitalism: reclaiming our wealth, our liberty, and our democracy. Wiley, Hoboken, p 336

Alperovitz G, Daly L (2008) Unjust deserts: how the rich are taking our common inheritance. The New Press, New York

[19] Wolfgang Sachs.

[20] Robert and Edward Skildelsky define advertising as "the organized creation of dissatisfaction." (Skidelsky and Skidelsky 2013: p. 40); Clive Hamilton proposes to "ban advertising and sponsorship from all public spaces and restricting advertising time on television and radio." And further change the tax laws that make it a deductible business expense. (Hamilton 2004, p. 219). Professor Hamilton is exceedingly polite. Dante would have consigned the makers, purveyors, and enablers to one of the lower tiers of Hell.

[21] Blaise Pascal, Pensees.

Bacon, F (1620) The new organon. Bobbs-Merrill Co., Indianapolis 1960
Barnosky AD, Hadly EA, Bascompte J, Berlow EL, Brown JH, Fortelius M, Getz WM, Harte J, Hastings A, Marquet PA, Martinez ND, Mooers A, Roopnarine P, Vermeij G, Williams JW, Gillespie R, Kitzes J, Marshall C, Matzke N, Mindell DP, Revilla E, Smith AB (2012) Approaching a state shift in Earth's biosphere. Nature 486:52–58. doi:10.1038/nature11018
Barber BR (2007) Consumed: how markets corrupt children, infantilize adults, and swallow citizens whole. W.W. Norton, New York, p 406
Becker G (1976) The economic approach to human behavior. University of Chicago Press, Chicago
Bell D (1976) The cultural contradictions of capitalism. Basic Books, New York, p 398
Berry W (2015) Our only world. Ten essays. Counterpoint, Berkeley, p 196
Bollier D (2014) Think like a commoner: a short introduction to the life of the commons. New Society Publishers, Gabriola, p 192
Colander D, Kupers R (2014) Complexity and the art of public policy: solving society's problems from the bottom up. Princeton University Press, Princeton, p 320
Cook B, Ault TR, Smerdon JE (2015) Unprecedented 21st century drought risk in the american southwest and central plains. Science 1:e1400082
Daly H (2007) Ecological economics and sustainable development. Edward Elgar, Cheltenham
Daly H (2014a) Economics for a Full World. Blue Planet Prize speech, November, 2014
Daly H (2014b) From uneconomic growth to a steady-state economy. Edward Elgar, Cheltenham
Daly H, Cobb JB (1994) For the common good. Beacon, Boston
Diamandis P, Kotler S (2012) Abundance: the future is better than you think. The Free Press, New York, p 386
Diamond J (2004) Collapse: how societies choose to fail or succeed. Penguin Group, New York, p 592
Fromm E (1976) To have or to be. Bantam Books, New York
Galbraith JK (1971) Economics, peace, and laughter. Signet Books, New York, p 382
Georgescu-Roegen N (1974) The entropy law and the economic process. Harvard University Press, Cambridge, MA
Gill E (1983) A holy tradition of working. Golgonooza Press, Ipswich
Gleeson-White J (2015) Six capitals: the revolution capitalism has to have – or can accountants save the planet? Allen & Unwin, Crows Nest, p 368
Gowdy J (ed) (1997) Limited wants, unlimited means: a reader on hunter-gatherer economics and the environment. Island Press, Washington, DC, p 378
Hamilton C (2004) Growth fetish. Pluto Press, London, p 288
Harvey D (2005) A brief history of neoliberalism. Oxford University Press, Oxford, p 254
Harvey D (2010) The enigma of capital: and the crises of capitalism. Oxford University Press, New York, p 296
Havel V (1990) Living in truth: 22 essays published on the occasion of the award of the Erasmus prize to Vaclav Havel. Boston, London
Heinberg R (2005) The party's over: oil, war and the fate of industrial. Societies, Gabriola, p 288
Heinberg R (2011) The end of growth. New Society Publishers, Gabriola, p 336
Helm D (2015) Natural capital. Yale University Press, New Haven
Higgs K (2014) Collision couirse: endless growth on a finite planet. MIT Press, Cambridge, MA, p 416
Hoffer E (1951) The true believer: thoughts on the nature of mass movements. Harper & Brothers, New York, p 176
Homer-Dixon T (2000) The ingenuity gap: facing the economic, environmental, and other challenges of an increasingly complex and unpredictable world. Knopf, New York, p 496
Hosey L (2012) The shape of green: aesthetics, ecology, and design. Island Press, Washington, DC
Illich I (1976) Medical nemesis. Bantam Books, New York
Jackobs J (1984) Cities and the wealth of nations: principles of economic life. Random House, New York, p 257
Johnson B (2014) Carbon nation: fossil fuels in the making of american culture. University Press of Kansas, Lawrence, p 262

Judd T (2010) Ill fares the land. Penguin Press, New York
Kelley CP, Mohtadi S, Cane MA, Seager R, Kushnir Y (2015) Climate change in the Fertile Crescent and implications of the recent Syrian drought. PNAS 112(11):3241–3246
Klare M (2012) The race for what's left: the global scramble for the world's last resources. Metropolitan Books, New York, p 320
Klein N (2014) This changes everything. Capitalism vs. the climate. Simon & Schuster, New York
Kohr L (1957) The breakdown of nations. Dutton, New York, p 256
Korten D (2010) Agenda for a New economy: from phantom wealth to real wealth. Berrett-Koehler Publishers, San Francisco
Kunstler JH (2012) Too much magic: wishful thinking, technology, and the fate of the nation. Atlantic Monthly, New York, p 245
Kurzweil R (2005) The singularity is near: when humans transcend biology. Penguin Publishers, New York, 672 pp
Lepore J (2015) Richer and poorer. *The New Yorker*. March 16, 2015, pp 26–32
Lovins A (2011) Reinventing fire: bold business solutions for the new energy era. Rocky Mountain Institute, Snowmass, p 352
Lovins H, Cohen B (2011) Climate capitalism. Hill and Wang, New York
McKibben B (2012) Global warming's terrifying new math. Three simple numbers that add up to global catastrophe – and that make clear who the real enemy is. Rolling Stone. http://www.rollingstone.com/politics/news/global-warmings-terrifying-new-math-20120719#ixzz3lpZHgvhj
Meadows DH, Meadows DL, Randers J, Behrens WW III (1972) The limits to growth. Universe Books, New York, p 205
Meadows DH, Randers J, Meadows DL (2004) Limits to growth: the 30-year update. Chelsea Green, White River Junction, p 368
Mill JS (1848) Principles of political economy. Longmans, Green, and Co, London, pp 746–751, 1940/1848
National Research Council (2010) The hidden costs of energy. National Academy of Sciences, Washington, DC, p 473
Nordhaus W (2013) The climate casino: risk, uncertainty, and economics for a warming world. Yale University Press, New Haven, p 320
Odum HT, Odum EC (2001) A prosperous way down. University of Colorado, Boulder, p 326
Orrell D (2010) Economyths: ten ways economics gets it wrong. Wiley, Ontario
Picketty T (2014) Capitalism in the twenty-first century. Penguin, New York
Polanyi K (1944) The great transformation. Beacon, Boston, 1967
Price C (1993) Time, discounting, & value. Blackwell, Oxford, p 393
Reich R (2007) Supercapitalism: the transformation of business, democracy, and everyday life. Knopf, New York, p 288
Rist G (2011) The delusions of economics: the misguided certainties of a hazardous science. ZED books, London, p 224
Rowe J (2008) Our phony economy. Harpers. June 2008
Ruskin J (1860) Unto this last. Cornhill Magazine
Sachs W (1999) Planet dialectics: explorations in environment and development. Zed Books, New York, p 226
Sahlins M (1972) Stone age economics. Aldine, Chicago, p 348
Sale K (1980) Human scale. Coward, McCann, & Geoghegan, New York, p 558
Sandel MJ (2012) What money can't buy: the moral limits of markets. Farrar, Straus and Giroux, New York, p 335
Schor J (2010) Plenitude: the new economics of true wealth. Penguin, New York, p 258
Schumacher EF (1973) Small is beautiful: economics as if people mattered. Harper Torchbooks, New York
Schumacher EF (1979) Good work. Harper Colophon, New York
Schweickart D (2002) After capitalism. Rowman & Littlefield Publishers, New York, p 222
Shuman M (2015) The local economy solution. Chelsea Green, White River Jct

Sinsheimer R (1978) The Presumptions of Science. Daedalus 107:23–35
Skidelsky R, Skidelsky E (2012) How much is enough? Money and the good life. Other Press, New York, p 272, 2012
Smith A (1759) The theory of moral sentiments
Smith A (1776) An inquiry into the nature and causes of the wealth of nations. W. Strahan and T. Cadell, London
Speth JG (2012) America the possible: manifesto for a New economy. Yale University Press, New Haven, p 272
Srinivasan UT, Carey SP, Hallstein E, Higgins PAT, Kerr AC, Koteen LE, Smith AB, Watson R, Harte J and Norgaard RB (2008) The debt of nations and the distribution of ecological impacts from human activities. In: Proceedings of the National Academies of Science. 5 Feb 2008, pp. 1768–1773
Stern N (2007) The economics of climate change. Cambridge University Press, New York, p 712
Stern N (2015) Why are we waiting?: the logic, urgency, and promise of tackling climate change (Lionel Robbins lectures). MIT Press, Cambridge, MA, p 376
Tainter J (1988) The collapse of complex societies. Cambridge University Press, Cambridge, 262 pp
Trainer T (2007) Renewable energy cannot sustain a consumer society. Springer, Dordrecht, p 198
Turner G (2008) A comparison of the limits to growth with thirty years of reality. CSIRO, Canberra, p 52
Wagner G, Weitzman M (2015) Climate shock: the economic consequences of a hotter planet. Princeton University Press, Princeton, p 264
Wilkinson R, Pickett K (2009) The spirit level: why greater equality makes societies stronger. Bloomsbury Press, New York, p 352
Zehner O (2012) Green illusions: the dirty secrets of clean energy and the future of environmentalism. University of Nebraska Press, Lincoln, p 464

Chapter 2
How Economics Can Become Compatible with Democracy

Peter Söderbaum

Abstract This chapter introduces a Scandinavian and European tradition of institutional economics as an alternative non-orthodox branch of economics that claims relevance in addressing sustainable development issues. The challenge of sustainable development is complex. Existing development trends are unsustainable in more ways than one, for example concerning climate change, loss of biological diversity, depletion of fish stocks, risks of nuclear accidents. When attempting to deal with these issues, it is often assumed that we can rely on science and on economics in particular. Economics claims to supply a conceptual framework and theory for efficient resource allocation at various levels; at the level of individuals, of business corporations and of society.

University departments of economics educate students in one way, nationally and globally, so called neoclassical economics. This theory can offer some ideas about how to deal with sustainability issues. But neoclassical theory has been dominant in a period when serious problems related to sustainability have emerged. It is therefore probably wise to also consider alternatives to neoclassical theory, such as institutional economics. In the present essay I will – while pointing in the direction of institutional theory – suggest a way of opening up economics to make the field more compatible with democracy. Economics has to move from the present monism to pluralism and become more sensitive to value or ideological issues in present society.

Keywords Ecological economics • Democracy • Neoclassical economics • Institutional economics • Sustainable development

---

P. Söderbaum (✉)
School of Economics, Society and Engineering, Mälardalen University, Box 883, 721 23 Västerås, Sweden
e-mail: peter.soderbaum@mdh.se

S. Shmelev (ed.), *Green Economy Reader*, Studies in Ecological Economics 6,
DOI 10.1007/978-3-319-38919-6_2

2.1 Introduction

Reductionism (rather than holism) has been a tendency in social sciences over the years. More recently there have been various counter-movements and the tendency to separate one discipline from another and social science from politics has been challenged. Can economics be separated from political science? Democracy is for many connected with political science. Does this mean that democracy is 'outside' economics? Can science be separated from politics? Is there a value-neutral economics?

Introductory textbooks in economics are important because they are read by millions of students each year. There is a monopoly for neoclassical economic theory in the sense that textbooks do not differ much but are standardized to the same assumptions and the same theory. In the extreme case of this standardization one single book is used at many university departments of economics through-out the world. In the early period of this process toward a common and restricted understanding Paul Samuelson's *Economics* (Samuelson 1948) was very popular and more recently N.Gregory Mankiw's *Principles of Economics* (Mankiw 2009) has become the most well-known and respected introductory textbook. A new version of this book has appeared and we are back to *Economics* as the title (Mankiw and Tailor 2011).

The reader is invited to take a look at the textbook used at her/his university or a nearby university. I live in Uppsala and at the Department of Economics, Uppsala University, the mentioned new version of the Mankiw textbook is used. My present interest is "democracy in economics" and how value or ideological issues are considered in the text. I then as a first step turn to the "glossary" and "index" part of the 900 pages book and make the observation that neither the word 'democracy' nor 'ideology' is there. Democracy appears to be a non-issue for, or at least downplayed by Mankiw, the implicit recommendation being that we should rely on experts.

Mainstream neoclassical economics does not question or problematize the present political economic system. This theory – with connected ideology – can instead be described as being part of a defense for our present kind of capitalism and market economy. One feature of institutional economic theory is, as the term suggests, a focus on institutional arrangements and institutional change processes. While neoclassical economists may point to the option of minor institutional adjustment as in the case of "market failure" and "external impacts", i.e. impacts upon third parties that should be "internalized", K. William Kapp among institutional economists argued that environmental impacts are underestimated if the focus is on one external impact at a time (Kapp 1976). According to him there is a systematic tendency for business corporations to reduce their monetary costs by shifting environmental impacts on outsiders and society at large:

> Thus, a system of decision-making, operating in accordance with the principle of investment for profit, cannot be expected to proceed in any other way but to try to reduce its costs whenever possible by shifting them to the shoulders of others or to society at large.
>
> (Kapp 1970, p.18)

The institution of joint stock company is therefore not without problems according to Kapp. Another way of arguing is to point to the fact that joint stock companies are defined in financial or monetary terms in the sense that a monetary profit is needed for survival. But many of the present issues facing local, national, regional and the global society are non-monetary in kind. This is so for climate change, biodiversity loss, pollution, overfishing but also for unemployment and health issues. It can therefore not be excluded that joint stock companies are miss-constructed in relation to present needs. Also major institutional change, in the sense of considering other legal contexts for organizations need to be considered.

But this is an ideological and political issue one may object. Yes, this is so and perhaps economics is always political economics and this field of study should be subordinated to a general aspiration for a strengthened democracy in any nation or other community.

2.2 A Return to Political Economics

Classical economists were broad-minded and referred to 'political economics'. A more narrow and specialized version of economics appeared about 1870. Attempts were made to keep the political element at a distance and reference was made to 'economics' rather than 'political economics'. Economists compared their discipline with physics and accepted positivism as the theory of science. Standing outside, explaining various phenomena objectively and making predictions became the main ambition. These ideas are dominant today and one may speak of a close to monopoly position for neoclassical theory in introductory textbooks and to a somewhat lesser extent in the discipline more generally.

It is now argued that it was a mistake to abandon 'political economics' as a label for the discipline. Economics is closely connected with politics and value or ideo logical elements are necessarily present even in our research and educational activities. Gunnar Myrdal who started as a neoclassical economist interested in price theory but eventually declared himself an institutional economist (Myrdal 1978) argued as follows:

> Valuations are always with us. Disinterested research there has never been and can never be. Prior to answers there must be questions. There can be no view except from a viewpoint. In the questions raised and the viewpoint chosen, valuations are implied. Our valuations determine our approaches to a problem, the definition of our concepts, the choice of models, the selection of observations, the presentations of our conclusions – in fact the whole pursuit of a study from beginning to end.
>
> (Myrdal 1978, p. 778)

There is an unavoidable political and ideological element in economics. Neoclassical economics is a kind of political economics and other schools of thought in economics can also be described with respect to ideological features. Value issues behind specific research and educational programs should therefore be stated as clearly as possible and discussed openly. Similarly, individuals and organizations as actors in the economy, and in society more generally, can be interpreted in ideological and political terms.

2.3 The Concepts of Ideology, Ideological Orientation and Mission

Neoclassical theory starts from an assumption that behavior is exclusively related to markets and that market behavior is based on self-interest. As an alternative to such Economic Man assumptions, reference will instead be made to a Political Economic Person (PEP) as actor guided by her/his ideological orientation. As an alternative to the profit-maximizing firm of neoclassical theory, reference is similarly made to a Political Economic Organization (PEO) guided by its ideological orientation or 'mission'. The word 'mission' opens the door for ethical considerations and performance also in non-monetary terms for example concerns about Corporate Social Responsibility (CSR).

Reference to 'ideology' and 'ideological orientation' may at first appear a bit strange in relation to economics. But individuals are not only consumers, wage earners or related to capital markets. We are also citizens as argued by Mark Sagoff (1988) and in a democratic society this citizenship comes first. As I see it, this is however not a matter of 'either-or' but rather 'both-and'. In dealing with sustainability issues it appears relevant to broaden the perspective. An individual in her different roles (as actor in the market place, as professional, as parent, as citizen etc.) is assumed to be guided by her ideological orientation. The ideological orientation of the individual may be narrow or broad and is something to be investigated rather than assumed as given.

In the social sciences we deal with so called 'contested concepts' (Connolly 1993) and 'ideology' is such a concept. 'Institution', 'democracy' and 'power' are other examples. Ideology stands for a means-ends relationship. It is not exclusively about ends or final values, nor is it exclusively about means; it rather relates means to ends. An actor refers to her ideological orientation as her compass and guiding vision. It is about where you are (present position), where you want to go (desired future positions) and how to get there (strategy in bringing objectives and instruments together).

The word 'ideology' has primarily been used at a collective level. Political parties and social movements may refer to their ideologies. We may speak of Liberal (and even Neo-liberal) ideology, Social Democracy, Christian ideology or Green ideology. Social movements represent aggregates of individuals and if ideology exists at a collective level then something similar is relevant at the level of individuals. In political elections politicians try to convince individuals about the merits of their ideologies and individuals respond in one way or other depending upon their own ideological orientations.

Another example of ideology at a collective level is'economic growth ideology' in GDP-terms where growth is believed to solve almost all problems (employment, environment, health care, welfare). Those who embrace this ideology are not necessarily organized for this purpose but still have beliefs in common. Some of them may share the same background as in the case of (neoclassical) economics education. I will now argue that sustainable development is another example of an ideological orientation. Much like 'economics growth', it can be shared by many actors belonging to different political and professional categories.

Like other contested concepts 'ideology' and 'ideological orientation' can be interpreted in more ways than one. For some – even among those who claim to take sustainable development seriously – ideology is used in a negative sense about belief systems and ideas that are considered dysfunctional and something that should be kept at a distance. As an example Bryan G. Norton in his book *Sustainability. A Philosophy of Adaptive Ecosystems Management* (Norton 2005) starts out with a "Preface: Beyond Ideology" and later returns to this use of words in a subtitle "Avoiding Ideology by Rethinking Problems". Especially in US contexts the word ideology is sometimes used to invite negative connotations.[1]

In the present chapter an actor may like, as well as dislike, a specific ideological orientation. Ideology is a necessary fact of life and 'ideology' is a useful word in introducing thinking about 'sustainable development' and 'sustainability' in the sense that each interpretation of sustainable development exemplifies an ideological orientation. But let us first further discuss how ideological orientation can be understood as a guiding principle for decision-making and behavior. The ideological orientation of an individual may contain elements of qualitative, quantitative and visual kind. It is stable in many ways but also changes more or less with context; it is fragmentary and uncertain rather than complete and certain. It can still be used for decision-making purposes where the ideological orientation of an actor is matched against the expected multidimensional impact profile of each alternative considered. (Söderbaum 2008, p. 58). Some alternatives match the ideological orientation of our actor better than others. Reference can also be made to appropriateness as in the case of Friedrich Schumacher's early argument for "appropriate technology" (Schumacher 1973). James March similarly points to a "logic of appropriateness" to be distinguished from a "logic of consequence" (March 1994, p. viii). The idea of matching has also become popular when attempting to combine persons with specific qualifications with job requirements. In political elections an individual as citizen looks for that political party which best fits or 'matches' her ideological orientation and so on.

The complexity of the problems faced in relation to sustainability is such that one can refer to "wicked problems" and we have to live with "fuzzy" concepts as described by Silvio Funtowicz and Jerome Ravetz (1991, 1994) among others. Ideas about science and analysis need to be modified or reconsidered suggesting that there are reasons to open the door for a so called 'post-normal science'. Reference to

[1] Bryan Norton is not alone in using 'ideology' in a negative sense. For a history of the word ideology, see MacKenzie (1994).

mathematical objective functions to be optimized then become exceptional cases and subject to criticism from those who take complexity seriously.

2.4 Sustainable Development as Ideological Orientation

Sustainable development can be interpreted as an ideological orientation or rather a set of ideological orientations with some features in common. In attempts to depart from unsustainable trends and move towards trends that are more sustainable, different interpretations of sustainable development can be articulated and discussed.

Sustainable development became part of political discourse through the so called Brundtland report (World Commission on Environment and Development 1987). This report has been interpreted differently but nobody can miss-interpret its title *Our Common Future*. It is an exhortation to individuals as actors in different roles to extend perspectives in space and time; it is about considering the interests of other people living today as well as future generations. Nobody can be completely altruistic but all can take steps by considering ethical aspects of their behavior and decisions. Institutional arrangements and incentive systems can be changed to facilitate such changes in behavior.

As suggested by the name of the World Commission, there was also a focus on the environment and natural resources in the Brundtland report. Irreversible degradation of the natural resource base in relation to future generations is an issue of inequality and should be avoided or minimized. We should rather improve the natural resource base available to future generations wherever possible. In an early article for the FAO journal *Ceres*, I suggested four "ecological imperatives for public policy". Assuming that we are preparing decisions for a municipal political assembly concerning for example investments in infrastructure of some kind (housing, energy system, road) then

- Alternatives of choice that involve irreversible degradation of the natural resource base in the home region now and in the future should be avoided;
- Alternatives of choice that involve irreversible degradation of the natural resource base in other regions (and ultimately at a global level) now and in the future should be avoided
- A precautionary principle should be applied in the sense that risks of significant irreversible degradation of the natural resource base in the home region or abroad is enough reason to avoid the alternative
- Only alternatives that remain after this selection process can be considered as broadly compatible with sustainable development (as ideological orientation). If no alternative remains then the analyst and other actors involved should initiate a search process to develop and design such sustainable alternatives (Söderbaum 1982).

The examples mentioned above referred to public policy options. But the same ethical imperatives are of course relevant also for private decision-making concerning for example forestry and agricultural management. Organic or ecological agriculture has impacts that differ from conventional agriculture. And the degree of popularity of organic agriculture is partly a matter of decisions at the household level. Ethical considerations become potentially relevant for all kinds of market decisions.

Since sustainable development is an ideological orientation, it is not very meaningful to look for a single 'true' or 'correct' interpretation. We have to live with different interpretations and be open about how we relate to them. In a democratic society, each person has the right to refer to her/his ideological orientation and interpret various phenomena accordingly. As a political economic person the individual can furthermore argue in favor of one interpretation or set of interpretations as opposed to others. I will here point to three possible interpretations that are more or less frequent in public debate. Each interpretation is expressed in relation to present development trends:

- *Business as usual* (BAU). The actor thinks in terms of conventional monetary indicators such as GDP-growth and profits in business and believes that markets and entrepreneurial creativity will automatically solve any problems that exist or may appear. Focus should continue to be on 'sustainable economic growth' and 'sustained profits' in business. No intervention in the present functioning of markets is believed to be needed.
- *Ecological modernization* (c.f. Hajer1995). The actor makes the judgment that there are environmental and other development problems that require action and believes that ethical concerns should enter the picture to modify or "modernize" the behavior of various actors. Only minor institutional change will be enough. Examples include Corporate Social Responsibility, Environmental Management Systems and Environmental Impact Assessment.
- *Radical change of institutional arrangements*. The actor makes the judgment that minor institutional change in the right direction is important but that also major change in present political economic system has to be considered. As an example the joint stock company is defined in monetary and financial terms while some of the most important global and regional challenges today are non-monetary in kind. Another example is the World Trade Organization (WTO) which builds its recommendations on the basis of an over-simplified neoclassical international trade theory.

There is a tendency in all societies or communities to avoid sensitive matters and behave opportunistically in relation to those in power. In the community of economists for example life becomes easier if you join the mainstream and something similar is true in society at large. But if acceptance of mainstream perspectives and behavior will aggravate problems related to sustainability rather than solve them, then there is reason for concern.

In a democratic society public debate is encouraged rather than avoided. Tensions between advocates of different perspectives (and even conflicts of interest in a narrower sense) are respected and constitute the life-blood of a democracy. There are competing political parties rather than one single political party in our societies.

If ideology is involved in economics, then the monopoly position of neoclassical economic theory becomes a problem in relation to a functioning democracy. In a democratic society, open-mindedness in relation to opinions other than your own is a virtue.

2.5 Ideological Features of Neoclassical Economic Theory

What are the ideological features of mainstream neoclassical economics? Again the reader is invited to make his/her own observations. I will here point to three interrelated features of neoclassical economic theory as presented for example in introductory textbooks that are important in relation to the prospects of sustainable development locally and globally:

- Focus on markets and market incentives
- Emphasis on self-interest
- Emphasis on monetary performance indicators
- Belief in correct prices for purposes of efficient resource allocation

The market and market relationships are at the heart of neoclassical ideas of economics and the economy. Markets for commodities, labor and capital are taken into account in specific ways. Human beings and ecosystems (with human and non-human species) are there only in so far as they are part of markets and market relationships. This focus on markets means that ecosystems and natural resources are less visible. Human beings are present but only in market-related roles. Institutions other than firms and markets, for example 'democracy', play a peripheral role.

Environmental problems are connected with 'market failure' (as in the case of externalities) and 'government failure' (when the government is subsidizing environmentally harmful activities). Environmental policy then becomes a matter of getting prices and market incentives right by eliminating failures of the two kinds. Designing and implementing new markets such as markets for pollution permits is another often preferred option.

Neoclassical ideas of environmental policy as described above are of interest as part of a pluralistic strategy. But ecological economists for example want to broaden the view to include ecosystems and the natural resource base locally and globally in our understanding of the economy (Boulding 1966; Costanza, ed. 1991). Social economists and feminist economists suggest alternative ideas of individuals and how they relate to each other. The list of potential failures that may explain environmental degradation and human suffering can furthermore be expanded to include failure of science (theory of science and paradigm in economics for instance), failure of ideology, failure of life-styles, failure of technology, failure of democracy and other institutional arrangements (Söderbaum 2008, pp. 37–52). It need not be added that such lists opens the door for a richer and multi-facetted dialogue about problems and policies. Environmental policy is not only a matter of what governments can do; it is also about individuals as actors in their different roles.

The second ideological feature of neoclassical economics identified is its Economic Man assumptions and the emphasis on self-interest. Each actor is assumed to behave in ways that satisfies her/his self-interest. This assumption legitimizes behavior where the concern for others is limited or non-existent. Neoclassical economists tend to argue that it is unrealistic to expect individuals to depart from self-interest. A different economic theory that we can label institutional economics opens the door also for ethical and cooperative considerations. The individual as actor does not behave in one way on the basis of one kind of motivation. Instead it is assumed that the individual refers to her ideological orientation which changes more or less over time depending upon context, for example. In this way the individual gets closer to or departs more from some idea of sustainable development.

Our third statement about ideological specificity of neoclassical economic theory refers to an emphasis on monetary performance indicators. The only organization dealt with in neoclassical microeconomics is the firm and the firm is assumed to maximize monetary profits. Accounting practices of firms are equally monetary in kind. At the level of society, investment decisions are prepared using Cost-Benefit analysis (CBA) in monetary terms. This can be described as an attempt to carry out a societal profitability analysis. Gross Domestic Product (GDP) plays a central role in national accounting practices. Reference to GDP is part of macroeconomic policy and in spite of its known limitations for the purpose, GDP is used as an indicator of progress.

It is clear from the above examples that monetary indicators play an essential role in neoclassical theory. There is a preference for quantification, more precisely quantification in monetary terms. From a critical point of view, this can be referred to as 'monetary reductionism'. Interest in non-monetary aspects of business management or non-monetary performance indicators at the national level is limited.

Neoclassical theory of the firm or contributions to economic policy may still be of interest and useful for some purposes as part of a pluralistic approach to economics but the neoclassical approach by itself can hardly be sufficient in relation to complex sustainability issues. A different conceptual framework is needed to deal with Corporate Social Responsibility and Environmental Management Systems, for example. Accounting systems at the level of organizations and nations need to deal systematically with non-monetary performance.

Our fourth and final ideological feature of neoclassical economics is a belief in correct prices for purposes of efficient resource allocation at the national level. When neoclassical economists advocate the use of Cost-Benefit Analysis (CBA) to prepare investments in infrastructure (roads, dams, energy systems etc.), this is equal to a dictatorship in terms of values and ideological orientation. Actual market prices and sometimes hypothetical prices of different impacts should be used in a summation procedure to arrive at the best and optimal alternative.

This idea of correct market prices for purposes of efficient resource allocation is built upon an assumed consensus in society about the relevance of the approach to evaluation of CBA. As argued by Ezra Mishan, himself a textbook writer on CBA (Mishan 1971), it is no longer realistic to assume the existence of such a consensus, considering the wide difference of opinions about environmental issues in present society (Mishan 1980). Some other approaches are needed, approaches that are dealing with multidimensional impacts in multidimensional terms and that are compatible with normal ideas of democracy (Söderbaum 2006; Söderbaum and Brown 2010, 2011).

2.6 Reconsidering Economics in Relation to Democracy

In mainstream definitions of 'economics' the scarcity of resources is emphasized: "Economics is the study of how society manages its scarce resources" (Mankiw, 2011, p.2). This scarcity aspect is certainly relevant for example in relation to quality and quantity of land and water and other natural resources but I will here suggest a definition which emphasizes additional aspects:

"Economics is about multidimensional management of resources in a democratic society"

In complex decision situations of the kind discussed in this chapter multidimensionality should be respected. One-dimensional calculation in monetary or other terms is questioned. Qualitative and visual impacts are not less important than quantitative ones (this being a matter of your ideological orientation). Inertia of different kinds (commitments, path-dependence, lock-in effects and irreversibility) is present in monetary as well as non-monetary terms. In our present political-economic system there is normally a focus on monetary and financial impacts. Non-monetary impacts are too often down-played or simply forgotten as part of a trade-off philosophy in monetary or other terms (where all kinds of impacts can be traded against each other). A specific effort is therefore needed to identify various aspects of inertia, such as irreversibility. The costs of building a road on agricultural land is not limited to the financial costs of purchasing land and the construction costs. Also the fact of irreversible change in land-use has to be taken seriously as a non-monetary cost when preparing decisions. 'Positional thinking' in multiple stages and 'positional analysis' are ways of illuminating such non-monetary impacts (Söderbaum 2008).

Why should ideas about democracy be brought into economics? A first answer is that economics deals with issues of values, ethics and ideology in governance at various levels. Decisions have to be prepared in ways that to some extent reflect the diversity of ideological orientations among citizens in a region and members of the political assemblies of that region.

Democracy is often connected with freedom of speech, freedom of organization, human rights and how political elections are carried out. Those principles are fundamental and we have seen cases where they have not been respected. Reference can be made to a minimalist interpretation of democracy where the ability to carry out political elections according to normal rules is at issue.

But fair political elections are part of a general cultural climate in any nation or community. In any society there are tensions between different individuals, groups and political parties who refer to different ideological orientations in our language.

Such tensions may play both positive and negative roles in society. Public debate and tensions normally lead to reconsiderations of ideological orientations and represent a possibility for creative solutions to problems. Political actors who differ in ideological orientation have to respect each other. An actor A should respect and be tolerant in relation to another actor B as long as the ideological orientation of B does not go against democracy itself.

This statement suggests that democracy can be understood in relation to its opposite, dictatorship. Two kinds of dictatorship are relevant here:

- Technocracy in the sense of dictatorship by experts
- Political dictatorship by one group and one political party

Neoclassical Cost-Benefit Analysis (CBA) is an excellent example of technocracy. An optimal alternative is identified (that with the highest monetary present value) and politicians, although responsible to citizens, need not be involved. A technocracy, such as the neoclassical economists in their monopoly position, is in some ways comparable with political dictatorship in the form of one party system. But it should be remembered that the mentioned kind of technocracy only represents a segment of the larger society where there may be many openings for public debate and democracy.

2.7 Elements of a More Open Political Economics

How can one move from a close to monopoly position for neoclassical economic theory to a more pluralist position in economics education and research? One response is to argue that also other schools of thought than the neoclassical one should be represented. Institutional economics can be taught as a complement to the neoclassical view, social economics, feminist economics, ecological economics, Marxist economics are other options. Textbooks in the history of economics ideas can inform students about the fact that tensions between schools of thought in economics are not a new thing.

Something can be done however that touches upon all the various schools in economics. As one of the first steps we can deal openly with the political and ideological element in economics to make the discipline more compatible with democracy. What is needed is a conceptual framework for a political economics in a broad sense rather than unambiguous explanations about how consumers and firms behave.[2] The following conceptual framework is proposed:

- *Political Economic Person assumptions*. Individuals are guided by their ideological orientation. The ideological orientation varies among individuals and for each individual over time and is something to be investigated rather than taken as

[2]Any tendency to connect 'political economics' exclusively with Marxist economics is here rejected. Marxist economics is just one kind of political economics among other schools.

given. Those with a Neo-liberal orientation can be expected to behave differently from those with a Green political orientation

- *Political Economic Organization assumptions*. Organizations are guided by their ideological orientation or mission. All organizations adapt to the present political economic system. Some business corporations focus exclusively on monetary profits while others have a different attitude to their social responsibility. Whereas neoclassical theory of the firm legitimizes narrow profitability motives, the present approach makes the mission of a corporation an open issue
- *A political economic view of markets*. While not excluding the neoclassical mechanistic model of markets in equilibrium terms, our political economic approach makes the actors in the market place (PEP:s and PEO:s) more visible. Power issues as when actor A in some sense is exploiting actor B can be brought into the picture. Also stakeholder models of firms and markets and network models are offered as ways of understanding behavior. Different models can sometimes complement each other and the idea of one (neoclassical) model representing the only possibility is abandoned.
- *Positional analysis (PA) as an approach to decision making* at the societal level. Approaches to decision-making have to be compatible with a multi-dimensional idea of economics and with – as previously argued – democracy. The one-dimensional monetary analysis of CBA is replaced with an approach where non-monetary processes and impacts are not regarded as 'less economic' than monetary ones. The democracy aspect suggests that the analyst has to respect different ideological orientations relevant among citizens and politicians rather than pursue analysis on the basis of one single ideological orientation (or objective function). The task of the analyst becomes one of illuminating an issue by carrying out a many-sided analysis with respect to ideological orientations, alternatives of choice, impacts, inertia, uncertainties etc. Conclusions in the form of ranking alternatives then become conditional in relation to each ideological orientation considered.
- *Efficiency and rationality as a matter of ideological orientation*. In neoclassical theory it is assumed that individuals as consumers use their available monetary budget when they maximize utility in a self-interested way. Business companies are similarly efficient when they maximize monetary profits. These assumptions can be accepted or rejected depending on the ideological orientation (mission) of a specific actor. The important thing is to open the door for alternative conceptions of efficiency and rationality.
- *Institutional change as a matter of conceptual and ideological power games between actor categories*. Small or larger institutional change processes take place all the time in the national and global economy. The conceptual framework and ideology of neoclassical theory plays a role in stabilizing the present political economic system. Actually, the ideology of neoclassical theory is in many ways equal to the ideology of capitalism. Many actors use large parts of their resources to defend this particular ideology. Those of us who make the judgment that the present political-economic system is not performing well in financial

terms or in relation to sustainability should invite attempts to socially construct alternative conceptual frameworks in economics.

2.8 Conclusions for Sustainability Politics

Present development trends are unsustainable in important ways. How can we move toward a more sustainable development? Focus on 'market failure' and 'government failure' in neoclassical economic theory is a possibility but hardly enough. Something more is needed. A move away from the present close to monopoly position of neoclassical theory to pluralism would be an important step. Neoclassical economic theory and environmental economics of the neoclassical kind is not just theory but at the same time specific in ideological terms. Neoclassical theory as one among theoretical perspectives can still contribute in some ways but the present monopoly for neoclassical theory goes against our ideas about democracy.

How can democracy enter into economics education and research? Should we expect neoclassical economists to change their minds and become pluralists? Will the values of democracy in terms of respect and tolerance be embraced by neoclassical economists or will they defend the monopoly and the present technocracy?

Students of economics can play a role. An example is the recent "International Student Initiative for Pluralism in Economics" where economics students from 48 universities and 20 countries participates (International Student initiative 2014). It is not yet clear how mainstream professors will respond to this international manifestation.

Will politicians enter into this dialogue and perhaps intervene? Or will they rely on a value-neutrality in university education and research that does not exist? In any case ideological issues have to be discussed more openly not only among politicians but also in university circles. Dominant ideology in business and market terms is perhaps one of the most important explanations of present unsustainable trends.

Hopefully, the idea of separating economics from politics and democracy is losing ground internationally. Economics textbooks that do not take democracy seriously should be replaced and left to the study of economic history. The values and principles of pluralism and democracy have to be propagated in each country as well as globally.

References

Boulding KE (1966) The economics of the coming spaceship earth. In: Jarret H (ed) Environmental quality in a growing economy. John Hopkins Press, Baltimore, pp 3–14

Connolly WE (1993) The terms of political discourse, 3rd edn. Blackwell, Oxford

Costanza R (ed) (1991) The science and management of sustainability. Columbia University Press, New York

Funtowicz S, Ravetz J (1991) A new scientific methodology for global environmental issues. In: Costanza R (ed) Ecological economics: the science and management of sustainability. Columbia University Press, New York, pp 137–151
Funtowicz S, Ravetz J (1994) The worth of a songbird: ecological economics as a post-normal science. Ecol Econ 10:197–207
Hajer MA (1995) The politics of environmental discourse. Ecological modernization and the policy process. Clarendon Press, London
International Student Initiative for Pluralism in Economics (2014) Available at: www.isipe.net/open-letter/
Kapp KW (1976) The nature and significance of institutional economics. Kyklos 29(2):209–232, April
Kapp KW (1970) Environmental disruption: general issues and methodological problems. Soc Sci Inf (International Social Science Council) 9(4): 15–32
MacKenzie I (1994) Introduction. The arena of ideology. In: Eccleshall R, Vincent G, Jay R, Kenny M, MacKenzie I, Welford R (eds) Political ideologies. An introduction, 2nd edn. Routledge, London, pp 1–27
Mankiw NG (2009) Principles of economics. Cengage Learning, Andover
Mankiw NG, Taylor MP (2011) Economics. Cengage Learning, Andover
March JG (1994) A primer on decision making: how decisions happen. The Free Press, New York
Mishan EJ (1971) Cost-benefit analysis. Allen & Unwin, London
Mishan EJ (1980) How valid are economic evaluations of allocative changes? J Econ Iss 14(1):143–161
Myrdal G (1978) Institutional economics. J Econ Iss 12(4):771–783
Norton BG (2005) Sustainability: a philosophy of adaptive ecosystem management. The University of Chicago Press, Chicago
Sagoff M (1988) The economy of the earth: philosophy, law and the environment. Cambridge University Press, Cambridge
Samuelson P (1948) Economics. McGraw-Hill, New York
Schumacher EF (1973) Small is beautiful: a study of economics as if people mattered. Arcus, London
Söderbaum P (1982) Ecological imperatives for public policy. Ceres: FAO Rev Agric Dev 15(2):28–30
Söderbaum P (2006) Democracy and sustainable development. What is the alternative to cost-benefit analysis? Integr Environ Assess Manag 2(2):182–190, April
Söderbaum P (2008) Understanding sustainability economics: towards pluralism in economics. Earthscan/Routledge, London
Söderbaum P, Brown J (2010) Democratizing economics: pluralism as a path toward sustainability. Ann N Y Acad Sci, Ecological Economics Reviews 1185(1): 179–195
Söderbaum P, Brown J (2011) Pluralism and democracy in economics. Int J Pluralism Econ Educ 2(3):240–243
World commission on Environment and Development (1987) (The Brundtland Report) Our Common Future. Oxford University Press, Oxford

Chapter 3
Gaps in Mainstream Economics: Energy, Growth, and Sustainability

Robert U. Ayres

Abstract This chapter begins with a review of the state of neoclassical economics, insofar as it deals with energy. The main point is that useful energy (exergy) is an essential ingredient in the formula for economic activity, especially growth. Conventional neoclassical theory asserts that the only two "factors of production" are capital and human labor, and that raw materials are "produced" by some combination of capital and labor. Of course this is literally impossible, but the impossibility is avoided by asserting that the process of "resource extraction" is a "value-added activity", where the value is added by the combination of capital and labor. This is an accounting trick, because both capital and labor are unproductive without an exergy input. So energy (or work) must be a third factor of production. This being so, social stability depends on growth (for now, at least), future economic growth depends on the availability of exergy. But long-term environmental sustainability also depends on decarbonization. The good news is that decarbonization will employ a lot of capital and labor, which promote growth. But the investment must be seen to be profitable. Fortunately this is possible.

Keywords Growth • Energy • Sustainability • Thermodynamics • Exergy • Limits to growth • Decarbonization

3.1 Background

A recent (Sept. 2014) book review in the New York Times by no less than Nobel Laureate Paul Krugman, makes the argument for ecological economics unintentionally but forcefully. The book, by Jeff Madrick, is entitled "Seven Bad Ideas: How mainstream economists have damaged America and the World" (Madrick 2014). The list starts with Adam Smith's oft –but inappropriately -quoted sentence about

R.U. Ayres (✉)
Economics and Political Science and Technology Management,
Management and the Environment, Emeritus, INSEAD,
Boulevard de Constance, 77305 Fontainebleau, France
e-mail: Robert.AYRES@insead.edu

S. Shmelev (ed.), *Green Economy Reader*, Studies in Ecological Economics 6,
DOI 10.1007/978-3-319-38919-6_3

"the invisible hand" and proceeds, through Say's Law, a couple of gems from Milton Friedman, to #7, the bad idea that globalization and free trade are a good thing.

But Madrick, and Krugman both miss the worst idea of all: the bad idea that economists don't need to understand science, or the laws of thermodynamics. (That is a cousin of the bad idea permeating the courses in many Business Schools, that managers don't need to understand what their companies are making or doing).

Economic concepts, from foundational issues like markets, supply and demand and trade to money and finance, lack any systematic awareness of implications of the Laws of Thermodynamics for the physical process of production. A corollary, almost worthy of being a separate bad idea on its own, is that energy doesn't matter (much) because the cost share of energy in the economy is so small that it can be ignored e.g. (Denison 1984). The so-called "production functions" used by all schools of economic thought that build growth models omit any necessary role for energy, as if output could be produced by labor and capital alone—or as if energy is merely a form of man-made capital that can be produced (as opposed to extracted) by labor and capital.

The neoclassical defense for neglecting energy, as such, as a production input is that extraction activities such as mining or drilling for oil are "value added" activities, properly accounted for in terms of capital and labor employed. This is true, on its face. But it is also an accounting trick to confuse a monetary abstraction (value added) with a physical resource that is essential for the production process. (One can argue, the same way, that manufacturing bombs is a "value added" process, even though they are actually used to destroy value.) The point is that in the real world, production requires exergy inputs, either as fuels or electric power, (replacing human or animal muscle work) or in the form of capital goods that are also a kind of embodied exergy. Labor and capital are inert and unproductive without exergy inputs.

The essential truth missing from economic education is that energy is the stuff of the universe, that all matter is also a form of energy, and that the economic system is essentially a system for extracting, processing and transforming energy as resources into energy embodied in products and services. This is a thermodynamic process, as Georgescu-Roegen said (Georgescu-Roegen 1971). The economic process is subject to both the first law of thermodynamics (conservation of mass/energy; nothing can be created or destroyed) and the second law of thermodynamics (increasing entropy; all transformation processes are irreversible). The "first law" implies that the notion of "consumption" as applied to products is misleading: material transformation processes unavoidably generate large quantities of material wastes or residuals (Ayres and Kneese 1969) (Ayres 1989). Some of those wastes are merely inconvenient but others are harmful or toxic. The second law says that energy becomes less useful (exergy is destroyed) by every action.

There is a circular graph in many economics textbooks, showing monetary flows between a box labeled "producers" and another labeled "consumers". These flows represent wages and expenditures (and value added). Expenditures are conceptually equal to the vector product of product quantities times prices. But if the materials balance (first law) of inputs and outputs for production were shown explicitly, it would be clear that much – indeed most – of the output consists of wastes, some of which are pollutants. Those wastes have no market prices, because there is no market for them, but if there was a market, the prices would be negative.

The attractive Walrasian idea of supply-demand equilibrium for all goods and services simultaneously is meaningless in a market with negative prices (costs) and negative demand. Mainstream economists call this a "market imperfection" and assume that the imperfection is insignificant. Since the problem was first seen to be non-trivial, in 1967 (Mishan 1967; Kneese et al. 1970), various schemes have been tried to compensate for the market imperfections by "internalizing externalities". Unfortunately, the methodologies currently applied are, themselves, imperfect and inadequate. There is no economic theory to prevent large-scale externalities, such as the creation and collapse of asset bubbles or – worse—the exhaustion of essential resources for which there is no substitute.

The second (entropy) law of thermodynamics says that all physical transformation processes in the real world are irreversible.[1] The irreversibility has been called "time's arrow" (Eddington 1928). From a macro-perspective, we are interested in "order", one of those words that are hard to define, but which we recognize when we see it. In statistical mechanics orderliness is expressed in terms of *gradients* of temperature, pressure, density or material composition. More order, as in complex structures, means more and sharper gradients, although gradients alone do not define order. It is known that complex "structures" can be self-organized" and maintained far from equilibrium, by fluxes of free energy (exergy) from outside the system e.g. (Nicolis and Prigogine 1977; Prigogine and Nicolis 1989). It is now suspected by many physicists and biologists that complex structures, from weather systems to living organisms and human societies, exist as consequences of the entropy law, operating far from thermodynamic equilibrium (Prigogine and Stengers 1984; Prigogine and Nicolis 1989; Brooks and Wiley 1986–1988) (Schneider and Sagan 2005).

Such fluxes of free energy may be realized as sunlight, nuclear reactions in stars, chemical energy in fossil fuels, or – in the human context – as electric power. These fluxes drive heat engines (transforming heat into kinetic energy, or kinetic energy into electricity generation) and materials transformations – such as reduction of metals from ores, ammonia synthesis and are ultimately, after the products themselves reach the end of usefulness – converted into useless waste heat. Material products are the end-points of chains of such transformations. Evidently living organisms (and the economy) depend upon a continuing supply of useful energy (exergy). Evidently, too, "recycling" is a transformation process that requires useful energy, whence such ideas as "zero waste" and "the circular economy" are misleading if taken too seriously and potentially dangerous.[2]

[1] Textbooks refer to "reversible" processes, but only in idealized situations, for explanatory purposes.

[2] It should be noted that the idea, still popular among some ecologists, that the al nutrients are recycled by the ecosystem is false; fossil fuels, iron ore, limestone and phosphate rock are just

The earliest economists cannot be criticized for lack of awareness of the role of energy in economics since energy and the basic principles of thermodynamics were not even understood by physicists until late in the nineteenth century. The only mainstream economist to grasp the importance of thermodynamics in economics (though he did not fully understand it) was the Rumanian émigré, Nicholas Georgescu-Roegen who wrote a book "The Entropy Law and the Economic Process" in 1971 (Georgescu-Roegen 1971).This book had a significant effect on one branch of science (ecology), but it very little impact on mainstream economists. Moreover, it must be said that the "physics" in the G-R book was somewhat at odds with mainstream physics.[3] However, G-R did make a strong case for the idea that the earth contains a limited stock of "low entropy" resources that are being exhausted by human civilization, with predictably bad long-term consequences. He was thinking mainly of mineral (including hydrocarbon) resources) although others would certainly include biological resources as well.

G-R's idea was adopted (without attribution) by the authors of the famous Report to the Club of Rome "Limits to Growth". (Meadows et al. 1972). Their model predicted economic collapse within a few decades, brought about partly by resource exhaustion and partly by environmental disaster. The book, which sold literally millions of copies, was based on a dynamic modeling scheme pioneered by MIT professor Jay Forrester (Forrester 1971, 1961). Mainstream economists reacted very negatively both to its thesis and to its analytic methodology. The first critic was William Nordhaus of Yale, who called it "measurement without data" (Nordhaus 1973).

The next wave of critics argued that "Limits to Growth" did not explicitly allow for technological substitutions of apparently scarce resources. Based on standard neoclassical economic theory, using the tools of classical physics in the form of utility (or consumption) maximization[4] several mainstream theorists announced authoritatively that resource exhaustion was not a problem, provided only that the "elasticity of substitution" (between abstract labor and abstract capital) remained at a sufficiently high level. (Nordhaus 1973) (Solow 1974a, b) (Stiglitz 1974; Dasgupta and Heal 1974) (Goeller and Weinberg 1976). What the pundits had really assumed (without noticing) was that there are no resources without a substitute. Of course energy (exergy) has no substitute.

This remarkable result was from a simple abstract model of a *closed* economy, with two factors of production (labor and capital) controlled optimally, in the sense

some of the examples of biological waste products that have become "ores" for humans. They were not recycled biologically.

[3] G-R denied that energy and mass are inter-convertible and proposed an unnecessary "Fourth Law" of thermodynamics, conservation of matter. He also concluded, incorrectly, that solar power is too diffuse to be an effective substitute for fossil fuels. This error was due to his lack of understanding of technology.

[4] In those days, "optimal controls" (borrowed from missile technology and classical physics) were "hot" subjects for theoretical economists e.g. (Pontryagin and et al. 1962; Mirowski 1989).

of maximizing consumption over time. This micro-model has little resemblance to the real open economy, especially in regard to the role of material/energy resources. But the conclusion, from over-simplified theory and sophisticated mathematics, seemed to be confirmed by a case study of the numerous applications of one scarce metal, mercury, and its various substitutes. This study was published as an article "The Age of Substitution", in *Science* (Goeller and Weinberg 1976). It was later reprinted in the *Journal of Economic Literature* because of its apparent relevance to the issue of substitutability.

As to the environmental problems arising from imperfect markets, such as the famous "tragedy of the commons"(Hardin 1968), the mainstream critics of "Limits" assumed that they could be resolved via tort law (Coase 1960) or eliminated by "internalizing" the externalities i.e. by creating markets for environmental property rights. (For a discussion of this approach see (Kneese 1977) pp 123 et seq.) The "invisible hand" of free markets would then assure that environmental harms and services would be correctly priced in a Pareto optimal economy. In the 1970s, environmental economists like Kneese were more sanguine than their successors are today, about the feasibility of market-based solutions, such as the "cap and trade" proposals.

The energy crisis resulting from the OPEC oil embargo of 1973–1974 definitely raise a lot of consciousness, especially as it occurred only a year after the "Limits to Growth" was published. In the first place it confirmed the fact that geologist M. King Hubbert's 1956 prediction of US peak output in 1969–1970 was uncomfortably on the mark (Hubbert 1956). When the embargo came along, 3 years later, the US had no spare capacity in the oil industry. Shortages resulted. Oil pricing power moved from Texas to OPEC, more or less overnight.

The magnitude of the oil price increase, and the severity of the resulting recession caught the attention of politicians and business leaders, energy professionals and economists as well as consumers. Inflation, which had been creeping up throughout the 60's was already a serious problem.[5] The concept of net energy and net energy analysis (NEA) emerged more or less synchronously with the embargo. Two new journals, *Energy* and *Energy Policy* were born in 1974–1976. The American Physical Society sponsored a summer study in the same year (conceived at a meeting in Los Alamos in 1973), held at Princeton. The report, since published as a book, was entitled "Efficient Use of Energy: A Physics perspective" (Carnahan et al. 1975). That study was an intellectual *tour de force*, because for the first time it created a consistent terminology and framework for efficiency calculations, based on thermodynamic theory. It also provided exemplary efficiency numbers for a range of generic energy-consuming products and services, ranging from refrigera-

[5] Inflation rose further in the 70s and still further when the second oil crisis erupted as a consequence of the overthrow of the Shah of Iran in 1979 and the following Iran-Iraq war. The Federal Reserve, under Chairman Paul Volcker "killed" the US inflation by raising interest rates to more than 20 % in 1982. This resulted in another deep recession. But oil prices began to decline sharply, partly due to reduced demand, and that kicked off the Reagan boom.

tion and air-conditioning, to pumps, water heaters, storage batteries, lighting and automobiles.

The APS summer study also introduced the important distinction between "first-law" efficiency and "second-law" efficiency. The former, which is commonly used without identification as such, is a simple ratio of "useful" energy output to energy input, subject to the mass/energy balance criterion. An example would be the efficiency of a gas-fired water heater, in terms of the ratio of heat actually absorbed by the water to the heat produced by the flame. An efficiency of 90 % in this case would mean that only 10 % of the heat generated by the gas flame is lost as hot gases up the chimney and the rest is absorbed by the water in the tank. However, what this definition of efficiency misses is the fact that the hot water delivered to the bath or the radiator in the house is much lower in temperature than the hot gases generated by the flame. Low temperature heat is much less "useful" than high temperature heat. So the second law efficiency, in this case, would be a ratio of "useful" output (the temperature difference between the water in the bath and the ambient temperature in the room, *vis a vis* the difference between the flame temperature or the temperature of the exhaust gases, and the ambient temperature in the room.

For example, if the bath temperature is 20° C above ambient while the flame temperature is 1500° C above ambient, we have a second law efficiency of 20/1500 or 1.33 %. This is a fairly extreme example. The same hot water system would look better if the high temperature input were actually lower. For instance, one might argue that the temperature of the flame is not really relevant, because what counts is the temperature of the exhaust gases, which might be 1000° C. In that case the efficiency would be 2 %. Of course some of the temperature differential may be captured "on the way down", as it were, thereby increasing the overall efficiency of the system. In fact the basic idea of "combined heat and power" (CHP), is that high temperature heat may be utilized first in a heat engine of some sort, producing electric power, while the waste heat from the steam engine condenser is still warm enough to heat a bath (or a house).

It is important to note that second-law efficiencies are always lower than first law efficiency, except in the case of electric power generation, where they are the same (typically around 33 % for large central power stations). But, sadly, most people, except specialist engineers, use the more convenient, but less meaningful first law definition of efficiency. The term "exergy" used previously in this paper as a term for *useful* energy, takes into account the temperature, pressure or other differences between inputs and outputs. So, the ratio of exergy in to exergy out is exactly the same as the second law efficiency of any transformation.

As already mentioned net energy analysis (NEA) became popular in the early 1970s, also in response to the crisis. Presidents Nixon, Ford and Carter all grappled with the issue. During the 1973 crisis two long-lived measures were adopted in addition to an emergency petroleum allocation law (1973). They were a national speed limit, and daylight saving. In 1975 came the Energy Policy and Conservation Act under Ford. And in 1977 Carter created the Department of Energy (DOE) with significant regulatory and other powers, including conservation. (DOE absorbed the former Atomic Energy Commission and thus became the official sponsor of nuclear power in the US.)

One emergent problem was to answer questions such as: how much useful energy (exergy) is required to produce a given amount of "final" exergy e.g. how much exergy is needed to produce a kilo-joule of ethanol (for motor fuel)? Or how much exergy is needed to drive an electric car one km? The methodology of net energy analysis or NEA (which requires input-output analysis) can be applied to any given product or service (Bullard and Herendeen 1975; Bullard et al. 1978). For instance, how much exergy does it take to make a tire? How much exergy is needed to produce a ton of steel? How much to produce a bushel of wheat? How much goes into a bar of chocolate or a cup of coffee? Then, as the methodology was refined, more controversial questions could be addressed. For instance: which is more efficient exergetically: the use of paper diapers that can be thrown away or cloth diapers that have to be washed? Or, more recently, how does a hard-cover book printed on paper compare with an e-book? In principle, NEA is about exergy flows, and especially indirect inputs, but it is a specialized version of life cycle analysis. In recent years the complexity of the analysis for any given product is such that full-blown NEA studies are now rare.

There have been several attempts to argue that the GDP corresponds so closely with the energy consumption of modern societies that there is a fixed, or nearly fixed relationship between money value and energy consumption e.g. Costanza (Costanza 1980, 1982), Garrett (Garrett 2011). The implication of such a relationship would be that increased exergy efficiency is not a realistic policy goal. Some of the economists who emphasize the importance of the so-called "rebound effect" seem to agree (Khazzoom 1980, 1987; Brookes 1990; Saunders 1992). However I think this supposed fixed relationship is based on inadequate data, and that exergy consumption per unit of GDP can be reduced significantly in the future, with greater investment in efficiency and new technology.

The "bottom line", so to speak, is that net energy analysis stopped too soon. It is not enough to know how much exergy is "embodied" in a specific product made using yesterday's technology. It is more important to know how much exergy is embodied in the $ output of the economy as a whole, and how much exergy is embodied in each additional increment of output, at the margin. And, by the way, it is important to know how much $ output would decline if the availability of exergy were to decline. Finally it is important to know how these relationships may change in the future. That is what production theory should be telling us.

Having made a few points about theoretical deficiencies in economic growth theory, it may be pertinent to comment on some of the real-world implications. The rest of this chapter is a discussion of what went wrong in the prequel to the 1929 crash, how it was fixed in the 1930s and then unfixed in the '1980s and '1990s, leading to the crash of 2008 and the Great Recession that followed. I conclude with some thoughts on what another better fix might look like.

3.2 The Current Economic Malaise

To reiterate a key message from the above paragraphs, useful energy (exergy) is essential to all productive activity. Capital, whether physical or monetary, produces nothing until it is invested in physical plant and equipment plus economically useful information services. "Plant", in the sense of buildings (walls, floors, roofs), roads, bridges, tunnels, pipes, and wires is also unproductive without exergy for activation. The activating agent is useful energy: food for workers, electric power, mechanical power or high temperature heat to drive metallurgical or chemical processes. Energy is so fundamental, in fact, that all material products can be regarded as "congealed" energy.[6] This is because every step in the sequence of materials transformation processes from extraction through purification, reorganization at the molecular level, shaping and forming, and final assembly, not to mention distribution, involve the consumption (and destruction) of exergy.[7]

One cause of the economic malaise is that conventional economic theory seriously under-rates the economic importance of energy.[8] Energy (exergy) has also been under-valued (too cheap) in the real world for a long time. It is this undervaluation that has led to a consumption-based "throwaway" economy, now threatened by its own waste products. This is important for the present discussion because energy services were declining in cost (and price) from the eighteenth century until the late twentieth century except for a few interruptions. During those two centuries the declining cost of useful energy (and useful work) has been a major driver of economic growth. But that long-term decline is now in the process of ending.

Another crucial fact is that climate change driven by accumulation of greenhouse gases (GHGs) in the atmosphere is already happening, and all signs point to acceleration. Storms are becoming more violent, arctic and glacier ice are melting, the West Antarctic ice sheet is (slowly) sliding into the sea, and the sea-level is rising. Climate change is due to the so-called "greenhouse effect", which, in turn, is due to mainly to the atmospheric accumulation of combustion products (such as CO_2) of fossil fuels. (By-products of agriculture, especially methane, are also important). The possible policy responses to climate change cannot be summarized and compared in a few words. But the "bottom line", in business-speak, is that, to prevent climate catastrophe, fossil fuel combustion must be cut rather sharply in the coming decades.

But the two centuries of declining energy (and work) costs is now in the process of ending. The ending may not look like a sharp corner ("peak oil") but there are many indications that energy prices, and oil prices in particular, will rise in coming

[6] To reiterate a point made earlier in footnote 2, the statement that energy is essential (and under-valued) is not equivalent to an assertion that energy is a measure of economic value. It is not.

[7] The word "destruction" seems odd, but it is accurate. Energy is conserved (first law of thermodynamics) but the useful component (exergy) is not conserved. It is used up (destroyed) in every activity or process.

[8] The mathematical arguments are too complex for explanation here. For details see (Ayres et al. 2013; Kuemmel et al. 2010) (Ayres et al. 2013; Kuemmel et al. 2010).

decades, even though episodes of temporary decline are also likely. Yet, whether due to rapid increases in demand from China and the developing countries, to natural scarcity (of cheap petroleum) or due to legal restrictions on fossil fuel use, or all three, it is very probable that carbon prices will rise in the future, especially if economic growth outside the OECD accelerates.

On the other hand, the costs of energy efficiency technologies (EETs) and renewable energy technologies (RETs), has been declining and will continue to decline (Ayres 2014) pp 282–300. This decline is a natural consequence of R&D, economies of scale, and "learning-by-doing" (Arrow 1962) and the so-called "experience curve" (Wene 2000). (The experience curve says that every time the total cumulative number of "widgets" produced doubles, the cost per widget declines by a (relatively) fixed fraction.[9] This phenomenon is, of course, the major reason why mass-produced computer chips, flash drives, and cell phones are now so cheap. It applies equally to photovoltaic (PV) cells, and panels, wind turbines and batteries.

A second set of facts relates to the state of the world financial system and the world economy. One fact is that, since the Reagan era, worker protection has been eroded, unions have been weakened, and corporate profits have soared to record heights. Government tax revenues have also been cut, as a share of nominal GDP, and squeezed in real terms by declining corporate and personal tax rates. Deficits have been covered by borrowing. But government debt constitutes a barrier to increased discretionary spending on problems like climate change. (Military spending has taken the lion's share of the available discretionary tax dollars.) Since the financial collapse of 2008 and the deep recession that followed, economic growth has been slow and unemployment has been high, especially of young people. This situation is socially unsustainable; it needs to be changed.

Another disturbing fact is that the financial system is increasingly unstable and the scope and magnitude of the potential consequences of instability are growing. This is partly due to economies of scale – banks are getting bigger and bigger – and partly to increasing interconnectedness and globalization. The negative externalities caused by purely financial activities such as asset "bubbles", credit denial to small business and rising credit card and mortgage debt default have simultaneously increased enormously in scale and scope since the 1980s. The negatives due to financial instability constitute a serious challenge to the future of capitalism and democratic society.

The stock market crash in 1929 was mainly due to buying stocks in enterprises (many of which were largely fictional) with borrowed money using the stocks themselves as collateral. A series of legislative "fixes", including the bank reform (Glass-Steagall) act of 1933, stabilized the system. A key component of that Act was the separation of commercial banking from investment banking. However that (and other) elements of the post-1929 reforms, have since been undone by legislation sponsored by the financial industry. Inadequate regulation enabled banks to participate in a Ponzi game where they created home mortgages for unqualified buyers,

[9] This is a rule-of-thumb, not a law of nature, and there are many exceptions. However it is a helpful place to start thinking about likely future cost trends.

sold them to government-backed agencies (e.g. Fannie-mae) that provided guarantees, packaged them into bonds and resold the bonds to pension funds or insurance companies. The end result was another crash and a large amount of financial wealth (especially home equity) destroyed.

Economic growth has become even more imperative than it was in the past, because the expected "recovery" from the deep recession of 2008–2009 is much slower than mainstream economists had expected. The reasons for the slow recovery are multi-fold, but the key point is that the sovereign debt crisis is the excuse for austerity policies that are currently in effect, both in the US and Europe. Those austerity policies have increased unemployment and decreased government revenues, causing a downward spiral that may now be slowing but that shows no signs of vigorous reversal. This is mainly because under-capitalized (excessively leveraged) banks, operating under risk-management guidelines imposed by the Bank for International Settlements (BIS) in Basel, Switzerland, still favor bank lending to governments, government sponsored enterprises (GSEs) and mortgages, *instead of* lending to business, especially small business.

The new rules also favor short-term lending over long-term lending. But small and medium sized business (SMEs) accounts for a large fraction of the economies of all OECD countries, but especially the southern tier (Greece, Italy, Spain and Portugal). Yet small business, almost the only source of employment growth (given the cost-cutting proclivities of big business) is at the bottom of the lending priority list under current Basel (BIS) rules.

3.3 The Win-Win Strategy

The facts, taken together, suggest that a growing share of private and corporate) savings must be redirected away from private consumption and home ownership to "de-carbonization". By that, I mean breaking the "addiction" to carbon-based fossil fuels in general, and to oil and gas in particular. The need to break that addiction will soon become increasingly evident, even to bankers. Moreover, it is imperative that renewed (and accelerated) economic growth be achieved by saving and investing, without depending primarily on government subsidies, or borrowing. But accelerated growth without increased government borrowing is a lot easier to recommend than to achieve. The way out of this dilemma lies in another direction.

Suppose the role of financial innovation must be to direct private investment toward two goals (1) decarbonization (renewables) combined with (2) energy independence. (The current "fracking" boom came out of a misplaced emphasis on goal #2 without goal #1.) Luckily, some of the financial techniques that were used in pursuance of a well-intentioned but poorly executed policy of increasing home ownership prior to 2008, can now be employed to pursue a different and necessary objective. In fact, large scale investing in renewables may be the way to achieve an economic triple-win ("trifecta"): (1) cut green-house gas (GHG) emissions, (2) stimulate innovation-based economic growth and employment, and (3) offer long-term investment opportunities for the insurance companies and pensions funds that were so eager a decade ago to buy mortgage-based securities (CDOs) with fraudulent AAA ratings from Wall Street banks.

To make this happen sooner than later, one financial innovation from the 1980s is relevant: *securitization.* A far-sighted Wall Street investment bank may see advantages in creating securities (like CDOs) based on revenue streams from a wide variety of energy efficiency projects (EETs) and renewable energy technologies (RETs). Admittedly, some of them will fail while others will succeed spectacularly. *But these investments, in aggregate, will increase in value because of the combined effect of rising oil prices and declining renewable costs (rather than by rising real estate or financial asset prices).*

The financial tools to create investment vehicles for long-term investors like pension funds already exist, with one gap. Those tools, such as securitization, by bundling large numbers of diverse individual project investments, created the sub-prime mortgage-based bonds and CDOs that caused the financial collapse. But that collapse occurred because it was truly an unsustainable bubble: it was driven by rising real-estate prices and it collapsed when prices stopped rising. An investment boom financed by decarbonization CDOs could (and should) continue for many decades, until truly long-term base-power solutions are developed. Among them could be geothermal power, tidal power, fusion power, nuclear power based on thorium (not uranium), or even solar power from solar satellites or the moon. It is the medium term – the next 20 or 30 years that will require major efficiency investments in existing buildings and expanded utilization of technologies like wind farms, solar rooftops and solar concentrators that are currently available at scale. These technologies suffer from low capacity utilization due to intermittency. But the problem of intermittency can be ameliorated by increased grid inter-connection and advanced electricity storage technology.

There is one gap in the financial tool arsenal. The missing piece is a mechanism for assuring liquidity of the long-term bonds or other securities. A key role of investment banks and brokers is to ensure liquidity – to "make a market" – for new securities. This role was performed by the banks in 2003–07 and it was the large inventories of mortgage-based bonds kept for this purpose that caused the paper losses that nearly destroyed the banking system itself when trading stopped (because nobody knew what they were worth). Nevertheless, the commercial banks, or perhaps the World Bank or the regional Development banks, would need to perform this role. The difficulty, of course, will be to devise means of providing current valuation for traders, especially during the construction and startup phases of projects. This will be a major challenge, but not an impossible one.

There are some barriers to overcome. Skeptics are not in short supply. Some will say that there is no climate crisis, it's just a hoax or "groupthink" by "greens", so nothing should stop us from exploiting that ocean of oil (and gas) locked in shale that can and will be extracted profitably. Some other skeptics will argue that renewables like wind power and rooftop PV or solar concentrators, or tidal power, or oil from algae or new storage batteries are not profitable now; that they need to be subsidized to make in the market-place, and that such investments are necessarily too speculative. Others (the ones trained in neo-classical economics) will argue that the economy doesn't really depend much on energy, because energy has (has had) such a small cost share in the GDP. It follows from that assumption, that economic growth can continue indefinitely and that our grand-children will be a lot richer than we are. It also follows from that line of argument that we don't need to change anything, because "the economy wants to grow" and it will grow automatically.

Many of those skeptics are using arguments based on factual errors, and all of these skeptics are betting that the future will be like the past, even though the game has changed. They are betting on the wrong horse. When that happens, the opportunities for betting on under-rated "outsiders" can be very good. That is exactly the right time to bet. I think the horse-race metaphor is apt and that this is such a time.

3.4 Conclusion

The economic mainstream's profound ignorance of physical reality is unforgivable over a century after the ideas of thermodynamics were clarified by physicists, especially at a time when both the continuing availability of easily accessed (cheap) fossil fuel energy (specifically oil) is in question. And ignorance of physics by economists of all stripes – not only the "mainstream" – is especially dangerous at a time when the retention of chemical wastes from energy "production" (by fossil fuel combustion) in the atmosphere is now arguably humanity's most pressing problem. Unusually for economics, this is a criticism which applies to both the main rival schools of economic thought: the dominant "Neoclassical" school and the minority "Post Keynesian" school. The common practice in both schools is to treat output as being produced by labor and capital alone, and to not consider the waste that is necessarily generated by production. Both schools also perpetuate the comfortable (but false) idea that externalities are minor, unimportant distortions in markets that are normally in or very near equilibrium (Solow 1987).

In fact, the idea (that economic externalities are of negligible importance) was demolished in 1969 by pointing out that all goods and services produced in the economy depend on the materials extraction and processing activities that generate large waste residuals. Thanks to the first law of thermodynamics, the fraction of materials extracted, *but not embodied in final goods*, cannot simply disappear and therefore becomes a residual (Ayres and Kneese 1969). Actually the great preponderance of materials that are extracted from the Earth's crust, end up as wastes or pollutants. Thus, wastes and pollutants are unavoidable consequences of extraction and production. Yet the economics profession still ignores this key fact by constructing economic models that treat inputs and outputs as immaterial abstractions and presume equilibrium to start with.

Notwithstanding a general absence of scientific qualifications, economists claim to have much to say about the costs of responding to the challenge of climate change (e.g. (Nordhaus 1991a, b, c, 1992, 1993, 1998) (Stern 2008; Stern et al. 2006). They think of "costs" in terms of lost economic growth. But their approach is predicated on the hidden assumption that our grand-children will be much richer than we are, because economic growth would continue at historical rates in the absence of new government interventions (e.g. taxes or regulation). No mainstream economists have considered the possibility that economic growth cannot continue at historical rates in a world of constrained energy ("peak oil"). Strangely, the mainstream has recently recognized the notion of "secular stagnation" without any awareness of its probable link to decreasing energy availability.

Having said this, there is a light at the end of the tunnel, namely large-scale investment in decarbonization and energy efficiency. The rising price of oil will make this profitable, and the increased investment will stimulate growth, employment and profitable investment opportunities for pension funds and insurance companies that need higher returns.

References

Arrow KJ (1962) The economic implications of learning by doing. Rev Econ Stud 29:155–173

Ayres RU (1989) Industrial metabolism. In: Jesse A, Sladovich HE (eds) Technology and environment. National Academy Press, Washington

Ayres RU (2014) The bubble economy. MIT Press, Boston

Ayres RU, Kneese AV (1969) Production, consumption and externalities. Am Econ Rev 59(June):282–297

Ayres RU, van den Bergh JCJM, Lindenberger D, Warr B (2013) The underestimated contribution of energy to economic growth. Struct Chang Econ Dyn 27:79–88

Brookes L (1990) Energy efficiency and economic fallacies. Energy Policy 18(2):199–201

Brooks DR, Wiley EO (1986–1988) Evolution as entropy: towards a unified theory of biology. In: Hull DL (ed) Science and its conceptual foundations. University of Chicago Press, Chicago

Bullard CW III, Herendeen RA (1975) The energy costs of goods and services. Energy Policy 3(4):268–278

Bullard CW, Penner PS, Pilati DA (1978) Net energy analysis: handbook for combining process and input–output analysis. Resour Energy 1:267–313

Carnahan W, Ford KW, Prosperetti A, Rochlin GI, Rosenfeld AH, Ross MH, Rothberg JE, Seidel GM, Socolow RH (1975) Efficient use of energy: a physics perspective. American Physical Society, New York

Coase RH (1960) The problem of social costs. J Law Econ 3(October):1–44

Costanza R (1980) Embodied energy and economic valuation. Science 210(December):1219–1224

Costanza R (1982) Economic values and embodied energy. Science 216:1143

Dasgupta P, Heal G (1974) The optimal depletion of exhaustible resources. In: Symposium on the economics of exhaustible resources, 41: 3–28

Denison EF (1984) Accounting for slower economic growth: an update. In: Kendrick JW (ed) International comparisons of productivity and causes of slowdown. Ballinger, Cambridge, pp 1–45

Eddington A (1928) The nature of the physical world. Cambridge University Press, Cambridge

Forrester JW (1961) Industrial dynamics. MIT Press, Cambridge

Forrester JW (1971) World dynamics. Wright-Allen, Cambridge

Garrett TJ (2011) Are there physcal constraints on future anthropogenic emissions of carbon dioxide? Climate Change 3:437–455

Georgescu-Roegen N (1971) The entropy law and the economic process. Harvard University Press, Cambridge

Goeller H, Weinberg A (1976) The age of substitutability. Science 191(February):683–689
Hardin G (1968) The tragedy of the commons. Science 162:1243–1248
Hubbert MK (1956) Nuclear energy and the fossil fuels. Shell Development Corporation, Houston
Khazzoom JD (1980) Economic implications of mandated efficiency standards for household appliances. Energy J 1(4):21–39
Khazzoom JD (1987) Energy savings resulting from the adoption of more efficient appliances. Energy J 8(4):85–89
Kneese AV (1977) Economics and the environment. Penguin, New York
Kneese AV, Ayres RU, d'Arge RC (1970) Economics and the environment: a materials balance approach. RFF Press distributed by Johns Hopkins University Press, Baltimore
Kuemmel R, Ayres RU, Lindenberger D (2010) Thermodynamic laws, economic methods and the productive power of energy. J Non-equilib Thermodyn 35:145–181
Madrick J (2014) Seven bad ideas: how mainstream economics has damaged America and the world. Alfred Knopf, New York
Meadows DL, Meadows DH, Jorgen R, William B III (1972) The limits to growth, club of Rome reports. Universe Books, New York
Mirowski P (1989) More heat than light: economics as social physics; physics as nature's economics. In: Goodwin CD. paperback (ed) Historical perspectives on modern economics. Cambridge, Cambridge University Press. Reprint, 1999
Mishan EJ (1967) The costs of economic growth, 1st edn. Staples Press, London
Nicolis G, Prigogine I (1977) Self-organization in non-equilibrium systems. Wiley-Interscience, New York
Nordhaus WD (1973) World dynamics: measurement without data. Economic J 83(December):1156–1183
Nordhaus WD (1991a) The cost of slowing climate change: a survey. Energy J 12(1):37–66
Nordhaus WD (1991b) A sketch of the economics of the greenhouse effect. Am Econ Rev Pap Proc 81:146–150
Nordhaus WD (1991c) Unknown. Science (September):1206
Nordhaus WD (1992) The DICE model: background and structure of a dynamic integrated climate economy model of the economics of global warming. Cowles Foundation, New Haven
Nordhaus WD (1993) Rolling the'DICE': an optimal transition path for controlling greenhouse gases. Resour Energy Econ 15(1):27–50
Nordhaus WD (1998) Roll the DICE again; the economics of global warming. Yale University, New Haven
Pontryagin LS et al (1962) The mathematical theory of optimal processes. Wiley, New York
Prigogine I, Nicolis G (1989) Exploring complexity. W. H. Freeman and Company, San Francisco
Prigogine I, Stengers I (1984) Order out of chaos: Man's new dialogue with nature. Bantam Books, London
Saunders H (1992) The khazzoom-brookes postulate and neoclassical growth. Energy J 13(4):131–148
Schneider E, Sagan D (2005) Into the cool: energy flow, thermodynamics and life. University of Chicago Press, Chicago
Solow RM (1974a) The economics of resources or the resources of economics. Am Econ Rev 64(2):1–14
Solow RM (1974b) Intergenerational equity and exhaustible resources. Rev Econ Stud 41:29–45
Solow RM (1987) Growth theory and after. Lecture, Nobel Prize in Economics, Stockholm, December 8
Stern N (2008) The economics of climate change. Am Econ Rev Pap Proc 98(2):1–37
Stern N, Peters S, Bakhshi A, Bowen C, Cameron S, Catovsky D, Crane S, Cruikshank S, Dietz N, Edmondson S-L, Garbett I, Hamid G, Hoffman D, Ingram B, Jones N, Patmore H, Radcliffe R, Sathiyarajah M, Stock C, Taylor T, Wanjie Vernon H, Zeghelis D (2006) Stern review: the economics of climate change. HM Treasury, London

Stiglitz J (1974) Growth with exhaustible natural resources. efficient and optimal growth paths. Review of economic studies 41 (Symposium on the economics of exhaustible resources)
Wene C-O (2000) Experience curves for energy technology policy. International Energy Agency (OECD), Paris

Chapter 4
A Further Critique of Growth Economics

Herman Daly

Abstract *Is aggregate physical growth still economic, as it was in the empty world, or has it become uneconomic as we have moved into the full-world era?* First I discuss why this critical question has been ignored by mainstream economists and policy makers. Second I consider 11 fallacies or anomalies that confuse reasoning about "economic growth", and undercut the current policy presumption in its favor.

Keywords Growth • Sustainable development • Development • Wealth • GDP • World Bank • Physical limits • Resource throughput • Resource flows • ISEW • Capital • Labor • Unemployment • IMF • WTO • Trade

4.1 Introduction

Some years ago I wrote a critique of the "Growth Report", a 2-year study by the prestigious international Commission on Growth and Development, published by the World Bank.[1] Here I would like to reflect on the "reaction" to my review – specifically that it was ignored! Many issues and many people are deservedly ignored. But should we ignore the question of whether growth still increases wealth faster than "illth", as it did in the past empty world, or whether in the new full world it has begun to increase illth faster than wealth? Is growth still economic in the literal sense, or has it become uneconomic? This question was not asked in the Growth Report, and consequently was the main question raised in my review. Surely it is not a trivial question, and my discomfort at seeing it roundly ignored transcends the mere personal pique that one feels at being brushed off. So I will begin with a few remarks on why I think my critical review failed to initiate a dialog with the authors

[1] "Growth and Development: Critique of a Credo", in Population and Development Review, **34** (3), September 2008. (Review of Commission on Growth and Development, The Growth Report: Strategies for Sustained Growth and Inclusive Development. Washington, DC: The World Bank, 2008. xiii+180 p. $26.00 (pbk.). Available at «https://openknowledge.worldbank.org/bitstream/handle/10986/6507/449860PUB0Box3101OFFICIAL0USE0ONLY1.pdf»

H. Daly (✉)
Emeritus, School of Public Policy, University of Maryland, College Park, MD, USA
e-mail: hdaly@umd.edu

S. Shmelev (ed.), *Green Economy Reader*, Studies in Ecological Economics 6,
DOI 10.1007/978-3-319-38919-6_4

of the Growth Report, and why I think that is indicative of a deeper failing within the economics profession. Following that I will consider the 11 fallacies and confusions that in my experience most frequently obstruct reasoning about growth.

4.2 The Growth Report

The "Growth Report" was done by a blue ribbon panel of 18 members from 16 countries, including two Nobel laureates in economics. It had many august sponsors, the main one being the World Bank. It can fairly be taken to represent the prevailing orthodox view on growth. My review was quite critical. I expected a debate, or at least a reply from the authors of the report. As indicated, they ignored it. Is this fact insignificant, or like Sherlock Holmes' dog that failed to bark in the night, might it be the clue to solving a mystery?

A few sympathetic former colleagues at the World Bank made sure that a copy of my review was sent to authors connected with the World Bank, with the suggestion that a reply was in order. The editor of *Population and Development Review* renewed his offer to the authors to publish their reply, if they chose to make one. No reply. I realize, of course, that one could waste a lot of time replying to all critics. Some critics are morons. Forgive my immodesty, but for the time being I am assuming that I am not a moron.

Might there be other reasons for silence? Certainly the Commission did not lack intellectual firepower or financial backing for a reply. I think perhaps they made a political calculation of interest and advantage. What would be gained from their point of view by a reply? A blue ribbon panel of experts is presumed to be correct, (especially if defending growth!) and a single critic is presumed to be wrong. Why risk upsetting that default presumption with a reply? The Report, after all, was a political manifesto (that is why it had so many co-authors and sponsors), a hymn to growth in the guise of an objective study. It had been widely and favorably reported by the establishment media and therefore had already achieved it's goal – namely, to counter the emerging and threatening suspicion that the economic growth of the past empty-world era was morphing into uneconomic growth in the new full-world era. Scholarly debate about the correctness of the report, and the continued viability of growth as the supreme goal of all nations, was not on the agenda – it was very much off message. Probably the authors believed that the case for growth was so ironclad and obvious that any defense of it against criticism was unnecessary. But then, why did they bother to mount such a grand defense of growth in the first place?

I tell this story because it illustrates the unhappy state of public discourse on economic matters, and the lack of seriousness of many economists engaged in such discourse. *The Journal of Economic Perspectives*, for example, has a policy of not printing comments on articles they have published. Perhaps because they would get too many comments, exposing too much disagreement? Or so few comments because there is such a consensus among economists? Other economics journals do publish comments and replies, but it seems that this practice is less frequent than in the past. Why comment on someone else's work – there is not much academic credit in so doing. Correcting errors may be a necessary part of science, but since economics is not a science anyway, why waste time on it? Besides, you might make an enemy. Furthermore, consensus among experts is considered the hallmark of a mature science, so by prematurely declaring a consensus among "all *competent* economists", and avoiding public debate on fundamental questions, economists preemptively lay claim to the status of a mature science.

The advantage of a reputation as a "mature science" is that economists can profitably sell themselves as credibility-enhancing professional consultants to all sorts of interest groups. This was convincingly demonstrated in the documentary film, *"Inside Job"*, detailing the disgraceful behavior of some prominent economists leading up to the 2008 financial debacle.

Pointing to the silence of others when invited to reply to criticism, while a fair debating tactic, is a less than convincing argument against their position. One needs a more direct and specific critique. That was provided in my review, but limited to the specifics of the Growth Report, and will not be repeated here.

What I have called "silence" could just be lack of a response to my particular review, invited by the editor of the journal in which it was published. Perhaps the authors of the Growth Report responded to other critics in other venues who might have raised the same or different issues. Also the Commission may have responded in their own subsequent publications. A wider review of the literature is in order.

There have been two further publications by the Growth Commission since their main Report in May of 2008. In 2009 they published, Post-Crisis Growth in Developing Countries, which asked if the unforeseen financial crisis of September 2008 (4 months after the publication of their Report) required any important changes in their conclusions. Understandably the Commission was absorbed in considering a massive "critique" of growth-mania coming from the real world. Academic criticisms could wait. The Commission's vision of growth as *summum bonum* remained undiminished, however, and was even reinforced by the crisis. Their next publication, Equity and Growth in a Globalizing World, (2010), provided another opportunity to reply, but there was no direct reference, nor anything that might be construed as an indirect reply.

Google and Google Scholar searches of the Commission, the Report, and of the names of each of the Commission co-Chairmen (Danny Leipziger and Michael Spence) combined with my name, failed to turn up any replies. That did not surprise me as much as did the fact that a search for any reviews of the Report itself turned up only a few, and they were mainly just descriptive summaries. For example, *Amazon.com* urges prospective purchasers to "be the first to review this book". Help from a research librarian who surveyed other data bases failed to turn up critical reviews, replies, or rejoinders. The Commission was not overwhelmed with reviews, perhaps another reason, and an understandable one at that, for their belief that a reply was unnecessary. As lamented earlier, there is not much incentive to write reviews – especially critical ones. Alas, disagreements tend to remain unexpressed, doubtful claims un-debated, and errors uncorrected.

This unwillingness to engage in discussion, from both directions, leads me to reflect more broadly on the major fallacies of growth economics in the more general context of economic and environmental policy. In this larger context these fallacies also played a part in the 2012 US presidential election, and are still present in the 2016 campaigns. The one thing the Democrats and Republicans agree on is that economic growth is our number-one goal and is the basic solution to all problems. The idea that growth could conceivably cost more than it is worth at the margin, and therefore become uneconomic in the literal sense, is not considered, because if true, it would totally overturn the applecart.[2] But, aside from political denial, why do many people (especially economists) not understand that continuous growth of the economy (measured either by real GDP or resource throughput) could in theory, and probably has in fact, become uneconomic? What is it that might confuse them? The remainder of this essay considers eleven confusions or fallacies that frequently serve as "thought-stoppers" in discussions about growth.

4.3 Eleven Confusions About Growth

1. **One can nearly always find something whose growth would be both desirable and possible**. For example, we need more bicycles and can produce more bicycles. More bicycles mean growth. Therefore growth is both good and possible. QED.

However, this confuses aggregate growth with reallocation. Aggregate growth refers to growth in everything: bicycles, cars, houses, ships, cell phones etc. Aggregate growth is growth in scale of the economy, the size of real GDP, which is a value-based index of aggregate production and consequently of the total resource throughput required by that production. In the simplest case of aggregate growth everything produced goes up by the same percentage. Reallocation, by contrast, means that some things go up while others go down, the freed up resources from the latter are transferred to the former. The fact that reallocation remains possible and desirable does not mean that aggregate growth is possible and desirable. The fact that you can reallocate the weight in a boat more efficiently (and even redistribute it more equitably among passengers) does not mean that there is no Plimsoll Line. Too much weight will sink a boat even if it is optimally allocated and justly distributed.

Reallocation of production away from more resource-intensive goods to less resource-intensive goods ("decoupling") is possible to some degree and often advocated, but is limited by two basic facts. First, the economy grows as an integrated whole, not as a loose aggregate of independently changeable sectors. A glance at the input-output table of an economy makes it clear that to increase output of any sector

[2] For a cogent argument that ecological economics must be more willing to overturn applecarts, see, B. Andersen and M. M'Gonigle, "Does Ecological Economics Have a Future? Contradiction and Reinvention in the Age of Climate Change", *Ecological Economics*, **84** (2012) 37–48.

requires an increase in all the inputs to that sector from other sectors, and then a second round of increased inputs required by the first round of input increases, etc. Second, in addition to this supply interdependence of sectors there are demand constraints – people are just not interested in information services unless they first have enough food and shelter. So trying to cut the resource-intensive food and shelter part of GDP to reallocate to less resource-intensive information services in the name of decoupling GDP from resources, will soon result in a shortage of food and shelter, and a glut of information services.

Aggregate growth was no problem back when the world was relatively empty. But now the world is full, and aggregate growth likely costs more than it is worth in terms of sacrificed natural services, even though more bicycles (and less of something else) might still be possible and desirable.

2. Another confusion is to argue that **since GDP is measured in value terms it is therefore not subject to physical limits**. This is another argument given for easy "decoupling" of GDP from resource throughput. But growth refers to real GDP, which eliminates price level changes. Real GDP is a value-based index of aggregate quantitative change in real physical production. It is the best index we have of total resource throughput. The unit of measure of real GDP is not dollars, but rather "dollar's worth". A dollar's worth of gasoline is a physical quantity, currently about one-fourth of a gallon. The annual aggregate of all such dollar's worth amounts of all final commodities is real GDP, and even though not expressible in a simple physical unit, it remains a physical aggregate and subject to physical limits. The price level and nominal GDP might grow forever (inflation), but not real GDP, and the latter is the accepted measure of aggregate growth.
3. A more subtle confusion results from **looking at past totals rather than present margins**. Just look at the huge net benefits of past growth! How can anyone oppose growth when historically it has led to such enormous benefits? Well, there is a good reason: the net benefits of past growth reach a maximum precisely at the point where the rising marginal costs of growth equal the declining marginal benefits – that is to say, at precisely the point where further growth ceases to be economic and becomes uneconomic! Before that point wealth grew faster than illth; beyond that point illth grows faster than wealth, making us poorer, not richer. No one is against being richer. No one denies that growth used to make us richer. The question is, does growth any longer make us richer, or is it now making us poorer? If aggregate growth now makes us poorer, then it can no longer be appealed to as "necessary to end poverty". Ending poverty requires sharing – redistribution rather than more uneconomic growth.

To understand the question requires that we recognize that real GDP has a cost, that illth is a negative joint product with wealth. Examples of illth are everywhere and include: nuclear wastes, climate change from excess carbon in the atmosphere, biodiversity loss, depleted mines, deforestation, eroded topsoil, dry wells and rivers, sea level rise, the dead zone in the Gulf of Mexico, gyres of plastic trash in the oceans, the ozone hole, exhausting and dangerous labor, and the un-repayable debt from trying to push growth in the symbolic financial sector beyond what is possible in the real sector. Since no one buys these annually produced bads (that accumulate into illth) they have no market prices, and since their implicit negative shadow values are hard to estimate in a way comparable to positive market prices, they are usually ignored, or mentioned and quickly forgotten.

The logic of maximization embodied in equating rising marginal cost with declining marginal benefit requires a moment's thought for the average citizen to understand clearly, but surely it is familiar to anyone who has taken Econ 101.

4. **Even if it is theoretically possible that someday the marginal cost of growth will become greater than the marginal benefit, there is no empirical evidence that this has happened yet.** On the contrary, there is plenty of casual evidence for anyone who has not been anesthetized by the official party line of Madison Avenue and Wall Street. As for empirical evidence of the statistical type, there are two independent sources that give the same basic answer. First are the objective measures that separate GDP sub-accounts into costs and benefits and then subtract the costs from GDP to approximate net benefits of growth. The Index of Sustainable Economic Welfare (ISEW) and its later modifications into the Genuine Progress Indicator (GPI) both show that, for the US and some other wealthy countries, GDP and GPI were positively correlated up until around 1980, after which GPI leveled off and GDP continued to rise. In other words, increasing throughput as measured by real GDP no longer increased welfare as measured by GPI. A similar disconnect is confirmed using the different measure of self-evaluated happiness. Self-reported happiness increases with per capita GDP up to a level of around $20,000, per annum, and then stops rising. The interpretation given is that while absolute real income is important for happiness up to a sufficiency, beyond that point happiness is overwhelmingly a function of the quality of relationships by which our very identity is constituted. Friendships, marriage and family, social stability, trust, fairness, etc., not per capita GDP, are the overwhelming determinants of happiness at the present margin, especially in high-income countries. If we sacrifice friendships, social stability, family time, environmental services, and trust – for the sake of labor mobility, a second job, and quarterly financial returns, we often reduce happiness while increasing GDP. Relative income gains may still increase individual happiness even when increases in absolute income no longer do, but aggregate growth is powerless to increase everyone's relative income because we cannot all be above average. Beyond some level of sufficiency, growth in GDP no longer increases either self-evaluated happiness or measured economic welfare, but it continues to increase costs of depletion, pollution, congestion, stress, etc. Why is there such resistance to measuring the very magnitudes that could tell us if we have reached this point? A possible answer follows.
5. Many believe that the way we measure GDP automatically makes its growth a trustworthy guide to economic policy. To be counted in GDP, there must be a market transaction, and that implies a willing buyer and seller, neither of whom would have made the transaction if it did not make them better off in their own judgment. *Ergo*, **growth in GDP must be good or it would not have happened**. The problem here is that there are many third parties who are affected by many

transactions, but did not agree to them. These external costs (or sometimes benefits) are not counted in GDP. Who are these third parties? The public in general, but more specifically the poor who lack the money to express their preferences in the market, future generations who of course cannot bid in present markets, and other species who have no influence on markets at all.

In addition, GDP, the largest component of which is National Income, counts consumption of natural capital as income. Counting capital consumption as income is the cardinal sin of accounting. Cut down the entire forest this year and sell it, and the entire amount is treated as this year's income. Pump all the petroleum and sell it, and add that to this year's income. But income in economics is by definition the maximum amount that a community can produce and consume this year, *and still be able to produce and consume the same amount next year.*[3] In other words income is the maximum consumption that still leaves intact the capacity to produce the same amount next year. Only the sustainable yield of forests, fisheries, croplands, and livestock herds is this year's income – the rest is capital needed to reproduce the same yield next year. Consuming capital means reduced production and consumption in the future. Income is by definition sustainable; capital consumption is not. The whole historical reason for income accounting is to avoid impoverishment by inadvertent consumption of capital. By contrast our national accounting tends to encourage capital consumption (at least consumption of natural capital), first by counting it in GDP, and then claiming that whatever increases GDP is good!

As already noted we fail to subtract negative by-products (external costs) from GDP on the grounds that they have no market price since obviously no one wants to buy bads. But people do buy anti-bads, and we count those expenditures. For example, the costs of pollution (a bad) are not subtracted, but the expenditures on pollution clean up (an anti-bad) are added. This is asymmetric accounting – adding anti-bads without having subtracted the bads that made the anti-bads necessary in the first place. The more bads, the more anti-bads, and the greater is GDP – wheel spinning registered as forward motion.

There are other problems with GDP but these should be enough to refute the mistaken idea that if something is not a net benefit it would not have been counted in GDP, so therefore GDP growth must always be good.

6. **As natural resources become scarce we can substitute capital for resources and continue to grow**. Growth economists assume a high degree of substitutability between factors of production, including capital for resources.[4] But if one considers a realistic analytic description of production, as given in Georgescu-Roegen's fund-flow model,[5] one sees that factors are of two qualita-

---

[3] Hicks, J. R., *Value and Capital: An Inquiry into Some Fundamental Principles of Economic Theory*. Oxford: Clarendon Press, 2nd ed. 1946.

[4] Daly, H. E., Ecological Economics and Sustainable Development, Edgar Elgar, Publishers, Cheltenham, UK, 2007. See chapter 11.

[5] Georgescu-Roegen, Nicholas, The Entropy Law and the Economic Process, Harvard University Press, Cambridge, MA, 1972.

tively different kinds: resource flows that are physically transformed into flows of product and waste; and capital and labor funds, the agents or instruments of transformation that are not themselves physically embodied in the product. There are varying degrees of substitution between different resource flows, and between the funds of labor and capital. But the basic relation between resource flow on the one hand, and capital (or labor) fund on the other, is complementarity. You cannot bake a ten-pound cake with only one pound of ingredients, no matter how many cooks and ovens you have. Efficient cause (capital) does not substitute for material cause (resources). Material cause and efficient cause are related as complements, and the one in short supply is limiting. Complementarity makes possible the existence of a limiting factor, which cannot exist under substitutability. In yesterday's empty world the limiting factor was capital; in today's full world remaining natural resources have become limiting. This fundamental change in the pattern of scarcity has not been incorporated into the thinking of growth economists. Nor have they paid sufficient attention to the fact that capital is itself made and maintained from, as well as powered by, natural resources. It is hard for a factor to substitute for that from which it is made! And consider yet another oversight. Substitution is reversible – if capital is a good substitute for resources, then resources are a good substitute for capital. But then why, historically, would we ever have accumulated capital in the first place, if nature had already given us a good substitute? In sum, the claim that capital is a good substitute for natural resources is absurd.

In reply to these criticisms growth economists often point to modern agriculture, which they consider the prime example of substitution of capital for resources. But modern, mechanized agriculture has simply substituted one set of resource flows for another, and one set of funds for another. The old resource flows (soil, sunlight, rain, manure) were to a significant degree replaced by new resource flows (chemical fertilizer, fossil fuels, irrigation water), not by "capital"! The old fund factors of labor, draft animals, and hand tools were replaced by new fund factors of tractors, harvesters, etc. In other words new fund factors substituted for old fund factors, and new resource flows substituted for old resource flows. Modern agriculture involved the substitution of capital for labor (both funds), and the substitution of nonrenewable resources for renewable resources (both flows). In energy terms it was largely the substitution of fossil fuels for solar energy, a move with short-term benefits and long-term costs. *But there was no substitution of capital funds for resource flows.* The case of mechanization of agriculture does not contradict the complementarity of fund and flow factors in production, nor the new role of resources as limiting factor.

7. **Knowledge is the ultimate resource and since knowledge growth is infinite it can fuel economic growth without limit.** Like many, I am eager for knowledge to substitute physical resources to the extent possible, and consequently advocate severance taxes to make resources expensive, and patent reform to make knowledge cheap. But if I am hungry I want real food on the plate, not the knowledge of a thousand recipes on the Internet. Furthermore, the fact that

knowledge is naturally depleting while ignorance is naturally renewing makes me doubt that knowledge can save the growth economy. Ignorance is renewable, mainly because ignorant babies continually replace learned elders. In addition, vast amounts of recorded knowledge are destroyed not only by death, but also by decay, fires, floods, bombs, and bookworms. Modern digital storage does not seem to be immune to the teeth of time, or to that new bookworm, the computer virus. To be effective in the world knowledge must exist in someone's mind (not just in the library or on the Internet) – otherwise it is inert. And even when knowledge increases, it does not grow exponentially like money in the bank. Some old knowledge is disproved or cancelled out by new knowledge, and some new knowledge is discovery of new biophysical or social limits to growth.

New knowledge must always be something of a surprise – if we could predict its content then we would know it already, and it would not really be new. Contrary to common expectation, new knowledge is not always a pleasant surprise for the growth economy – frequently it is bad news. For example, climate change from greenhouse gases was recently new knowledge, as was discovery of the ozone hole. How can one appeal to new knowledge as the panacea when the content of new knowledge must of necessity be a surprise? Of course we sometimes get lucky with new knowledge, but should we borrow against that uncertainty? Why not count the chickens after they hatch?

8. **Without growth we are condemned to unemployment.** The Full Employment Act of 1946 declared full employment to be a major goal of US policy. Economic growth was then seen as the means to attain the end of full employment. Today that relation has been inverted – economic growth has become the end, and if the means to attain that end – automation, off-shoring, excessive immigration – result in unemployment, well that is the price "we" just have to pay for the supreme goal of growth. If we really want full employment we must reverse this inversion of ends and means. We can serve the goal of full employment by restricting automation, off-shoring, and easy immigration to periods of true domestic labor shortage as indicated by high and rising wages. In addition, full employment can also be served by reducing the length of the working day, week, or year, in exchange for more leisure, rather than more GDP.

Real wages have been falling for decades, yet our corporations, hungry for cheaper labor, keep bleating about a labor shortage. They mean a shortage of cheap labor in the service of growing profits. Actually a labor shortage in a capitalist economy with 80 % of the population earning wages is not a bad thing. How else will wages and standard of living for that 80 % increase? What the corporations really want is a surplus of labor, and falling wages. With surplus labor wages generally do not rise and therefore all the gains from productivity increase will go to profit, not wages. Hence the elitist support for automation, off-shoring, and lax enforcement of democratically enacted immigration laws.

9. **We live in a globalized economy and have no choice but to compete in the global growth race.** Globalization was a policy choice of our elites, not an inevitability. Free trade agreements had to be negotiated. Who negotiated and signed

the treaties? Who has pushed for free capital mobility and signed on to the WTO? Who wants to enforce US intellectual property rights worldwide with trade sanctions? The Bretton Woods system was a major achievement aimed at facilitating international trade after WWII. It fostered trade for mutual advantage among separate countries. Free capital mobility and global integration were not part of the deal. That came with the WTO and the effective abandonment by the World Bank and IMF of their Bretton Woods charter. Globalization is the engineered integration of many formerly relatively independent national economies into a single tightly bound global economy organized around absolute advantage, not comparative advantage which assumes capital immobility internationally. Once a country has adopted free trade and free capital mobility it has effectively been integrated into the global economy and is no longer free not to specialize and trade. Yet all of the theorems in economics about the gains from specialization and trade assume that trade is voluntary. How can trade be voluntary if countries are so specialized as to be no longer free not to trade? Countries can no longer account for social and environmental costs and internalize them in their prices unless all other countries do so, and to the same degree. To integrate the global omelet you must disintegrate the national eggs. While nations have many sins to atone for, they remain the main locus of community and policy-making authority. It will not do to disintegrate them in the name of abstract "globalism", even though we certainly require some global federation of national communities. But when nations disintegrate there will be nothing left to federate in the interest of legitimately global purposes. "Globalization" (national disintegration) was an actively pursued policy, not an inertial force of nature. It was done to increase the power and growth of transnational corporations by moving them out from under the authority of nation states and into a non-existent "global community". It can be undone, as is currently being contemplated by some in the European Union, formerly heralded as the forerunner of more inclusive globalization.

10. **Space, the high frontier, frees us from the finitude of the earth, and opens unlimited resources for growth**. – In a secular age where many have lost faith in the spiritual dimension of existence, and where the concept of "man as creature" is eclipsed by that of "man as creator", it is to be expected that science fiction might be called on to fill the dead void of space with a happy population of the "scientifically raptured". The spiritual insights of centuries are replaced by technocratic projections of the "Singularity" in which mankind attains the final goal of (random?) evolution and becomes a new and immortal species, thanks to the salvific power of exponential growth in information processing technology. Eternal silicon-based life awaits the new elect who can stay alive until the Singularity; oblivion for those who die too soon! And this comes from materialists who think that they have outgrown religion!

Of course many technical space accomplishments are real and amazing. But how do they free us from the finitude of the earth and open up unlimited resources for growth? Space accomplishments have been extremely expensive in terms of terrestrial resources, and have yielded few extraterrestrial resources – mainly those useless moon rocks that incited thievery by a NASA intern. As for new services, space tourism has provided orbital joy rides to a few billionaires. On the truly positive side of the ledger we can list communications satellites, but they are oriented to earth, and while they provide valuable services, they do not bring in new resources. And apparently some orbits are getting crowded with satellite carcasses.

Robotic space exploration is a lot cheaper than manned space missions, and may (or may not) yield knowledge worth the investment to a society that has not yet provided basic necessities and elementary education for most of its inhabitants. In such a world political willingness to finance the expensive curiosity of a scientific elite might be less, were it not for the heavy military connection (muted in the official NASA propaganda). Cuts in NASA's budget have led to the hyped reaction by the "space community" in proclaiming a pseudo religious technical quest to discover "whether or not we are alone in the universe". Another major goal is to find a planet suitable for colonization by earthlings. The latter is sometimes justified by the claim that since we are clearly destroying the earth we need a new home – to also destroy?

The numbers – astronomical distances and time scales – effectively rule out dreams of space colonization. But another consideration is equally daunting. If we are unable to limit population and production growth on earth, which is our natural and forgiving home, out of which we were created and with which we have evolved and adapted, then what makes us think we can live as aliens within the much tighter and unforgiving discipline of a space colony on a dead rock in a cold vacuum? There we would encounter limits to growth raised to the hundredth power.

11. **Without economic growth all progress is at an end.** On the contrary, without growth, now actually uneconomic growth if correctly measured, true progress finally will have a chance. As ecological economists have long argued, **growth** is quantitative physical increase in the matter-energy throughput, the metabolic maintenance flow of the economy beginning with depletion and ending with pollution. **Development** is qualitative improvement in the capacity of a given throughput to provide for the maintenance and enjoyment of life in community. Growth means larger jaws and a bigger digestive tract for more rapidly converting more resources into more waste, in the service of frequently destructive individual wants. Development means better digestion of a non-growing throughput, and more worthy and satisfying goals to which our life energies could be devoted. Development without growth beyond the earth's carrying capacity is true progress. The main ways to develop are through technical improvement in resource efficiency, and ethical improvement in our wants and priorities. Resource efficiency must be an adaptation to lower resource throughput. So far we have sought efficiency independently of limiting throughput and have consequently run into Jevons' Paradox – better efficiency in using a resource tends to increase the total amount used. If we first limit throughput then we will get efficiency increase as a secondary adaptation; if we first seek efficiency increase we secondarily get Jevons' paradox. Limiting physical growth is necessary to force the path of progress on to development. Since

physical growth has become uneconomic one might think that limiting it would not be so controversial! But of course most economists do not admit that growth is, or even could be, uneconomic. They seem determined to avoid discussion of arguments or evidence to the contrary.

4.4 Conclusion

If growth economists will make an effort to overcome these 11 fallacies, and break their guild's stonewalling silence, then maybe we can have a productive dialog about whether or not what used to be economic growth has now become uneconomic growth, and what to do about it. It was too much to hope that the issue of uneconomic growth would make it into the 2012 or 2016 elections, but maybe 2020, ….or sometime?

One can hope. But hope must embrace not just a better understanding regarding these confusions, but also, at a deeper level, more love and care for our fellow humans, and for all of Creation. I say Creation with a capital "C" advisedly, and not in denial of the facts of evolution. If our world and our lives are not in some sense a Creation, but just a purposeless happenstance – a random statistical fluke of multiplying infinitesimal probabilities by an infinite number of trials – then it is hard to see from where we will get the will and inspiration to care for it. Indeed, our decision-making elites may already tacitly understand that growth has become uneconomic. But apparently they have also figured out how to keep the dwindling extra benefits for themselves, while "sharing" the exploding extra costs with the poor, the future, and other species. Why not, if it is all just a purposeless happenstance? The elite-owned media, the corporate-funded think tanks, the kept economists of high academia, and the World Bank – not to mention Gold Sacks and Wall Street – all sing hymns to growth in harmony with class interest and greed. The public is bamboozled by technical obfuscation, and by the false promise that, thanks to growth, they too will one day be rich. Intellectual confusion is real, but moral nihilism, abetted by the prevailing ethos of naturalistic scientism, may be the bigger problem.

Chapter 5
Multidimensional Assessment of Sustainability: Harmony vs. the Turning Point

Stanislav Shmelev

Abstract Multidimensional assessment of sustainability is a way to reconcile the need for simultaneous consideration of various indicators of progress beyond GDP growth with a policy focused visualization of multi-dimensional trends in a clear and transparent manner. The various composite measures used for sustainability assessment often hide the trade-offs between economic, social and environmental dimensions of sustainability. This chapter discusses indicators used for sustainability analysis at the macro scale and offers a multi-criteria sustainability assessment framework. It discusses results that were obtained in sustainability assessments for the USA, Brazil, China, France, Germany, Britain and Russia. The Multicriteria Decision Aid tool, Aggregated Preference Index System (APIS) is used for the assessment with the following three headline indicators: GDP per capita; CO_2 emissions and Life Expectancy at birth. The indicators represent economic, environmental and social dimensions respectively. The multidimensional assessment is designed with two different policy priorities: priority of economic over environmental and social dimension versus priority of environmental and social dimensions over economic. Results help to identify countries, where economic development happened at the expense of environmental and social dimension and lead to policy conclusions.

Keywords Sustainability assessment • Sustainable development • GDP • Multi-criteria decision aid • MCDA • Policy priority • Progress • Criteria • Indicators • Well-being • Energy • Multidimensional

5.1 Introduction: Sustainable Development

The new macroeconomic theory is not possible without the theoretically sound method for assessing progress. There is a substantial literature on assessing progress in ecological economics, which has largely been focusing on the Index for

S. Shmelev (✉)
Environment Europe Ltd, 38 Butler Close, Woodstock Road, Oxford OX2 6JG, Oxfordshire, UK
e-mail: director@environmenteurope.org

S. Shmelev (ed.), *Green Economy Reader*, Studies in Ecological Economics 6,
DOI 10.1007/978-3-319-38919-6_5

Sustainable Economic Welfare (Daly, Cobb and Cobb 1989) introduced for the USA and later applied for Austria (Stockhammer 1997), Chilie (Castaneda 1999) and other countries. With the general mainstream debate becoming more sympathetic to the 'Beyond GDP' idea, which has now been embraced by OECD, European Commission and politicians COM (2001), macroeconomic models will need to include a module for the assessment of progress in a more multidimensional sense, than GDP alone. (Shmelev 2012). In our previous works we have offered a new methodology for assessing sustainability over time with the help of Multi-Criteria Decision Aid (MCDA) methods (Shmelev 2011) (Shmelev and Rodríguez-Labajos, 2009) . This methodology has a number of advantages: it doesn't require weighting different criteria, allows consideration of alternative policy priorities, presents the results in a framework, which allows comparisons between countries and at the same time integrates information on social, environmental and economic dimensions of sustainability. It allows considerations of various criteria sets, which could reflect national priorities and reflect the recent developments in Sustainable Development Goals (UN 2012).

Sustainable Development is understood as a process with simultaneous improvements in at least three areas: economic, social and environmental. The Brundtland Report (WCED 1987) defined it as a development, where the needs of the future generations are not compromised by current consumption. This implies solving the pressing problems in the environmental dimension: climate change, water, biodiversity, air pollution, solid waste and social dimension: unemployment, crime, poverty, health, and well-being issues. The United Nations prepared a set of indicators, which could be used to assess sustainability (United Nations 1996, 2007) and currently there is a major international process going on Sustainable Development Goals (SDGs), which was started at the Rio + 20 United Nations Conference on Sustainable Development (UN 2012). At the moment SDGs are discussed within the following Thematic Clusters:

1. Poverty eradication;
2. Food security and nutrition, sustainable agriculture
3. Desertification, land degradation and drought
4. Water and sanitation
5. Employment, decent work and social protection
6. Youth, education and culture
7. Health and population dynamics
8. Sustained and inclusive economic growth
9. Macroeconomic policy questions
10. Energy
11. Sustainable Development Financing
12. Means of implementation
13. Global partnership for achieving sustainable development
14. Needs of countries in special situations
15. Human Rights
16. Regional/Global governance
17. Sustainable cities and human settlements

18. Sustainable transport
19. Sustainable consumption and production
20. Climate change
21. Disaster risk reduction
22. Oceans and seas
23. Forests and biodiversity
24. Promoting equality, including social equity, gender equality and women
25. Conflict prevention, post-conflict peace building and the promotion of durable peace
26. Rule of law and governance

The Sustainable Development Goals framework will clearly need to develop more structure by grouping thematic clusters into a smaller set of dimensions and, most importantly, an element to treat interdisciplinary linkages among various dimensions and the relevant feedback loops (Shmelev, Shmeleva, 2009).

Extensive list of the dimensions presented above points towards the fact that a paradigm shift in macroeconomics is required to compliment traditional macroeconomic aggregates, such as GDP, inflation, interest rates, unemployment, consumption, savings, investment and international trade measures with data on associated resource and environmental flows and social implications of development.

Given that many dimensions are to be taken into account simultaneously when discussing sustainable development, Multi-Criteria Decision Aid tools are a strong candidate to compare various states in the development process and present the multidimensional trends. The methodology of multi-criteria tools will be discussed in the next section.

5.2 Multicriteria Tools

The method of Multi-Criteria Decision Aid (MCDA) was introduced by Prof. Bernard Roy of SEEMA, and later Universite Paris Dauphine (France) (Roy and Vincke 1981; Roy 1985, 1991, 1996). This method emerged in response to the need to deal with many criteria when justifying expenditure on large infrastructure projects and management consultancy tasks. Usually, MCDA is presented in terms of alternatives or courses of actions (scenarios) that need to be compared; criteria that are used to assess performance of these actions; multi-criteria aggregation procedure (MCAP), and the recommendations that could be given. Prof. Roy suggested using several types of 'problematique' – classes of problems that could be solved with the help of multicriteria methods.

Within the description problematique (δ) the aid helps to answer the following questions: In what terms should we pose the problem? What type of results should we try to obtain? How does the analyst see himself/herself fitting into the decision process to aid in arriving at these results? What kind of procedure seems the most appropriate for guiding his/her investigation?

Within the choice problematique (α): the aid is oriented towards a selection of a small number (as small as possible) of "good" actions in such a way that a single alternative or a subset may finally be chosen; the subset N of the selected actions could contain all the most satisfying actions, which remain non comparable between one another. The sorting problematic (β) focuses on an assignment of each action to one category (judged the most appropriate) among those of a family of predefined categories: e.g. the family of four categories could contain: (i) actions for which implementation is fully justified; (ii) could be advised after only minor modifications, (iii) can only be advised after major modifications; (iv) is inadvisable.

Within the ranking problematique (γ): the aid is oriented towards creation of a complete or partial preorder on the decision set A, which can be regarded as an appropriate instrument for comparing actions between one another. A ranking from best to worst is one of the possible outcomes of the γ type of methods. Currently there are dozens of MCDA methods available in the form of standalone software packages or libraries of tools. There is a wide spectrum of application of MCDA for sustainability related problems (Shmelev 2012, 2010), with the some of the first (Shmelev 2011) works in this area attributed to (Rath-Nagel and Stocks 1982, Anselin et al. 1989, Munda 1995, Martinez-Alier et al. 1998).

It is important to distinguish between the continuous (numerous alternatives, large system optimization) and discreet (several alternatives, the problem of choice) versions of MCDA. The MCDA tools could be of help when searching for compromise solutions and dealing with multiple dimensions of the economy-environment interactions and are an important building block of ecological economics used for the assessment of ecosystem services and their role in the macroeconomy example of which is the IUCN project described in Chapter 9 of (Shmelev 2012).

Alternative in MCDA usually represents a scenario, a course of actions or an object to be compared. In our chapter alternatives will be represented by various years in the development process of a country in question. Criterion is a measure defined in a single-dimensional space, capable of differentiating the performance of various alternatives, e.g. GDP per capita, CO_2 emissions, Gini index of income inequality, when comparing the sustainability performance of countries with each other or over time.The next section will discuss indicators of sustainable development, the Driver-Pressure-State-Impact-Response framework and the policy implications of selecting particular indicators as development goals.

5.3 Indicators for Sustainability Assessment

Since the adoption of Agenda 21 (United Nations 1992), the product of the United Nations Rio Summit in 1992, National Sustainable Development Strategies started to emerge in various countries. Their aim was to achieve better policy coordination horizontally (across policy sectors), vertically (across levels of governance), temporarily (across time), and across societal sectors (public, private, academia, civil society) and they became the prime tools for realizing governance for sustainable development (Pisano et al. 2013). Agenda 21 stated that "governments […] should adopt a national strategy for sustainable development" which should "ensure socially responsible economic development while protecting the resource base and the environment for the benefit of future generations" (United Nations 1992). In the European Union as a result of the publication of EU Sustainable Development Strategy (EC 2006), all Member States were required to develop their National Sustainable Development Strategies by 2007 and commit to bringing them in harmony with the EU Strategy. Some of the first National Sustainable Development Strategies were developed in Sweden (1994), United Kingdom (1994), Switzerland (1997), Finland (1998), Belgium (2000), Germany (2002) and Austria (2002). Figure 5.1 presents the map, where countries, which developed their National Sustainable Development Strategies by 2010 are shown. There have been 106 such countries in 2010.

The United Nations Conference on Sustainable Development, known as the Rio + 20 Summit took place from 20 to 22 June 2012 in Rio de Janeiro. It had three central objectives (1) to secure renewed political commitment for sustainable development; (2) to assess the progress to date and the remaining gaps in the implementation of the outcomes of the major summits on sustainable development; and (3) to address new and emerging challenges. The two main themes of the conference have been: (i) a green economy in the context of SD and poverty eradication; and (ii) the institutional framework for SD. One of the major conference outcomes was an agreement by Member States to launch a process to develop a set of sustainable development goals (SDGs). (Pisano et al. 2013).

Institutionally, different countries implemented diverse approaches to address bringing sustainable development into the political fore. For instance, UK intro-

Fig. 5.1 Countries that developed their National Sustainable Development Plan. *Green* – NSDS being implemented, *Brown* – NSDS development in progress, *Violet* – no NSDS, *Beige* – no information available (Source: United Nations)

duced departmental sustainable development action plans to make sure that SD principles are implemented across government departments; Germany introduced a State Secretaries' Committee on Sustainable Development, a political body fostering cross-sectoral integration of sustainable development policies into all branches of government. Finland introduced a National Sustainable Development council, which brought in important stakeholder groups. Austria was very successful in bridging the gap between the national and regional levels and introduced binding strategies for both levels.

Sustainable Development Indicators describe various parts of the system of economy-environment interactions. The phenomena they describe are well explained by The Driver-Pressure-State-Impact-Response (DPSIR) framework, which could be particularly useful in classifying sustainable development indicators from the point of view of the logic of feedback loops that exist among the elements of the system. The DPSIR framework was initially developed by OECD (OECD 1993).

In the Pressure-State Response Framework, depicted in Fig. 5.2, Pressures are human activities focused on energy generation, industrial production, transport, agriculture, and other areas. The State of the environment is normally registered by studying the quality of the air, water, soil, ecosystems and biodiversity, availability of resources and human health. Response is represented by the actions of households, enterprises and government acting at sub-national, national and international level. Within the framework presented in Fig. 5.2, Percentage of Renewable Energy in the Energy Mix can be seen as an indicator of Response; Total Primary Energy Supply as an indicator of Pressure and concentrations of PM_{10} or CO_2 in the atmosphere – indicators of the State.

This methodology has been revised and adopted by the United Nations Environment Programme (UNEP 2007) and the European Environment Agency

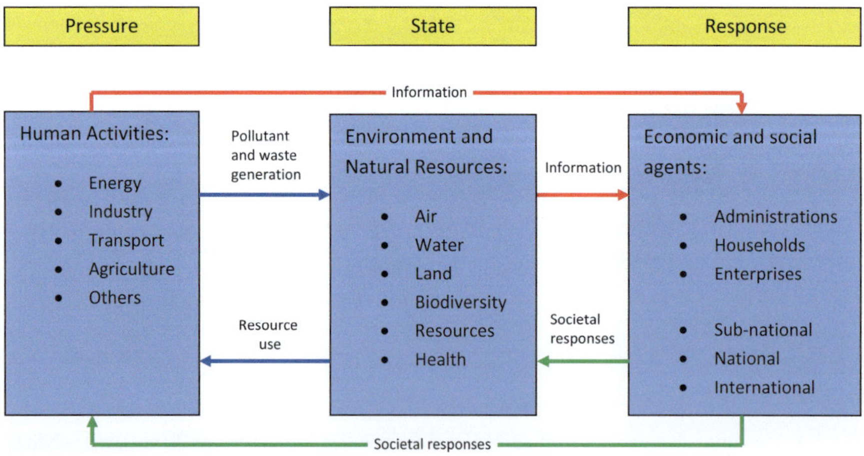

Fig. 5.2 Pressure-state-response framework (OECD 1993)

(EEA 1999). It represents a systemic view on economy-environment interactions from the point of view of policy interventions and the correction of adverse impacts (Fig. 5.3). The framework illustrates multiple directions of possible interventions, focusing on addressing the drivers, pressure, state and impacts through policy: resource and environmental taxation, advanced quality standards, system-wide interventions (public transport; planning and architecture, technological change).

When it comes to the analysis of indicators from the point of view of policy interventions, it is particularly insightful to assess the level of consistency between the declared top level and high level sustainable development priorities and related indicators in various European countries (Table 5.1). Only two countries – Germany and Estonia have a perfect 100 % match between the two with 4 top level indicators and 21 and 12 high level indicators respectively. This indicates the level of importance, focus and transparency of the policies oriented towards sustainable development in these countries. Other countries have lower levels of consistency, e.g. Finland 100 % at the top level and 69 % at the high level, Czech Republic 100 % at the top level and 88 % at the high level. France 78 % at the top level and 0 % at the high level, Slovenia 100 % at the top level and 84 % at the high level, Italy 100 % at the top and 82 % at the high level. The UK has only 67 % of consistency at the top level and 0 % at the high level, Austria 80 % at the top and 0 % at the high level, wich might indicate insufficient attention to the issues of sustainable development, when it comes to the details. In Sweden and the Netherlands the Sustainable Development Priorities were found not to match with Sustainable Development Indicators. As will be seen later in this chapter, the high level of cross-departmental

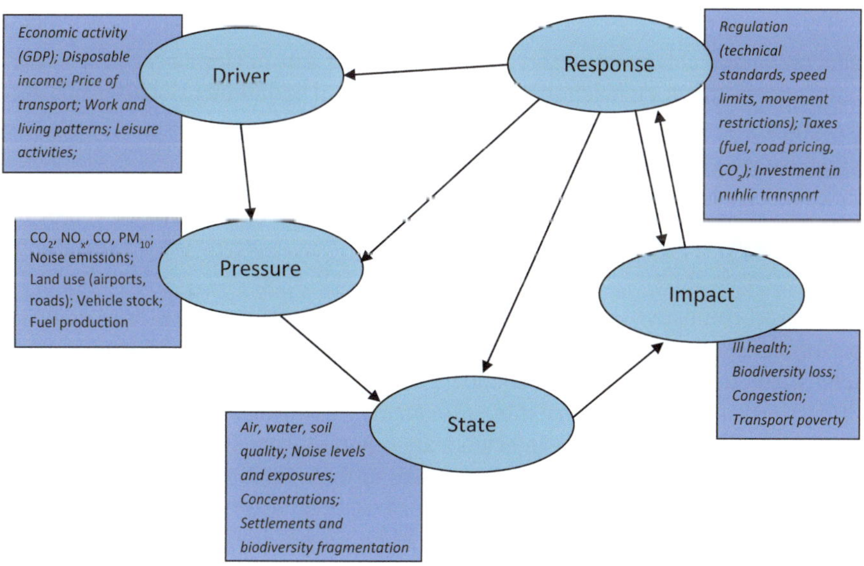

Fig. 5.3 Driver-pressure-state-impact-response framework (EEA 2003)

Table 5.1 Three levels of sustainable development priorities, indicators and consistenc

	Top level			High level			Key Issues
Country	Consistency	SDP	SDI	Consistency	SDP	SDI	SDP
Austria	0.80	5	4	0.00	23	0	131
Belgium	0.00	7	0	0.00	31	0	193
Czech Republic	1.00	6	6	0.88	17	15	144
Denmark	1.00	21	21	0.44	87	38	92
Estonia	1.00	4	4	1.00	12	12	16
Finland	1.00	6	6	0.69	26	18	154
France	0.78	9	7	0.00	50	0	16
Germany	1.00	4	4	1.00	21	21	
Greece	0.40	5	2	0.40	25	10	26
Ireland	0.29	7	2	0.38	16	6	170
Italy	1.00	4	4	0.82	28	23	110
Latvia	0.62	26	16	0.00	79	0	214
Lithuania	0.00	27	0	0.00	48	0	535
Malta	1.00	4	4	0.50	28	14	214
Netherlands	0.00	13	0	0.00	22	0	54
Slovakia	0.00	11	0	0.00	28	0	238
Slovenia	1.00	5	5	0.84	19	16	145
Sweden	0.00	8	0	0.00	19	0	92
UK	0.67	6	4	0.00	33	0	121

Pisano et al. (2013)

integration and coherence observed in the indicator sets in Germany could be partially responsible for a much stronger performance in multidimensional sustainability terms.

5.4 Data

In order to compare the performance of a country in various years, or to compare performance of various countries or regions, one needs to have comparable data sets, which use identical methodologies, cover all necessary years. This is why a discussion on data in the context of sustainability assessment is essential. The fact that only a limited number of indicators are reported from 1960 (GDP, life expectancy, CO_2 emissions) determined our choice of these three measures representing economic, social and environmental dimensions as an express set throughout this chapter and in the subsequent assessments. More 'exotic' indicators, e.g. unemployment, or the share of renewable energy in the energy are available for a majority of

Table 5.2 Sustainable development indicators

Economic	Social	Environmental
GDP per capita	Life expectancy at birth	CO_2 emissions
R&D expenditure, % GDP	GINI index	Recycling rate, %
Government Debt	Unemployment rate, %	Share of renewables, %

countries through World Bank, IMF or UN Stat databases. However, it should be noted that African countries are normally not covered in as much detail as OECD countries. In this chapter we will discuss the results of the multidimensional sustainability assessment using the express criteria set, which could be seen in the first line of Table 5.2. It was decided that the indicator set should be balanced in terms of social, economic and environmental priorities, that is why the large set includes three indicators from each group. Availability of data has been a factor for such countries like China, where data on recycling rates is not available. Other indicators from this table will be incorporated into the sustainability assessment at later stages.

In the past decade the countries made a huge leap forward in terms of collating and documenting their sustainable development indicator sets. For all of the countries discussed in this chapter, the data on express sustainability indicators: GDP per capita, Life Expectancy and CO_2 emissions were available for 1995–2011. Unfortunately, even at the time of publication in 2015, data covering 2012 and 2013 was not available, which could be a point of criticism of the delays in the currently existing statistical data provision procedures.

5.5 Applications

5.5.1 The APIS Method

The multicriteria decision aid tool Aggregated Preference Indices System (APIS), developed by Nikolai Hovanov (Hovanov et al 2009), presents a useful way to compare alternatives on multiple criteria in the situation of uncertainty regarding the relative importance of criteria in a given situation. It is a single decision maker tool as opposed to the group decision making tools, however APIS has a built-in capacity to test alternative visions by considering different sets of priorities. The method requires explicit specification of alternatives, criteria and the decision-making matrix. The Multi-Criteria Aggregation Procedure in this method uses the principle of the Monte Carlo method and generates admissible distributions of weights using the information on relative policy priorities (e.g. increase in GDP is more important than the CO_2 emissions reductions) obtained from the decision maker. The particular attractive aspect of the method is that it is capable of presenting the same development process as seen from the point of view of different stakeholders (e.g. an industrialist vs. a green activist). This method has been used to formulate a new approach to the sustainability assessment in (Shmelev 2011).

5.5.2 *Interpretation of the Charts*

The charts that are going to be presented in the following sections can be read in the following way.

1. The Multidimensional Sustainability Performance Index is represented by midpoints of the red bars;
2. The length of the red bars represents the degree of uncertainty;
3. The blue bar denotes the probability (between 0 and 1) of the fact that the alternative on the left of the blue bar dominates the one on the right.

The multidimensional dynamics of the development process comes to life when one explores the changes in sustainability scores. It is peculiar that in countries like Germany, France and the UK there seems to be much more coherence between economic, social and environmental policies, which is illustrated as a positive sustainability trend regardless of the priorities (GDP vs. CO_2 and Life expectancy). On the other hand in China, Brazil, Russia and somewhat in the USA the development process happens with less integration between sustainability dimensions and a pronounced social or environmental cost of development.

Interpretation of the subsequent charts has been carried out with the help of stakeholders possessing the knowledge of the economic systems in question. This made the preliminary results presented in this chapter more relevant. All charts in this chapter are based on a three-criteria assessment covering GDP per capita, CO_2 emissions and Life Expectancy.

5.5.2.1 United States of America (Fig. 5.4)

Sustainable development ideas have been prominently advocated in the United States in the works by Rachel Carson (Carson 1962), Kenneth Boulding (Boulding 1966), Herman Daly (Daly 1968, 1972, 1974, 1977), Robert Ayres (Ayres and Kneese 1969; Ayres R.U. et al. 1970; Ayres R. U. 1978) and the 'Limits to Growth' systems dynamics modeling project (Meadows and Club of Rome. 1972). Influenced by the environmental movement, the United States adopted a range of important environmental legislation: Clean Air Act (1963), Solid Waste Disposal Act (1965), Water Quality Act (1965), Air Quality Act (1967), National Environmental Policy Act (1969), Noise Control Act (1972), Endangered Species Act (1973). Environmental Protection Agency has been formed in 1970. In 1998 the USA has signed the Kyoto Protocol, which would have committed the USA to reducing GHG emissions 7 % below 1990 level between 2008 and 2012, however in 2001 disengaged and opted not to ratify it (Encyclopaedia of the Earth 2006). Recently, as a result of the US-China deal, USA committed to 26–28 % GHG reduction below 2005 levels by 2025, which will be brought to the COP21 in Paris in 2015.

The sustainable development trend based on three criteria of GDP per capita, Life expectancy at birth and CO_2 emissions in the USA is depicted in Fig. 5.5. It is

Fig. 5.4 Map of the USA

clear that there is a steady upward trend in sustainability (understood in terms of three criteria chosen for this assessment), which seems to reflect the overarching principles in US policymaking: economic growth is a clear objective to which other policy goals must conform. When at the time of the Financial Crisis in 2008 GDP per capita fell in the USA in absolute terms from 44,872 to 43,234 (World Bank, 2014) the decrease in CO_2 emissions by 340 million t, which was the result of reduced economic production, made up for this loss, so that even under GDP priority the Multidimensional Sustainability Performance Index continued to increase. Because the goal of all fiscal and monetary policy of the US government seemed to have been to maximize GDP growth or minimize the fall in GDP at the time of the recession, the three criteria assessment is skewed to show more improvement than a more inclusive analysis, incorporating additional dimensions, especially thanks to the incidental decrease in CO_2 emissions due to an economic downturn.

The Multidimensional Sustainability Performance Index chart under CO_2 minimization and life expectancy maximization priority (Fig. 5.6) shows an overall positive trend starting in 2005 and a substantial improvement in 2009, where the recession substantially reduced CO_2 emissions. At the same time the fuel switching from coal to the lower-carbon natural gas, largely obtained as a result of a fracking boom of 2005–2006, might also explain the causes of the largely positive trend observed since 2005. On the other hand, new research suggests that often-

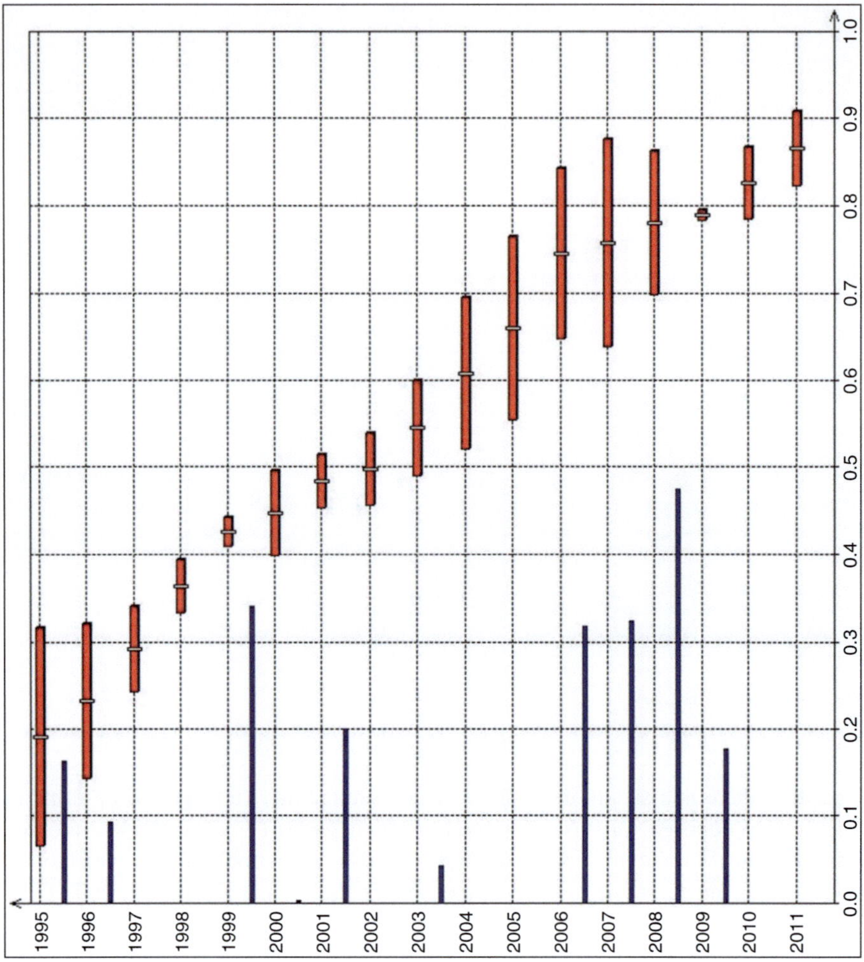

Fig. 5.5 Multidimensional sustainability performance index, USA: 3 criteria (GDP, CO_2 and life expectancy); GDP maximization priority

unaccounted methane leakages from gas-fracking are large enough to negate any climate advantage of burning fracked gas instead of coal.

5.5.2.2 Brazil (Fig. 5.7)

The Multidimensional Sustainability Performance Index with a GDP priority in Brazil (Fig. 5.8) shows an upward tendency, where even the 1 % reduction in GDP per capita after the crisis in 2009 was accompanied by a 4 % reduction in CO_2 emissions and lead to an overall improvement in sustainability terms. The subsequent 6 % increase in GDP was enough to offset the increases in carbon intensity and lead

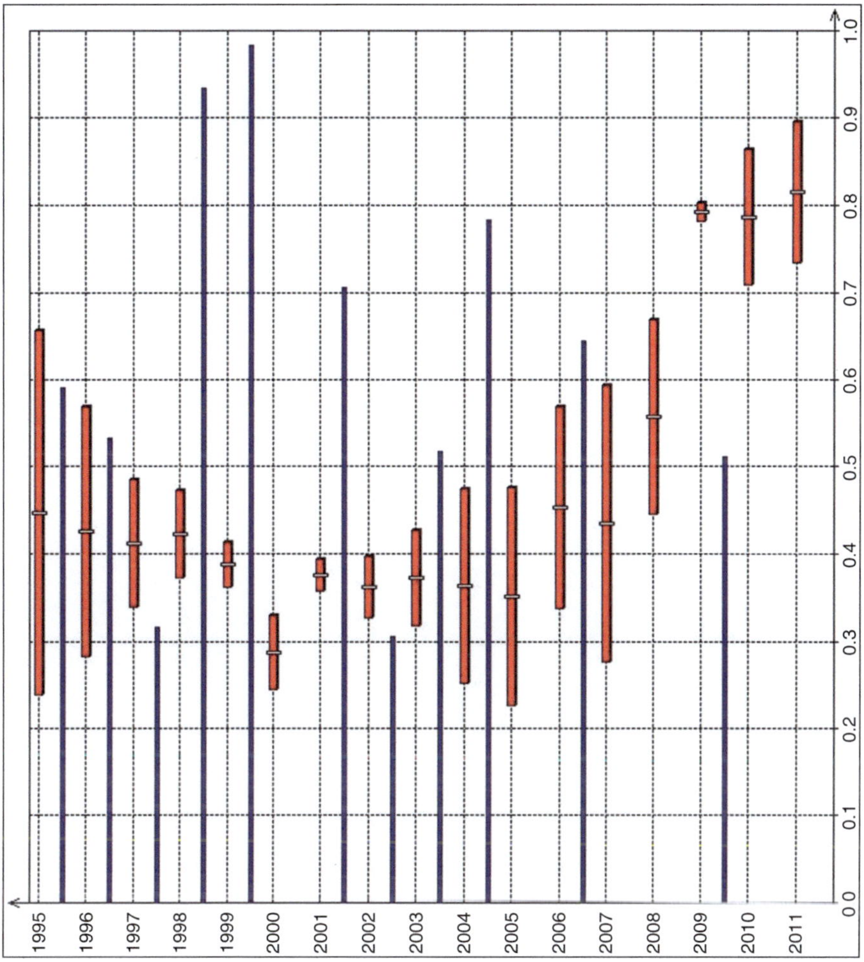

Fig. 5.6 Multidimensional sustainability performance index, USA: 3 criteria (GDP, CO_2 and life expectancy); CO_2 minimization and life expectancy maximization priority

to further sustainability improvements under GDP priority assumption. The fact that this trend differs substantially especially from the post-crisis trend under CO_2 and life expectancy priority illustrates relative unsustainable tendencies in post-crisis Brazil.

The slow but steady sustainability improvement observed in Brazil since 1995 under social and environmental priorities (Fig. 5.9) is likely to have multiple causes. Among them are higher than average share of renewables in the energy mix (over 80 %) with a substantial proportion of hydro and biofuels; steady trend in quality of life improvements resulting in life expectancy increases and a lower reduction in GDP per capita after the financial crisis than that in the USA (1 % vs. 3 %). Biofuel economy in Brazil was shown to have a favourable energy balance of 8.3 (1 unit of

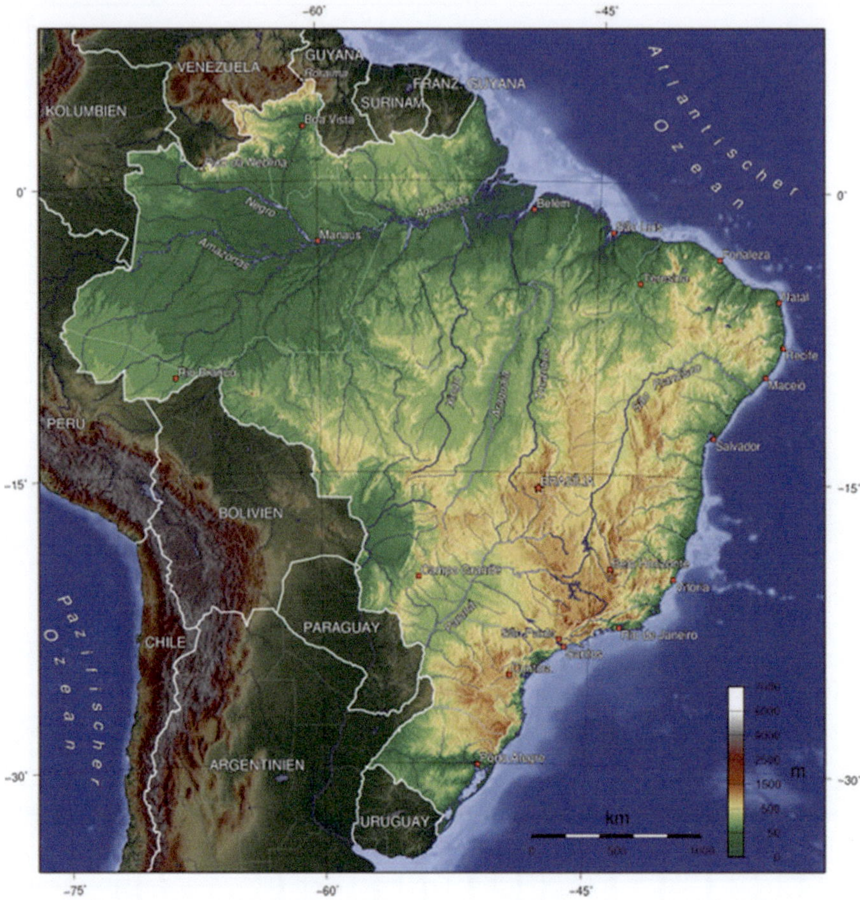

Fig. 5.7 Map of Brazil

energy is required to produce 8.3 units of biofuel), making Brazilian transport sector the least carbon-intensive in the world. On the other hand the share of renewable energy in the energy mix has been declining likely due to delays in developing the new hydropower capacity in the Amazon. The latter could explain the reversal of the trend between 2009 and 2011 (Fig. 5.9), when Brazil was becoming less sustainable according to the chart. The turning point in the Multidimensional Sustainability Performance Index in the year 2009 in Brazil is in line with the 'threshold hypothesis', proposed by Max Neef (1995): 'for every society there seems to be a period in which economic growth (as conventionally measured) brings about an improvement in the quality of life, but only up to a point – the threshold point – beyond which, if there is more economic growth, quality of life may begin to deteriorate.'

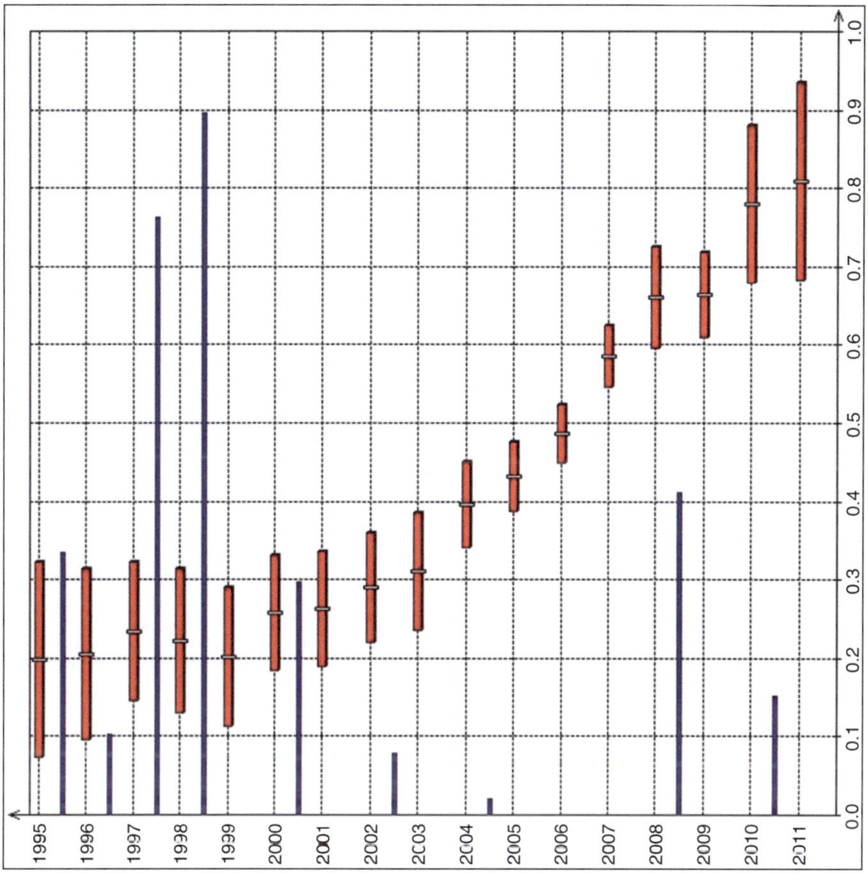

Fig. 5.8 Multidimensional sustainability performance index: Brazil, 3 criteria (GDP, CO_2 and life expectancy); Priority to GDP

5.5.2.3 France (Fig. 5.10)

France has been one of the few countries where economic, social and environmental priorities are in relative harmony. Under the conditions of GDP priority (Fig. 5.11), four distinct periods with explicitly positive trends were observed: 1996–1997, 1998–2000; 2003–2007 and, finally, 2009–2011. The fact that France obtains 75 % of its electricity from nuclear power stations determines the relatively low levels and low elasticity of CO_2 emissions in response to GDP changes. The 3.6 % GDP per capita drop in 2008–2009 financial crisis is clearly visible here, however the situation improves soon afterwards and 2011 performs better in sustainability terms than 2007, the best pre-crisis year.

Under the conditions of CO_2 minimization and life expectancy maximization priority (Fig. 5.12), France experienced several positive trends, the longest positive one lasting from 2005 to 2011. Over the last 6 years under consideration the positive

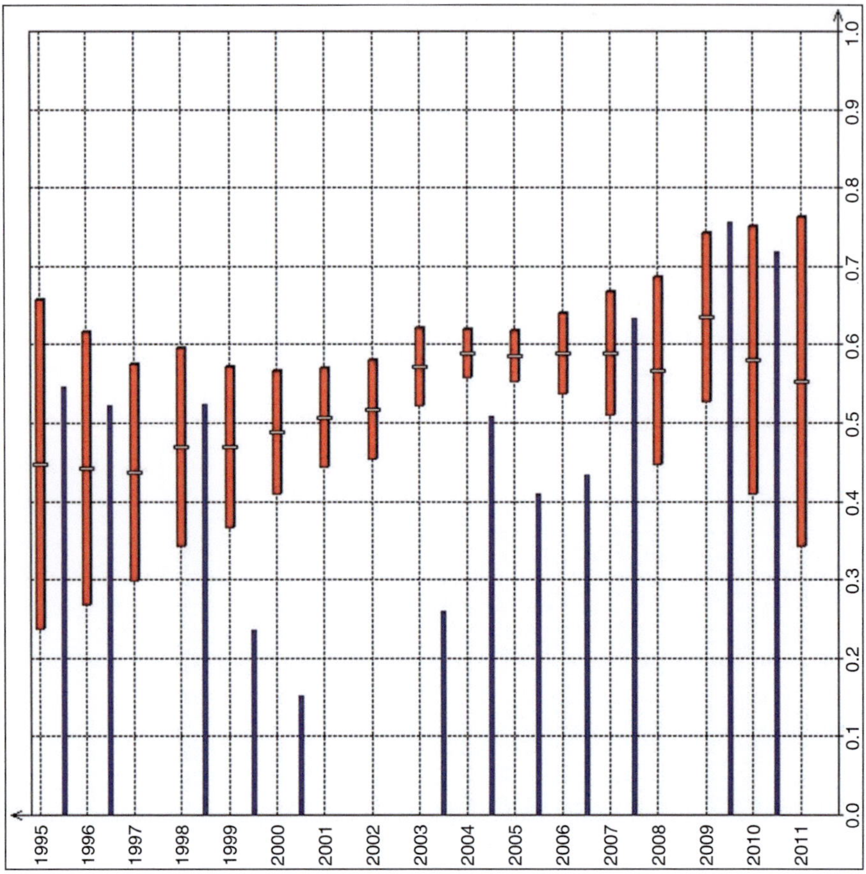

Fig. 5.9 Multidimensional sustainability performance index: Brazil, 3 criteria, GDP, CO_2 and life expectancy: priority to CO_2 and life expectancy

trend in sustainability coupled with the positive trend under GDP priority in France represents a relatively higher degree of harmony between social, economic and environmental priorities. Part of the reason is historic attention to the social dimensions of sustainability and the quality of life: the Ministry of the Ecology, Sustainable Development and Energy in France was called Ministry of the Quality of Life between 1974 and 1977.

5.5.2.4 Germany (Fig. 5.13)

Germany presents a very coherent picture with very low levels of uncertainty and a general positive trend under GDP priority assumption (Fig. 5.14). The exception to the general trend appeared at the time of the financial crisis in 2009, when GDP per

Fig. 5.10 Map of France

capita went down by almost 5 % but CO_2 emissions reduced by even more substantial 6.5 %. Under the GDP priority this resulted in a decline in sustainability. After 2010 the German economy returned to a sustainable trend without increasing CO_2 emissions. It should be said that German share of renewable energy in the energy mix has been steadily increasing and reached 20 % in 2011. At the same time Germany was one of the very few countries where CO_2 emissions were reduced in absolute terms largely due to the policy of technological innovation.

The sustainability trend under the CO_2 minimization and Life expectancy maximization priority in Germany (Fig. 5.15) looks very positive, which illustrates the success in technological innovation, deployment of renewables and the policy coherence between economic, social and environmental dimension. Germany introduced the State Secretaries' Committee on Sustainable Development, a political body fostering cross-sectoral integration of sustainable development policies into all branches of government. Such an active consultative body is rare and signifies a major step in reconciliation between economic, social and environmental priorities. The German sustainability trends look very different from those of the USA, Brazil, China, Russia but somewhat similar to those of France and the UK.

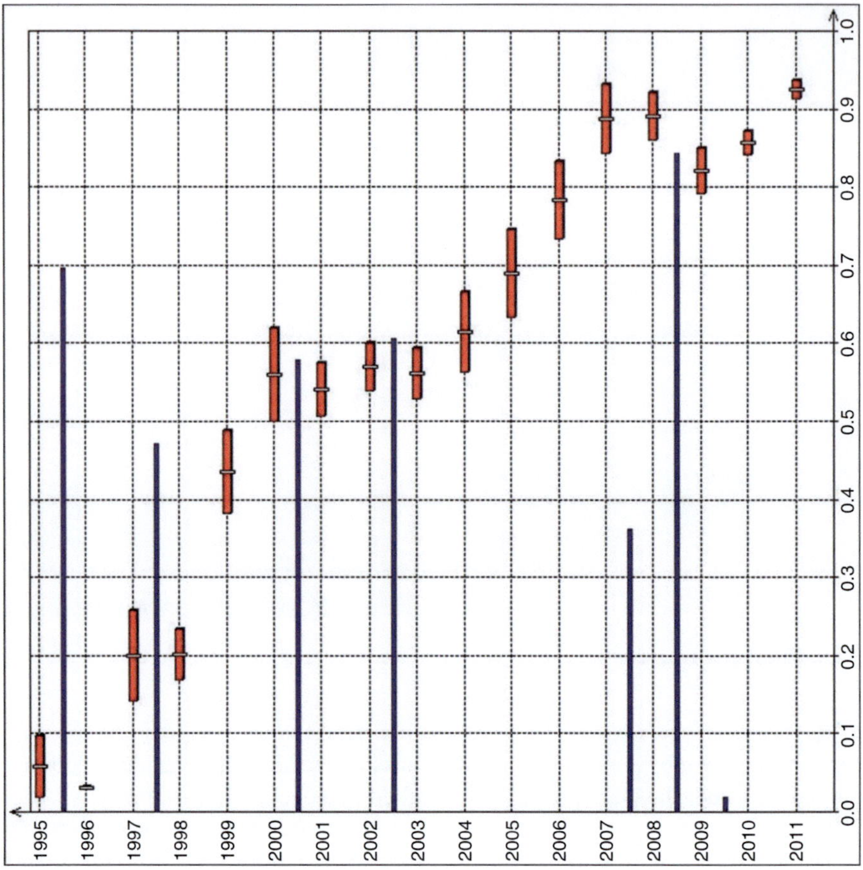

Fig. 5.11 Multidimensional sustainability performance index: France, 3 criteria, GDP, life expectancy and CO_2, priority to GDP

5.5.2.5 United Kingdom (Fig. 5.16)

The UK Multidimensional Sustainability Performance Index under a GDP priority (Fig. 5.17) shows a steady improvement between 1996 and 2008 and continues to improve between 2010 and 2011. The trend is characterized by a slightly higher degree of uncertainty than in France and Germany, which makes it closer to the USA case. The trend is still positive except the decline between 2009 and 2010, when per capita GDP declined 5.8 %. CO_2 emissions in the UK have declined due to the increasing share of renewables (up to 14 % in 2012), introduction of the fuel duty escalator in 1993 and climate change levy in 2001, and possibly outsourcing production to China and other countries.

After the financial crisis GDP in the UK has not returned to its 2006 level by 2012, at the same time emissions started to grow again after 2009. With the general trend of CO_2 emissions to decline and the significant progress that was made in the

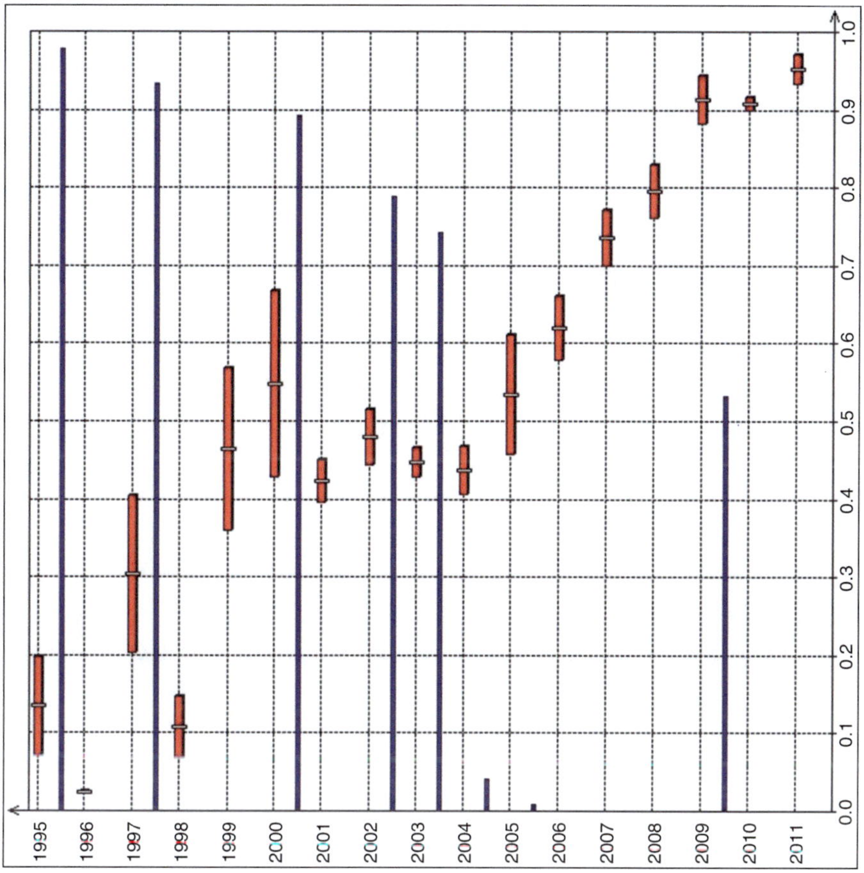

Fig. 5.12 Multidimensional sustainability performance index: France, 3 criteria, GDP, life expectancy and CO_2: priority to life expectancy and CO_2

UK with renewables deployment (from nearly zero to over 14 % in 2012), UK has been in a strong position to overcome unsustainable tendencies related to post-crisis reductions in GDP. Unfortunately, there has not been so much investment in sustainable economic sectors: education, health care, recreational services, which received a reductions of their budgets as part of the austerity package. My earlier paper (Shmelev 2010) showed how investment in such sectors could provide a neo-Keynesian push at the same time not requiring excessive resource use and emissions both directly and indirectly. Similarities expressed in the low uncertainty, positive tendencies in Multidimensional Sustainability Performance Index under two sets of priorities in Figs. 5.17 and 5.18 point towards the relative degree of harmony between social, economic and environmental dimensions and allows us to place the UK in the same group as Germany and France.

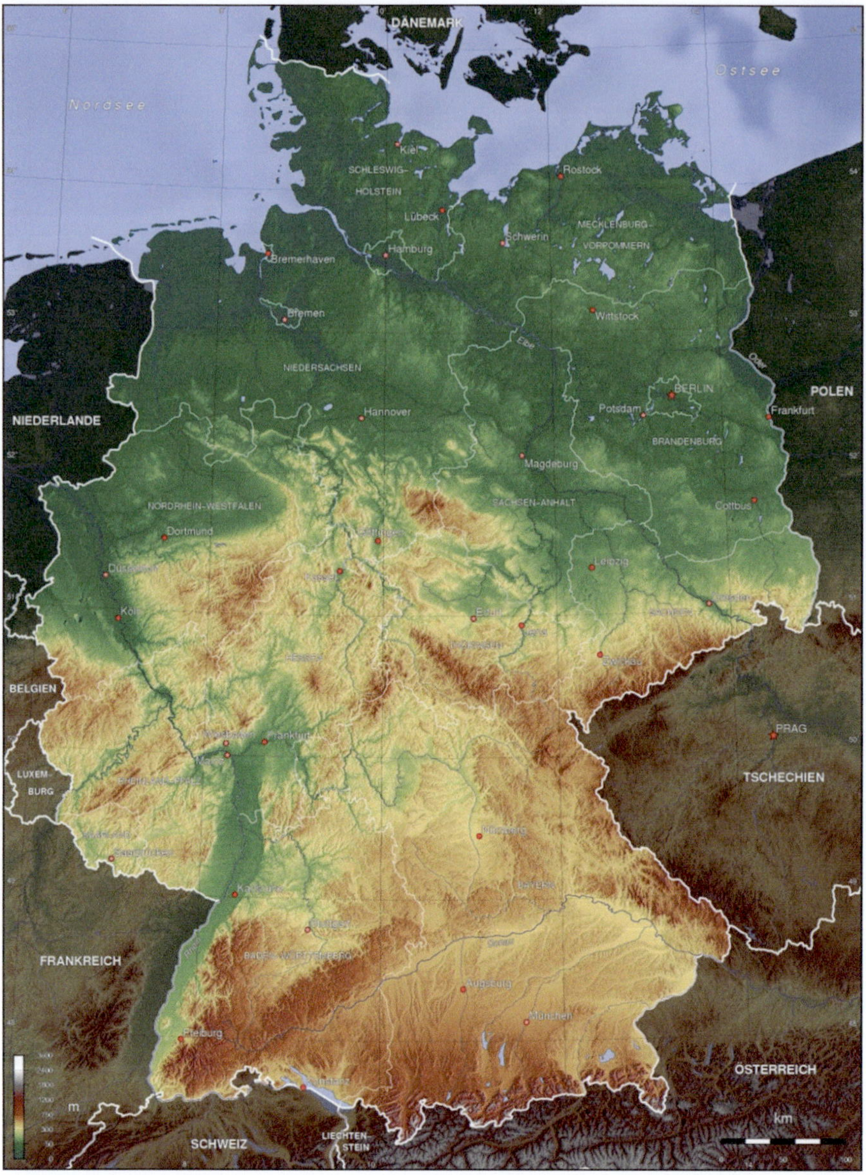

Fig. 5.13 Map of Germany

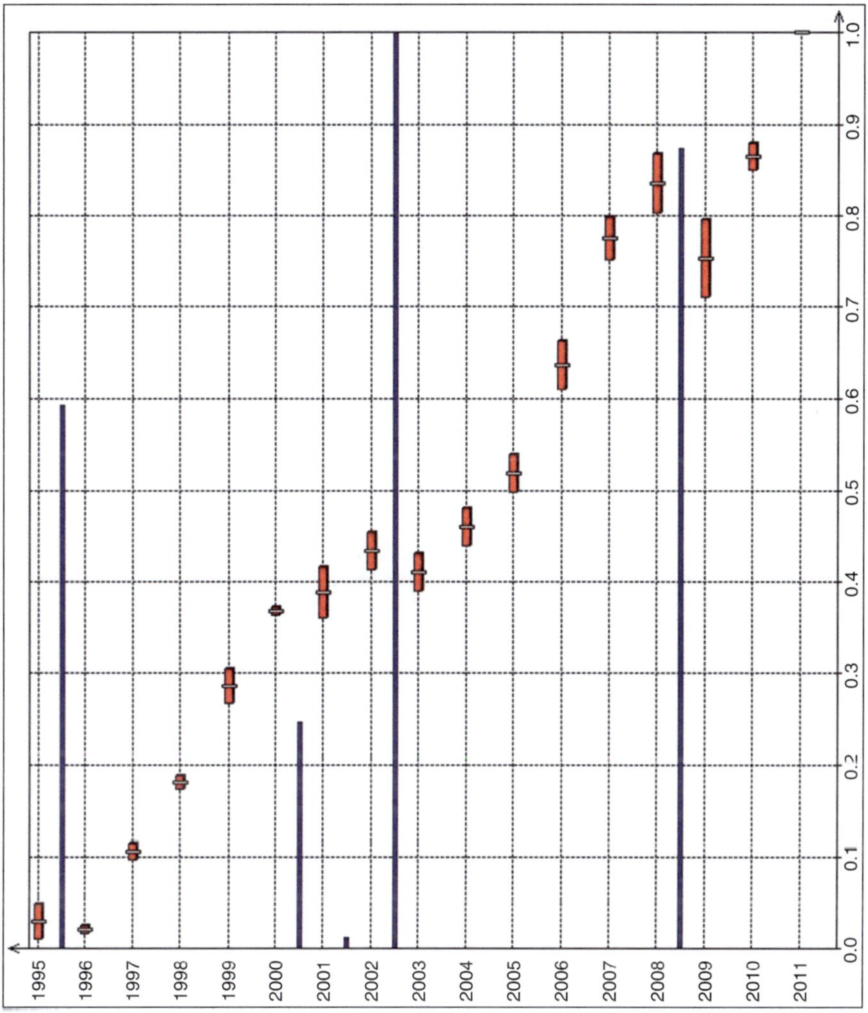

Fig. 5.14 Multidimensional sustainability performance index: Germany, 3 criteria, GDP, life expectancy and CO_2: priority to GDP

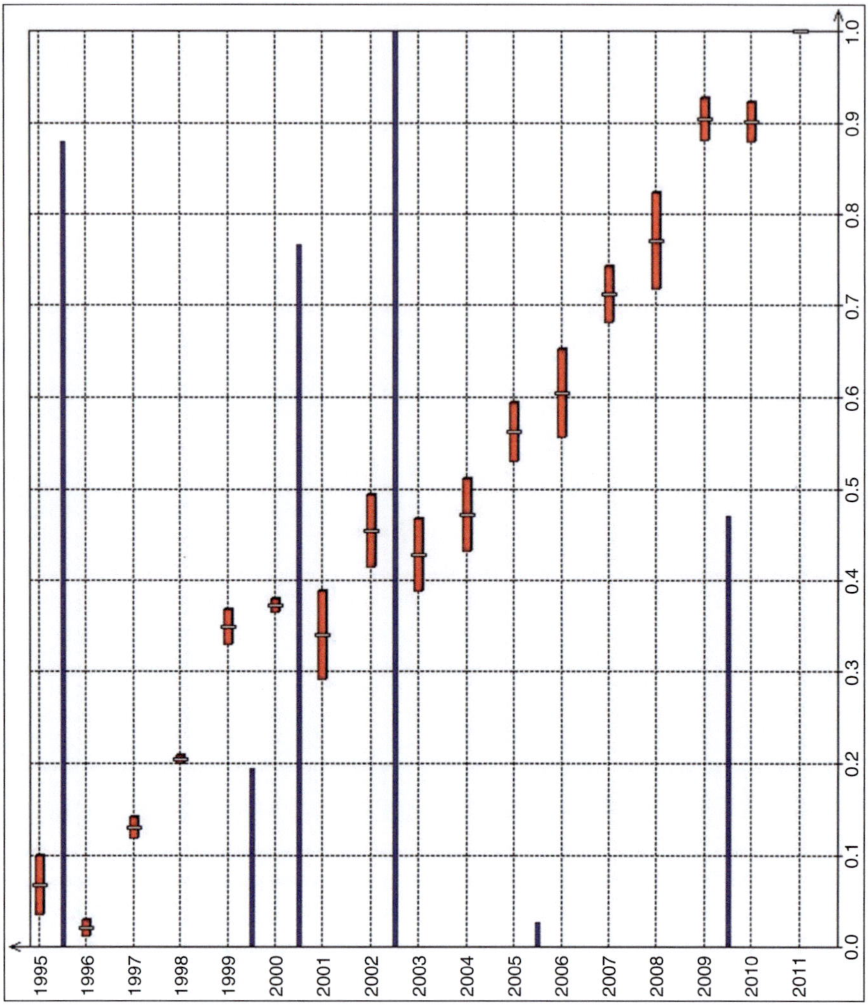

Fig. 5.15 Multidimensional sustainability performance index: Germany, 3 criteria, GDP, life expectancy and CO_2: priority to life expectancy and CO_2

5.5.2.6 China (Fig. 5.19)

China presents a fascinating economic and environmental case and has been the subject of much ecological-economic research (Polenske and Lin 1993; Lin and Polenske 1995; Guan et al. 2008). Exporting around a third of its production, China is experiencing severe deterioration in air quality, including the largest cities of Beijing and Shanghai.

Although the GDP priority case shows a strong positive trend in the Multidimensional Sustainability Performance Index (Fig. 5.20), the Chinese case

Fig. 5.16 Map of the UK

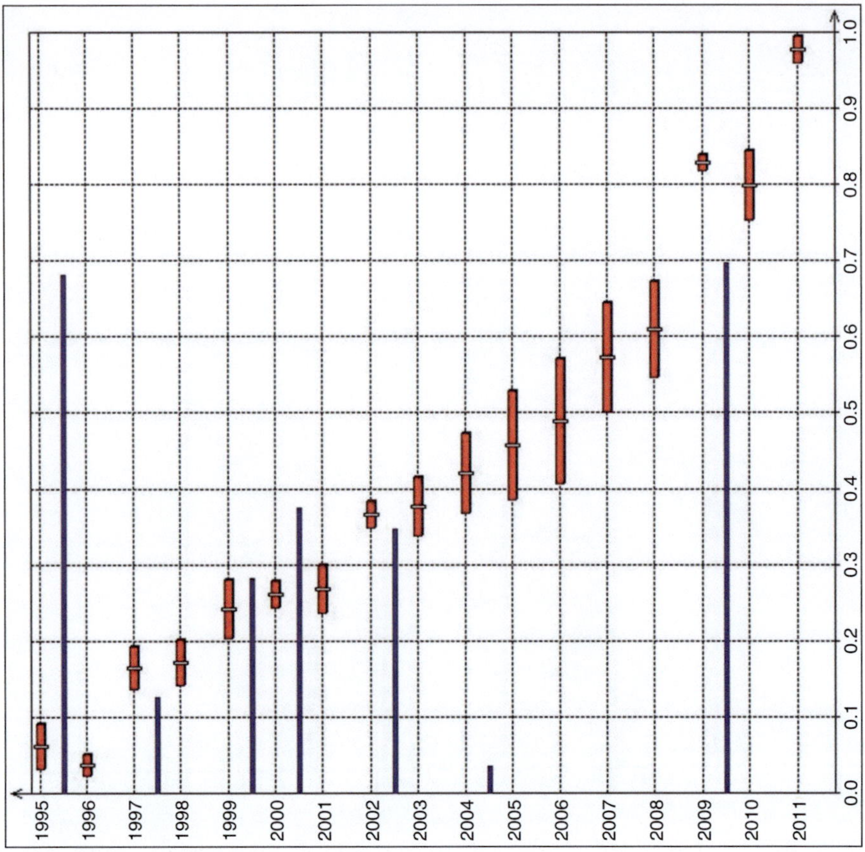

Fig. 5.17 Multidimensional sustainability performance index: UK, 3 criteria, GDP, life expectancy and CO_2: priority to GDP

demonstrates a higher level of uncertainty, than cases of post-industrial countries of France, Germany and the UK. In becoming the world's centre of production, China has relied on coal energy, which brought considerable levels of CO_2 emissions and PM air pollution. At the same time, the recent years saw construction of high speed railways and underground systems, largest hydropower installations in the world and a rapid improvement in the Solar PV technology. The recent US-China climate change deal announced the intention of China to allow for the GHG emissions peak in 2030 and increase non-fossil fuel share of all energy to 20 % by 2030. USA and China recently agreed to expand joint efforts on clean energy research and facilitate Low-Carbon Cities initiative among other measures.

The CO_2 priority case (Fig. 5.21) demonstrates a very distinct tendency: sustainability improves until 2002, after which the trend is reversed and sustainability declines to around 1998 levels. Massive industrialization, development of the coal energy and the expansion of the car fleet in a country with 1.357 bln, put China into

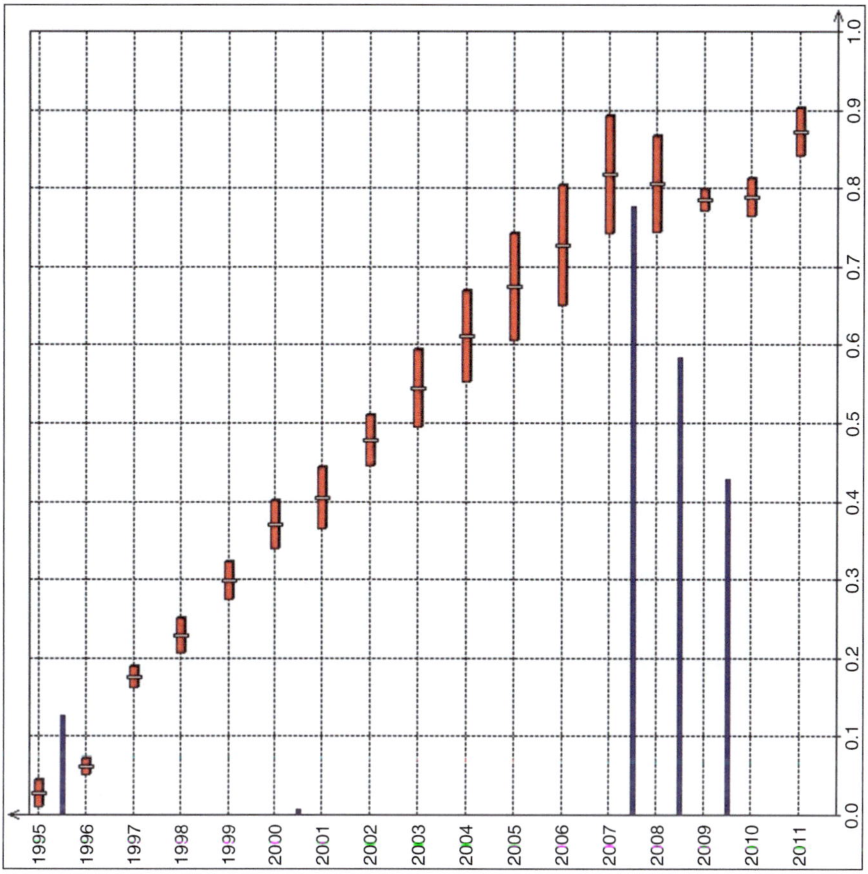

Fig. 5.18 Multidimensional sustainability performance index: UK, 3 criteria, GDP, life expectancy and CO_2: priority to life expectancy and CO_2

the 'turning point' group, where economic growth was achieved at the expense of social and environmental deterioration. This corresponds to the Max-Neef's 'threshold hypothesis' discussed earlier.

5.5.2.7 Russia (Fig. 5.22)

When it comes to social and economic dimensions of development, Russia presents, perhaps, the most complex case of all. Increases in GDP after the disastrous hyperinflation and economic reforms of the early 1990s, which brought the life expectancy of men down from 68.84 in 1987 to 57.55 in 1993, where largely determined by favourable natural resource prices. This export-oriented trend created a substantial gap between the rich and the poor (Gini increased from 23.8 in 1987 to 48.38 in 1993). Despite the positive GDP trend, the financial crisis of 2008–2009 had

Fig. 5.19 Map of China

substantial implications (GDP per capita declined 7.8 %) and the Multidimensional Sustainability Performance Index trend even under GDP priority reversed and started to decline after 2009 (Fig. 5.23).

In the CO_2 minimization and life expectancy maximization case (Fig. 5.24), the positive trend in the Multidimensional Sustainability Performance Index of the years signifying the end of hyperinflation and relative stablization of 1995–1997 was followed by a reduction in sustainability between 1998 and 2000. The sustainability improvement of 2000–2001 was outweighed by a decrease in 2001–2003 followed by an improvement between 2003 and 2009. The positive tendency was followed by a decline of 2009–2011, which brought Russia back to the level of 2008 in sustainability terms. The relatively high level of nuclear (16 %) and hydro energy (16 %) in the electricity mix, relatively compact cities, development of Combined Heat and Power, which represent the positive achievements in Russia, coexist with a rapid expansion of the car fleet and overreliance on the extractive economy, which put together results in placing Russia in the 'turning point' group. This perhaps could be an illustration of a case with several 'thresholds' in Max-Neef's terminology.

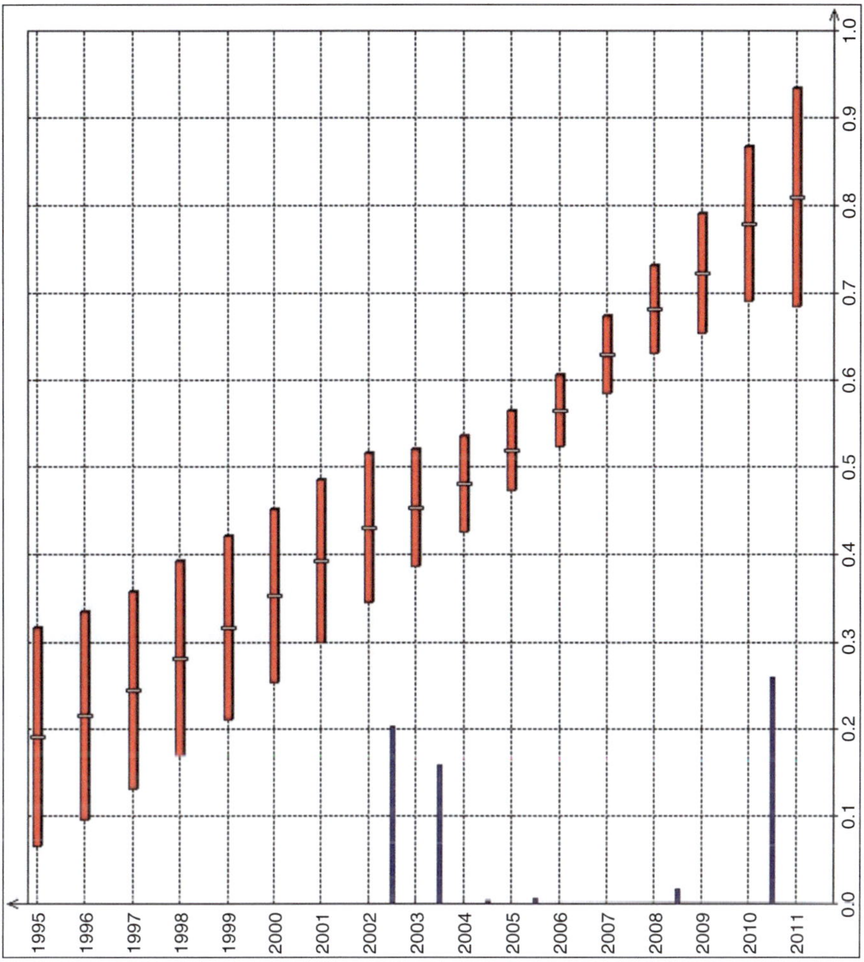

Fig. 5.20 Multidimensional sustainability performance index: China, 3 criteria, GDP, life expectancy and CO_2; Priority to GDP

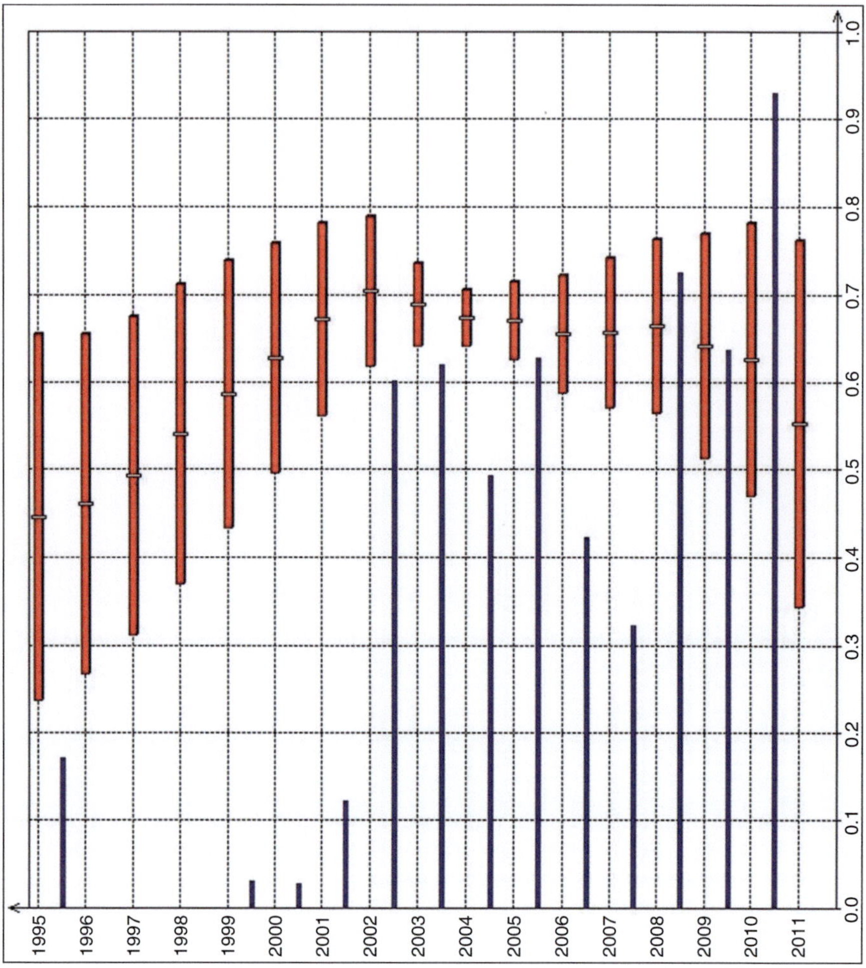

Fig. 5.21 Multidimensional sustainability performance index: China, CO_2 and life expectancy priority with 3 criteria

5.6 Conclusions

This chapter focused on the dynamic sustainability assessment with the help of Multidimensional Sustainability Performance Index for several countries. First, we discussed the issues of selecting indicators for sustainability assessment, presented the methodology of Multi-Criteria Decision Aid and compared indicator and priorities in various countries' sustainability strategies. We then explored Multidimensional Sustainability Performance trends with an express set of three indicators: GDP per capita, Life Expectancy and CO_2 emissions in USA, Brazil, France, Germany, UK, China and Russia. The analysis presented demonstrated a wide spectrum of

Fig. 5.22 The map of Russia

development trajectories in different countries. The level of coherence between economic, social and environmental dimensions varies from the most sustainable cases of France and Germany to slightly less sustainable UK and US and the 'turning point' countries of China, Russia and Brazil. The observed tendencies correspond well with the Manfred Max-Neef's 'Threshold Hypothesis', which states that quality of life begins to deteriorate after a certain point along the economic growth trajectory.

The results presented in this chapter have been framed around two policy priorities: priority of GDP growth and priority of CO_2 emissions reduction and Life expectancy increase. Different policy priority settings, which we were able to test, clearly showed that some countries evolve in a sustainable direction (positive trends both with GDP and CO_2 and Life expectancy priorities), whereas in some countries, development happens at the expense of social and environmental dimensions. The reasons differ from country to country, however cross-departmental interaction and sustainability orientated government bodies working across various ministries (Economy, Energy, Environment, Transport, etc) are more likely to deliver increased levels of harmony and coherence, which the German example amply illustrates. It remains to be seen, how bringing in environmental and social impacts embodied in international trade would change the Multidimensional Sustainability Performance trajectories.

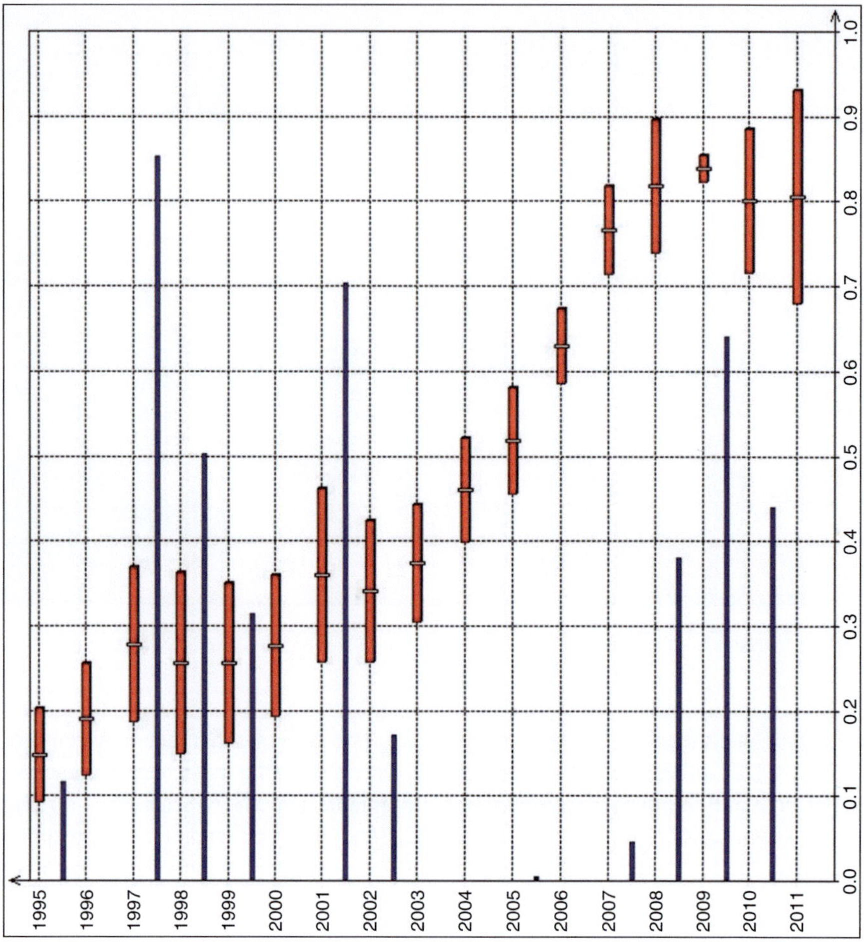

Fig. 5.23 Multidimensional sustainability performance index: Russia, GDP priority with three criteria

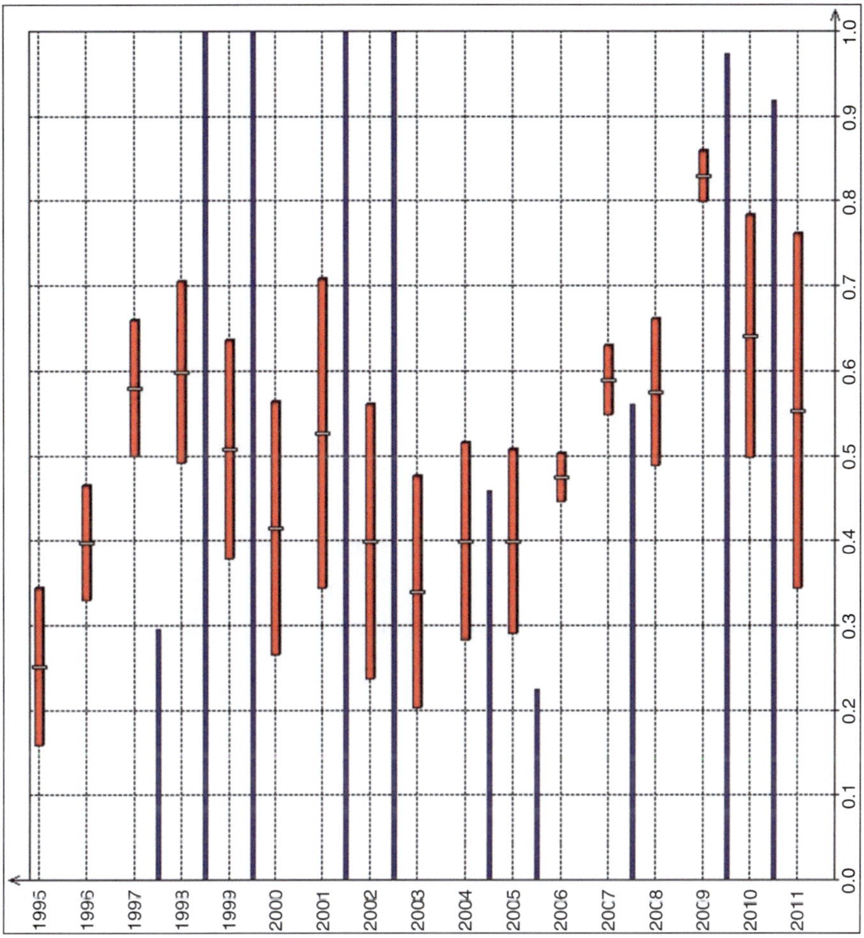

Fig. 5.24 Multidimensional sustainability performance index: Russia, CO_2 minimization and Life expectancy maximization priority with 3 criteria

References

Anselin A, Meire PM, Anselin L (1989) Multi-criteria techniques in ecological evaluation: an example using the analytical hierarchy process. Biol Conserv 49(3):215–229

Castaneda B. (1999) An index of sustainable economic welfare (ISEW) for Chile. Ecol Econ 28:231–244

COM (2001) 264. 15.5.2001. Communication from the commission. 'A Sustainable Europe for a Better World: A European Union Strategy for Sustainable Development' (Commission's proposal to the Gothenburg European Council). http://eur-lex.europa.eu/LexUriServ/site/en/com/2001/com2001_0264en01.pdf

EEA (1999) Environment in the European Union at the turn of the century

EEA (2003) Environmental indicators: typology and use in reporting
Encyclopaedia of the Earth (2006) Kyoto protocol and the United States. http://www.eoearth.org/view/article/154065/
Hovanov NV, Yudaeva M, Hovanov K (2009) Multicriteria estimation of probabilities on basis of expert non-numeric, non-exact and non-complete knowledge. Eur J Oper Res 195(3):857–863
Martinez-Alier J, Munda G, O'Neill J (1998) Weak comparability of values as a foundation for ecological economics. Ecol Econ 26:277–286
Max-Neef M (1995) Economic growth and quality of life: a threshold hypothesis. Ecol Econ 15:115–118
Munda G (1995) Multicriteria evaluation in a fuzzy environment. Theory and applications in ecological economics, Contributions to economics Series. Physica-Verlag, Heidelberg, 255 pp
OECD (1993) OECD core set of indicators for environmental performance reviews. A synthesis report by the group on the state of the environment. OECD, Paris, 35 pp.Polenske, K.R. & Lin, X., 1993. Conserving energy to reduce carbon dioxide emissions in China. Structural Change and Economic Dynamics, 4(2), pp.249–265.
Pisano U, Lepuschitz K, Berger G (2013) National sustainable development strategies in Europe 2013. Taking stock and exploring new developments. ESDN Q Rep 29:32
Rath-Nagel S, Stocks K (1982) Energy modelling for technology assessment: the MARKAL approach. Omega 10(5):493–505
Roy B (1985) Méthodologie multicritère d'aide à la décision. Economica, Paris, 699 pp
Roy B (1991) The outranking approach and the foundation of electre methods. Theor Decis 31:49–73
Roy B (1996) Multicriteria methodology for decision aiding. Kluwer Academic Publishers, Dordrecht, Boston, London, 293 pp
Roy B, Vincke P (1981) Multicriteria analysis: survey and new directions. Eur J Oper Res 8:207–218
Shmelev S (2010) Environmentally extended input–output analysis of the UK economy: key sector analysis
Shmelev S (2011) Dynamic sustainability assessment: the case of Russia in the period of transition (1985–2008). Ecol Econ 70(11):2039–2049
Shmelev S (2012) Ecological economics: sustainability in practice. Springer, Dordrecht, 248 pp
Shmelev SE, Rodríguez-Labajos B (2009) Dynamic multidimensional assessment of sustainability at the macro level: the case of Austria. Ecol Econ 68(10):2560–2573
Shmelev SE, Shmeleva IA (2009) Sustainable cities: problems of integrated interdisciplinary research. Int J Sustain Dev 12(1):4–23
UN (1992) Agenda 21, Programme of action for sustainable development, adopted at the United Nations conference on environment and development, Rio de Janeiro, Brazil, 1992
UN (1996) United Nations, indicators of sustainable development framework and methodologies. United Nations Sales Publication No.E.96.II.A.16, New York, August 1996
UN (2007) Indicators of sustainable development, guidelines and methodologies, 3rd edn. United Nations Department for Economic and Social Affairs, Division for Sustainable Development, http://www.un.org/esa/sustdev/csd/review.htm
UN (2012) Sustainable development goals. http://sustainabledevelopment.un.org/
UNEP (2007) GEO 4. Global environmental outlook report. http://www.unep.org/geo/GEO4/report/GEO-4_Report_Full_en.pdf
WCED (1987) Our common future. Oxford University Press, Oxford, 416 pp

Chapter 6
System of Accounts for Global Entropy-Production (SAGE-P): The Accounting in the Topological Domain Space (TDS) of the Econosphere, Sociosphere, and the Ecosphere

Anthony Friend

Abstract This chapter describes the accounts of the Econosphere, Sociosphere and Ecosphere governed by the Second Law of thermodynamics. Applied symmetries of 'production' (i.e., neg-entropy), 'consumption' (i.e., entropy) and capital accumulation (i.e., net-valued low entropy fund available for human consumption). These entropy production accounts permits holistic assessment of sustainability in the following state condition: (a) steady-state, (neg-entropy = entropy), (b) surplus-state, (neg-entropy > entropy) and (c) deficit-state (neg-entropy < entropy). Examined are the root of entropy accounting in the labour theory of values and the Ricardian (long-term) production equations of distribution of wages, profits and rent. This is contrasted to the neoclassical general equilibrium hypothesis of unlimited productivity per unit of consumption and where distributions are determined by state conditions of the supply and demand curve governed by consumer choice and/or preference functions. We introduce the equations of the entropic process described by G-R Flow-Fund Model whereby quantitative (material) production functions are transcribed into qualitative (immaterial) consumption functions, (i.e., enjoyment of Product). This powerful logic of entropy production as a limit function of Production, Consumption and Capital Accumulation, reinstates the Ricardian hypothesis of the longterm end result of Capitalism: *Wages* fall to subsistence, *Profits* fall to zero, and *Rents* rise to a maximum. The SAGE-P provides the accounts of entropy production essential for redirecting the trajectories of unsustainable growth, towards a sustain-

We have chosen the concept of Topological Domain Space (TDS) because of its generalisation of a mathematical object defined as any set of points that satisfy a set of postulates. These are postulates applied here are those expounded in the G-R Flow Fund Model, see Appendix I.

A. Friend (✉)
OIKOS, Ottawa, Canada
e-mail: afriend@sympatico.ca

S. Shmelev (ed.), *Green Economy Reader*, Studies in Ecological Economics 6,
DOI 10.1007/978-3-319-38919-6_6

able economy reduced to maintaining the stock (i.e., wealth) of any well-defined Low Entropy Fund (LEF) available for human consumption, (see Fig. 6.4). The object of policy is thus redefined as the minimum socially acceptable rate of entropy production per unit of consumption. The Appendices provides a window on the formalism underpinning the accounting concepts of SAGE-P.

Keywords Urban sustainability • Sustainable cities • Multi-criteria decision aid • MCDA • Benchmarking, etc.

6.1 Prologue

6.1.1 Entropy Production and the Econosphere

> *The law that entropy always increases -second law of thermodynamics-holds, I think, the supreme position among the laws of Nature. If someone points out to you that your pet theory of the universe is in disagreement with Maxwell's equations -then so much worse for Maxwell's equations. If it is found to be contradicted by observation -well, these experimentalists do bungle things sometimes. But if it is found to be against the second law of thermodynamics I can give you no hope; there is nothing for it but to collapse in deepest humiliation.* (Sir Arthur Eddington, The Nature of the Physical World, 1927)

The value conserved in the System of Accounts for Global Entropy-Production (SAGE-P) is the flow of services drawn from the consumption of the Low Entropy Fund (LEF) composed of economic, social and ecological objects. Boulding (1949) pointed to the desirability of governments to design policies which maximise the service flows (benefits) and minimise the entropy production (cost) of the national economy, to wit:

> *It is not the increase in consumption or production that makes us rich, but the increase in capital, and any invention which enables us to enjoy a given stock of capital and a smaller amount of consumption and production, out-go or income, so much the gain.*

Boulding clearly saw that the objective function of economic policy is to maximise the flow of service income from capital as opposed the current accounting of GDP, (i.e., money-income generated from the gross domestic product). Efficiencies applying Boulding is the maximization of values conserved in use of economic, social, human and natural capital. While recognised in the discourse on sustainability, it is rarely practiced in a market economy. On the contrary, conserved values-in-exchange dominate the assessments of economic performance as per capita GDP. The protocols of longterm sustainability (i.e., intergeneration equity), cannot be developed from the quantitative metric of GDP. For this to be done, a mapping of the qualitative variables of the human welfare measure on GDP is a *sine qua non*.

Mayumi (2001) proposed that sustainability is best measured as given 'socially acceptable' minimum rate of entropy production for any well-specified measure of GDP expanded to a composition of capital – economic, social, human and ecological. In effect, a cultural choice of the social efficiency matrix of capital, or, in Georgescu-Roegen (1971) terms the rate of replenishment of the consumed stock of any well-define LEF.

Ecological Economic Theory permits the mapping of quantitative data, (i.e., cardinal-valued metrics) on qualitative. (i.e., ordinal-valued metrics) to measure qualitative change of the human welfare function, (Georgescu-Roegen 1971; Daly and Cobb. 1989). Further, the theory infers an *a priori* judgment of preferred future states of the system,. In other words, inferred is a pre-analytical vision statement, based on ethical principles that enter, in important way in the evolving discourse of the globalisation of the economic process which exhibit unequal distribution, and unequal access, to the Global Low Entropy Fund, (Martinez-Alier 1987).

6.2 Entropy Production

The rate of entropy production of GDP is in proportional to, and symmetric with, the material/energy inflows/outflows to/from any well-defined Low Entropy Fund (LEF). Production is translated in the language of thermodynamics as a neg-entropic process (i.e., increasing order in the *Econosphere* measured by use-intensity of the factors of production, land, labour and capital), **Consumption** is stripped down to the properties associated with the entropic process [i.e., decreasing order in the *Econosphere* measured by the expenditures of households (private goods and services), government (public goods and services) and the net-value of trade with the rest-of-the-world]. **Capital** in GDP is residual value of surplus generated by the entropic process and translated into (new) investment which accumulate in the physical stock of the LEF, (i.e., houses, factories, farms etc.) GDP is can be viewed and he annual inflow from the Econosphere to the LEF where values are conserved-in-exchange. Other annual inflows to the LEF are the goods and services from t*he Sociospher*e where values are conserved in use the Ecosphere where values are conserved in themselves, or existential (see Fig. 6.4).

While symmetry remains, the proportionality assumption is dropped with respect to the immaterial economy.[1] The rate of entropy production is represented in the GDP accounts as a set of physical objects consumed in the I/O matrix **A** described by production, consumption and capital accumulation functions, as well as a net-valued trade with the rest-of-the-world. **GDP** thus represents **LEF** at market prices, (i.e., values conserved-in-exchange) available for human consumption. For every object, or element '**x**' in the matrix **A** there exists a corresponding inflow of negentropy (i.e., production = **a**) and outflow of entropy (i.e., consumption = **a'**) permitting the correspondence mapping of $a \rightarrow a'$. The arrow depicts the algorithm of a structure-preserving mapping (homomorphism): $\mathbf{A} \rightarrow \mathbf{A'}$ (i.e. entropy production). Literally, entropy production is the rate of heat dissipation (i.e., unavailable energy

[1] The products of the immaterial economy, (i.e., abstract objects like financial services), are not directly subject to the Entropy Law. Abstract objects are nonetheless agents, like investment decisions, for either increasing or decreasing the rate of entropy production in the system. Abstract objects acting on other abstract objects have no discernible rate of entropy production.

for further work) in the transformation of the set of low entropy ***a*** objects into an equal valued set of high entropy ***a'*** objects, (i.e., waste).

Applying the mapping LEF → GDP we are provided with measure of sustainability based on the rate of entropy production per unit of consumption, (Mayumi 2001). This measure is generalised in the System of Accounts for Global Entropy Production (SAGE-P) in the hierarchical-structure of the topological domain spaces (TDS) of the ***Econosphere,*** (i.e., values are conserved-in-exchange), which is a proper-subset of the topological domain of the ***Sociosphere,*** (i.e., values are conserved-in-use or participation), which in turn is a proper subset of the topological domain of the ***Ecosphere,*** (i.e., values are conserved-in-themselves or intrinsic[2]).

SAGE-P may be presented to decision-makers, and the pubic-at-large, as a continuous space-time mapping of net-valued entropy production and the results recorded in the balance sheet of the LEF accounts: (i) surplus-state: (i) $\mathbf{A} > \mathbf{A'} = \mathbf{A}^{(+)}$, (i.e., sustainable + growth); (ii) deficit-state $\mathbf{A} < \mathbf{A'} = \mathbf{A}^{(-)}$, (i.e., unsustainable + degrowth); and (iii) steady-state, $\mathbf{A} = \mathbf{A'} = \mathbf{A}$, (i.e., sustainable + no-growth). Since the equations represent inequalities, the assessment of sustainability are time differential of the rate of replenishment of the **a** objects greater than, or equal to, the rate of entropy production of the **a'** objects. Non-renewable resources (e.g., fossil fuels) are at all times in the state $\mathbf{a} < \mathbf{a'}$. Also, in entropy production accounting substitution represents objects with different qualitative properties, and in the strict sense of the word, are not equivalence in values.

The LEF represents the I/O matrix of objects/functions available for human use and enjoyment produced in: (i) the ***Ecosphere*** ($\mathbf{A}^{\mathbf{n}}$), (ii) ***Sociosphere*** ($\mathbf{A}^{\mathbf{s}}$) and the ***Econosphere*** ($\mathbf{A}^{\mathbf{e}}$). Note the conventions of the System of National Accounts (SNA) the LEF is a incomplete set of accounts of (a) natural capital $\mathbf{K}^{\mathbf{n}}$ (i.e., inventory of natural resources), (b) human capital $\mathbf{K}^{\mathbf{h}}$ (i.e., inventory of human resources), (iii) social capital $\mathbf{K}^{\mathbf{s}}$ (i.e., inventory of institutions and public goods[3]) and (iv) economic capital $\mathbf{K}^{\mathbf{e}}$, (i.e., inventory of production capacity for goods and services for sale[4]) and natural capital $\mathbf{K}^{\mathbf{n}}$.

The innate properties of physical and abstract objects are mutually exclusive to the categories in any well-defined TDS, (i.e., economic, social or ecological function). The same objects mapped from one category to another, while maintaining their innate properties, conserve values in accordance to function. Thus, like the Bauhaus architectural and urban planning design where form follows function, entropy accounting design is where form follows 'entropy efficiencies' identified with the categories of production, consumption and capital accumulation. The cha-

[2] Intrinsic values are in principle infinite. For the purpose of accounting it seems more rationale to convert intrinsic to to human-calculated existential values. What is the value of existence of rare and endangered species? Or the cost of conservation of ecosystem functions? Existence-values converge into cultural values and may thus be considered as equivalences.

[3] Public goods imply free access to the LEF but are frequently privileged to income, as in airports and/or national parks. Public goods also include military expenditure to protect the Nation, but are the locus of destruction of capital and wealth.

[4] Centrally-planned economies $\mathbf{K}^{\mathbf{e}}$ represents the public good, but like market economy, the access to economic LEF is privileged to apparatchiks who run the economy and the state.

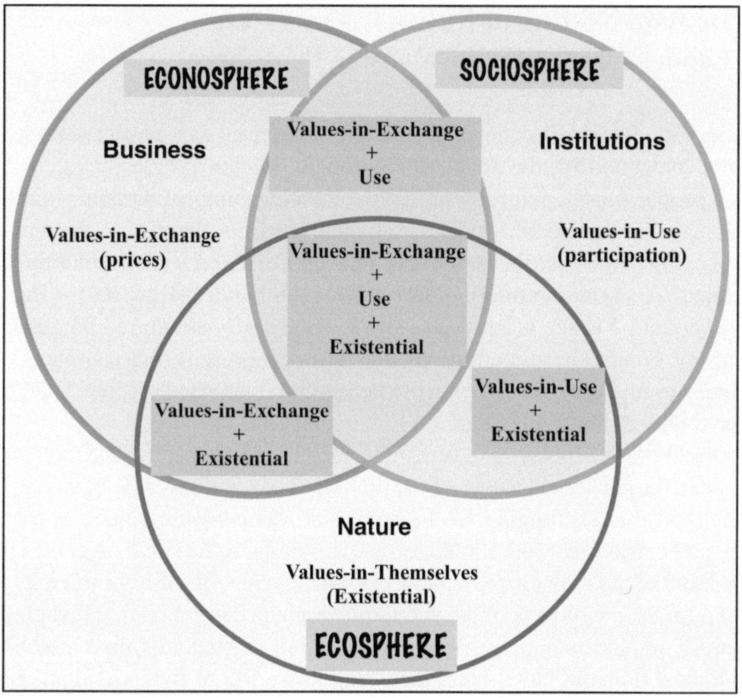

Fig. 6.1 Pluralism of values

meleon colour change of values may be viewed as analogous values follow function referred to as Pluralism of Values, (see **Fig. 6.1: Pluralism of Values**).

For instance a living tree is valued for its ecological function in the Ecosphere Accounts, (e.g., photosynthesis & habitat), its social function in the Sociosphere Accounts, (e.g., recreational & aesthetic) and its economic function the Econosphere Accounts, (e.g., wood products). Thus, the same object conserves values unique to its function: (i) conserved value-in-exchange (Econosphere); (ii) conserved value-in-use (Sociosphere); and (iii) conserved value in themselves, or exstential,[5] (Ecosphere).

[5] Intrinsic values are either infinite or zero. While this may hold true for the individual, the society may express through culture, myth, memory etc., a value for objects and function unrelated to usefulness or economic gain. This value is best summed up as *existential value* which must be conserved for its own sake. There is no direct matrix for existential value, the society nonetheless expresses these values indirectly through allocation of funds, such as conservation of assets (i.e., low entropy fund), like establishment of a system of national parks, protection of historical monuments, museums etc., and through mass protests and politics. In the case of the latter wars have been fought to protect the Nation's abstract values like integrity, honour and freedom.

6.3 Domain Nesting of Entropic Processes: Econosphere → Sociosphere → Ecosphere

The economic domain (Econosphere) is conceptualised as a proper subset of, and thus fully integrated in, the social-demographic domain (Sociosphere), which, in turn, is a proper subset, and fully integrated in, the ecological domain (Ecosphere). Each domain is represented by the statistical datasets describing the quantities, qualities, and spatial distribution where relevant, of fixed, and circulating capital: Econosphere (economic capital) → Sociosphere (human/social capital) → Ecosphere (natural capital). **Figure 6.2** presents the hierarchically-structured datasets of the low entropy Fund. The directions of the arrows represent material-energy flows from lower-to higher-order (trophic) structures, and reversed arrows, from higher- to lower-orders of the trophic chain.

The accounting objects and functions in the low entropy Fund are uniquely expressed in the conserved values of the domain. Thus, we have all objects expressed in exchange-values belonging to the economic domain, use-values in the social domain, and existential-values in the ecological domain. As such, objects in the low entropy Fund of the Ecosphere, with an exchange-value, like the commercial timber value of a forest, are objects of the economic domain. The recreational values of the same forest are objects in the social domain, and its existential-values are objects of the ecological domain. Note, here, that that human life of the individual, and *ipso facto*, the population, are existential-valued objects, belonging, like any other spe-

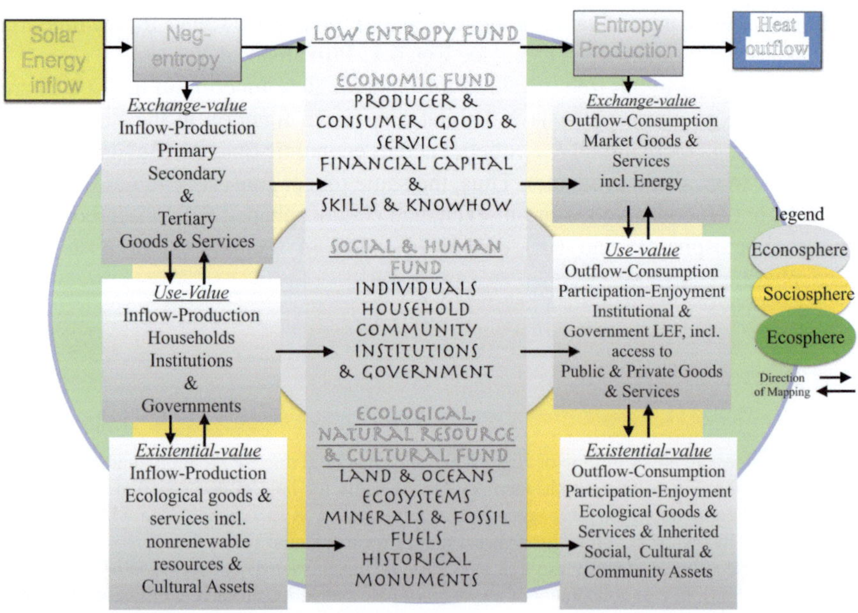

Fig. 6.2 SAGE-P hierarchically-structured topological space

cies, to the ecological domain. However, if wages are paid in exchange for work, the objects (i.e., the labour force) belong to the economic domain. Similarly, population participation in activities where no money is exchanged, are objects of the social domain.

The highest order value, in which all other entropic events are dependent functions, is the inflow of solar radiation, and the outflow of heat dissipation from the Earth. Vitousek *et al.* (1986) estimate that humans currently appropriate approximately 25 % of the potential total global Net Primary Productivity (NPP), and 40 % of the terrestrial potential.[6] It is important to recognise that the overwhelming quantity of solar energy is heat absorbed/reflected and only a minuscule amount of solar energy fuels 'living organisms' (i.e., photosynthesis/respiration). The solar energy balance sheet includes the NPP indirectly appropriated by humans (agriculture, forestry, fisheries etc.,) and the direct solar energy captured for heating and useful work (photovoltaic cells, wind, hydro, tidal, etc.).

Georgescu-Roegen argued that the economy is an entropic process of transformation of low to entropy objects – the ultimate end-product of the house is a pile of rubble! Thus, entropy production must be purposeful and has an ordinal-value. To wit:

> *Since the economic process materially consists of a transformation of low entropy into high entropy, i.e., into waste, and since this transformation is irrevocable, natural resources must necessarily represent one part of the notion of economic value. And because the economic process is not automatic, but willed, the services of all agents, human and material, also belong to the same facet of that notion. For the other facet, we should note that it would be utterly absurd to think that the economic process exists only for producing waste. The irrefutable conclusion is that the true product of that process is an immaterial flux, the enjoyment of life. This flux constitutes the second facet of economic value. Labor, through its drudgery, only tends to diminish the intensity of this flux, just as a higher rate of consumption tends to increase it.* (Georgescu-Roegen 1971, p. 18)

Central to the valuation of the entropic process is defining the efficiency measure of the immaterial flux. Irving Fisher (1906), in *The Nature of Capital and Income*, saw income (flow) and capital (stock) as two facets of the same object – income being the accumulated flow of abstract services measured over a period of time, and capital being a (low entropy) fund measured as material wealth at an instant in time:

> *[Income] … is a flow through a period of time and not, like capital, as a fund at an instant in time, … consisting of abstract services and not, like capital, of concrete wealth. The income from any instrument is thus the flow of services rendered by the instrument. The income of a community is the total flow of services from all its instruments. The income of an individual is the flow of services yielded to him from his property.* (Fisher 1965, p. 102)[7]

[6] NNP = solar energy captured by plants and other photosynthetic organism minus that used by the organisms themselves for respiration.

[7] In the Fisher analysis, abstract objects in SAGE-P, like bank accounts, assume a material object of 'concrete wealth'. What this really means is that wealth is an instrumental means to produce more wealth (i.e., capital) or an instrumental means to sustain or enjoy life (i.e., consumption). For this condition to hold, economic instruments must have attached property rights, which are owned either by an individual or a collective, such as a community. Note that in SAGE-P 'property rights' represent the higher order (abstract) institutional objects of the sociosphere.

Kenneth Boulding expanded the Fisher service-flow income from stocks to include the entropy efficiency of capital consumption itself:

> *I shall argue that it is capital stock that we derive satisfaction, not from additions to it (production) or subtractions from it (consumption): that consumption far from being a desideratum, is a deplorable property of the capital stock which necessitates the equally deplorable activities of production: and that the objective of economic policy should not be to maximize consumption or production, but rather minimize it, i.e. to enable us to maintain our capital stock with as little consumption and production as possible. It is not the increase in consumption or production that makes us rich, but the increase in capital, and any invention which enables us to enjoy a given stock of capital and a smaller amount of consumption and production, out-go or income, so much the gain.'* (Boulding 1949)

Georgescu-Roegen argued that understanding the nature of the economic process is contextual to historic experience of peoples and the capacity to adapt to technological change (i.e., exosomatic evolution). To wit:

> *And paradoxical as though it may seem, it is the Entropy Law, a law of elementary matter, that leaves us no choice but to recognize the role of cultural tradition in the economic process. The dissipation of energy, as the law proclaims, going on automatically everywhere. This is precisely why the entropy reversal as seen in every line of production bears the indelible hallmark of purposive activity. And the way this activity is planned and performed certainly depends upon the cultural matrix of the society in question. There is no other way to account for the intriguing difference between some developed nations endowed with a poor environment, on the one hand, and some underdeveloped ones surrounded by an abundance of natural riches. The exosomatic evolution works it way through cultural tradition, not only through technological knowledge.'* (Georgescu-Roegen 1971, p. 18)

6.4 SAGE-P: Roots in the Classical Model of Fixed and Circulating Capital

The key, and unique feature of SAGE-P, is the articulate accounting principles of the LEF. The ethical principles concerning the distribution and access to Global Low Entropy Lund is echoed in the opening sentence of the Report of the World Commission on the Environment and Development, a.k.a. the Brundtland Report, to wit:

> *The Earth is one but the world is not. We all depend on one biosphere for sustaining our lives. Yet each community, each country, strives for survival and prosperity with little regards for the impact on others. Some consume the Earth's resources at a rate that would leave little for future generations. Others, many more in number, consume far too little and live with the prospect of hunger, squalor, disease, and early death.* (WCED 1987, p. 27)

We shall argue that the issue raised by the Brundtland report has deep roots in the distribution issue of classical economics, and well articulated in the Malthusian Spectre considered a critical factor in David Ricardo's end state of Capitalism, (Malthus 1959; Ricardo 1973; Robinson 1980.). LEF in classical economics is composed of fixed (i.e.,structures) and circulating capital (i.e., money). Further distinctions were made between the stock of objects available for immediate consumption (e.g., food) and the stock of objets (i.e., machines) necessary to replenish LEF at its rate of consumption.

Sraffa's (1960) essay: The Production of Commodities by Means of Commodities is a self-definining production, consumption and capital accumulaion system. The Labour Fund (i.e., wages) is reduced to its base subsistence, the Capital Fund (i.e., profits) is reduced to the appropriated part of the Labour Fund, (i.e., surplus from work), and Natural Capital (i.e., rent) to its capacity to repoduce itself.[8] The latter, while a limiting factor for economic growth, is not in itself, a factor of the share of the National Prduct between wages and profits. All values are determined by the thermodynamic relationships of factors of production: land, labour, and capital. Joan Robinson admired the logic of Sraffa's I/O Model of Production reduced to circulating capital, to wit:

> Evidently we are in a capitalist economy, but to avoid ambiguity which have clustered around the word, capital is never mentioned. There is profit, but no enterprise; wages, but no pay-packets; prices, but no markets. Nothing is mentioned but but the equations of production and the necessary conditions of exchange, Robinson 1980: 144–150.

While rooted in Ricardian economics and the labour (or work) theory of value, the Sraffa I/O Model of circulating capital may be viewed as a precursor of G-R Flow-Fund Model of the entropic process, (Robinson 1980; Mayumi 2001; Friend 2005).

With the advent of the general equilibrium system, circa the 1870s, the labour theory of value was replaced by a subjective, willingness-to-pay theory of value Thus freeing economic theory from the Mathusian Spectre of the 'dismal sciences'. The subsistence laour theory of value was unbounded by the ever expanding frontier of technology, productivity and wages. Where constraints to reproductive capacity of natural resources appeared, the substitution theory prevailed, (Simon 1981). Alfred Marshall, was a little more sanguine on this point:

> *Even when cultivation has reached a stage after which each successive dose applied to a field would get less return than the preceding dose, it may be possible for an increase in the population to cause a more than proportional increase in the means of subsistence. It is true that the evil day is only deferred: but it is deferred. The growth of population, if not checked by other causes, must ultimately be checked by the difficulty of obtaining raw produce; but in spite of the law of diminishing return, the pressure of population on the means of subsistence may be restrained for a long time to come by the opening up of new fields of supply, by the cheapening of railways and steamship communication, and by the growth and organization of knowledge. Against this must be set the growing difficulty of getting fresh air and light, and in some cases fresh water, in densely peopled places. The natural beauties of a place of fashionable resort have a direct money value which cannot be overlooked; but it requires some effort to realize the true value to men, women and children of being able to stroll amid beautiful and varied scenery.* (Marshall 1947, p. 166)

While Alfred Marshall's warning was left unheeded by the economic growth theorists, the entropy production feedback loop industrialization, if left unchecked, would lead to both the ruination of the reproductive capacity the ecosystem as well

[8] Nonrenewable resources like fossil fuels and minerals are not circulating capital, other than recycling, and treated as appropriated in distributional relationship of the wage-profit cycle.

as the aesthetic qualities of the planet Earth.[9] Indeed, as noted by some social analysts (e.g., Ellul 1964; Koestler 1967; Schumacher 1973; and Bataille 1988),[10] the Machine + Man Age harbour the inner logic of the entropy feedback loop, (i.e., Tragedy of the Commons), (Hardin 1968).

Today, the exponential rate of global entropy production associated the material consumption is fully recognized in the academic discourse on sutainability with the biophysical limits of the Earth and openly discussed in terms of sustainability of high consumer societies, (Arrow et al. 2004; Stern 2007). The concern of material consumption identified with environmental degradation, at least in Britain in the 1840s, elegantly expressed in John Stuart Mill's essay on the Stationary State, to wit:

> *It is scarcely necessary to remark that a stationary condition of capital and population implies no stationary state of human improvement. There would be as much scope as ever for all kinds of mental culture, and moral and social progress; as much room for improving the Art of Living, and much more likelihood of its being improved, when minds ceased to be engrossed by the art of getting on* (Mill 1985, p. 116)

The labour theory of value includes the notion that Nature work along with Man but in context of the free gift of the Global Commons, to wit:

> *In agriculture, too, Nature labours along with man; and though her labour costs no expense, its produce has its value, as well as that of the most expensive workman.* (Smith 1994, p. 393).

Services, while recognised as the necessary condition to maintain order and structure of a Society, as well as pleasure and enjoyment, could not add to the Wealth of Nations because services, the tertiary sector of the National Accounts"… *perish the very instance of its production.*" The oft quoted passage in the Wealth of Nations makes this point clearly and unambiguously as follows:

> *The labour of some of the most respectable orders in society is, like that of the menial servants, unproductive of any value, and does not fix or realize itself in any permanent subject, or vendible commodity, which endures after labour is past, and for which an equal quantity of labour can be procured. The sovereign, for example, with all the officers both of justice and war who serve under him, the whole army and navy, are unproductive labourers....In the same class must be ranked, some both of the gravest and most important, and some of the most frivolous professions; churchmen, lawyers, physicians, men of letters of all kinds;*

[9]Theodore Roosevelt (1906) said: 'We have become great because of the lavish use of our resources. But the time has come to inquire seriously what will happen when our forests are gone, when the coal, the iron, the oil, and the gas are exhausted, when the soils have still further impoverished and washed into the streams, polluting the rivers, denuding the fields and obstructing navigation.'

[10]Georges Betaille's philosophical concern is the logic that 'surplus product' must ultimately 'destruct'. This led to the examination of the human experience with respect to the distribution (power) and destruction (choice) of the 'economic surplus'. The annual flood of the Nile River aquatic ecosystem (i.e., a low entropy Fund) created the 'surplus funds' to construct the tombs of the Pharaohs – a useless object. In the more literal sense of destruction, the economic surplus provides the capacity for nation's to wage wars or to build-up arms for the politically-motivated (imaginary) pseudo-wars, like the Cold-War.

> *players, buffoons, musicians, opera-singers, opera-dancers, etc. ...Like the declamation of the actor, the harangue of the orator, or the tune of the musician, the work of all of them perish the very instance of its production.* (Smith 1994, p. 364)

While abstract objects, referred to as non-tangible, are denied contribution to wealth, the physical objects that make-up this stock of wealth offer the paradox, ical difference between use-value and exchange-value, to wit;

> *The word VALUE, it is to be observed, has two different meanings, and sometimes expresses the utility of some particular object, and sometimes the power of purchasing other goods which the possession of that object conveys. The one may be called "value in use"; the other, "value in exchange". The things which have the greatest value in use have frequently little or no value in exchange; on the contrary, those which have the greatest value in exchange have frequently little or no value in use. Nothing is more useful than water: but it will purchase scarce anything; scarce anything can be had in exchange for it. A diamond, on the contrary, has scarce any use-value; but a very great quantity of other goods may frequently be had in exchange for it.* (Smith 1994, p. 57)

Adam Smith in the opening passage of the Wealth of Nations has an uncanny resemblance to the G-R Flow Fund Model when the economy is described in terms of the annual product of labour (negentropy) as inflow to, and the annual consumption of product (entropy) as outflow from, the stock of LEF available for human consumption, to wit:

> *The annual labour of every nation is the fund which originally supplies it with all the necessaries and conveniences of life which it annually consumes, and which consists always either in immediate produce of that labour, or what is purchased with that produce from other nations.* (Smith 1994, p. lix)[11]

6.5 Emergence of Dual Valuation of Land and Labour in Classical Economics

The *Political Arithmetic* of William Petty (1623–1687) calculated the value of the nation's wealth on a per capita income of £6 13 s 4d per annum and a population of six million souls, thus yielding a national income for England of £40 m. On value, Petty continued the debate begun by Aristotle, and chose to value the Wealth of the Nation in its natural denominations, land and labour. Both of which are prime source of tax, *land* (i.e., Nature's Œconomy) and *labour*, (i.e., Human Œconomy). Petty's famously quoted concept of value: "LAND is the MOTHER of WEALTH and LABOUR is the FATHER of WEALTH."

The Physiocrates, in the *Tableau Economique* identified the augmentation of value in the soil, which clearly suggests that out of nothing, nature creates value

[11] Economists have often pointed to the contradiction between 'annual labour', a flow concept, and a 'Fund', a stock concept. This can be resolved by replacing the word 'annual' with the word 'potential'. The labour force is thus a stock (a quantitative value) representing qualities of the potential work, in conjunction with nature, to supply the nation '...with all the necessaries and conveniences of life which it annually consumes.'

(i.e., *the primary product* of the Planet Earth).[12] Labour conserves value prescribed by the magnitude of fixed and circulating capital and the quantity of the labour exchange in the market (i.e., wages). In other words, the circulation of primary product in any well-defined physical process of production, consumption and capital accumulation. Adam Smith disagreed, while recognising that Nature is the primary producer, (i.e., natural capital), wealth creation (i.e., augmentation of value) is coming into existence, so to speak, the action of industry and labour. to wit:,

> *The materials of all wealth originates primarily in the bosom of the earth; but it is only by the aid of labour they can truly constitute wealth. The earth furnishes the means of wealth; but wealth cannot have any existence, unless through industry and labour which modifies, divides, connects and combines the various production of the soils so as to render them fit for human consumption.* (Smith 1809, p. xxxvii).[13]

Mirowski (1989) detected in the reasoning of the Physiocrates the essence of the entropic process inferring that the primary producer of photosynthesis (i.e., solar energy) is the origin of value. The logic of recursive, hierarchically-structured, circulating capital seem to confirm the hypothesis of conserved values in the following order: existential-value > use-value > exchange-value. This is how Mirowski put it:

> *If one chooses, as did the Physiocrates, to locate the augmentation of value in a single sector, then it follows that trade between sectors can readily be defined as trade of equivalents: this is the real meaning of the Tableau Economique. In this schema, primary production is well defined as the locus of increase of the value substance; secondary production, trade or circulation as where value substance is conserved, and finally, consumption as the locus of final destruction.* (Mirowski 1989, p. 159).

6.6 Georgescu-Roegen's Flow-Fund Model (FFM)

6.6.1 Time as a Factor of Production

At Harvard University in the 1930s, Georgescu-Roegen studied under Joseph Schumpeter,. His Theory of Economic Development (1934) was framed in the formalism of the dialectics of the creative and destructive process, a precursor of the indeterminism entailed by the Entropy Law (i.e., arrow of time). While not explicit, the General Equilibrium Model is framed in statistical time series which predict the future from the probability of past events, (i.e., frequency distribution). Determinism in economic analysis is based on the Newtonian equation of motion and degree of belief in the Laplacian Demon.[14]

---

[12] Primary production is the synthesis of organic compounds from atmospheric or aqueous carbon dioxide. It principally occurs through the process of photosynthesis, which uses light as its source of energy.

[13] The quote is from M. Garnier who wrote a introductory essay entitled "A Short View of the Doctrine of Smith with that of the French Economists" included in the 1809 edition of "An Inquiry into the Nature and Causes of the Wealth of Nations."

[14] We may regard the present state of the universe as the effect of its past and the cause of its future. An intellect which at a certain moment would know all forces that set nature in motion, and all

The Entropy Law introduces to economic analysis the ultimate anti-Laplacian Demon, that of pure randomness of events, both past and future. In the pure state of chaos there is no linkages and connections among the observed events with a statistical probability of zero. Ilya Prigogine, who studied the Entropy Law in chemical reactions and synthesis, concluded that what might appear to the observer as a random event is the manifestation of the disequilibrium state of the system at the limit, (i.e., far from equilibrium) where anticipated trajectories enter the zone of bifurcation, like the mutation in evolutionary theory. Prigogine (1997) refers to this state condition of the system in terms of nonlinear processes characterised as "*emergent properties of dissipative structures far from equilibrium*," (Prigogine and Glandorf 1971; Prigogine and Stengers 1984; Prigogine 1997).

Schumpeter, studied under Eugen von Böhm-Bawerk, (1891) whose capital theory founded on historical time assumed the dynamics of recursive functions of the Austrian School.[15] While the original 'roundaboutness' theory of capital assumed interdependency of current consumption to past capital formation, the Neo-Austrian School expanded the interdependencies to include complex pathways through the social and natural capital, (Malte et al. 1999).

Time recursion is a characterised in the FFM in the algorithms of hysteresis (i.e., lagged effect of past state conditions on the present state conditions).[16] This factor of time in evolutionary processes requires information of the memory of the system, which, *inter alia*, assumes knowledge of the initial state conditions of the system. Georgescu-Roegen argued that (nonlinear) evolutionary processes cannot be predicted from *a priori* assumptions of equilibrium state conditions entailed by the (Walrasian) General Equilibrium System

The logic of the behaviour of economic agents framed in the analytical domains of complex, adaptive, systems are irreducible to the linear regressions analysis of econometric models, no matter how many endogenous variables are included. Predictions of the next event in the time series, exception of near future events, are better framed in the theory of a priory distribution or anticipatory events identified with Bayesian Statistics, (see **Appendix II: Bayesian Statistical Methods**, (Rosen 1985; Malte and Proops 1992).

A key distinction between the Neoclassical School and the Austrian School is that the latter describes qualitative change dependent of time, whereas the former, qualitative change is independent of time,(*de gustibus non est disputandum*) Indeed qualitative change, are empirically observed in consumer preferences in the market

positions of all items of which nature is composed, if this intellect were also vast enough to submit these data to analysis, it would embrace in a single formula the movements of the greatest bodies of the universe and those of the tiniest atom; for such an intellect nothing would be uncertain and the future just like the past would be present before its eyes. (translated from French), – Pierre Simon Laplace, A Philosophical Essay on Probabilities.

[15] Von Böhm-Bawerk was a student of Karl Menger, who, along with Stanley Jevons and Leon Walrus, are considered the founders of the Neo-classical School.

[16] For instance the 2008 financial meltdown may be viewed as lag effect of deregulation of the Banking sector in the 1980s and the subsequent explosion of financial products created by traders in the 1990s, like derivatives, hedge-funds, credit/default swaps etc. The secondary lag effect of the 2008 financial market are reverberating in the debt crises of sovereign States.

place. Government intervention, with respect to public goods, is similarly assumed as exogenous variables in any statistical regressions of consumer and/or producer behaviour.

Schumpeter's (1934) analytical method of creative-destruction assumes discontinuities in production/consumption and capital accumulation cycles. Holling (1994) adapted the method to the Figure Eight Model (see **Fig. 6.3**) to describe dynamics of ecosystem restructuring at the planetary-scale of the Anthropocene, to wit:

> The one overall conclusion is that discontinuous change is an internal property eco systems. In a sense, key structural parts of the eco system become "accidents waiting to happen"… There is both a destructive feature to such changes (entropy) and a creative one (negentropy). Organisms self-destruct in part because of their very success in competing with other organisms and in appropriating and accumulating the prime resources of energy, space and nutrients. The accumulated resources, normally bound tightly and unavailable, are suddenly released by forces of change. Such forces therefore permit creative renewal of the system. I call this … ecosystem function "creative destruction", a term borrowed from Schumpeter's economic theory…The full dynamic behaviour of the system at an aggregate level can therefore be represented by the sequential interaction of four ecosystem functions: exploitation, conservation, creative destruction, and renewal. The progression is such that these functions dominate at different times: from exploitation…slowly to conservation… rapidly to creative destruction…rapidly to renewal…and rapidly back to exploitation. (Holling 1994).

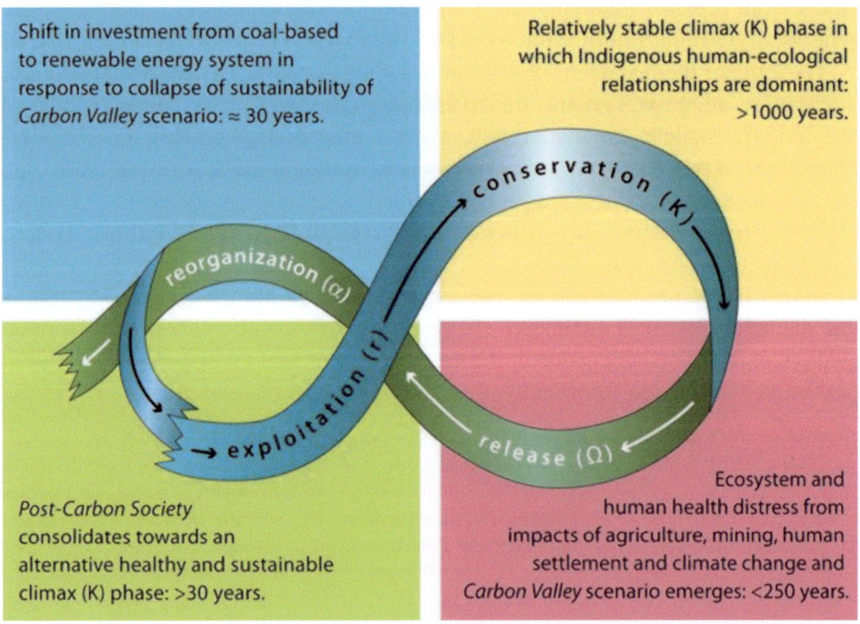

Fig. 6.3 Holling figure eight model

6.7 Some Observations of SAGE-P and Adaptive Logics

SAGE-P is a system of accounting of complex, nonlinear, evolutionary systems, where the (accounting) objects and functions evolve and change over time and space. The algorithmic mappings of category sets, therefore, require special adaptive logics to describe bifurcation points of qualitative properties of objects, such that they are no longer the same object. This question was explored in a paper on nonlinear accounting presented at the European Society for Ecological Economics Conference, Ljubljana, Slovenia, in 2009:

> Here, it should be noted, we make no assertions about which logic is the best fit for nonlinear accounting, but rather offer adaptive logics as a possible first base from which to work. Adaptive logics are sensitive to, and can account for, evolutionary processes. Further, the objective is to develop a formal logic framework to construct the analytical space for an ordinal structured system of non-ergodic processes (Georgescu-Roegen 1971; Friend 2009). One of the key features which we can reflect in adaptive logics is the idea that objects change properties over time, which at some indefinite point, are no longer the same objects defined by their initial conditions. In other words, objects, themselves evolve and change. (Friend and Friend 2009)

In theory, each event recorded in a non-ergodic statistical database is deemed unique. In practice, events in the *Ecosphere*, described by cycling systems, (i.e., atmosphere, hydrosphere and lithosphere) and the reproductive processes of the ecosystems are of sufficient similarity in coarse-grained data to be treated as recurring events. However, a measure of recurrence is a problem in the statistical databases of the fast moving *Econosphere* and *Sociosphere*.

The quantitative/qualitative properties of physical objects visibly change form and function over space-time., While the statistical objects change qualities, the common denominator of entropy production is a conserved value in any well-defined object and/or function over time. Abstract objects, while changing volume and qualities over time, are not directly subject statistics of entropy production. However, SAGE-P permits an algorithm of correspondence mapping of abstract objects on the associated physical objects.

Adaptive logics permit nonlinear relationships to be expressed, and normalised, in nonlinear events of Bayesian Statistics where, for instance: (i) inputs $\neq$ outputs[17]; (ii) the critical event horizon $\rightarrow$ limit function $\rightarrow$ bifurcation points and (iii) resilience capacity to absorb stress (e.g., degree of robustness). Bayesian Statistics may be expressed in stress-response vectors leading to system collapse (i.e., chaos $\rightarrow$ release $\rightarrow$ reorganisation) and the emergent properties of dissipative structures far from equilibrium, (see, **Fig. 6.3**). The Stress-Response Model was proposed as a framework for the development of environment statistics following the the first UN Conference on the Global Environment in Stockholm, 1972, (Rapport and Friend 1979).

---

[17] This means that abstract objects do not obey the *superposition principle* which states that, for all linear systems, the net response at a given place and time caused by two or more stimuli is the sum of the responses which would have been caused by each stimulus individually.

Feedback loops are constructed from algorithms of statistical correspondence of objects/functions that either amplify or dampen the rate of entropy production in the SAGE-P., (Rosen 1985, 1991; Lawvere and Schanuel 1997). Of interest, especially in the long-term analytical frame for climate change policies, are the accounts of entropy production of new technologies, such as Hydrolic Fracturing of shale rocks to release sequestered gas and oil. The long term amplifying feedback loop of previous technologies are well-documented, (Ellul 1964; Bataille 1988).

While technological change is portrayed in the literature as progress in the human condition, the underlying premise of SAGE-P is that exosomatic evolution (i.e., the invention of the tool) is paradoxical with respect to the measure of 'entropy efficiency' in the human welfare function. At the micro-scale of the individual, household and community, entropy efficiency generally increases in proportion to the diffusion of technological progress, (i.e., measure of entropy production per unit of consumption) but at the macro scale of (global) economic production, consumption and capital accumulation the rate of entropy production increases with the growth of the (material) GDP, (i.e., measure of entropy production for the whole economy), (Jevons 1965; Polimeni et al. 2008). Entropy production is an outcome of the Second Law of thermodynamics exclusive to the material world. The immaterial economy of production (e.g., financial services), consumption (e.g., entertainment) and capital accumulation, (e.g., knowledge) entropy production is not proportional to its output, although indirectly it may be instrumental in accelerating the rate of entropy production through financial investment in production, consumer expenditure in tourism and the power consumption of the emergent (immaterial) digital economy.

In theory, each event recorded in a non-ergodic statistical database is deemed unique. In practice, events in the *Ecosphere,* described by cycling systems, (i.e., atmosphere, hydrosphere and lithosphere) and the reproductive processes of the ecosystems are of sufficient similarity in coarse-grained data to be treated as recurrence of the same event. However, a measure of recurrence does not hold in the statistical databases depicting the entropic process in the fast moving events of the *Econosphere* expressed in the hourly movement of the stock markets and somewhat slower household, community and institutional changes in Sociosphere, expressed in decades of demographic and institutional databases.

6.8 The Flow-Fund Model (FFM) of the Entropic Process[18]

The FFM is an accounting system of the LEF available for human consumption in the form of the metabolic process. Observed is the *inflow* of low entropy objects and *outflow* of high entropy objects (i.e., waste). The Fund coefficients represent the metabolic agents of transformation, (e.g., I/O coefficients of manufacturing). The flow elements (E_i) enter as inputs to, and exit as outputs from, the Fund. The

[18] For a Technical description see **Appendix I: Georgescu-Roegen's Flow-Fund Model**.

elements (E_p), represent the factors of production and exist simultaneously in two states: (i) the 'agents' of the process (i.e., the technical coefficient of the I/O matrix), and (ii) the elements used and/or acted upon (i.e., material-energy inflow/outflow). The E's are defined, and empirically observed, at the I/O boundary of any well-define dentropic process. What goes on inside the process remains a black box. The entropy production, or the rate of consumption of the Fund, represents the efficiencies of the human participation in the Fund elements, expressed as categories of consumption, C_1, C_2, C_3 ..., $C_{n.}$ (i.e., the net-value entropy production per unit of consumption), (see Appendix 1:Georgescu-Roegen's Flow-Fund Model).

The net-value of entropy production is the difference between the quantity of input flow (i.e., low entropy objects, like natural resources) to the fund – which is given a positive sign (negentropy) – and the quantity of the output flow, (i.e., high entropy objects, like waste residuals) – which is given a negative sign (entropy). Sustainability implies that the balance accounts of SAGE-P are non-negative over some well-specified time period and/or spatial entity. If the result is positive, the fund assumes a 'surplus' of low entropy stocks available for investment and/or future consumption. If negative, the fund assumes a 'deficit', which needs to be replenished at a rate greater than the present rate of consumption. An example of 'deficit' is the rate of greenhouse gas emissions (i.e., the present consumption of fossil fuels), which is greater than the absorptive capacity of the Ecosphere carbon cycle, resulting in an inevitable build-up of the concentration of greenhouse gases in the atmosphere.

FFM represents the dynamics continuous change in form and function. While recurrence of entropic processes are observed in natural cycles and reproductive processes, Georgescu-Roegen is careful to rule out illusion of recursion from the irreversible entropic process. Thus, the logic of the Entropy Law decrees that all events in a statistical time series are non- recurring (Friend 2009). In practice, and in the short-term, however defined, it matters little to assume recurrence of events, as the error term is too small to be of any consequence for policies. This is not true for the long-term as the small error (i.e., slight perturbation) amplifies in any algorithm of recurrence in any positive feedback loop.

Georgescu-Roegen distinguished statistical probability in the abstract, (i.e., analytic *a priori*), from the statistical probability of causal relationships, (i.e., synthetic *a posteriori*). His view of the two formulations of probability not only apply to different domains, but are fundamentally antagonistic. While the latter assumes an irreducible 'randomness' in nature itself, the former assumes incompleteness in knowledge, viz:

> *If probability is an ultimate element of nature, then forcibly its definition must rest on probability. And if the centuries-old struggle with the problem of finding an analytical definition of probability has produced only endless controversies between the various doctrines, it is, in my opinion, because too little attention has been paid to the singular notion of random. For the dialectical root, in fact, lies in this notion: probability is only an arithmetical aspect of it.* (Georgescu-Roegen 1971, pp. 56).

Erwin Schrödinger captures the essence of the entropic process in his 1944 essay on *What is Life?* to wit:

> *A living organism continually increases its entropy – or, as you may say, produces positive entropy – and thus tends to approach the dangerous state of maximum entropy, which is death. It can only keep aloof from it, i.e. alive, by continually drawing from its environment negative entropy – which is something very positive as we shall immediately see. What an organism feeds upon is negative entropy. Or, to put it less paradoxically, the essential thing in metabolism is that the organism succeeds in freeing itself from all the entropy it cannot help producing while alive.'* (Schrödinger 1967)

The entropic process exhibits the behavioural parameters of complex, non-equilibrium, thermodynamic systems. The causal realationship of bio-econmics Ecosphere assumes the qualitative properties and the quantitaive values of the inflow of solar radiation, the production of of biomass, (i.e., primary productivity), and the outflow of heat to outer space. Stuart Kaufman, (1995) put forward the idea that, like the Newtonian Linear laws of Motion, there must be an equivalent corresponding set of formalised general laws to enable the prediction of the behaviour (material) systems far from equilibrium, (Prigogine 1997). While there is progress in defining formalism and the mathematics of non-equilibrium systems, the predictive cause-effect are far from robust. Morel and Fleck (2006) propose the material Fourth Law of Thermodynamics as a formal foundation of 'emmergent properties of dissipative strutures far from equilibrium, (i.e., how order is created out of chaos).:

> The Fourth Law of thermodynamics…brings the vast area of reacting systems within the domain of thermodynamics. By stressing the consistency of system behavior, it explicitly incorporates the concept of causality into the formal foundations of thermodynamics. The significance and manifestations of the Fourth Law are dramatic in far-from-equilibrium systems where spontaneous investments in local ordering – in dissipative structures – inevitably increase the rates at which systems increase entropy. (Morel and Fleck 2006)

The Morel-Fleck proposition formalises the Georgescu-Roegen material Fourth Law. Whether this will hold up to scientific scrutiny is another matter, but the proposition is logically consistent with the Aristotelean causes defined in SAGE-P in the following manner: (a) *material cause* (i.e., observed event), (b) *formal cause* (i.e., identified with a higher order structure), (c) *efficient cause* (i.e., identified with the instrumental means of change) and (d) *final cause* (i.e., identified with policy epressed in social-cultural valuest.

These definitions do not correspond to our modern understanding of causal relations in the natural and social sciences. Rosen (1991) redefined the four Aristotelean causes into a hierarchical structure: (a) *material cause* corresponds to the change of object observed (e.g., tree cut down) → *efficient cause* corresponds to the instrumental means (e.g., chain saw) for the change of state of the object → *formal cause* corresponds to the human-designed structure (e.g., forest industry) that caused (a) and (b), → final cause corresponds to an abstract object (e.g., government policy regulating the forest industry) that caused (a), (b), and (c).

While the Entropy Law governs direction and volume of entropy production in a closed thermodynamic system, the Fourth Law is exclusively applied to open thermodynamic systems (i.e., temporary reversal) governing the causal relationship of (material) object and functions arising out of complex, hierarchically-structured systems that are not predicted from extrapolations of the past into the future characterised as "*emergent properties of dissipative structure far from equilibrium.*"

The great idea of Georgescu-Roegen was to recognise that Newtonian Laws of Motion are applied to material production, consumption and capital accumulation, In entrapp entraps economic analysis to the one-dimensional mechanistic universe. While the mechanism of quantitative analysis is important for low income countries, (i.e., relatively low entropy per unit of consumption), for high income economies the central question is the qualitative change in the consumption function (i.e., to reduce to a minimum the rate of entropy production per unit of consumption). The primary science of matter is non-Newtonian, and the indeterminism and complexities of quantum mechanics[19] is universal and applicable to the macro-world of the social sciences, to wit:

> *The significant fact for the economist is that the new science of thermodynamics began as a physics of economic value and, basically, can still be regarded as such. The Entropy Law itself emerges as the most economic of all natural laws. It is in ...the primary science of matter that the fundamental nonmechanistic nature of economic process fully reveals itself.* (Georgescu-Roegen 1971, pp. 3)

Georgescu-Roegen viewed the neoclassical project to reduce the behaviour of *Homo economicus* into a set of deterministic laws of statistical mechanics, such as the Hamiltonian, an error of the first order. This mania of the mechanical regression of statistics, (i.e., econometrics), he coined the neologism "arithmorphism." By this, he meant reducing the infinity of all other possible numbers to a logical class of discrete and distinct objects represented by a single number. Indeed, this is the neoclassical project of applying 'Economics as Social Physics, and Physics as Nature's Economics' – the secondary title to Mirowski's (1989) book entitled: *More Heat than Light*, which is a discourse on how the conservation of energy theories were hijacked to demonstrate the mathematical proofs of the conservation-of-value in (linear) econometric models.

Solow (1974) in reviewing the Entropy Law and the Economic Process wrote:

> *Georgescu-Roegen central message is a devastating attack on mathematization in the social sciences in general and economics in particular, and a wise critique of the values and limitations of mathematics in the analysis of human behavior and social phenomena. He agues as follows: Mechanistic model building, and the arithomorphism of theory, the consequences of an effort in by modern economics to conform to the formal cannons of classical physics, has destroyed the relationship between statement (premise) and experience (observation),* (brackets added).

While the First Law of Thermodynamics is an expression of the conservation of energy, it is the Second Law that expresses the fundamental nature of irreversibility, indeterminism, and complexity in Nature. Thus, unlike the neoclassical concern of reducing complexity to a set of (static) equalities of conserved value,

---

[19] A 'quantum' is a quantity of something, a very specific amount. 'Mechanics' is a study of motion. Therefore, quantum mechanics is the study of motion in quantities. Quantum theory says that nature comes in bits and pieces (quanta), and quantum mechanics is the study of this phenomena (Zukav 1979, p. 45).

Georgescu-Roegen chose to examine the change of forms and qualities of economic objects and functions by applying the (holistic) dialectic method.[20] In other words, it involved an analysis of an evolutionary economic process as an ordinal progression in continuous real time/space.[21]

Georgescu-Roegen's exploration of the 'physics of value' led to the discovery of the new discipline of 'Bioeconomics', which combined elements of evolutionary biology, ecology, and conventional economics (Mayumi 2001). Alfred Marshall anticipated Georgescu-Roegen's Bioeconomics [22]:

> *The Mecca of the economist lies in economic biology rather than economic dynamics. But biological conceptions are more complex than those of mechanics; a volume on Foundations must therefore give a relatively large place to mechanical analogies; and the frequent use of the term "equilibrium," which suggests something of a statical analogy. This fact, combined with the predominant attention paid in the present volume to the normal conditions of life in the modern age, has suggested the notion that its central idea is "statical", rather than "dynamical". But in fact it is concerned throughout with the forces which cause movement: and its key-note is that of dynamics, rather than statics.* (Marshall 1947, pp. xiv)

Bioeconomics entails the parameters of complexity. Georgescu-Roegen proposed the dialectical methods to study 'emergence' of novelty and to apply the entropic process to evolutionary change that cannot be predicted by extrapolating the past into the future. The validity of statistical projections of the past into probable states of the unknowable future was challenges as follows:

> *Social Scientists…apply mathematical operations on paper to any number they can can get hold of, or think of, without stopping for one moment to consider whether these operations have any meaning at all. Do we not frequently see economists adding discounted utilities to future dates -i.e., discounted future fluxes – as if they were annuities paid in money (a cardinal variable)?*[23] *… As I have intimidated, quantity cannot be regarded as a notion prior to quality, either in the logical or evolutionary order*, (Georgescu-Roegen 1971.p.99).

The dialectical theory was employed, largely by continental social scientists after Hegel (i.e., thesis → antithesis → synthesis), to identify the salient vectors of chronological trajectories of future sates of the system. The alternative theory employed,

---

[20] Fichtean/Hegelian Dialectics postulate that: (i) everything is transient and finite, existing in the medium of time; (ii) everything is made out of opposing forces/opposing sides (contradictions); (iii) gradual changes lead to turning points, where one force overcomes the other (quantitative change leads to qualitative change); and (iv) change moves in spirals (or helices), not circles (sometimes referred to as 'negation of the negation').

[21] For instance administrative boundaries are abstract space and can be changed from one instant to the next by decree. Geographical space require that the boundary conditions be specified by observed differences in the qualities of the biome or in hydrological/geological formation, river basins, mountain ranges etc.

[22] Alfred Marshall similarly rejected the notion of a general equilibrium system. While the individual markets may tend towards equilibrium, (i.e., supply = demand), the relationship among markets may tends towards disequilibrium states, like the world oil market, international trade, and any market with political interference, subsidies, rationing, currency, etc.

[23] The *Stern Report* (2007) present-value discount of income flows, albeit at 2–3% of global GDP, in order to avoid an even greater loss of income in the future, is an example of the neoclassical arithomorphism applied to the wellbeing of future generation of unknowable values.

largely by analytical social scientists after Darwin, chose 'random mutation' as the salient vector of evolutionary processes. Unlike the Hegelian dialectic, in which 'emergence' identified with the synthesis of paired opposites (thesis/antithesis), randomness can built in the algorithm of any well-defined entropic process. The search for'Synthesis' of conflictual, multi-dimensional, parameters identified in complex adaptive systems, can be pre-specified (i.e., Baysian a priori) in the vector of convergence towards some well-identified 'attractor.' The stabilisation of tha attractor draws the algorithm to a conclusion, (Prigogine 1997).

Rosen (1991) proposed the method of anticipatory modelling of formal systems parameters (i.e., algorithms of the Econosphere) mapped on the observed parameters of natural system (i.e., algorithms of the Ecosphere). Of interest is the method to claculate deviations of formal systems (i.e., models), upon which policy is based, from the observed trajectory of natural systems.

6.9 Entropy Production and the UN Agenda 21

Entropy production, expressed as a rate of depletion of the low entropy ecological Fund (i.e., natural capital), came onto the world stage at the UN Conference on the Human Environment (Stockholm 1972). Twenty years later, the UN Conference on Environment and Development (Rio de Janerio 1992) produced an action plan to reduce the rate of depletion of this Fund under the rubric of sustainable development. The plan, entitled *Agenda 21*, produced a comprehensive blueprint for the conservation of the global fixed (i.e., Lithosphere) and circulating natural capital (i.e., hydrosphere and atmosphere)– in other words, an action plan to reduce the rate of entropy production at the global, national, and local level through government policy, international agreements and NGO citizen's actions.

While *Agenda 21* was rich in semantics, there was a singular lack of a comprehensive, syntactical structure to integrate the 'action plans' into a higher-order complex systems analytical framework. The result, perhaps inevitable considering the number of experts involved, was to advise on how to mitigate the material cause of events detailed in the programme areas of desertification, deforestation, biodiversity, atmospheric pollution, and so forth, in the hope that the sum of the parts = the whole. In an integrated framework, the whole would be addressed before the particulars. In other words, the analysis, and the *a priori* questions, would focus on formal causes, such as national economic and social policies; final causes, such as social and cultural values; and time-delay feedback loops of historical events in recognition that the sum of the parts ≠ the whole (Gunderson and Holling 2002).

Chapter 40 of *Agenda 21* addresses the question of information and data. It is here that the lack of syntax in *Agenda 21* is most evident (Friend and Rapport 1991). While macro-accounting of material-energy flows and balances is referred to as important and necessary, there is no mention whatsoever on statistical methods by which the social and natural science datasets are integrated into a single framework. Nor is there any reference to the valuation methods that distinguish objects/functions in the Econosphere (i.e., economy) from those of the Sociosphere (i.e., households, communities and institutions) and the Ecosphere (i.e., living organisms, habitata and the cycling systems of the lithosphere, hydrosphere and atmosphere). The advice on data was primarily on the development of ad-hoc indicators to measure the degree of sustainability as follows:

40.4. Commonly used indicators such as the gross national product (GNP) and measurements of individual resource or pollution flows do not provide adequate indications of sustainability. Methods for assessing interactions between different sectoral environmental, demographic, social and developmental parameters are not sufficiently developed or applied. Indicators of sustainable development need to be developed to provide solid bases for decision-making at all levels and to contribute to a self-regulating sustainability of integrated environment and development systems.

40.6. Countries at the national level and international governmental and non-governmental organizations at the international level should develop the concept of indicators of sustainable development in order to identify such indicators. In order to promote the increasing use of some of those indicators in satellite accounts, and eventually in national accounts, the development of indicators needs to be pursued by the Statistical Office of the United Nations Secretariat, as it draws upon evolving experience in this regard.

6.10 SAGE-P: Category Theory and the Homomorphism of Entropy Production

Category Theory, also referred to as conceptual mathematics, is a formalism for object/functional mapping of: (i) *object → objects*; (ii) *object → functions*; (iii) *function → objects*; and (iv) *function → functions* (Lawvere, and Schanuel 1997; Friend and Friend 2009). The formalism assumes 'structure preserving' operations termed *homomorphism* (i.e., a transformation of one set into another that preserves in the second set the relations between elements of the first). The *category set* is a collection of elements defined by the boundary condition of the set, such as an economy defined by national boundaries. Technically, the objects of the set are identified by 'common properties' and/or 'common end-objectives'. The latter defines the function of objects classified to the three fundamental category sets of SAGE-P – namely: (i) **Econosphere**; (ii) **Sociosphere**; and (iii) **Ecosphere**.

The category sets are *isomorphic* structural accounts of the low entropy Fund (i.e., empirically-observed data). In mathematics, isomorphism involves structure preserving mapping between objects that shows a relationship between two properties or operations. If there is an isomorphism between two structures, we can call the two structures *isomorphic*. In a certain sense (i.e., if you choose to ignore finer-grained differences that may arise from how they are defined), isomorphic structures are structurally identical.

The (analytical) datasets are produced through the algorithms of correspondence mappings (i.e., the *homomorphism*), which maps the rate of entropy production on the elements of the low entropy Fund. The structure preserving algorithm is a set of *isomorphism of conserved values* identified with specific objects and/or functions described in the datasets.

The fundamental category sets of the System of National Accounts (SNA) are: (i) production accounts; (ii) consumption accounts; (iii) capital accumulation accounts; and (iv) trade with the rest-of-the-world accounts (Stone 1968). The Keynesian definition of the boundary conditions of the institutional-market economy assumes an isomorphic structure of the System of National Accounts (Keynes 1960). These are the identifiers (or algorithm) of conserved values of Income (*Y*), Consumption (*C*), Saving (*S*) and Investment *(I*). Note, here, that SAGE-P, is a more general and larger-scale description of the economic process where the System of National Accounts is a proper subset. However, the accounting methodology remains as given, being a special case of the institutional-market economy.

Unlike the pluralism of values in SAGE-P the System of National Accounts assumes single-value methods denoted in prices. (Norgaard 1989; Friend 1993).[24] While prices may be applied to the Polluter Pays Principle, such as a carbon tax, the reliance on a single valuation method is dangerous, not only for policy decisions (witness the resistance to the Kyoto Protocols due to the economic cost of reducing greenhouse gas emissions), but because social and natural science databases are a *sine quo non* for understanding complexity, qualities of human welfare, and the limit functions of the (material) economy.

The SAGE-P isomorphism of entropic processes are of two types: the algorithm to construct (i) statistical time series of entropy production, and (ii) geo-coded spatial databases of entropy production. While (ii) are statistical constructions of Geographical Information Systems (GIS), (ii) assume the information systems of entropic-evolutionary processes.

A further distinction that needs to be made is between the correspondence mapping between abstract and physical objects. Clearly, abstract objects in themselves have no spatial coordinates. However, abstract properties, like beauty of place, may have an ordinal relationship to physical objects. It is nonetheless the mapping of abstract objects on phyical objects is tricky, insofar as abstract objects are, for the most part, ephemeral and, above all, not subject to the Second Law of Thermodynamics. In other words, abstract objects are not directly observable, like ground cover from satellite imagery, but must be identified by association, such as the price of a specific object or be institutionally defined, like geographical administrative spaces.[25]

---

[24] The non-market valuations drop out of the System of National Accounts, as these data are represented in the money-valued elements in the social and environmental accounts. Thus, the System of National Accounts is freed from imputations of equivalences between market and non-market objects.

[25] Administrative boundaries can sometimes be clearly observed from air by observing the patterns of land-use. The US-Mexican border is observed sharply in contiguous urban and agricultural zones, but not in with the wild desert zone.

SAGE-P assumes nonlinearity in algorithms and associated vector analysis of entropic processes and transformation functions.[26] Category sets of the entropic process are: (i) the set of objects that enter the Fund (i.e., the rate of negentropy production); (ii) the set of objects that exit the Fund (i.e., rate of entropy production); and (iii) the set of objects which constitute the low entropy Fund. While (i) and (ii) are measures of flows over a period of time and equivalent to production and consumption in the System of National Accounts, (iii) is a measure of stock at an instant-in-time and equivalent to the surplus/deficits in the balance sheets of the System of National Accounts. The third category set thus represents the net-value of the Fund in three sub-sets: (i) fixed capital/fund (i.e., non-renewable objects, like fossil fuels); (ii) circulating capital/fund (e.g., reproductive objects, like food and manufactured goods); and (iii) cycling capital/fund (i.e., atmosphere, hydrosphere, and lithosphere).

SAGE-P proposes hierarchical, valued-structured, topological domain spaces of three basic categories of objects and functions, viz:

- values conserved-in-exchange = economic objects/functions → Econosphere;
- values conserved-in-use = social objects/functions → Sociosphere;
- values conserved-in-themselves (existential) = ecological objects/functions → Ecosphere.

The low entropy Fund is an ordinal-valued dataset of objects and functions mapped from lower- to higher-order categories as follows: (i) {[(Econosphere) → Sociosphe re] → Ecosphere}; (ii) the hierarchical structured mapping of efficiencies: {[(economic money-valued efficiency) → social participatory-valued efficiency] → existential -valued eco-efficiency}; and (iii) the hierarchical structured mapping of the welfare function: {[(economic well-being) → social well-being] → ecosystem health and integrity}, (see Fig. **6.4**).

The objects of the SAGE-P fundamental category sets are distinguished by being either material or non-material objects/functions, the latter are abstract, (i.e., occupy time but not space) and the former, physical objects (i.e., occupy space/time). Since our central concern is to account for the "rate of entropy production" in the system, we need to distinguish objects that are renewable and the LEF can be replenished at its rate of consumption, and those that are non-rewable, and dininish at its rate of consumption, such as fossil fuels.

Abstract objects do not occupy space. Therefore, while they may appear, disappear, grow, and decay, they are not directly subject to the Second Law of Thermodynamics. Nonetheless, abstract objects may be clearly identified as inflows,

[26] Entropy production functions in complex systems need not be identified in the strict sense of causal relationships, but rather in the weaker sense of probabilities. The precautionary principle is based on uncertainty, and the risk factor is merely a probability function assumed by *a priori* knowledge. See the discussion on Bayesian statistical methods.

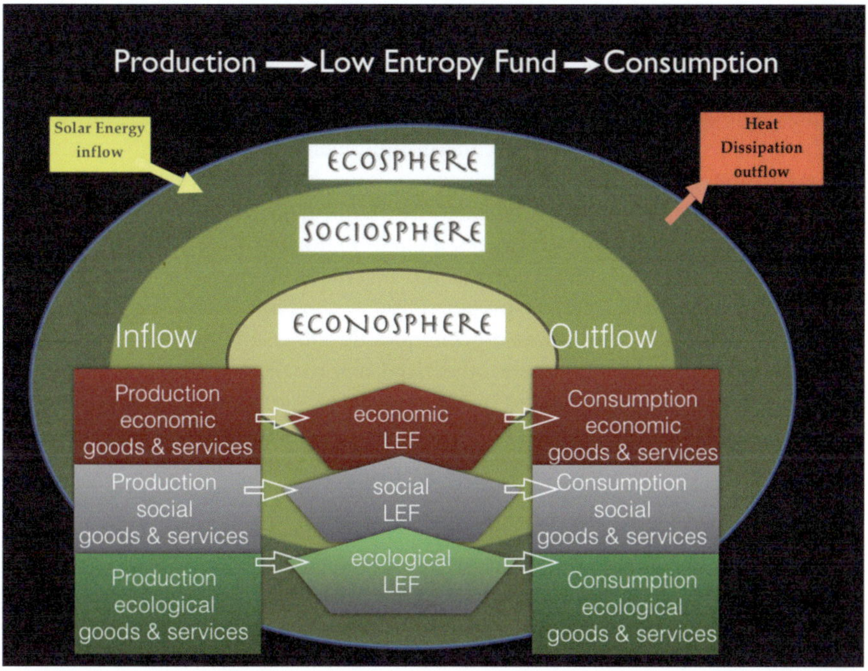

Fig. 6.4 Construction of the low entropy fund statistics for the system of accounts for global entropy production, SAGE-P.

outflows, at the boundary of any well-defined entropic processes and given an net-entropy value.[27]

Replenishing physical objects may be categorised as circulating capital that: (i) occupy space; (ii) are subject to the Second Law of Thermodynamics; (iii) are conditionally reproducible[28]; and (iv) cardinally measured in quantitative properties and ordinal measured in qualitative properties.

Abstract objects may be categorised as the collection of objects in the immaterial world. The boundary condition – for example, financial investment – may be identified either directly, or indirectly, with causal factors associated with any well-defined entropic process. Abstract objects are clearly identified in the economic

[27] In information theory, entropy is a measure of the uncertainty associated with a random variable. In this context, the term usually refers to *Shannon entropy*, which quantifies the expected value of the information contained in a message, usually in units such as bits. Equivalently, the Shannon entropy is a measure of the average information content one is missing when one does not know the value of the random variable. The concept was introduced by Claude E. Shannon in his 1948 paper, 'A Mathematical Theory of Communication'.

[28] These are state conditions where the object, while replenished, changes properties. In economic analysis, substitution assumes qualitative change of objects for equivalent values. In SAGE-P analysis, substitution reflects the human acceptance of inferior qualities of objects in the consumption function, because the superior quality object is unavailable or unaffordable.

process as the inflows and outflows of 'services' in material production, consumption, and capital accumulation. Abstract objects may also be mapped on other abstract objects to produce a new set of abstract objects. This, indeed, is the familiar operational functions of the financial sector of the economy, but includes the boundary conditions (i.e., inflow/outflow), the institutional structure of the Sciosphere – e.g., education, governance, jurisprudence, religion, health services, entertainment, etc. – but not the epidemiological dataset. The latter represent the 'health status' of physical objects of the low entropy (population) Fund.

Ecological Economics theory assume the uncertainty of post-normal science and that the economy is reduced to a proper subset of the larger-scale Ecosystem. The agent upon which the theory on bio-economics develops is *Homo sapiens ecologicus*. Neoclassical Economics, on the other hand, assume the certainties of Newtonian Physics and an economy bounded by the System of National Accounts where the larger-scale ecosystem are reduced to the inflows and outflows of *primary production,* Agriculture, Fisheries, Forestry, Mining and so forth. The agent upon which the neoclassical theory develops is *Homo sapiens economicus* governed by the laws of the (Walrasian) General Equilibrium Model. The question raised is whether rational choices of the individual is compatible with the rational choice of the collective in order to arrive at a socially acceptable rate of entropy production per unit of consumption. Leading economists, including the doyen of 'environmental economists' wrote a collective article entitled, "Are we consuming too much?," (Arrow et al. 2004). The conclusion that this to be true in sense of the limits in the Planet's capacity to satisfy the growing material needs of the seven billion *Homo-economicus,* but could be avoided by: (a) increase productivity per unit of output, (b) substitution of human for natural capital, and (c) establishing the protocols for the 'Green Economy.' The proviso for a non-declining welfare function is to keep productivity of natural capital intact.

The powerful isomorphism of SAGE-P that links physical with abstract objects permits the mapping of ordinal values on the state (position), and the change of state (movement), of well-defined accounting objects and functions. While immaterial things do not occupy space by definition, they nonetheless may be presented by volume, density, or other metrics in imaginary space – for example, global information pathways in the internet, the physical location of bank accounts in Switzerland, or the cyclical movement in the New York stock exchange, etc.[29] Abstract objects mapped on other abstract objects provides the datasets of the higher-order effect

[29] That does not mean that the abstract objects do not have spatial identifiers, like the enjoyment of scenery in the Alps, or the enjoyment of the ballet in the Bolshoi Theatre in Moscow. The essential idea is that of mapping abstract objects on abstract objects, like financial flows and balances, or the mapping of abstract objects on physical objects, like prices on goods, but not services. Indeed, GDP may be viewed as a mapping of prices on all the (final) goods (physical objects) and services (abstract objects) produced in the economy in one year. It should be noted that the System of National Accounts is a method to reduce physical objects (material economy) and abstract objects (immaterial economy) into the single denominator of market prices. In other words, it is a method to reduce complex physical processes into a pure abstract value-object disconnected from the physical universe.

(i.e., formal cause) on the rate of entropy production in the material economy. For example, a mapping of approvals of sub-prime mortgages → debit/credit swaps → toxic debts in bank balances → 2009 Financial Crisis ← physical location and number of house foreclosures in American suburbs.

Conserved values-in-exchange constitute the homomorphism of the System of National Accounts (i.e., supply=demand). The correspondence mapping of GDP on the SAGE-P datasets results in the production matrix = negentropy and the consumption matrix = entropy. The mapping neg-entropy on entrpy results in three states of the (material) GDP: (i) sutainable steady-state where production = consumption, (ii) unsustainable where production < consumption, and (iii) sustainable surplus-state where production > consumption.

Daly and Cobb (1989) have observed that when price assumes efficiency in the allocation of scarce resources results in the logical fallacy of 'misplaced concreteness'. In other words, price efficiencies≠entropy efficiencies. In this case, entropy-efficiency is the ordinal-valued utility function to be optimised (i.e., minium rate of entropy production per unit of consumption)., (Mayumi 2001).

6.11 Bayesian Statistical Methods: Nonlinear Statistics and Evolutionary Processes

SAGE-P database permits (non-discounted) present day evaluations of risk and uncertainty of large-scale events up to 30–50 years hence based on the qualitative state conditions of LEF available for human consumption, (see Fig. 6.4). The pluralistic valuation method entails conservation values of the entropy production to the algorithm of the TDS of *the Econosphere, the Sociosphere* and *the Ecosphere*. Rosen (1985) 'anticipatory system' provides a template for the mapping of formal systems (models) on natural systems (observations). The common denominator of algorithms is the I/O of any well-defined entropic process. The Rosen's anticipatory system assumes the Bayesian method of 'prior probability' in unfolding event process, where a future state is latent in the present structure of the system, and thus detectable, if one knows what one is looking for, (see **Appendix II: Bayesian Statistical Methods**).

The distinction needs to be made between statistical 'uncertainty', concerning bias in the sample for instance, and scientific uncertainty, concerning evidence and proof. The issue of scientific uncertainty embedded in complex systems was well-explored by Funtowicz and, Ravetz, (1991) in "Uncertainty and Quality in Science Policy." The conclusion drawn is that investigative science of complex systems can no longer be independent of ethical and moral questions. A secondary conclusion is importance of qualitative assessment of the data itself. In other words, transparency and openness of statistics applied to long-term policy at the the scale of climate change and transformation to sustainable, green, economies.

The algorithms for mapping objects → objects; objects → functions; functions → objects; and functions → functions in SAGE-P while constructed from *a priori propositions* (i.e., pre-set instructions of the programmer), the data entered into the algorithm are *a posteriori proposition* (i.e.,observations) of the real world,[30] (Berlinski 1999).

Kant's analytic-synthetic distinction, also called the analytic-synthetic dichotomy, is helpful in understanding the distinctions between scientific uncertainty. (i.e., *a posteriori* propositions) and statistical uncertainty, (i.e., *a priori* propositions).

- *analytic proposition*: a proposition whose predicate concept is contained in its subject concept;
- *synthetic proposition*: a proposition whose predicate concept is not contained in its subject concept.

These two propositions can be further classified with respect to:

- *a priori proposition*: a proposition whose justification does not rely upon experience;
- *a posteriori proposition*: a proposition whose justification does rely upon experience.

Combining the above definitions, we now have four types of proposition to construct the accounting algorithms of SAGE-P:

1. analytic, *a priori* propositions;
2. synthetic, *a priori* propositions;
3. analytic, *a posteriori* propositions;
4. synthetic, *a posteriori* propositions.

Applying the above four proposition to the construction of the accounting algorithms for the System of National Account (SNA) and SAGE-P we have:

SNA = *analytic propositions,* (i.e.,General Equilibrium System) with a correspondence mapping of *a priori propositions,* (i.e., statistics pre-defined by *analytic propositions* for Production, Consumption, Capital and Trade with the rest-of-the-world).

SAGE-P = *synthetic propositions,* (i.e., Entropic Process) with a correspondence mapping of *a posteriori propositions,* (i.e., statistics defined by *a posteriori propositions* of the boundary conditions for Negentropy, Entropy and Low Entropy Fund).

[30] There is some ambiguity of what actually is being observed in statistical databases. Economic and social statistical databases are largely constructed from surveys, with pre-defined categories. Demographic data is a direct observation of the population and can be counted in numbers of people with specific time and location. Environmental data is perhaps the least ambiguous observation if taken from satellite imagery, the location and time of the data are exact. The data on spot checks and sampling of physical objects run into the same degree of uncertainty as those of socio-economic surveys.

Note that the *synthetic propositions* of SAGE-P are symmetric (i.e., translation) to the SNA but not the reverse. The reason: the SNA (Econosphere) is a proper subset of the SAGE-P (Ecosphere).

6.12 SAGE-P: Construction of Synthetic, *a Posteriori*, Statistical Databases

Bayesian statistics, *a priori* knowledge, also known as the prior distribution (θ) of the observed events (Θ) is permissible under some ordinal set of well-defined causal protocols. SAGE-P adapts the Aristotelean causal protocols in the mapping of probability and limit functions on the datasets of: (i) material cause (π^1); (ii) efficient cause (π^2); (iii) formal cause (π^3); and (iv) final cause (π^4). These are mapped pairwise to produce joint prior distribution estimators of expected events (i.e., $\pi^2\ \theta \rightarrow \pi^1\ \Theta = \theta'$; $\pi^3\ \theta \rightarrow \pi^2\ \Theta = \theta''$; $\pi^4\ \theta \rightarrow \pi^3\ \Theta = \theta'''$), where:

- θ represents a statistical distribution (e.g., observed events) of the material cause (π^1);
- θ' represents the prior distribution (e.g., technology) of the efficient cause (π^2);
- θ'' represents the prior distribution (e.g., policy) of the formal cause (π^3);
- θ''' represents the prior distribution (e.g., socio/cultural values) of the final cause (π^4);

Note that the material cause represents the observed data of the normal distribution of events with accuracy of any particular prediction being informed by its standard deviations (σ) from the mean (μ). Thus, Bayesian Methods provide the decision-maker (and the public) with the *estimators of probability distribution of events* ($\boldsymbol{x}^1$ …. $\boldsymbol{x}^n$) representing the initial state conditions of the material cause, $\pi^1\ \Theta = \theta^1$ (data observed on the ground). The SAGE-P material dataset may be modified under well-defined conditions representing probabilities of the change of state of the system due to efficient cause $\pi^2\ \Theta = \theta^2$ (technological conditions); formal cause $\pi^3\ \Theta = \theta^3$ (policy conditions); and *final cause* $\pi^4\ \Theta = \theta^4$ (socio-cultural value conditions)

Bayesian statistics fix the normal distribution of the observed event $\boldsymbol{x}$ *a priori*. While traditional statistical average probability of an event $\boldsymbol{x}$ happening over the state conditions of the observed sample itself: $\boldsymbol{x} \in \Theta$, the result is a (linear) extrapolation of probability function, assuming a statistical correlation of past events into the future. Bayesian probability of the event $\boldsymbol{x}$ happening is modified by 'degrees of belief' of a potential and/or hypothetical distribution of (probable) events θ: $\boldsymbol{x} \in \Theta$. This is known as a generalized Bayes rule with respect to $\pi\ \theta$. There maybe more than one generalized Bayes rule, since the model may require multiple and complex conditional relations between $\boldsymbol{x}$ and $\boldsymbol{y}$. This may be generalized as mapping of different formal systems, or models, on natural systems: $\theta \in \Theta$ (e.g., What if? scenario modeling). Applying the Rosen 'anticipatory system', the prior distribution (θ) predicts the probability of event $\boldsymbol{x}$ as a mapping: $\theta \rightarrow \Theta$, [i.e., mapping of a formal system (θ) on a natural system (Θ) to predict the next event in Θ] (see Appendix 3.2 on Bayesian Statistical Methods).

6.13 Further Observation of the Bayesian Method

Macro-accounting assumes, as given, the institutional structure of society and its social values. While construction of the datasets may be scientific, applying sound statistical methods, the social value structure, including political intervention (e.g., the definition of 'unemployment'), may not only bias the selection of data collected (i.e., administrative *vis-à-vis* scientific data), but also the qualitative assessments of the state and the change of state of the observed datasets.[31] The latter is particularly sensitive to the order of the evaluation protocols, where, for example, commercial values override cultural values (Maslow 1968). The Bayesian Method permits values to be modified, and indeed evolve, with the change of social values.

In reality, the accounts are constructs of the accountant who assumes the role of the neutral observer of 'facts'. Bayesian parameters of 'degrees of belief' are thus ruled out of any consideration for the objective function for the accounts. This, in the Bayesian language, is illusionary, as degrees of belief are found in every fact, since there is no absolute certainty. This becomes glaringly obvious in the assumption that willingness-to-pay equals willingness-to-sell, the code-word for conserved values-in-exchange, which, under given conditions, represents in the neoclassical model an unbiased sample of consumer preferences. Even under the condition of 'perfect competition', consumer preferences are not only manipulated by advertising, or pure lack of information, but by time preferences of (discounted) future streams of consumption. The belief in future state conditions of consumption is pure Bayesian and finds representation in an index number of 'consumer confidence' following an economic crisis.

Even more problematical are the (inflexible) institutionally-defined objects and functions of accounts, exacerbated by the penumbra of the boundary conditions of categories. For instance, is 'housework' value-added, or value-subtracted to the national product. Clearly the inflows material products, like tools, cleaning fluids, household appliances and the human labour performed, is no different labour performed by the producer and added to the low entropy fund available of for consumption, albeit by same members of the household. While clearly not an economic product for sale, with a definite exchange-value, it has indeed definite use-value and

[31] This was clearly recognized by the UN Statistical Office by permitting the social value bias to enter the System of National Accounts. Here, the Marxist versus the capitalist bias entered the valuation of the 'National Product.' The latter valued the annual goods and services at 'willingness-to-pay' basis, or market prices, while the former valued at the cost of production of the material product – thus, assuming a labour theory of value. However, the UN Statistical maintained its own bias with respect to the non-monetised sector of the national economies and thus the 'product' of the household and the non-monetised 'natural product' was left out of the accounts. This can be corrected by a 'service-flow' income accounting.

to the extent that the family is a micro-ecosystem, existential value. If government policy is to increase the national product, it would make sense add housework.

The non-institutional SAGE-P there is no distinction made between household and market production functions, and indeed, except for its factors of production, is a distinction made between human and Nature's production production, consumption and capital accumulation functions. What is observed, and measured, is the direction of material inflows towards order, disorder, or steady-state of dissipative structures.

Appendices

Appendix I

Georgescu-Roegen's Flow-Fund Model

Factors of production are divided into two categories: the Fund elements, which represent the agents of the process, and the flow elements, which are used or acted upon. Each flow element is represented by one coordinate $E_i(t)$. The fund element enters and leaves the process with its efficiency intact. Specifically, we can represent the participation of a Fund C_σ by a single function $S_\sigma(t)$ showing the amount of services of C_σ up to the time t, where $0 \leq t \geq T$.

The analytical presentation of a Process can thus be written $[[E_{i0}{}^t(t), S_0{}^t(t)]]$, where the i subscript represents input or output and σ represents both inputs and outputs.

As for the analytical coordinates of a partial process, analysis must renounce the idea of including in the description of a process, either inside or outside it, the problem associated with the happenings with a process reducing to recording only what crosses the boundary. For convenience, we may refer to any element crossing the boundary from the environment into the process as an input, and to any element crossing it from the opposite direction as an output. At this juncture, analysis must make some additional heroic steps all aimed at assuming away dialectical quality.

Discretely distinct qualities are still admitted into the picture as long as their number is finite and each one is cardinally measurable. If we denote the elements that may cross the t boundary of a given process by C_1, C_2, C_3, … C_n, the analytical description is complete if, for every C_i, we have determined two non-decreasing functions $F_i(t)$ and $G_i(t)$, the first showing the cumulative input, the second, the cumulative output of C_i up to the time (t). Naturally, these functions must be defined over the entire duration of the process which may be always represented by a closed time interval such as [0, T]. The question of whether this analytical model is operational, outside paper-and-pencil operations, cannot be decided without an examination of the nature of the elements usually found in actual processes. Such an examination reveals that there exists numerous elements for which either $F_i(t)$ or $G_i(t)$ is identically null for the entire duration of the process. Solar energy is a typical example, which is only an input for any terrestrial process. The various materials ordinarily covered by the term 'waste' are clear examples of elements which are only outputs. In all these cases, we may simplify the analytical picture by representing each element by one coordinate only – namely, by:

$$E_i(t) = G_i(t) - F_i(t)$$

For an output element, $E_i(t) = G_i(t) \geq 0$; for an input element, $E_i(t) = -F_i(t) \leq 0$. The sign of the suffices indicate which is actually the case (Georgescu-Roegen 1971, p. 215).

Georgescu-Roegen further distinguishes $E_i(t)$, which are (basic) elements necessary to maintain the production cycle at a steady-state (e.g., seeds → crops) and $E_i(t)$, which are (non-basic) elements that are surplus available for consumption, $E_i(t) = G_i(t) - F_i(t) \geq 0$.

Appendix II

Bayesian Statistical Methods[32]

Thomas Bayes addressed both the case of discrete probability distribution of data and the more complicated case of continuous probability distribution. In the discrete case, Bayes' theorem relates the conditional and marginal probabilities of events *A* and *B*, provided that the probability of *B* does not equal zero. Thus: $P(A|B) = P(B|A) P(A)/P(B)$

In Bayes' theorem, each probability has a conventional name:

- $P(A)$ is the prior probability (or 'unconditional' or 'marginal' probability) of *A*. It is 'prior' in the sense that it does not take into account any information about *B*; however, the event *B* need not occur after event *A*. In the nineteenth century, the unconditional probability $P(A)$ in Bayes' rule was called the 'antecedent' probability; in deductive logic, the antecedent set of propositions and the inference rule imply consequences. The unconditional probability $P(A)$ is called '*a priori*'.
- $P(A|B)$ is the conditional probability of *A* given *B*. It is also called the posterior probability because it is derived from or depends upon the specified value of *B*.
- $P(B|A)$ is the conditional probability of *B* given *A*. It is also called the likelihood.
- $P(B)$ is the prior or marginal probability of *B*, and acts as a normalising constant.

[32] Wikipedia entry on Bayesian Statistics.

Bayes' theorem, in this form, gives a mathematical representation of how the conditional probability of event *A* given *B*, is related to the converse conditional probability of *B* given *A*.

Bayes' Theorem With Continuous Prior and Posterior Distributions

Suppose a continuous probability distribution with probability density function $f\Theta$ is assigned to an uncertain quantity Θ. In the conventional language of mathematical probability theory, Θ would be a 'random variable'. The probability that the event *B* will be the outcome of an experiment depends on Θ; it is $P(B \mid \Theta)$. As a function of Θ, this is the following likelihood function:

$$L(\theta) = P(B \mid \Theta = \theta)$$

The posterior probability distribution of Θ (i.e., the conditional probability distribution of Θ, given the observed data *B*), has the probability density function:

$$f_{\Theta}(\theta \mid B) = \text{constant} \bullet f_{\Theta}(\theta) L(B \mid \theta)$$

where the 'constant' is a normalising constant so chosen as to make the integral of the function equal to one, so that it is indeed a probability density function. This is the form of Bayes' theorem actually considered by Thomas Bayes. In other words, Bayes' theorem says: 'To get the posterior probability distribution, multiply the prior probability distribution by the likelihood function and then normalise.' More generally, still, the new data *B* maybe the value of an observed continuously distributed random variable *X*. The probability that it has any particular value is therefore zero. In such a case, the likelihood function is the value of a probability density function of *X* given Θ, rather than a probability of *B* given Θ:

$$L(\theta) = f_x(x \mid \Theta = \theta)$$

Notation and Definitions

In the notation $P(A|B)$, the symbol *P* is used only as a reference to the original probability. It should not be read as the probability *P* of some event $A|B$. Sometimes the more accurate notation $P_B(A)$ is used, but the use of complex events as index of the symbol *P* is cumbersome. Notice that the line separating the two events *A* and *B* is a vertical line.

Conditional probability is the probability of some event *A*, given the occurrence of some other event *B*. Conditional probability is written $P(A|B)$, and is read 'the (conditional) probability of *A*, given *B*' or 'the probability of *A* under the condition *B*'. When in a random experiment, and the event *B* is known to have occurred, the possible outcomes of the experiment are reduced to B, and hence the probability of the occurrence of A is changed from the unconditional probability into the conditional probability given B.

Joint probability is the probability of two events in conjunction. That is, it is the probability of both events occurring together. The joint probability of A and B is written as $P(A|B)$, $P(AB)$, or $\mathrm{P}(A, B)$. Marginal probability is then the unconditional probability $P(A)$ of the event A; that is, the probability of A, regardless of whether event B did or did not occur. If B can be thought of as the event of a random variable X having a given outcome, the marginal probability of A can be obtained by summing (or integrating, more generally) the joint probabilities over all outcomes for X.

The conditioning of probabilities (i.e., updating them to take account of (possibly new) information), may be achieved through Bayes' theorem. In such conditioning, the probability of A given only initial information I, $P(A|I)$, is known as the prior probability. The updated conditional probability of A, given I and the outcome of the event B, is known as the posterior probability, $P(A|B, I)$.

A continuous probability distribution is a probability distribution which possesses a probability density function. Mathematicians also call such a distribution absolutely continuous, since its cumulative distribution is absolutely continuous with respect to the Lebesgue measure λ. If the distribution of X is continuous, then X is called a continuous random variable. There are many examples of continuous probability distributions: normal, uniform, chi-squared, and others.

The probability density function, or density of a continuous random variable, is a function that describes the relative likelihood for this random variable to occur at a given point. The probability for the random variable to fall within a particular region is given by the integral of this variable's density over the region. The probability density function is non-negative everywhere, and its integral over the entire space is equal to one.

Appendix III

SAGE-P Datasets on the State and Change of State of the Econosphere, Sociosphere, and the Ecosphere

Propositions

Proposition I: An entropic Process in SAGE-P is an algorithm to map the transformation of (statistical) objects from lower into higher entropic states (i.e., consumption), and its inverse from higher into lower entropic states (i.e., production).

The proposed methods to construct the algorithm are prescribed Bayesian Rules for scalar operators in any well-defined, hierarchical-structured, complex adaptive system in Topological Domain Space (TDS), defined as:

Econopshere $(E')\subseteq$ Sociosphere $(E'')\subseteq 0\ (E''')$

where $\subseteq =$ subset.

Scalar operators of the ecological flow-fund are a slow moving, but much larger scale to the socio-demographic Fund, which, in turn, is larger, but slower moving to the economic Fund (i.e., $E''' > E'' > E'$).

- The elements of the E' dataset belonging to the Econosphere TDS are characterised by fast moving state, and change of state, variables described by emergent properties of dissipative economic structures far from equilibrium: (i.e., domain properties of the *low entropy economic fund*);
- The elements of the E'' dataset belonging to the Sociosphere TDS are characterised by slow moving state, and change of state, variables described by emergent properties of dissipative social-institutional structures near equilibrium (i.e., domain properties of the low entropy socio-demographic Fund);
- The elements of the E''' dataset belonging to the Ecosphere TDS are characterised by very slow moving state, and change of state, variables described by emergent properties of dissipative ecosystem structures very near equilibrium (i.e., domain properties of the low entropy global ecosystem Fund).

Proposition II: The homomorphism of SAGE-P datasets are conserved value mappings of objects → objects; objects → functions; functions → objects; and functions → functions among any well-defined TDS:

- Econosphere: the homomorphism of economic objects/functions are the values conserved-in-exchange;
- Sociosphere: the homomorphism of socio-demographic objects/functions are the values conserved-in-use;
- Ecosphere: the homomorphism of socio-demographic objects/functions are the values *conserved-in-themselves* (i.e., intrinsic value).

Proposition III: SAGE-P functions are defined by the boundary conditions of processes unique to any well-defined TDS. Objects are defined by the boundary conditions of the objects-in-themselves, but change values with respect to function:

- Objects where values are conserved-in-exchange ∈ Econosphere
- Objects where values are conserved-in-use ∈ Sociosphere
- Objects where values are conserved-in-themselves ∈ Ecosphere

Proposition IV: SAGE-P qualitative properties of *Objects* change with *Function*:

- Social and ecological objects where qualities are values conserved-in-exchange ∈ Econosphere;
- Economic and ecological objects where qualities are values conserved-in-use ∈ Sociosphere;
- Economic and social objects where qualities are values conserved-in-themselves ∈ Ecosphere.

Structure of Datasets

SAGE-P material datasets are observed phenomena (i.e., statistical database); all other datasets are constructed from correspondence mappings of:

(i) objects → objects;
(ii) objects → functions;
(iii) functions → objects;
(iv) functions → functions.

Working Definitions

Objects are a collection of statistical elements of the dataset, ($E_{1\text{-}n} = \Theta$), and represent numerical cardinal/ordinal values of the quantitative/qualitative properties of physical/abstract objects.

Functions are a collection of algorithmic operators where the elements of the set are the instructions, [$f\ (E_{1\text{-}n}) = \pi\ \Theta$], (i.e., vector mapping of objects on object, objects on functions, functions on objects, and function on functions).

SAGE-P algorithmic operators are formalisms expressed in terms of entailment properties of objects and functions. These may be classified to Aristotelean hierarchical structure of causes, viz: material cause (π^1) → efficient cause (π^2) → formal cause (π^3) ← final cause (π^4). The reverse arrow of π^4 reflects the (abstract) socio-cultural values mapped on sustainable development policy in terms of an intensity value measure of the socially acceptable rate of entropy production.

Homomorphism is a structure preserving algorithm permitting one-to-one correspondence mapping of the elements in the SAGE-P datasets – an example being the mapping of prices (values conserved-in-exchange) on economic objects to construct the System of National Accounting dataset (Note the symmetries conserved in linear datasets: inputs = outputs, and broken in nonlinear datasets, inputs ≠ outputs).

SAGE-P datasets are hierarchically-structured matrices of entropic processes representing the notion of:

- *Production* (P_e) (i.e., negentropic processes);
- *Consumption* (C_e) (i.e., entropic processes);
- *Capital Accumulation* (K_e) (i.e., the low entropy Fund available for future consumption, or $K_{e(t)=}\ P_{e(t\text{-}n)} - C_{e(t\text{-}n)}$.

Category Sets of the Statistical Database

SAGE-P datasets are divided into two separate and distinct categories, viz:

- Physical Objects/Function (Category I): material statistics subject to the Second Law of Thermodynamics, where Category I = (Θ^p) and;

- Abstract Objects/Function (Category II): immaterial statistics not subject to the Second Law of Thermodynamics, where Category II = (Θ^a).

Category I: Quantitative values of inflow/outflow of Physical Objects in any well-defined entropic process and a parallel set of balance sheet accounts of the low entropy (physical) Fund:

- $E' I \rightarrow$ Econosphere $\rightarrow \Theta^{p'}$
- $E'' I \rightarrow$ Sociosphere $\rightarrow \Theta^{p''}$
- $E''' I \rightarrow$ Ecosphere $\rightarrow \Theta^{p'''}$

Category II: Quantitative values of inflow/outflow of Abstract Objects in any well-defined entropic process and a parallel set of balance sheet accounts of the low entropy (abstract) Fund:

- $E' II \rightarrow$ Econosphere $\rightarrow \Theta^{a'}$
- $E'' II \rightarrow$ Sociosphere $\rightarrow \Theta^{a''}$
- $E''' II \rightarrow$ Ecosphere $\rightarrow \Theta^{a'''}$

Category III: Mapping of Qualitative Values Algorithms (π) on Category I and II, π' = exchange, π'' = use, π''' = intrinsic

- $E' III \rightarrow (EI, EII) \rightarrow \pi' \Theta$
- $E'' III \rightarrow (EI, EII) \rightarrow \pi'' \Theta$
- $E'' III \rightarrow (EI, EII) \rightarrow \pi''' \Theta$

Category IV: Mapping of Spatial Co-ordinate Algorithms (π^s) on Category I (Note: Category II objects, by definition, have no geographical co-ordinates), $\pi^{s'}$ = economic co-ordinate space, $\pi^{s''}$ = social coordinate space, $\pi^{s'''}$ = ecosystem coordinate space.

- $E' IV \rightarrow$ Econosphere TDS $\rightarrow \pi^s \Theta^{p'}$
- $E'' IV \rightarrow$ Sociosphere $\rightarrow \pi^{s'''} \Theta^{p''}$
- $E'' IV \rightarrow$ Ecosphere $\rightarrow \pi^{s'''} \Theta^{p'''}$

References

Arrow K, Dasgupta P, Goulder I, Daily G, Ehrlich P, Heal G, Levin S, Maler K, Schneider S, Starrett D, Walker B (2004) Are we consuming too much? J Econ Perspect 18(3):147–172

Bataille G (1988) The accursed share: an essay on general economy, vol. I. Consumption (trans: from French by Hurley R). Zone Books, New York

Berlinski D (1999) The advent of the algorithm: the idea that rules the world. Harcourt, New York

Boulding K (1949) Income or wealth. Rev Econ Stud 17:77–86

Daly H (1977) Steady-State economics. W.H. Freeman, San Francisco

Daly H, Cobb J (1989) For the common good: redirecting the economy toward community, the environment, and a sustainable future. Beacon, Boston

Ellul J (1964) The technological society, (trans: from French by Wilkinson J). Vintage Books, New York

Fisher I (1906) The nature of capital and income. Macmillan, New York

Fisher I (1965) 'The nature of capital and income', in Reprints of Economic Classics, Augustus M. Kelley Publishers, New York

Friend A (1993) Towards pluralism in national accounting systems. In: Franz A, Stahmer C (eds) Approaches to environmental accounting. Physica Verlag, Heidelberg

Friend A (2005) Unearthing ecoprices in conservation accounts. In: Sahu N, Choudary A (eds) Dimensions of environmental and ecological economics. Hyderabad University Press, Hyderabad

Friend A (2009) The ergodic hypothesis and its relationship with the Clausius and the Blotzmann formulation of the Entropy Law: Discussion Paper I. unpublished manuscript

Friend A, Friend M (2009) Non-linear accounting methods and adaptive logic: an exploration of the meta-language of the Georgescu-Roegen Flow-Fund Model. Paper presented at the European society for ecological economics conference, Ljubljana, Slovenia, 30 July-2 August 2009

Friend A, Rapport D (1991) Evolution of macro-information systems for sustainable development. Ecol Econ 3:59–76

Funtowicz SO, Ravetz JR (1991) Uncertainty and quality in science policy. Kluwer Academic Publishers, Dordrecht

Georgescu-Roegen N (1971) The entropy law and the economic process. Harvard University Press, Cambridge, MA

Gunderson L, Holling C (eds) (2002) Panarchy: understanding transformations in human and natural systems. Press, Islands

Hardin G (1968) Tragedy of the commons. Science 162(3859):1243–1248

Holling C (1994) What barriers? What bridges? In: Gunderson L, Holling C, Light S (eds) Barriers and bridges in renewing ecosystems and institutions. Columbia University Press, New York

Jevons WS (1965) The coal question: an inquiry concerning the progress of the nation and the probable exhaustion of our coal-mines, (1865), 3rd edn. rev. by Flux AW, New York, Augustus M. Kelly

Keynes JM (1960) The general theory of employment, interest and money. Macmillan, London

Koestler A (1967) The ghost in the machine. Hutchinson, London

Kuaffman S (1995) At home in the universe: the search for laws of self-organization and complexity. Oxford University Press, New York

Lawvere F, Schanuel S (1997) Conceptual mathematics: a first introduction to categories. Cambridge University Press, Cambridge

Malte F, Proops J (1992) Evolution, time, production and the environment. Springer, Heidelberg

Malte F, Proops J, Speck S, Jöst F (1999) Capital and time in ecological economics: neo-austrian modeling. Edward Elgar, Cheltenham

Malthus T (1959) Population: the first essay. The University of Michigan Press, Ann Arbor

Marshall A (1947) Principles of economics, 8th edn. Macmillan, London

Martinez-Alier J (1987) Ecological economics: energy, environment, and society. Basil Blackwell, Oxford

Maslow AH (1968) Toward a psychology of being. D. Van Nostrand Company, New York

Mayumi K (2001) The origins of ecological economics: the bioeconomics of Georgescu-Roegen. Routledge, London/New York

Mill JS (1985) Principles of political economy: books 1V & V. Penguin, London (org. pub. 1848)

Mirowski P (1989) More heat than light: economics as social physics: physics as nature's economics. Cambridge University Press, Cambridge

Morel R, Fleck G (2006) A fourth law of thermodynamics. Chemistry 15(4)

Norgaard R (1989) The case for methodological pluralism. Ecol Econ 1:37–57

Polimeni J, Mayumi K, Giampeetro M, Alcott B (2008) Jevons' Paradox and the Myth of resource efficiency improvements. Earthscan Publications, Oxford

Prigogine I (1997) The end of certainty: time, chaos, and the new laws of nature. Free Press, New York

Prigogine I, Glandorf P (1971) Thermodynamic theory of structure, stability, and fluctuations. Wiley, New York
Prigogine I, Stengers I (1984) Order out of chaos: man's new dialogue with nature. Bantam Books, New York
Rapport D, Friend A (1979) Towards a comprehensive framework for environmental statistics: A stress-response approach, Catalogue #11-510. Statistics Canada, Ottawa
Ricardo D (1973) The principles of political economy and taxation. Everyman's Library, London
Rio de Janerio (1992) UN world conference on the environment
Robinson J (1980) Further contributions to modern economics. Basil Black Well, Oxford
Rosen R (1985) Anticipatory systems: philosophical, mathematical and methodological foundations. Pergomon Press, New York
Rosen, R. 1991. Life Itself: A Comprehensive Inquiry into the Nature, Origin, and Fabrication of Life, New York, Columbia Press
Schrödinger E (1967) What is life and mind and matter? Cambridge University Press, London
Schumacher EF (1973) Small is beautiful: economics as if people mattered. Harper and Row, New York
Schumpeter J (1934) The theory of economic development. Harvard University Press, Cambridge
Shannon CE (1948) A mathematical theory of communication. Bell Syst Tech J 27(3):379–423
Simon J (1981) The ultimate resource. Princeton University Press, Princeton
Smith A (1809) An Inquiry into the nature and causes of the wealth on nations, vol I. Mundell, Doig, and Stevenson, Edinburgh
Smith A (1994) An inquiry into the nature and causes of the wealth of nations. The Modern Library, New York
Solo R (1974) Arithmorphism and entropy, book review, Georgescu-Roegen, "The Entropy Law and the Economic Process". Econ Dev Cult Chang 22(3):510–517
Sraffa P (1960) Production of commodities by means of commodities. Cambridge University Press, Cambridge
Stern N (2007) Review report on the economics of climate change. UK Government Press, London
Stockholm (1972) UN world conference on the 'human environment.' Theodore Roosevelt refers to speech made to congress "lessons from China"
Stone R (1968) A system of national accounts: studies in methods, Series F, 2(3), United Nations, New York
Vitousek P, Ehrlich P, Ehrlich A, Matson P (1986) Human appropriation of the products of photosynthesis. Bio-science 36:368–373
von Böhm-Bawerk E (1891) The positive theory of capital. Macmillan, London
World Commission on Environment and Development (WCED) (1987) Our common future. Oxford University Press, Oxford
Zukav G (1979) The dancing Wu Lim masters: an overview of the new physics. William Morrow, New York

Part II
Ecological Economics of Physical Balance: Resources, Climate Change, and Renewable Energy

Chapter 7
The Green Economy in Europe: In Search for a Successful Transition

Stefan Speck and Roberto Zoboli

Abstract A green economy has initially been adopted as a policy concept and tool to address the financial and economic crisis. Then it became a strategy now extensively endorsed by the EU and international organisations to guide a transition for a fairer society living in a better environment. Although a broadly agreed definition of a green economy does not exist, three objectives are commonly reflected in the green economy paradigm: improving resource efficiency, ensuring ecosystem resilience, and enhancing social equity. The central question is how to implement a transition to a green economy and what can be the role of policies in this broad scale process that no one single policy can implement. In general, the major economic transformations, including the increasing share of services in the EU economies, are not leading to a green economy transition. Enabling policies and factors at the crossroads between policies and the real-economy dynamics are required. Among enabling factors are eco-innovation, the open circulation of green knowledge, availability of financial resources for investing in the long-term transition and fiscal reforms, in particular economic instruments, such as carbon pricing schemes. Achieving a green economy requires long-term thinking and actions, the widespread application of a coherent framework that drives profound changes in dominant structures and thinking. Coherent integration of objectives across all policy areas is required, treating economic, social and environmental performance objectives as equal.

Keywords Green economy • Policy • Fiscal instruments • Carbon tax • ETS • Eco-innovation

---

The chapter is based on the background works of the EEA report 'Resource-efficient green economy and EU policies' (EEA 2014a).

S. Speck (✉)
European Environment Agency, Copenhagen, Denmark
e-mail: stefan.u.speck@gmail.com

R. Zoboli
IRCrES Institute of Research on Sustainable Economic Growth, Catholic University, Milan, Italy
e-mail: r.zoboli@ircres.cnr.it

S. Shmelev (ed.), *Green Economy Reader*, Studies in Ecological Economics 6,
DOI 10.1007/978-3-319-38919-6_7

7.1 The Green Economy as a Policy Concept

The idea and concept of a green economy gained momentum at the beginning of the global economic and financial crisis in late 2007 as a strategy able to deliver a double divided: fast recovery from the economic and financial crises as well as solutions for environmental crises. This emerging strategic concept was quickly embraced by international organisations and different policy fora before culminating as one of the two main themes, 'Green economy in the context of sustainable development and poverty eradication', of the United Nations Conference on Sustainable Development in Rio de Janeiro in 2012.

The concept of a green economy is nothing new as the idea and its main features were already discussed in academia more than 20 years ago (Pearce et al. 1989; Jacobs 1991). These early works attracted only minor interest by policy-makers but covered many of the ideas and policies which are today discussed in the green economy context. There is no general agreed definition of a green economy but there is a set of 'working definitions' by international organisations that all include[1]: (i) a clear connection with the concept of sustainable development; (ii) the idea of green economy as a selection device for multiple-dividend growth paths combining economic benefits with environmental and social outcomes.

Economy-environment interactions are a central point of attraction at times of environmental pressure and economic slowdown (Ekins and Speck 2011). However, there are differences between the earlier debates of these interactions and the current discussion of the green economy because this was driven by the occurrence of multiple crises and a further developed understanding of systemic challenges (EEA 2010). The crisis of the late 2000s went along with a sharp reduction in economic activities in the majority of EU Member States associated with a financial crisis leading to severe challenges of the banking sector. The financial crisis can be symbolised with the failures of banks in several EU member states, massive declines in stock exchange indices and the reduction of the market values of equities and commodities. Furthermore, budgets of EU member states were reaching a high level of public debt in relation to GDP during this time and therefore lagging sustainable government finances, i.e. the budget of many EU member states breached the obligations of the Maastricht Treaty of keeping sound fiscal policies, with debt limited to 60 % of gross domestic product (GDP) and annual deficits no greater than 3 % of GDP.

The economic and financial crisis took place at a time of increased environmental and climate crisis and steep increases in commodity prices. The triple crisis – economic, financial and environmental –is also accompanied by growing disparities in income and wealth within European countries and globally as exemplified that by 2013 almost 25 % of EU's population is at risk of poverty or social exclusion as compared to 23.8 % in 2008 (Eurostat 2014). Moreover, this development led to 'a

[1] See EEA 2014a, for an overview of the different definition of green economy and green growth.

weakening of democratic institutions and legitimacy of governance systems and political parties' (Vergragt 2013).

The interest in the green economy concept became therefore evident in the late 2000s when government across the developed world implemented fiscal stimulus packages as an attempt to tackle the economic crisis and thereby aiming to counterbalance the fall in economic growth through green economy initiatives and measures (Bowen et al. 2009; Geels 2013). An example of these policy packages was the 'European Economy Recovery Plan (EREP)' launched by the European Commission (EC 2008). It was estimated that fiscal stimulus package in EU member states amounted to 5 % of GDP (EC 2009).

The fiscal recovery strategies seemed to lose momentum after 2010 and were coming to an end in 2012 (Geels 2013) primarily as governments were under pressure to consolidate public budgets and to implement debt reducing policies. Besides these pressures the macro-economic impact of green economy related strategies were questioned as they probably by far exaggerated expectations on income multiplier of green economy investments were not fulfilled. This has also to be seen in the overall context of the mismatch between short-term thinking and long-term expectations on which the green economy concept is based. The green economy concept was born to address multiple crises but the causes of them are inherent in the overall unsustainable system and structure of the economies and therefore requiring a fundamental shift in the dominant economic, financial and environmental systems which can only be achieved in the long-term. However, the green economy paradigm intrinsically entered the operational strategies of major international organisations, such as UN, World Bank, OECD and it seems bound to assume a stable role for these organisations.

7.2 The Green Economy as a Paradigm of Transition

The mounting importance of the green economy framework in the policy discourse shifts the interest from the short-term static thinking of the environment being a cost factor that constrains economic output growth and impairs competitiveness towards the idea that economic development can take place decoupled from environmental damage and within environmental limits (SRU 2012). The green economy must therefore be seen as a social, economic, and ecological transition towards better living within environmental limits.

7.2.1 Past Trends and the Economic Structure of the EU

One of the central objectives of EU policy and the green economy concept is improving resource efficiency. This objective is obviously necessary but it may not be sufficient in guaranteeing long-term sustainability (EEA 2012). The major EU environment and climate objectives and targets for 2020 and beyond are unlikely to be met without additional effort and more radical re-orientation of the European economic system. It is therefore obvious that the economic structure is important in the whole discussion of the green economy transition process.

Economic activity is the principal driver of pressure on resources and the environment. The structure of the economy, as defined by the sectoral industry mix, the technological system, and the combination of domestic production and consumption and international trade, is a major underlying factor for an economy to be 'green' or not. For example, differences in industrial structures in terms of energy intensity across countries can result in huge and fairly stable country differences in environmental indicators. Changes in the economic structure are slow but resilient. Normal business cycles and even crises do not usually change dramatically the macro-sectoral composition of the economy, which however is always evolving towards new configurations. Although trends in environmental indicators are closely related to the overall economic activity, structural changes of the economy may be even more crucial for steadily affecting environmental and resource efficiency performances.

It is therefore interesting to assess the environmental implications of the economic and financial crisis of the late 2000s in Europe. During the years of the crisis environmental pollution fell and resource consumption decreased. All these encouraging trends are measured in absolute terms but they did not lead to any improvements in resource efficiency indicators because economic indicators, like GDP, dropped too. Therefore, the economic and financial crisis did not lead to substantial improvements in resource efficiency, in some areas it has led to a decrease in efficiency indicators (EEA 2014b).

Structural changes of the economy, in particular the shift towards services is not per sè a factor leading to radical changes in resource efficiency of EU economies: services are less emitting directly in production, but when looked at final consumption they require input of other material-intensive and energy-intensive sectors – along the supply-chain – keeping their total need of resources very high. As economic productivity growth is much higher in manufacturing sectors than in services, the shift towards services may reduce overall growth (World Bank 2013 and OECD 2012a). Therefore, manufacturing will have a role to play in the green economy transition.

In addition, associated with the changing structure of the EU economy is an increasing role of international trade, especially of goods. This has implications for evaluating improvements in resource efficiency in the EU: there have been resource efficiency improvements in EU domestic production but EU consumption also results in emissions outside the EU through imported goods. Keeping an integrated environmental-economic perspective, although the relocation of production activities outside the EU shifts the associated emissions to these countries, it may bring them significant economic and social benefits. This must be borne in mind when considering trade-related aspects of the green economy.

Social and employment implications are also important to take into account when assessing the transition towards a service-oriented economic structure. Recent

UK data reveal that the increase of employment in the service sector and the simultaneous reduction in the manufacturing sector led to an overall reduction in earnings as the manufacturing sector has higher-paid jobs as compared to the lower-paid jobs of the service sector (Taylor et al. 2014). This development is highly relevant in the current discussions of social equity and precarious working conditions.

7.2.2 *Projections – Continued Economic Growth and Resource Use*

The future GDP and resource use developments are projected following more or less current trends with the exception that the projected growth rates are lower than the past ones, i.e. an annual average growth rate of 2.9 % between 2010 and 2050 as compared to 3.5 % between 2001 and 2010 (Fig. 7.1). The financial and economic crisis affected the GDP growth path in developed countries. These growth trajectories underline the basic change in the economic significance of some of today's emerging and developing countries as China is projected to become the biggest economy in the world by GDP around 2020 (based on 2005 USD Purchases Power Parity (PPP)) and India is projected to be the third biggest economy in 2050. This development comes along with a steep increase in the global middle class according to the OECD increasing from 27 % of the world population of 6.8 billion in 2009 to 58 % of a predicted world population of 8.4 billion in 2030 (Kharas 2010).

The growth of resource extraction is also projected to slow down at the global scale from 3.7 % during the period 2000 and 2008 to 2 % between 2010 and 2050

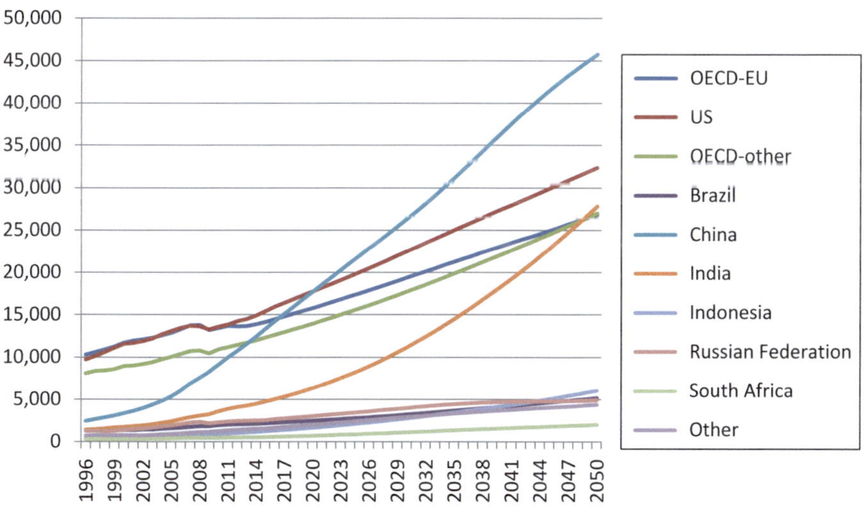

Fig. 7.1 Past and projected GDP trends (in billion 2005 USD PPP). Note: OECD-EU covers 21 EU Member States (Source: OECD 2014)

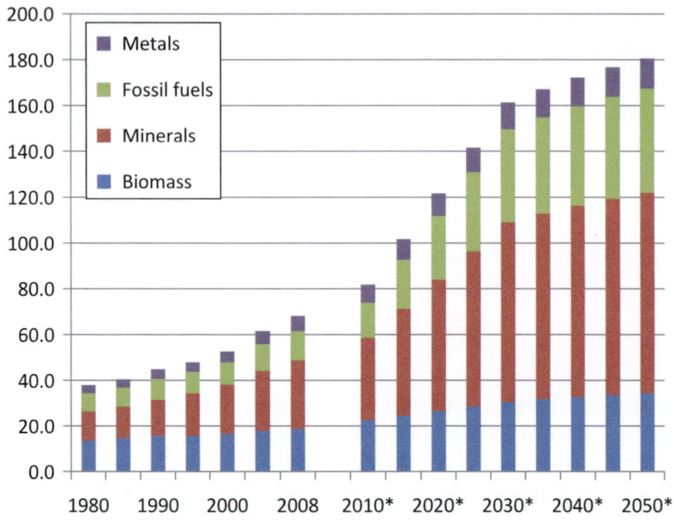

Fig. 7.2 Past and projected trends of global resource extraction. Note: * estimated (Source: Dittrich et al. 2012)

(Fig. 7.2). The projected data illustrates a relative decoupling between economic performance and resource use during the period 2010–2050 at the global scale. It is worthwhile to mention that the biggest increase is projected for fossil fuels (average annual increase of 2.8 %) followed by minerals (2.2 %).

The uncertainties of making these projections are well-known and discussed extensively.[2] The projections are reviewed periodically including the underlying assumptions of macro-economic and fiscal policies, commodity prices, exchange rates, etc. However, these projections are undoubtedly showing that the overall trends, which are based on the business-as-usual economic paradigm, are not directed to a resource-efficient and low-carbon green economy.

The EU has made progress in some areas of environmental and climate policy in the past, but when it comes to a longer-term outlook, the EU is far from being on the right track toward the transition to a green economy (EEA 2013a, b and 2014b). An example of the long-term challenges the EU is facing are the results of the EU energy and climate reference scenario presenting 'the development of the EU energy system under current trends and adopted policies' and 'includes all binding targets set out in EU legislation regarding of renewable energies and reductions of greenhouse gas (GHG) emissions, as well as the latest legislation promoting energy efficiency' (EC 2014a). Under the then adopted policies the scenario shows a decreasing trend in GHG emissions beyond 2020 which, however, is not sufficient to achieve the long-term EU objectives of reducing GHG emissions by 40 % in 2040 and by 80 % in 2050 relative to the 1990 level: 'the projections in the Reference 2013 scenario are 32 % reduction in 2030 and 44 % reduction in 2050' (EC 2014a).

[2] See http://www.oecd.org/eco/outlook/forecastingmethodsandanalyticaltools.htm

The necessity to align the short-term targets with long-term objectives is fundamental.[3] Furthermore, it seems to be quite obvious that new, innovative policy initiatives as well as reinforcing existing ones must be put in place with the aim of establishing a coherent policy trajectory for the achievement of these long-term objectives. Innovative policies and policy measures are required which enable the transition to a green economy in Europe as the transformation process does not happen on its own. A general policy framework as well as policy measures and instruments enabling the transition process are therefore essential.

7.3 A Transition Strategy: Enabling Factors

The current experience shows that the structure of an economy is changing over time but these changes are only incremental and will not promote the transition towards a resource-efficient, green economy and thereby accomplishing environmental and economic policy targets. Policies and factors enabling the transition process can be quite diverse but eco-innovation, international green technology transfer, the financing of the green economy transition, i.e. the provision of innovative investment vehicles, and environmental taxes as well as fiscal reforms can stimulate and accelerate the transition. These enabling factors have to provide a fundamental shift and not incremental changes in order to have a successful transition process.

7.3.1 Eco-Innovation and Green Technology Transfer

The role of eco-innovation and green inventions is crucial in the development of a greener and more competitive economy as stressed by international institutions, such as OECD (2008, 2010a, 2012a, 2013). The positive role of eco-innovation is also recognised by a large stream of recent economic research, which concludes that, while the economic and environmental benefits of eco-innovation are not always achievable as low-hanging fruit, they should be included in environmental policy making.

Consistently with these views, when looking at the indicators of the real economy it is clear that changing technology has been a key factor behind resource efficiency gains in Europe. For example, structural decomposition analysis of CO_2 emissions in European countries in 1995–2009 suggests that the reduction in emission intensity, which is the result of the diffusion of more emission-efficient

---

[3] A notable policy in this context is the Climate Change Act of the UK which establishes the GHG reduction target of at least 80 % from 1990 by 2050. The Act sets out five-yearly carbon budgets serving as interim targets and therefore fundamental building blocks for measuring overall progress towards achieving the 2050 target.

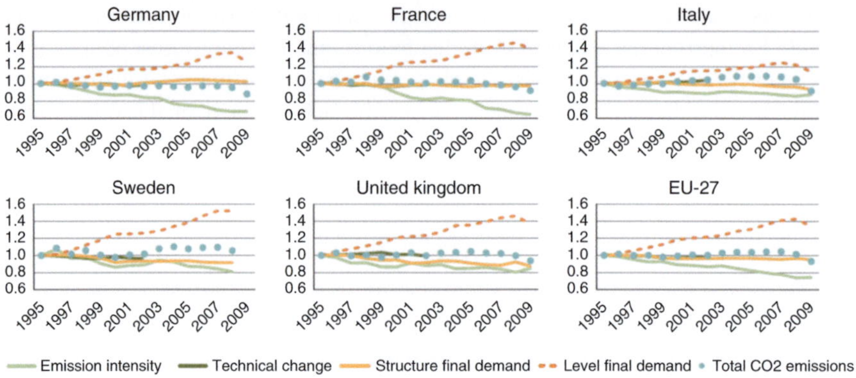

Fig. 7.3 Structural decomposition analysis for the production footprint of CO_2 emissions (index, 1995 = 1) (Source: EEA 2014a)

technologies, has been strong (France, – 36 %, Germany, – 32 % and the EU-27 as a whole, – 25.5 %) thus representing a crucial factor in compensating for the increase in emissions that would have resulted from the increase in final demand (see Fig. 7.3).

Not surprisingly, therefore, the role of innovation as an enabling factor for a green economy transition is recognised in EU specific and general policies, for example the adoption of the EU Eco-Innovation Action Plan and the Innovation Union flagship initiative as part of the Europe 2020 Strategy as well as by EU's Seventh Framework Programme for Research (FP7) and Horizon 2020, the EU research and technology programme.

It is important to recognise, however, that most eco-innovation policies generally focus on the early stage of eco-innovation (R&D, invention) but the critical steps in increasing resource efficiency is the diffusion and adoption of green technologies on a large scale. In fact, European data suggest that improved environmental conditions in a country are correlated with the rate of eco-innovation adoption by companies located in the country itself (EEA 2014a). In spite of a large portfolio of eco-inventions (e.g. patents), a relatively low share of firms adopt eco-innovation in many countries, according to the Community Innovation Survey (CIS).[4] To achieve a green economy transition, however, there is a need for more eco-innovation adoption and diffusion, especially in laggard countries.

One of the reasons for the slow adoption of eco-innovations is that companies face a set of barriers, from financial to knowledge barriers, which may be company-specific or related to the local industrial environment (Marin et al. 2014). Some of these barriers are common to all types of innovation and some others, being linked to the intrinsic features of the company, cannot be overcome by eco-innovation policies. Other barriers, instead, are specific to eco-innovation. For example, several companies in different sectors state, in responding to surveys, that a weak

[4] See http://epp.eurostat.ec.europa.eu/portal/page/portal/microdata/cis

environmental regulation can prevent them to adopt eco-innovations, which is an argument consistent with the so called ‘Porter hypothesis’, i.e. a well-designed environmental policy can spur innovation and create competitive advantages for companies in the local and global market (Porter 1991 and Porter and Van der Linde 1995).

The result is that the great potential for green innovation in Europe is partially untapped. In addition, as with all types of innovation, green technologies interact with organisations and individuals. Green innovation therefore requires some degree of social innovation if it is to facilitate a green economy.

The possibility of triggering eco-innovation and achieve its recognised benefits is also linked to the capacity of the EU to promote a global-scale green economy. The 2012 Rio + 20 conference (UN 2012) accorded a special role to the green economy in international cooperation for development. This development requires a process of green technology and knowledge transfer through technology trade, the circulation of green intellectual property rights and can also require an influence on environmental regulation at the international scale. These forms of knowledge transfer can provide developing countries with the potential for leapfrogging. Furthermore, it may offset the possibly higher costs faced by ‘forerunner industries’ in the EU which are often politically driven to implement more expensive but more energy and environmental efficient new technologies.

The total global export value of environmental goods (technologies) more than doubled during the 2000s, with developed and developing countries showing similar trends. Growth has been stronger for some technologies like renewable energies. International patent applications in green technologies are growing, especially in some areas of innovation in which the EU is also the recipient of patent applications from other countries. After the Kyoto Protocol, the number of international (European coverage) patent applications filed at the European Patent Office (EPO) increased 2.5 times for all green technologies. Trends in international green patenting across European countries are presented in Table 7.1, through a ranking of the residence of the main inventors of green patents, renewables and waste-related technologies for 2001–2010. The data show that green innovation as represented by international patents is highly concentrated in a few countries. The four countries with the most patent applications account for about 75 % of total green inventions, although this dominance is slightly less pronounced for renewables and waste. However, even in these two categories the top four countries account for more than the 60 % of total patented innovation.

The EU is also a recipient of green technological knowledge from other countries. Japan and the United States are the two countries that filed the most patents in Europe. In some technological fields the share of patents from these two countries together is more than the 90 % of the total patents filed in that field. The fields with high levels of US and Japanese innovation include energy-efficiency technologies in buildings and lighting and technologies that reduce emissions (EEA 2014a). This multi-directional circulation of green technological knowledge is always positive because it brings to a global improvement of resource efficiency of which Europe can benefit directly and indirectly.

Table 7.1 Residence of the main inventors of international green patents in the EU (% share, 2001–2010)

Country	Green technologies	Cumulative green technologies	Renewables	Cumulative renewables	Waste	Cumulative waste
Germany	48.2 %		41.9 %		29.4 %	
France	13.8 %	62.0 %	9.6 %	51.4 %	11.7 %	41.1 %
United Kingdom	6.9 %	68.9 %	7.7 %	59.1 %	9.7 %	50.8 %
Italy	6.5 %	75.4 %	5.6 %	64.8 %	13.3 %	64.1 %
Netherlands	6.0 %	81.4 %	5.8 %	70.6 %	8.0 %	72.1 %
Sweden	3.8 %	85.2 %	2.8 %	73.4 %	3.0 %	75.0 %
Denmark	3.5 %	88.7 %	9.7 %	83.1 %	2.2 %	77.2 %
Austria	2.7 %	91.4 %	3.1 %	86.2 %	5.2 %	82.4 %
Spain	2.1 %	93.5 %	5.8 %	92.0 %	4.1 %	86.6 %
Finland	2.0 %	95.4 %	2.0 %	94.0 %	4.8 %	91.4 %
Belgium	1.7 %	97.2 %	2.4 %	96.4 %	2.8 %	94.3 %
Luxembourg	1.0 %	98.2 %	0.6 %	96.9 %	1.2 %	95.4 %
Ireland	0.6 %	98.7 %	1.1 %	98.0 %	1.2 %	96.6 %
Others	1.3 %	100.0 %	2.0 %	100.0 %	3.4 %	100.0 %

Source: EEA 2014a, based on OECD statistics

A similar effect can be achieved by 'export' of EU environmental regulation, i.e. the adoption of EU environmental standards in non-EU countries. Examples are the standards for vehicles emissions and the standards in hazardous substances in electrical and electronic equipment. But there are also cases where a foreign standard has influenced European standards, for example in energy efficiency labelling. The diffusion of EU's environmental standards promotes eco-innovation in different ways, especially when EU standards are more stringent than the national ones, and they can also promote EU export as EU companies are already aligned to those level of stringency.

7.3.2 Financing the Transition

Financial resources can be a constraint to the transition to a green economy. Estimated financial needs at the European and global scales are huge –at least €270 billion per year is needed for the coming 40 years to support the transition to a low carbon economy – current investment trends fall far short of this (EEA 2014a). According to the Climate Policy Initiative (2013), achieving clean energy by 2020 needs USD 5 trillion of additional investment. However in 2012, only USD 359 billion was spent on climate change investment, even lower than the USD 364 billion invested in 2011. Of the USD 359 billion spent, USD 224 billion came from the private sector and USD 135 billion from the public sector through incentives such as low-cost loans, risk-coverage mechanisms, direct project investments and technical support.

Financing the green economy transition is a process at the macro-economic level which requires public policy initiatives to act as catalysers and the consideration of financial features in environmental policies. While it is quite noticeable that finance emerges as a constraint for the transition to a green economy, a range of innovative financing schemes were developed and implemented during the last decades that can support green investments on a significant scale. Although they are unlikely to fill the entire gap, mobilising them will help to cover a significant part of it. The relevant instruments, financial vehicles and initiatives are not necessarily focused on Europe alone, some are on a global scale. However, EU bodies, European governments and private economic operators in Europe are or could be involved in these financial opportunities, for example through climate funding to less developed countries, thus also contributing to the international transfer of green knowledge.

An example of private instrument is represented by pension funds, which own USD 30 trillion in assets at the global level, and insurance companies, which have USD 25 trillion in assets. Most pension funds are very interested in lower-risk investments that provide a steady, inflation-adjusted income stream in the long term. Green investments may meet these criteria, but pension funds do relatively little green investment (Della Croce et al. 2011). There are different reasons for this including a lack of appropriate investment vehicles, and the lack of knowledge and expertise among pension funds about these investments and their associated risks. However, data suggest that, given a total of about USD 15 trillion of assets under management, even a small fraction – 0.5 to 1 % – of this invested in climate change-related assets, would amount to USD 75 billion to USD 150 billion of investments in the green economy sectors.

Another example linked to institutional investors are Socially Responsible Investments (SRIs), i.e. those investments chosen by fund managers based on criteria related to the social and environmental attributes. The total SRI assets under management in Europe increased from EUR 2.7 trillion in 2007 to EUR 5 trillion at the end of 2009. This growth was, stronger than broader asset management growth. Furthermore, capital invested in SRIs increased by another 34 % between 2009 and 2011 (Eurosif 2012).

In the case of mixed public and private investors, green bonds represent a case of fast growing vehicles to finance the green economy. Including corporate and international organisations, green bond issues showed an impressive development as they increased more than fivefold in 2013 compared to 2012. Expectations are for USD 50 billion of issuance by 2015 implying that green bonds could account for 10–15 % of global bond issuance within 5–7 years.

Other important public actors, like the European Investments Bank (EIB), and State-owned instruments, like Sovereign Wealth Funds (SWF), can be added (Miceli 2013) that have the potential to channel huge amounts of financial resources to the green economy transition.

To realise these opportunities and avoid competition with conventional strategies being adopted by the financial system, a high level of commitment, persistence and risk-reducing strategies are needed.

7.3.3 Environmental Taxation and Fiscal Reforms

Economic instruments, such as environmental taxes and emission trading schemes, are policy tools for achieving environmental policy goals in a cost-effective manner thereby giving economic actors the flexibility to act individually as they are changing the pricing system, i.e. the relative prices. These price changes are essential for triggering the transition process toward a green economy (OECD 2011; UNEP 2011; World Bank 2012 and Ekins and Speck 2011). The basic motivation for their use is to correct market failures leading to an overexploitation of natural resources and environmental pollution thereby levying huge costs and risks on future generations. The internalisation of these externalities or external costs in the prices of goods and services can be done by utilising economic instruments, such as environmental taxes. This policy approach is the implementation of the Polluter Pays Principle (PPP), which is adopted in the Treaty on the Functioning of the European Union (Article 191(2) TFEU). Closely linked and probably a pre-condition for the usefulness of environmental taxation is the call for reforming and phasing-out of environmentally harmful subsidies. Some progress can be reported and multilateral co-operation to support the reform of environmentally harmful subsidies, in particular with regard to fossil fuel subsidies, is ongoing.

The topic of reforming fossil fuel subsidies is critical in the transition to a green economy as the provision of fossil fuel subsidies is in conflict with any climate policy actions as well as having a negative impact on public budgets. Figures published by the International Energy Agency (IEA) record an increase of subsidies on fossil fuel consumption from USD 300 billion (approx. €215 billion) in 2009 to USD 544 billion in 2012 (approx. €390 billion) (IEA 2011 and 2013).

The economic literature discusses the pros and cons of environmental taxes and comes to an understanding that these fiscal policy measures, such as carbon pricing, are effective instruments for tackling a range of environmental problems (OECD 2001 and 2006 and de Mooij et al. 2012). The major advantages of environmental taxes are to achieve static and dynamic efficiency gains as well as to generate revenues for public budget (Baranzini et al. 2000 and OECD 2001).

As Europe is still facing the consequences of the economic and financial crisis it is of no surprise that the overall policy focus is on growth, competitiveness and jobs. Past experiences and theoretical results are illustrating that the application of economic instruments for environmental policy is not an obstacle for achieving the economic objectives of growth, competitiveness and job creation (Andersen and Ekins 2009). Properly designed economic instruments can foster the achievements of them in a cost-effective manner. Furthermore, environmental taxes can help

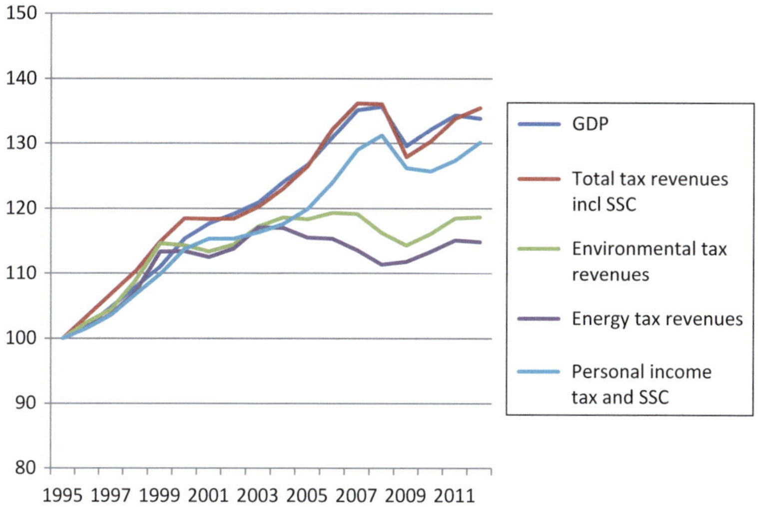

Fig. 7.4 Development of GDP and tax revenues of the EU-27 between 1995 and 2012 (in 2005 prices) Note: SSC – social security contributions; data are converted into 2005 prices by using the GDP deflator published by Eurostat (Source: author's calculation based on Eurostat data)

countries to increase the overall tax take and reduce debt and borrowing and release countries with the need to increase other taxes, such as income taxes or capital taxes.

Shifting taxation from labour to pollution, energy and resource use in a budgetary neutral way is a policy approach promoted by international institutions and in particular by the European Commission. Environmental taxation schemes make them especially well-suited to the post-financial crisis context, where countries wish to continue to grow, while also raising revenues to plug budget gaps. Environmental taxes are classified as taxes which are least detrimental to employment and growth (OECD 2010b and 2012b). Although the call for a more extensive use of environmental taxes is put forward for many years, the actual development looks rather different as environmental tax revenues increased alongside GDP and total tax revenues until 2000 and since then the overall trend decoupled with the exception of the period of the economic and financial crisis as tax revenues and GDP dropped during this period (Fig. 7.4). Therefore it is not surprising that the former Commissioner of DG Taxation and Customs Union spoke of environmental taxes as being under-used in many European countries.[5]

[5] See Statement by Commissioner Šemeta on Taxation Trends in Europe at http://europa.eu/rapid/press-release_STATEMENT-14-196_en.htm?locale=en [accessed on 4 November 2014]

7.3.3.1 Carbon Pricing – The Current Situation

Currently, the most discussed policy approach related to economic instruments is carbon pricing either introduced via carbon taxation or emission trading schemes. Moreover it is regularly labelled as a prerequisite for the transition to a low-carbon economy as high carbon prices are decisive to promote investments in clean, low-carbon technologies.

During recent years carbon pricing policies were introduced in numerous countries so that in 2014 this policy approach is in place in about 40 countries covering about 12 % of annual global greenhouse gas (GHG) emissions (World Bank 2014). This trend does not stop as the introduction of further carbon pricing schemes are planned to be implemented in some countries in due course, such as a national trading scheme in China, and in other countries the date for the introduction of new schemes are already fixed, such as CO_2 taxes in Chile and South Africa.

Both types of carbon pricing schemes are implemented in EU countries. The European Union Emission Trading System (EU ETS), the largest trading GHG emission trading scheme of the world, is portrayed as a cornerstone of the EU's policy to combat climate change and as a key tool for the reduction of industry's GHG. The EU ETS is a mandatory cap and trade scheme and latest data shows that the 2013 ETS emissions are below the cap. The overall GHG reduction of the sectors covered by the EU ETS was about 19 % compared to the 2005 level and close to the reduction target of 21 % for 2020 also compared to the 2005 level (EEA 2014b).

The allowance price of the EU ETS amounts to about €6 per tonne of CO_2 in the autumn of 2014 and is well below the carbon price seen to be needed for the investments in emission reduction technologies to become competitive. Business estimates are revealing that a carbon price between €20 and €40 is required at least (Business Green 2012). Other estimates are projecting a carbon price in the range of between €81 and €162 per tonne of CO_2 as necessary (WBCSD 2012).

Several EU Member States introduced CO_2 taxes more than 20 years ago and have been the first countries worldwide to make use of them. In the meantime this type of carbon pricing schemes is also implemented outside Europe (Table 7.2). The carbon taxation schemes differ widely in terms of the tax base and tax rates between countries, which is not too surprising as CO_2 taxes are implemented in connection with other fiscal schemes, such as energy taxes and emission trading schemes. The latter is of particular significance for European countries as the economic sectors covered by the EU ETS are in general exempt from CO_2 taxes and are only subject to energy taxes (Speck 2013). The range of CO_2 tax rates implemented in different countries is very wide as shown in Table 7.2.

Economic theory states that for having a cost-effective and efficient carbon taxation scheme, all energy sources and users would have to be subject to the CO_2 tax and that the tax rate 'would be set equal to the marginal benefits of emission reduction, represented by estimates of the social cost of carbon' (Aldy and Stevens 2012). However, the political reality looks very different as the uniformity of tax rates is

Table 7.2 The CO_2 tax rates implemented in selected countries

Country (year)	CO_2 tax rate - € per tonne CO_2	coverage – in % of GHG emissions	Comments
Finland (2014)	35–70	15	Tax differentiation between transport and heating fuels
Norway (2014)	28–51	50	Tax rates differ between energy products – rates for diesel and natural gas/offshore use are presented
Sweden (2014)	119.2	25	
Denmark (2013/4)	22.4	45	
Switzerland (2014)	49.2	30	Heating fuels are subject to the CO2 tax; further increases are planned for the coming years if reduction targets are not achieved. CO2 tax rate may increase to €98 in 2018
Canada – British Columbia (2014)	20.1	70	
Iceland (2014)	7	50	
Ireland (2014)	20	60	
UK – CCL (2014)	5.2 – 18.6	25	Tax rates differ between energy products – rates are shown for LPG and electricity
UK – CRC (2014)	14.8	25	Electricity; natural gas
UK – CPF (2014)	12 (22)	25	Carbon tax amounts to £9.55 as carbon price floor is set at £18 and is fixed until 2019/2020
Japan (2012)	1.5	70	Increase to €2.3 (JPY 289) in 2016
France (2014)	7.0	35	Further increases are planned to €14.5 in 2015 and to €22 in 2016.
Mexico (2014)	2.6	40	Tax can be offset by CDM projects; natural gas is not subject to the CO2 tax and coal is subject to a reduced tax rate (€0.8)
Australia	15.5	60	The carbon tax was abolished on 1 July 2014
South Africa (planned from 2016)	10.3	80	Annual increase of 10 % is planned until 2019. An offset scheme to reduce tax liability is planned
Chile (planned from 2017)	3.7		

Note: UK: CCL – climate change levy: was introduced in 2001 and is a tax on energy used by non-domestic users; CRC – Energy Efficiency Scheme (formerly known as the Carbon Reduction Commitment) was introduced in 2010 and since then revised. It covers all organisations in the UK using more than 6000 MWh per year of electricity and are not part of the EU ETS; CPF – carbon floor price: it came into effect in 2013 and is a tax on fossil fuels used to generate electricity. This carbon price support mechanism aims to support investment in low-carbon electricity generation. The CPF is set with the aim of achieving a predetermined price trajectory and therefore depends on the EU ETS allowance price. South Africa: a range of special tax exemptions are planned so that the effective tax rates are between €0.8 per tonne CO_2 and €3.3 per tonne CO_2 depending on economic sector

Source: Speck (2014)

not given because of different reasons. For example, EU member states are making use of the two different carbon pricing schemes.

Besides these legal carbon pricing policies, voluntary schemes are also in place as an increasing number of companies introduced internal carbon prices within planning processes aiming to capture prospective policy development and regulatory risks of CO_2 emissions into their business strategies (Sustainable Prosperity 2013, CDP 2013 and 2014). Companies employing internal carbon prices belong to all sectors of the economy, including manufacturing, energy and fossil-fuel companies, and are located in all world regions, such as the US, Canada, Brazil, UK, Netherlands, Australia, South Korea, India, China and Japan. Some of the highest internal carbon prices are being used by energy-intensive companies, such as Exon Mobil Corporation and National Grid, are amounting of more than USD 60 per tonne CO_2 (CDP 2014). The difference between the high internal carbon prices used by companies and the EU ETS allowance price is striking and may be interpreted that companies are expecting higher carbon prices for the future.

7.4 Main Findings and Conclusions

The strategic policy push for a green economy transition in the EU is extensive and, for a large part, already binding. The considerable number of legally binding targets and non-legally binding objectives in the areas if energy, GHG emission, ozone depleting substances (ODS), waste, water, sustainable consumption and production (SCP) and resource efficiency, chemicals, biodiversity and land use to be achieved form 2010 and 2050 (EEA 2013c) can be looked at as fundamental building blocks of a green economy in Europe as they reveal potential points of arrival of a desired transition.

However, the transition demands much more fundamental changes in resource efficiency and resource use trends compared to what can be observed at present. The factors behind these trends are of a macro-economic scale and largely depend, as in the case of industrial specialisation and dynamics, on factors that can be hardly governed by those same environmental policies and strategies that established ambitious objectives.

The green economy is a macro-structural transformation. Different directions of change for a successful transition can be envisaged.

- Eco-innovation: Innovation is a pillar of the Europe 2020 Strategy and major EU policies and can be a fundamental lever for the green economy transition by guiding the EU energy and material-use system towards a radical transition of efficiency and into the direction of a circular economy. The structural transformations of the EU economy, in particular the shift to services with special focus given to the renaissance of manufacturing industry[6] will require a significant attention to innovation to achieve a green re-manufacturing able to satisfy competiveness and environmental objectives together.

- Green technology transfer: The EU is a major actor of the world economy and international trade. Through trade, the green knowledge of the EU, be it embodied in traded technologies or disembodied (patents and other immaterial knowledge goods), can be transferred internationally. On the other side, green economy knowledge from non-EU regions can be imported. Overall, technology transfer is beneficial for countries as they can gain benefit and for the EU as it enlarges the markets and the innovation dynamics of green technologies/knowledge.
- The role of finance for the green economy: Finance is emerging as a constraint for the transition to a green economy. The estimated financial needs are huge and the crisis weakened the availability to invest. There are important instruments by which financial resources can be channelled to the green economy (pension funds, green bonds, the European Project Bond initiatives, social responsible investments, sovereign wealth funds, etc.).
- Environmental fiscal reforms and the general reconsideration of market-based instruments for environmental protection: While environmental fiscal reforms seemed to come back into the policy agenda at the beginning of the crisis, its political acceptability seems to have weakened with the persistency of the crisis and its adverse fiscal policy implications. Without a right pricing of resources and the environmental pressures it can be hardly expected that the necessary changes in the trends of resource efficiency will be achieved. However, the use and the design of market-based instruments can be reconsidered also in the light of a possibly changing responsiveness of the economy to high resource prices, as emerged in the recent years.

The potential benefits for the shift to a green economy are great. The on-going economic and financial crisis emphasised the imbalance between environmental and climate policies as these policies are addressing areas with a longer time horizon as compared to the society's short-term expectations towards policy making of addressing more pressing economic and social topics, such as the creation of new jobs. The effects of the crisis can be seen as impairing factors in the transition process as the current situation may delay the necessary programmes for accelerating the process. A secondary effect of the implications of the still prevailing financial crisis is a change in the overall priorities in EU member states' macro-policies as stringent fiscal and financial conditions must be met.

This overall policy development must be considered when assessing the current speed of the transition process. However, the green economy approach is structurally embodied in major EU strategies and policies outside the environmental domain. The business sector as well as consumers also adheres to concepts closely connected to aspects of the green and circular economy, such as improvement in

[6] EU industrial policy is calling for a 'industrial renaissance' as manufacturing industry should achieve a 20 % share in EU's GDP by 2020, from about 16 % in 2011 (EC 2014b).

resource efficiency. Incremental changes of the economic performance will be required but more profound changes are needed for the transition to fundamentally sustainable economy.

A green economy is not the creation of a new economy but rather the greening of the existing one. But this obviously requires a more widespread use of a whole policy framework including the backing of eco-innovation, in particular the extensive diffusion of new technologies, a broader application of economic instruments and innovative financing activities aiming to transform the current economy into the greener one. But this cannot be successful without the integration of all areas of policies and treating economic and environmental performances as dual objectives by regarding social inclusion as an additional pillar.

Disclaimer
The views expressed in this article are those of Stefan Speck and may not in any circumstances be regarded as stating an official position of the European Environment Agency or its Management Board.

References

Aldy JE, Stavins RN (2012) The promise and problems of pricing carbon: theory an experience. J Environ Dev 21(2):152–180

Andersen MS, Ekins P (eds) (2009) Carbon-energy taxation: lessons from Europe. Oxford University Press, Oxford

Baranzini A, Goldemberg L, Speck S (2000) A future for carbon taxes. Ecol Econ 32(3):395–412

Bowen A, Fankhauser S, Stern N, Zenghelis D (2009) An outline of the case for a 'green' stimulus. The Grantham Research Institute on Climate Change and the Environment and Centre for Climate Change Economics and Policy, Policy Brief, LSE's Grantham Research Institute on Climate Change and the Environment, London

Business Green (2012) Energy giants renew calls for action on EU carbon price, 28 June 2012. Available via http://www.businessgreen.com/bg/news/2187590/energy-giants-renew-calls-action-eu-carbon-price. Accessed 25 Oct 2014

CDP (2013) Use of internal carbon price by companies as incentive and strategic planning tool a review of findings from CDP 2013 disclosure, A white paper from Carbon Disclosure Project (CDP) North America, New York

CDP (2014) Global corporate use of carbon pricing Disclosures to investors. CDP North America. September 2014

Climate Policy Initiative (2013) Global landscape of climate finance 2013. Available viahttp://climatepolicyinitiative.org/publication/global-landscape-of-climate-finance-2013/. Accessed 30 Nov 2014

Della Croce R, Kaminker C, Stewart F (2011) The role of pension funds in financing green growth initiatives. OECD working papers on finance, insurance and private pensions, No. 10. Organisation for Economic Co-operation and Development, Paris

DeMooij R, Parry IWH, Keen M (eds) (2012) Fiscal policy to mitigate climate change a guide for policymakers. International Monetary Fund, Washington, DC

Dittrich M, Giljum S, Lutter S, Polzin C (2012) Green economies around the world? The role of resource use for development and the environment. Sustainable Europe Research Institute (SER), Vienna

EC (2008) A European Economic Recovery Plan. Communication from the Commission to the European Council, COM (2008) 800final. http://ec.europa.eu/economy_finance/publications/publication13504_en.pdf
EC (2009) Economic crisis in Europe: causes, consequences and responses. European Economy No.7/2009. Directorate-General for Economic and Financial Affairs, Brussels
EC (2014a) EU energy, transport and GHG emissions trends to 2050 reference scenario 2013. Directorate-General for Energy, Directorate-General for Climate Action, Directorate-General for Mobility and Transport. Publications Office of the European Union, Brussels
EC (2014b) For a European Industrial Renaissance. Communication from the Commission to the European Parliament, the Council, the European Economic and Social Committee and the Committee of the Regions, COM(2014) 14/2, Brussels
EEA (2010) The European environment – state and outlook 2010: synthesis. European Environment Agency, Copenhagen
EEA (2012) Environmental indicator report 2012 ecosystem resilience and resource efficiency in a green economy in Europe. European Environment Agency, Copenhagen
EEA (2013a) Air quality in Europe – 2013 report. EEA report No 9/2013. European Environment Agency, Copenhagen
EEA (2013b) Managing municipal solid waste – a review of achievements in 32 European countries. EEA Report No 2/2013. European Environment Agency, Copenhagen
EEA (2013c) Towards a green economy in Europe- EU environmental policy targets and objectives 2010–2050. EEA Report No 8/2013. European Environment Agency, Copenhagen
EEA (2014a) Resource-efficient green economy and EU policies. EEA report no 2/2014. European Environment Agency, Copenhagen
EEA (2014b) Trends and projections in Europe 2014 – Tracking progress towards Europe's climate and energy targets until 2020. EEA report no 6/2014. European Environment Agency, Copenhagen
Ekins P, Speck S (eds) (2011) Environmental tax reform: a policy for green growth. Oxford University Press, Oxford
Eurosif (2012) European SRI study 2012. Brussels
Eurostat (2014) More than 120 million persons at risk of poverty or social exclusion in 2013. Eurostat newsrelease 168/2014
Geels FW (2013) The impact of the financial-economic crisis on sustainability transitions: financial investment, governance and public discourse. Working paper no 39, WIFO – Austrian Institute of Economic Research, Vienna
IEA (2011) World energy outlook 2011. International Energy Agency, Paris
IEA (2013) World energy outlook 2013. International Energy Agency, Paris
Jacobs M (1991) The green economy environment, sustainable development and the politics of the future. Pluto Press, London
Kharas H (2010) The emerging middle-class in developing countries. OECD Development Centre. Working paper No 285. Organisation for Economic Co-operation and Development, Paris
Marin G, Marzucchui A, Zoboli R (2014) SMEs and barriers to eco-innovation in EU: a diverse palette of greens. INGENIO (CSIC-UPV) working paper series 2014-04. Valencia
Miceli V (2013) Do sovereign wealth funds herd in equity markets? Quant Finan 13(9). Taylor and Francis, Abingdon
OECD (2001) Environmentally related taxes in OECD countries: issues and strategies. Organisation for Economic Co-operation and Development, Paris
OECD (2006) The political economy of environmentally related taxes. Organisation for Economic Co-operation and Development, Paris
OECD (2008) Environmental policy, technological innovation and patents. Organisation for Economic Co-operation and Development, Paris
OECD (2010a) Environmental policy design characteristics and technological innovation: evidence from patent date. Organisation for Economic Co-operation and Development, Paris

OECD (2010b) Tax policy reform and economic growth. OECD Tax Policy Studies, No. 20. Organisation for Economic Co-operation and Development, Paris
OECD (2011) Towards a green growth. Organisation for Economic Co-operation and Development, Paris
OECD (2012a) The jobs potential of a shift towards a low-carbon economy. Final report for the European Commission DG employment. Organisation for Economic Co-operation and Development, Paris
OECD (2012b) Energy and climate policy: bending the technological trajectory. Policy brief – energy and climate policy: bending the technological trajectory of key findings. Organisation for Economic Co-operation and Development, Paris
OECD (2013) International cooperation for climate change innovation: a problem shared is a problem halved. Organisation for Economic Co-operation and Development, Paris
OECD (2014) Economic outlook No 95 – May 2014 – long-term baseline projections. Available via http://stats.oecd.org/Index.aspx?DataSetCode=EO95_LTB#. Accessed 3 Nov 2014
Pearce D, Markandya A, Barbier ED (1989) Blueprint of a green economy. Earthscan Publications, London
Porter M (1991) America's green strategy. Sci Am 264(4):168, New York
Porter M, Van der Linde C (1995) Toward a new conception of the environment- competitiveness relationship. J Econ Perspect 9(4):97–118, American Economic Association, Washington, DC
Speck S (2013) Carbon taxation: two decades of experiences and future prospects. Carbon Manage 4(2):171–183
Speck S (2014) Carbon taxation schemes – an overview of design schemes throughout the world. In: Paper presented at the 15th Global conference on environmental taxation, Copenhagen, 24–26 September 2014
SRU (2012) Environmental report 2012 responsibility in a finite world. Sachverständigenrat für Umweltfragen – German Advisory Council on the Environment), Berlin
Sustainable Prosperity (2013) Shadow carbon pricing in the Canadian energy sector. Policy brief, March 2013. University of Ottawa, Ottawa
Taylor C, Jowett A, Hardie M (2014) An examination of falling real wages, 2010–2013. Office for National Statistics (ONS), Newport
UN (2012) The future we want. Resolution adopted by the General Assembly on 27 July 2012, A/RES/66/288. United Nations, New York
UNEP (2011) Towards a green economy: pathways to sustainable development and poverty eradication. United Nations Environment Programme, Nairobi
Vergragt PJ (2013) A possible way out of the combined economic-sustainability crisis. Environ Innov Soc Trans 6:123–125
WBCSD (2012) The energy mix low-carbon pathways to 2050. World Business Council for Sustainable Development, Geneva
World Bank (2012) Inclusive green growth the pathway to sustainable development. World Bank, Washington, DC
World Bank (2013) China 2030: building a modern, harmonious, and creative society. World Bank, Washington, DC
World Bank (2014) State and trends of carbon pricing 2013. World Bank, Washington, DC

Chapter 8
Measuring Natural Resource Use from the Micro to the Macro Level

Stefan Giljum, Stephan Lutter, and Martin Bruckner

Abstract Many of today's most urgent environmental problems are related to the increasing volumes of worldwide production and consumption and the associated use of natural resources. Solid indicators to measure different dimensions of anthropogenic resource use are essential for designing appropriate policy measures for a sustainable management of these resources. Based on a brief review of the current state of the art of resource use indicators, this chapter describes a set of complementary environmental indicators, combining existing measures for the use of materials, water and land as well as emissions of greenhouse gases. This set can be applied consistently from the micro level of products and companies up to the macro level of countries and world regions, where all suggested indicators take a life cycle perspective on production and consumption activities. The set of indicators deals with the issue of the overall scale of the human production and consumption system and can be regarded as a framework of pressure indicators, based on which indicators on different environmental impacts can be derived. Moreover, these pressure indicators are considered as appropriate proxies for the human impact on the environment. The described set of indicators thus covers natural resource use in a comprehensive and complementary manner and can serve as a basis for setting resource-specific targets and evaluating specific resource policies.

Keywords GHG emissions • Land use • Material consumption • Resource footprints • Water use • Environmental accounting

---

S. Giljum (✉) • S. Lutter • M. Bruckner
Institute for Ecological Economics, Vienna University of Economics and Business (WU), Vienna, Austria
e-mail: stefan.giljum@wu.ac.at; stephan.lutter@wu.ac.at; martin.bruckner@wu.ac.at

S. Shmelev (ed.), *Green Economy Reader*, Studies in Ecological Economics 6,
DOI 10.1007/978-3-319-38919-6_8

8.1 Introduction[1]

Due to growth of world population, continued high levels of consumption in the developed world, and the rapid industrialisation of emerging economies such as China, India and Brazil, worldwide demand on natural resources and related pressures on the environment are steadily increasing. Renewable resources, and the ecological services they provide, are at great risk of degradation and collapse (UNEP 2012). The depletion of these ecological assets is serious, as human society is embedded within the biosphere and depends on ecosystems for a steady supply of the basic requirements for life: food, water, energy, fibre, waste sinks, and other services. At the same time, extraction of many renewable and non-renewable resources is already reaching or has already passed a peak (Seppelt et al. 2014); some authors even described today's situation as "peak everything" (Heinberg 2007).

The past 30 years saw a change in complexity and scope of environmental problems. Early environmental policy was mainly concerned with the reduction of local or regional environmental degradation through pollution of certain environmentally harmful substances, such as air pollutants, sewage effluents, and hazardous wastes. In this regard, developed countries and world regions, such as Europe, have achieved significant improvements due to technological innovations and substitution of harmful substances and products. This has resulted in better environmental quality of rivers and lakes, decreasing concentrations of pollutants in ground water, successful reduction of acid rain, and improved air quality in many cities.

However, since the mid-1980s, another type of environmental problem became increasingly important, which is associated with global changes in production, trade and consumption patterns. These problems are more closely related to the overall volume (or scale) of economic activities than a result of the specific potential for environmental harm of single substances (Giljum et al. 2005; Schmidt-Bleek 1992); and they are more difficult to address, as they are complex, international or even global in scope, and involve multi-dimensional cause-effect-impact relationships and time-lags. Issues such as climate change, loss of biodiversity, land cover conversion and high levels of energy and resource consumption are part of this new type of environmental problems. As evidence illustrates, world regions such as Europe have performed much worse in this regard: many species are threatened by extinction, fish stocks are depleted, water reserves shrink, overall waste volumes have been growing, urban sprawl transforms fertile land into sealed areas, valuable soil is lost through erosion, energy consumption grows, and Europe is far away from achieving a significant reduction in greenhouse gas emissions (GHGs) (EEA 2015).

Given this serious situation, it is necessary to develop solid and comprehensive systems which measure human resource use applying appropriate indicators. With

---

[1] This chapter is an update of the paper by Giljum et al. (2011): A comprehensive set of resource use indicators from the micro to the macro level. Resources, Conservation and Recycling 55, 300–308.

the help of such indicators the extent of human resource appropriation can be measured and monitored, and effective political management strategies can be designed. While standards for measuring greenhouse gas emissions (GHG) were developed within the UN framework convention on climate change (Solomon et al. 2007), for the issue of measuring resource use harmonised methodologies are only currently being introduced on the European and international level (EUROSTAT 2013; Fischer-Kowalski et al. 2011; Giljum et al. 2013).

Based on a brief review of existing indicators, this chapter describes a consistent and comprehensive set of resource use indicators. The indicator set comprises the resource input categories of abiotic and biotic materials, water, and land. In addition, it includes the output category of GHG emissions. Although being an output indicator, it was decided to include GHG emissions in this indicator set on resource use, as in terms of mass, GHGs are the biggest outflow from industrial systems back to the natural system. Furthermore, GHG emissions are a more policy relevant category than the use of air, which provides the oxygen for combustion processes on the input side.

The set of indicators suggested in this chapter can be applied on all levels of economic activity: from the micro level of products and companies, via the meso level of economic sectors to the macro level of countries and world regions. It can therefore be regarded as a framework of pressure indicators, based on which other indicators can be derived. For example, solid indicators measuring the different environmental impacts of resources and products (see van der Voet et al. 2009) require data on physical amounts as their basis. Moreover, the proposed aggregate pressure indicators are considered as appropriate proxies for the human impact on the environment.

The chapter is structured as follows. In Sect. 8.2 a review of existing resource use indicators is provided and their interrelations are illustrated. Section 8.3 lists criteria for the identification of comprehensive and sound sets of resource use indicators. Section 8.4 describes the suggested resource indicator set, explains the reasons for selecting these indicators and describes how this indicator set should be applied in practice. Section 8.5 concludes.

8.2 Existing Measurement Systems and Indicators

The past 15–20 years saw rapidly increasing interest in the quantitative assessment of the interrelations between the socio-economic system and nature. This is due to the fact that many of our environmental problems are regarded as the result of the quantity and quality of the resource throughput of our societies. This chapter provides a short review of existing measurement systems and resource use indicators.

8.2.1 Methodologies to Measure Resource Use

Five basic categories of natural resources serve as inputs to production and consumption processes: biotic and abiotic materials, energy, air, water and land (United Nations 2012b). For each of these categories, different methodologies have been developed.

8.2.1.1 Material Flow Accounting and Analysis (MFA)

Material flow accounting and analysis (MFA) is an approach, which focuses on the use of different materials by human activities. MFA on the economy-wide level (EW-MFA) builds on concepts of material and energy balancing, which were introduced already more than 40 years ago (for example, Kneese et al. 1970). MFA calculations are carried out in mass units (kilogrammes or tonnes). MFA in general accounts for material flows, i.e. the turnover of mass during a defined period of time, usually 1 year. Based on national or international statistical data, EW-MFA accounts for the domestic extraction of resources, as well as physical imports and exports. Biotic materials cover production from agriculture, forestry, fishery, and hunting; abiotic materials cover minerals (metal ores, industrial and construction minerals) and fossil energy carriers (coal, oil, gas, peat). Since the beginning of the 1990s, when first material flow accounts on the national level were presented, MFA has been a scientific field of rapidly growing interest, and major efforts have been undertaken to harmonise methodological approaches developed by different research teams (Adriaanse et al. 1997; Fischer-Kowalski et al. 2011; Krausmann et al. 2015; Matthews et al. 2000). In international working groups on MFA, a harmonisation of methodologies for accounting and analysing material flows on the national level was achieved and published in methodological guidebooks by EUROSTAT (2001, 2013) and the OECD (2007). In many EU and OECD countries, MFA is already part of the official environmental statistics reporting system. MFA data are also available for an increasing number of emerging and developing countries (Giljum et al. 2014; Schandl and West 2010; WU 2015; West and Schandl 2013).

In addition to the accounting of material flows on the economy-wide level (global, national, regional), resource use indicators have also been developed and applied for products, based on life cycle assessments (LCA). For instance, the concept of “Material Input per Service Unit (MIPS)” was developed at the Wuppertal Institute for Climate, Environment and Energy in Germany and aims at illustrating material, water and air inputs required along the whole life cycle of a product: from resource extraction (e.g. mining) and refining via manufacturing and trade to consumption and finally treatment or disposal. At all stages of the life cycle, MIPS accounts not only direct material use, but the equivalent of primary materials taken from nature (Saurat and Ritthoff 2013; Schmidt-Bleek 1992; Schmidt-Bleek et al. 1998). These life cycle-wide material inputs (also known as the ecological rucksack of a product) visualise the cumulated environmental pressures, which are in general invisible to final consumers.

8.2.1.2 Energy Flow Accounting (EFA)

Based on data from material flow accounts on the macro-level, a so-called "energy flow analysis" can be carried out. In contrast to standard energy balances published by statistical offices and covering only energy carriers, all material inputs are considered in an energy flow analysis. Materials are transformed into energy equivalents through their gross calorific values, reflecting the full amount of energy that could potentially be gained by the combustion of the materials. Thereby, also food products and manufactured products can be captured with their energetic dimension, allowing to comprehensively assess the energy system of societies (Schandl et al. 2002). Together with MFA, EFA can form a comprehensive concept for sustainability assessments (see Haberl et al. 2004).

8.2.1.3 Air Accounts

Air is a key resource input to combustion and other processes and serves as a balancing item to establish material balances e.g. for the use of fossil fuels, producing CO_2 from O_2 in the air and carbon in the fuels. For indicators such as the Ecological Footprint (Borucke et al. 2013) or the Carbon Footprint (Peters 2010), this category of resource input is therefore of importance in the underlying accounting method.

8.2.1.4 Water Accounts

The use of water is an issue with increasing policy relevance. Water accounts have already been included in statistical systems on the national level (for example, DESTATIS 2013). These statistics mostly represent the domestic uptake of water. However, current developments such as on the European level by Eurostat envisage accounting not only for water withdrawn but also actually consumed (i.e. water lost throughout the production process via incorporation into the product, evaporation, transpiration by plants, etc). Also, the United Nations set up a standard for compiling water accounts in their System of Environmental-Economic Accounting for Water (United Nations 2012a).

The so-called "Water Footprint of Nations" (Hoekstra and Mekonnen 2012) also takes a consumption perspective and accounts for the (domestic and foreign) uptake of water necessary to satisfy the national consumption of products. It thus includes the so-called "virtual water", i.e. embodied water, of internationally traded products. The Water Footprint Manual describes how to apply the methodology at different levels of economic activity (Hoekstra et al. 2009). An increasing number of

"Water Footprints" are also calculated on the product level (for example, Mekonnen and Hoekstra 2013) and for companies (Ruini et al. 2013).

Increasingly, water accounts distinguish between withdraws of water from rivers, lakes and aquifers (surface and ground water, the so-called "blue water") that is used in agriculture, industry and for domestic purposes, as well as water from rainfall ("green water") that is used to grow crops. The impact of water withdrawals depends largely on where and when water is extracted. A link to the renewable water stocks for the specific geographic region or country is particularly useful for an appropriate interpretation of water flow-based indicators (Boulay et al. 2015).

8.2.1.5 Land Cover and Land Use Accounts

Land cover accounts are generally established from satellite images applying a certain resolution (grid system). For example, the EU Corine land cover (CLC) system, which is used by the European Environment Agency (EEA) for producing and reporting land cover change accounts, is based on satellite images in a 100 m × 100 m grid (EEA 2006). Such systems aim at describing the geographical patterns of different land cover types across a country or region, the way they change over time and the processes that drive these transformations. Furthermore, the LUCAS survey carried out every 3 years by EUROSTAT assesses the state and the dynamics of changes in land use and cover in the European Union based on a point survey (EUROSTAT 2010). The latest LUCAS survey (2012) covers in-situ observations on more than 270,000 survey points in 27 EU countries. Recently, there is increasing interest in quantifying the land embodied in internationally traded products and in derived indicators such as the land footprint (Bruckner et al. 2015; Kastner et al. 2012; Weinzettel et al. 2013; Yu et al. 2013).

8.2.1.6 Environmental-Economic Accounting

In the last decades efforts have been made to set up internationally agreed standards for producing internationally comparable statistics on the environment and its relationship with the economy. The most prominent example is the UN System of Environmental-Economic Accounting (SEEA; United Nations 2012b). The European pendant to SEEA is the European Strategy for Environmental Accounts (ESEA; ESSC 2014). The aim is to align environmental accounts with the System of National Accounts (SNA) and to use concepts, definitions and classifications consistent with the SNA in order to facilitate the integration of environmental and economic statistics. Both accounting systems, SEEA and ESEA, are aligned to ensure consistency, and encompass different environmental categories, such as material, land, water, energy, emissions, and others.

8.2.2 Indicators Based on the Core Categories of Resource Inputs

Based on this system of five main categories of resource inputs, a number of indicators can be calculated (see Fig. 8.1). Two main types of indicators can be distinguished: input indicators (left side of the diagram) and indicators, which refer to outputs or combine inputs and parts of the generated outputs, in particular GHG emissions (right side of the diagram).

Input-oriented indicators include indicators derived from MFA accounts. Material flow-based indicators on the economy-wide level comprise input, consumption, trade and productivity indicators and are expressed in mass units or in mass units related to GDP (Bringezu et al. 2003). Material flow-based indicators (such as Domestic Material Consumption (DMC) in relation to GDP) have already been integrated in various European indicator sets. Most notably, GDP/DMC is the current headline indicator for monitoring the implementation of the "Roadmap to a resource efficient Europe" (European Commission 2011). Recently, significant efforts have been devoted to establish methodologies for calculating material flows embodied in internationally traded products (Giljum et al. 2015; Lutter and Giljum 2014; Wiedmann et al. 2015). On the level of single products, the indicator MIPS is most widely applied, which sums up primary material requirements along the whole

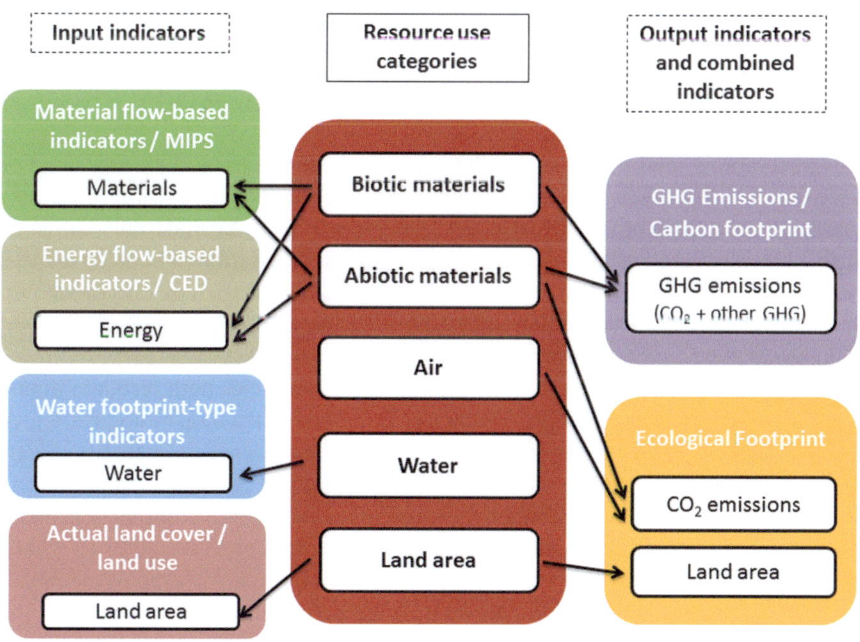

Fig. 8.1 The system of resource use indicators derived from the core resource input categories (Source: adapted from Giljum et al. 2011)

life cycle of a product (Saurat and Ritthoff 2013; Schmidt-Bleek et al. 1998). There are also attempts to link quantitative data on the amounts of resources consumed (from material flow accounts) with characterization factors from life cycle impact assessments informing about the specific environmental harm (global warming, toxicity, land intensity, etc.) of different types of materials, see for example the indicator of Environmentally weighted Material Consumption (EMC) (van der Voet et al. 2009; van der Voet et al. 2005).

Based on data from MFA, indicators of energy flow analyses can be derived. "Domestic Energy Consumption" equals the DMC indicator in the MFA framework and is calculated as the sum of domestic energy inputs plus imported energy minus exported energy (Schandl et al. 2002). The indicator regarding energy consumption on the product level is the so-called Cumulative Energy Demand (CED), which has been used as headline indicator for energy system since the 1970s. CED equals the sum of primary energy inputs along the whole product life-cycle (Huijbregts et al. 2005).

Water footprinting indicators account for the water input in production and consumption processes in the unit of litres or cubic metres. While on the micro level the water input along the production chain is quantified, on the macro level a methodology often applied is to allocate water abstraction to specific economic sectors. By subtracting water discharges from the respective sectors water consumption can be calculated. In order to obtain a comprehensive picture of human impact on the hydrological system, it is of key importance to account for both blue and green water consumption (see above). As measurement techniques especially for the latter are complex and data are often not available, these values have to be estimated using specific models. Examples for such models or estimation concepts are the Water Footprint (Hoekstra et al. 2011; Hoekstra and Mekonnen 2012) or biophysical models (Flörke et al. 2013).

Indicators on land use illustrate the land area required to produce a product or service (micro level) or all the goods produced or consumed in a region or country (macro level). Particularly valuable are indicators, which illustrate the change of land cover and land use from one year to another (e.g. expansion of built-up land on the cost of agricultural land) (EEA 2010). LCA databases such as Ecoinvent (see www.ecoinvent.org) also include information on the land inputs required for specific products and processes.

The Ecological Footprint is an indicator, which combines both resource inputs and waste outputs. The Ecological Footprint is defined as the total biologically productive land and water areas required to produce the resources a population consumes, and to assimilate the waste (i.e. CO_2 emissions) it generates. Its purpose is to answer the question of how much regenerative capacity of the biosphere is occupied by the resource consumption of the inhabitants of different countries (Borucke et al. 2013). The Ecological Footprint provides a bookkeeping system of biocapacity: by comparing the land appropriation of the population of a country with the ecological capacity available within a country or world-wide, national (or global) ecological deficits or ecological reserves can be quantified. National Ecological Footprint accounts build to a large extent on data from national material flow and land use accounts (see above). The compilation of such accounts starts from a population's resource consumption (domestically harvested resources plus imports minus exports) expressed in mass flows (tonnes per year). These physical flows are then converted into area equivalents, expressed in the unit of so-called "global hectares" (hectares with world-average biological productivity). This approach is repeated for six major land use types: cropland, pastures, fishery areas, forestland, built-up land, and so-called energy land. Built-up area is typically calculated based on land cover and land use accounts. The category of energy land illustrates which amount of biologically productive land (i.e. forests) would be required to absorb the CO_2 emissions released by these nations. It is important to note that through the inclusion of the energy land category, the Ecological Footprint aggregates actual and hypothetical land areas. Ecological Footprint calculations have been carried out for almost all countries of the world by the Global Footprint Network (WWF et al. 2014).

Finally, the Carbon Footprint on the product level assesses greenhouse gas emissions (CO_2 and other GHGs) throughout the complete supply chain of goods and services consumed in a region or country utilising a lifecycle approach (normally measured in grams or kilograms of CO_2 equivalents) (Wiedmann and Minx 2007). The methodology underlying the Carbon Footprint of products has been standardised in the ISO 14067.2 standard (ISO 2012). Carbon Footprints related to the total consumption of the inhabitants of countries including the GHG emissions embodied in imported and exported products have been calculated by a number of different research groups (Hertwich and Peters 2009; Steen-Olsen et al. 2012; Tukker et al. 2014).

8.3 Criteria for Deriving a Set of Resource Use Indicators

A set of resource use indicators should comply with a number of criteria (for a comprehensive discussion on criteria for deriving indicator sets, see Giljum et al. 2009) to ensure its applicability. The most important criteria shall be discussed below and will be reflected in the suggestion of the indicator set in Sect. 8.4.

8.3.1 Comprehensive and Complementary Coverage of All Relevant Resource Use Categories

A measurement and indicator system should account for all relevant categories of resource use and must ensure that possible shifts of environmental pressures between different types of resources can be identified and illustrated. This criterion calls for a set of indicators, which covers the different resource use aspects and delivers directionally safe information on the amounts of different types of resources used by human production and consumption activities.

8.3.2 Policy Relevance

One crucial criterion is the policy relevance of a measurement system and derived set of resource use indicators. Such a set should enable monitoring and evaluating macro policies (e.g. the implementation of a tax reform, border adjustments or licence trading systems to increase energy and resource productivity) as well as more specific (sectoral and cross-sectoral) policies related to resource use (e.g. energy, transport, trade, agriculture policies). This implies that it should be possible to disaggregate indicators by economic sectors.

8.3.3 Easy Communication

Resource use indicators should be easy to communicate in order to provide relevant information not only to experts but to a large number of policy makers as well as actors from civil society. They should show whether a country or world region is moving towards reductions in natural resource use and increases in resource productivity; or whether a given product, process or technology is using resources more efficiently than a comparable one or has some indirect ecological drawbacks. An example for an environmental issue including indirect drawbacks is the debate on biofuels. Substitution of fossil fuels by biofuels might improve the CO_2 performance, but can result in significant increases in the demand of water and land (UNEP 2009).

8.3.4 Application of a Life Cycle Perspective

Any resource measurement system that is intended to support decisions at a national, sectoral or product level should apply a life cycle perspective. This requires including the resource use along the whole life cycle of a product, i.e. in the production chain, during the use of the product as well as in waste treatment and recycling. In studies on natural resource use at the national level, a life-cycle perspective implies taking into account the indirect resource requirements of imported and exported products,[2] in order to capture possible shifts of environmental pressures related to domestic production or consumption to other countries and world regions (Giljum et al. 2013).

[2] Indirect flows of imported products have also been termed upstream or embodied flows.

8.3.5 Avoiding Double Counting

Resource use indicators should – to the extent possible – be additive across products, sectors and countries. Applying this principle restrictively, indicators could either reflect domestic environmental pressures related to domestic production (such as indicators on domestic material or water extraction).[3] Or they could be constructed according to a consumption principle, where resource requirements are *allocated to final consumption*; examples from MFA include the indicator Raw Material Consumption (RMC).

8.3.6 Measuring Resource Use at Different Scales

Depending on the issue of concern, measurements of resource use should be applicable at different levels of economic activities. Assessments at the micro level focus on resource use of products and organizations and can be related to monetary values, such as the price of a product or the material and energy costs spent by a company. Resource use and resource productivity of specific economic sectors (mining, chemicals, iron and steel, etc.) is assessed at the meso level. Economy-wide studies measure resource use and resource productivity of sub-national regions,[4] countries or world regions (macro level). Measurement systems and derived indicators of resource use and resource productivity should be designed in a consistent manner across these different scales (see also Bringezu et al. 2009). It should therefore be possible to consistently aggregate or disaggregate resource use indicators from products via sectors to countries. For example, macro-indicators on resource consumption related to final demand in one country should, conceptually, equal the sum of all final products and services consumed in that country.

8.3.7 Compatibility and Consistency with the System of National Accounts

A system for measuring resource use should be compatible with the economic System of National Accounts (SNA) (Radermacher 1992) as implemented in the UN SEEA system (United Nations 2012b) or the European NAMEA (National Accounting Matrix including Environmental Accounts) approach (EUROSTAT 2008) and ESEA (see above). This allows a consistent analysis of the interaction

[3] This would follow a production-oriented accounting principle, as it is, for example, implemented in the Kyoto protocol.

[4] Regional studies are sometimes also termed applications on the meso level.

between the economy and the environment and the assessment of the environmental implications of different patterns of production and consumption.

8.4 A Consistent Set of Resource Use Indicators

Taking into account the criteria listed in the previous section, the following set of complementary resource use categories and related indicators can be derived. The set covers the core resource input categories of materials, water and land plus the output category of GHG emissions, since GHG emissions represent the largest outflow from the economic system back to the environment causing climate change, which is generally regarded as the most pressing environmental problem.

In Table 8.1, we illustrate the suggested set of indicators for two levels: the product level and the national level. It is possible to cover also other levels of activities with this indicator set (for example, companies, economic sectors, etc.). It shall be emphasised that all indicators take a life cycle perspective (also termed footprint perspective), thus capturing possible shifts of environmental pressures related to domestic production or consumption to other countries and world regions. For the assessment of resource use on product level, the life cycle approach comprises all (direct and embodied) resources used from cradle to grave. The suggested methods and indicators are therefore oriented towards Life Cycle Assessment / LCA, a standardised method to assess resource use and the related environmental impacts along products life cycles from cradle to grave/cradle (UNEP 2003).

For the categories of biotic and abiotic materials on the product level, the concept of material input is suggested, for instance following the calculation guidelines of the MIPS concept (see above; Saurat and Ritthoff 2013). Assessment of material inputs of products should always separate biotic and abiotic materials, as they have very different economic and environmental implications. On the macro level, we suggest using material footprint-type indicators such as Raw Material Consumption (RMC; see above) as the main headline indicator. Again, calculations of MFA-based

Table 8.1 The suggested system of resource use indicators on the product and the national level

Resource use category	Product level	National level
Materials	Material footprint of products	Material footprint (RMC) of countries
Water	Water footprint of products	Water footprint of countries
Land area	Land footprint of products	Land footprint of countries (including land embodied in imports and exports)
GHG emissions	Carbon footprint of products	Carbon footprint of countries (including GHG emission embodied in imports and exports)

Source: adapted from Giljum et al. (2011)

indicators such as RMC should separately illustrate biotic and abiotic materials, with the latter further split up into the main categories of fossil fuels, metal ores and industrial/construction minerals. Applying the RMC indicator allows aggregation across countries without double counting, and incorporation of indirect flows from product import and export, unlike the simpler Domestic Material Consumption (DMC) which does not incorporate these indirect flows. DMC is already part of the EU indicator set underlying the "Roadmap to a resource efficient Europe" (European Commission 2011), while RMC is the targeted indicator for the future (European Commission 2014).

Regarding the category of water use and water consumption we suggest applying the concept of water footprinting, encompassing not only water input (withdrawal) but also water consumption (input minus output) as well as a separation between blue and green water. This concept is in general applicable for both the micro as well as the macro level. However, as stated above, data availability particularly regarding data on green water can be a constraint and require a reduced approach. In such cases, we suggest focusing on blue water, i.e. accounting for all water inputs which were withdrawn from groundwater, deep groundwater and surface water.

The land footprint of products reflects the life cycle-wide demand on land area for the production of goods or services. National land use inventories allow illustrating the state and changes of land use on a specific territory. Again it is necessary to consider the upstream land use of imported and exported products in order to calculate a national land use indicator from a consumption perspective. For a review of available land footprint accounting methods see Bruckner et al. (2015). As with the material indicator, it is recommended to report major categories of land use separately, including particularly cropland, grassland (i.e. permanent pastures and meadows), forestland, and built-up land.

The category of GHG emissions refers to the concept for calculating carbon footprints, a life cycle-wide inventory of emissions of different GHGs at the product level. On the national level, the current system of Kyoto GHG inventories represents a production (or territory) accounting principle. Also regarding this category, consumption-based indicators can be calculated through considering GHG emissions embodied in internationally-traded products.

8.4.1 *Explanation for Suggesting this Set of Indicators*

8.4.1.1 Complementary Coverage of Different Resource Use Aspects

Key criteria in the evaluation were the coverage of all relevant categories of resource use in order to monitor shifts of environmental pressure and a well-founded basis for policy making and target setting. These criteria can be better fulfilled by applying a set of indicators instead of only one indicator (e.g. Carbon Footprint). A set of indicators covers resource use in a complementary manner and allows setting resource-specific targets and evaluating specific resource policies. This approach has also been applied in the original Environmental Space (ES) studies in the 1990s, which assessed ES separately in different categories of resource use (non-renewable raw materials, wood, energy, water, land use) (for example, Spangenberg 1995). The suggested set of indicators avoids counting the same resources twice; with the exception of inputs of fossil fuels producing CO_2 and biotic material inputs, which produce GHG emissions other than CO_2, which are accounted in the material flow indicators and the carbon footprint.

8.4.1.2 Resource Use Indicators as the Basis and Complements to Environmental Impact Indicators

In a world which increasingly faces limits of ecosystem capacities and resource scarcities, reducing the amounts of used natural resources becomes a key determining factor for a sustainable global development. This set of indicator deals with the issue of the overall scale of the human production and consumption system and thus differs from environmental impact indicators, which are less suitable to address problems stemming from the overall scale of economic activities. The suggested indicators thus point to the need of reduction rather than substitution of specific environmentally harmful materials and substances.

Current policy strategies for achieving sustainable resource use in Europe, such as the EU Roadmap for a resource efficient Europe (European Commission 2011) define as their overall objective to decouple environmental impacts of resource use from GDP. To achieve this objective, a strategy of "double decoupling" should be implemented: decoupling GDP from resource use amounts and decoupling resource use amounts from the generated environmental impacts. On the country level, Indicators on resource use quantities as suggested in this paper show some correlation with the overall environmental impacts related to resource use. However, on the level of single materials, this correlation does not hold, as different materials have very different impacts per kilogram (van der Voet et al. 2005).

Sets of indicators have been suggested to measure these environmental impacts. For example, Best and colleagues (2008) suggested a basket of four indicators for monitoring the EU Resource Strategy: the Ecological Footprint illustrating the impacts on biocapacity and (global) carrying capacity, Environmentally-weighted Material Consumption (EMC) reflecting the specific environmental impacts of materials and products, Human Appropriation of Net Primary Consumption (HANPP) indicating the intensity of ecosystem use and Land and Ecosystem Accounts (LEAC) illustrating the drivers for land cover and land use changes, which have implications for biodiversity and ecosystem services. Other impact-oriented indicators of resource use are currently being developed, for example by the Joint Research Center of the European Commission (JRC 2010, 2012).

The indicator set suggested in this chapter complements such baskets of impact indicators through providing the information on the underlying volumes; in fact, in several cases, the indicator set is the physical basis for properly calculating such impact indicators. For example, solid accounts of material consumption of products or countries are among the main data bases for calculating the Ecological Footprint or the EMC of countries.

8.4.1.3 Strong Link to the Statistical System

We suggest including measurement methods and indicators, which have a strong link to the statistical system on the Member State and EU level. The example of MFA-based indicators illustrates that indicators with a solid statistical background are more accepted in policy spheres as are other indicators, which have been developed outside the statistical system of environmental accounting. The set of indicators therefore includes indicators, which can be calculated on the basis of real statistical data and do not require transformation and modelling of data.

8.4.1.4 The Indicator Set and the Ecological Footprint

The Ecological Footprint is the most prominent example for a macro environmental indicator, which integrates land and energy use and carbon emissions primarily accounted in different units (such as kilograms, kilojoules and square meters). Macro environmental indicators such as the Ecological Footprint enable an easier communication of overall results, as a large number of complex interrelations between the economy and the environment are illustrated in easily understandable terms. However, at the same time, this approach entails a number of important disadvantages, which shall be discussed using the example of the Ecological Footprint:

- It is difficult to cover all resource categories with only one indicator. In the case of the Ecological Footprint, GHG emissions other than CO_2 are currently not accounted and abiotic materials are only indirectly accounted through the demand for energy and land for extraction and processing.
- Strong assumptions need to be applied, in order to transform different types of primary data (e.g. material flows, land use, CO_2 emissions) into one common unit of calculation. For example, the Ecological Footprint transforms CO_2 emissions into forest areas required to absorb the CO_2 emissions. This approach is being criticised, as the sequestration does not actually happen (see Best et al. 2008 for a summary of this critique).

We therefore suggest measuring and illustrating different aspects of resource use in the original units (e.g. material consumption and carbon emissions in tonnes, water use in litres, land use in hectares), without transforming them into a single artificial unit of measurement.

8.4.2 Research Needs for the Suggested Indicators

8.4.2.1 Further Development of Accounting Standards

Some of the accounting methods underlying the suggested set of indicators already exist in an internationally standardised format: this holds true for material flow-based indicators on the product and the country level, the Carbon Footprint or Kyoto inventories of GHGs. The accounting method for other categories, in particular for water and land are currently being developed. As the measurement systems covering different types of natural resources have been developed separately, further methodological harmonisation is still required, in order to improve the comparability of the results. This requires in particular defining common system boundaries for accounting resource use.

8.4.2.2 Improving Data Availability

Data on material consumption in the EU countries is already collected by EUROSTAT (2014) and also calculations on the product level exist for a variety of products (for instance, Ritthof et al. 2002; Saurat and Ritthoff 2013). Comprehensive water accounts do not exist for a large number of countries. Hence, water footprint calculations based on real data can be carried out only to a certain extend or using modelled data as in the case of the Water Footprint indicator (see the case studies at www.waterfootprint.org). Data on actual land cover and land use is available for Europe through the European Environment Agency (EEA 2006), while data on land demand of products is very patchy, with the exception of biomass products, for which the UN Food and Agricultural Organisation maintains a data base (see http://faostat.fao.org). Since 1990 data on the territorial GHG emissions of countries are available from United Nations Framework Convention on Climate Change (UNFCCC), provided in the annual GHG inventory submissions by Annex I Parties and in the national communications under the Convention by non-Annex I Parties. First data sets on the Carbon Footprint of countries have also been presented (see www.carbonfootprintofnations.com).

8.4.2.3 Defining Sustainability Limits

The identification of sustainability limits for each of the resource use categories is one of the key issues for further development of the suggested set of indicators. For GHG emissions, a per-capita target of around two tonnes of CO_2 (equivalents) per inhabitant has been formulated by the IPCC and taken up by policy initiatives, such as the European Union's strategy for tackling climate change (European Commission 2015). Additional targets for other types of natural resources need to be defined on a scientific basis (see Bringezu 2015, for an approach to define targets for sustainable material consumption). These targets could refer to the maximum amount of biomass extraction from a given area of crop land and forests or the maximum (sustainable) uptake of fresh water, given the limited capacity for water renewal. A system of limits for each of the categories is crucial to evaluate trade-offs between different (resource) policy options (see below).

8.4.3 Applying the Set of Indicators in Practice

8.4.3.1 Illustration of Data in Aggregated and Disaggregated Form

The suggested indicators can be applied as an aggregated number per resource category (headline indicators), but also be disaggregated into components, such as different abiotic materials in the material flow-based indicators or different categories of land areas (cropland, forestland, built-up land, etc.). Disaggregation is often necessary, in order to link the resource use indicators closer to specific environmental problems and ensure a proper evaluation of results and trends. Also the links to related impact indicators can be established more easily on a disaggregated level. Such environmental problems include e.g. extinction of native species, reduction of biodiversity, etc. as a consequence of the expansion of built-up land for transport infrastructure or the expansion of agricultural land for production of bioenergy and biomaterials.

8.4.3.2 Consideration of the Regional/Local Context in the Interpretation of the Indicators

When interpreting resource use indicators, the regional or local context should be taken into account to the extent possible. In particular indicators on water use depend critically on the local or regional availability of renewable water; a certain water footprint of a product can be problematic in one region, but sustainable in another. Approaches such as the Water Stress Index (Pfister et al. 2009) combine the appropriation and the impact aspect of quantitative water use. Also, the Water Footprint Manual (Hoekstra et al. 2011) developed methodological approaches to calculate the scarcity of green and blue water at the level of river basins by comparing the green and blue water footprint with the availability of green and blue water in the river basin. However, due to limited water availability data at the river basin level, such calculations are still difficult to be performed in practice.

8.4.3.3 Analysing Trade-Offs

In a system of indicators illustrating the different types of resource use plus sustainability limits for each of the categories, trade-offs between different options can be analysed. For example, higher production of biofuels would likely decrease the abiotic resource indicator (less fossil fuels) and, depending on the type of biofuels, also the related GHG emissions. On the other hand, this would translate into increased demand for land and water. The set of indicators and related limits can illustrate, whether an improvement in one category leads to an unsustainable situation in another category.

8.4.3.4 Production Versus Consumption Indicators

In the traditional environmental accounting frameworks (such as the one applied in the Kyoto protocol), environmental pressures are accounted according to a territory principle (production principle), i.e. accounted where the pressures occur. In addition, a consumption perspective is necessary to illustrate the global environmental pressures related to the final consumption of goods and services by a given population. However, methods to account for the global resource use related to consumption in one country are still under development and refinement due to the lack of data on resources embodied in international trade (see above).

8.4.3.5 Policy-Oriented Application

On the EU level, the suggested measurement system already plays an important role in the implementation of the EU Roadmap for a resource efficient Europe (European Commission 2011), where a dashboard of macro-indicators on carbon, materials, water and land is foreseen to monitor European resource use both in a territorial and a global perspective. As discussed before, the indicator set can also help setting concrete targets for different types of resources. Empirical evidence generated with this indicator set should also support the statement that a substantial reduction of the negative environmental impacts related to resource use will need to address also the issue of overall *levels* of resource use in Europe. The indicator set could also play a role in EU impact assessments and sustainability impact assessments, when a number of policy options are compared. Other policy areas, where this indicator set could be useful are green public procurement, structural and cohesion funds and development aid.

8.5 Conclusions

This chapter described a comprehensive set of quantitative resource use indicators, which addresses primarily environmental problems related to the overall scale of production and consumption in Europe and its global implications. The suggested set of indicators therefore focuses on the absolute amounts of resource use instead of specific environmental impacts. For most of the indicators, data and methodologies are already available for both products and countries. The suggested set of indicators can therefore be implemented already now. However, resources should be devoted to improve data availability particularly for indicators related to land and water use as well as for natural resources embodied in internationally traded products. In order to allow a proper evaluation of these indicators and resulting necessary trade-offs between them from a sustainability point of view, the identification of sustainability limits for each of the different resource categories should be a high priority in the near future. The set of indicators should then feed into a number of EU and international policy processes and help to better assess the impact of policies on natural resource use.

References

Adriaanse A, Bringezu S, Hamond A, Moriguchi Y, Rodenburg E, Rogich D, Schütz H (1997) Resource flows: the material base of industrial economies. World Resource Institute, Washington

Best A, Giljum S, Simmons C, Blobel D, Lewis K, Hammer M, Cavalieri S, Lutter S, Maguire C (2008) Potential of the ecological footprint for monitoring environmental impacts from natural resource use: analysis of the potential of the ecological footprint and related assessment tools for use in the EU's thematic strategy on the sustainable use of natural resources. Report to the European Commission, DG Environment, Brussels

Borucke M, Moore D, Cranston G, Gracey K, Iha K, Larson J, Lazarus E, Morales JC, Wackernagel M, Galli A (2013) Accounting for demand and supply of the biosphere's regenerative capacity: the national footprint accounts' underlying methodology and framework. Ecol Indic 24:518–533

Boulay A-M, Motoshita M, Pfister S, Bulle C, Muñoz I, Franceschini H, Margni M (2015) Analysis of water use impact assessment methods (part A): evaluation of modeling choices based on a quantitative comparison of scarcity and human health indicators. Int J Life Cycle Assess 20:139–160

Bringezu S (2015) Possible target corridor for sustainable use of global material resources. Resources 4:25–54

Bringezu S, Schütz H, Moll S (2003) Rationale for and interpretation of economy-wide materials flow analysis and derived indicators. J Ind Ecol 7:43–64

Bringezu S, Sand IVD, Schütz H, Bleischwitz R, Moll S (2009) Analysing global resource use of national and regional economies across various levels. In: Bleischwitz R, Bringezu S (eds) Sustainable resource use. Greenleaf Publishing, Sheffield, pp 10–51

Bruckner M, Fischer G, Tramberend S, Giljum S (2015) Measuring telecouplings in the global land system: a review and comparative evaluation of land footprint accounting methods. Ecol Econ 114:11–21

DESTATIS (2013) Umweltnutzung und Wirtschaft. Bericht zu den Umweltökonomischen Gesamtrechnungen. Statistisches Bundesamt, Wiesbaden

EEA (2006) Land accounts for Europe 1990–2000. Towards integrated land and ecosystem accounting. European Environment Agency, Copenhagen

EEA (2010) The european environment. State and outlook 2010. Land use. European Environment Agency, Copenhagen

EEA (2015) The european environment. State and outlook 2015. Synthesis report. European Environment Agency, Copenhagen

ESSC (2014) European strategy for environmental accounts. ESEA 2014. European Statistical System Committee, Luxembourg

European Commission (2011) Roadmap to a resource efficient Europe. European Commission, Brussels

European Commission (2014) Analysis of an EU target for resource productivity. European Commission, Brussels

European Commission (2015) The Paris protocol – a blueprint for tackling global climate change beyond 2020. European Commission, Brussels

EUROSTAT (2001) Economy-wide material flow accounts and derived indicators. A methodological guide. Statistical Office of the European Communities, Luxembourg

EUROSTAT (2008) Revised European strategy for environmental accounting. Statistical Office of the European Communities, Luxembourg

EUROSTAT (2010) LUCAS – a multi-purpose land use survey. Statistical Office of the European Communities, Luxembourg

EUROSTAT (2013) Economy-wide Material Flow Accounts (EW-MFA). Compilation guide 2013. Statistical Office of the European Communities, Luxembourg

EUROSTAT (2014) Material flow accounts. Statistical Office of the European Communities, Luxembourg. Available at http://appsso.eurostat.ec.europa.eu/nui/show.do?dataset=env_ac_mfa&lang=en

Fischer-Kowalski M, Krausmann F, Giljum S, Lutter S, Mayer A, Bringezu S, Moriguchi Y, Schütz H, Schandl H, Weisz H (2011) Methodology and indicators of economy-wide material flow accounting. J Ind Ecol 15:855–876

Flörke M, Kynast E, Bärlund I, Eisner S, Wimmer F, Alcamo J (2013) Domestic and industrial water uses of the past 60 years as a mirror of socio-economic development: a global simulation study. Glob Environ Chang 23:144–156

Giljum S, Hak T, Hinterberger F, Kovanda J (2005) Environmental governance in the European Union: strategies and instruments for absolute decoupling. Int J Sustain Dev 8:31–46

Giljum S, Hinterberger F, Lutter S, Polzin C (2009) How to measure Europe's resource use. An analysis for friends of the Earth Europe. Sustainable Europe Research Institute, Vienna

Giljum S, Burger E, Hinterberger F, Lutter S, Bruckner M (2011) A comprehensive set of resource use indicators from the micro to the macro level. Resour Conserv Recycl 55:300–308

Giljum S, Lutter S, Bruckner M, Aparcana S (2013) State of play of national consumption-based indicators. A review and evaluation of available methods and data to calculate footprint-type (consumption-based) indicators for materials, water, land and carbon. Report for DG Environment of the European Commission. Sustainable Europe Research Institute (SERI), Vienna

Giljum S, Dittrich M, Lieber M, Lutter S (2014) Global patterns of material flows and their socio-economic and environmental implications: a MFA study on all countries world-wide from 1980 to 2009. Resources 3:319–339

Giljum S, Martinez A, Bruckner M (2015) Material footprint assessment in a global input–output framework. J Ind Ecol 19(5):792–804

Haberl H, Fischer-Kowalski M, Krausmann F, Weisz H, Winiwarter V (2004) Progress towards sustainability? What the conceptual framework of material and energy flow accounting (MEFA) can offer. Land Use Policy 21:199–213

Heinberg R (2007) Peak everything: waking up to the century of declines. New Society Publishers, Gabriola Island

Hertwich EG, Peters GP (2009) Carbon footprint of nations: a global, trade-linked analysis. Environ Sci Technol 43:6414–6420

Hoekstra AY, Mekonnen MM (2012) The water footprint of humanity. Proc Natl Acad Sci 109:3232–3237

Hoekstra AY, Chapagain AK, Aldaya M, Mekonnen M (2009) Water footprint manual. State of the art 2009. Water Footprint Network, Enschede

Hoekstra AY, Chapagain AK, Aldaya MM, Mekonnen MM (2011) The water footprint assessment manual. Setting the global standard. Earthscan, London/Washington D.C

Huijbregts MAJ, Rombouts LJA, Hellweg S, Frischknecht R, Hendriks AJ, van de Meent D, Ragas AMJ, Reijnders L, Struijs J (2005) Is cumulative fossil energy demand a useful indicator for the environmental performance of products? Environ Sci Technol 40:641

ISO (2012) Carbon footprint of products – requirements and guidelines for quantification and communication. ISO/DIS 14067.2. International Organization for Standardization
JRC (2010) Decoupling indicators, basket-of-products indicators, waste management indicators: framework, methodology, data basis and updating procedures. European Commission Joint Research Centre, Institute for Environment and Sustainability, Ispra
JRC (2012) Life cycle indicators for resources, products and waste. European Commission Joint Research Centre, Institute for Environment and Sustainability, Ispra
Kastner T, Rivas MJI, Koch W, Nonhebel S (2012) Global changes in diets and the consequences for land requirements for food. Proc Natl Acad Sci 109:6868–6872
Kneese AV, Ayres RU, d'Arge RC (1970) Economics and the environment: a material balance approach. John Hopkins Press, Baltimore/London
Krausmann F, Weisz H, Eisenmenger N, Schütz H, Haas W, Schaffartzik A (2015) Economy-wide material flow accounting. Introduction and guide. Social Ecology Working Paper. Institute for Social Ecology, Vienna
Lutter S, Giljum S (2014) Demand-based measures of material flows. A review and comparative assessment of existing calculation methods and data options. OECD / Working Party on Environmental Information, Paris
Matthews E, Bringezu S, Fischer-Kowalski M, Huetller W, Kleijn R, Moriguchi Y, Ottke C, Rodenburg E, Rogich D, Schandl H, Schuetz H, van der Voet E, Weisz H (2000) The weight of nations. Material outflows from industrial economies. World Resources Institute, Washington
Mekonnen M, Hoekstra A (2013) Water footprint benchmark for crop production. A first global assessment. Ecol Indic 46:214–223
OECD (2007) Measuring material flows and resource productivity. The OECD guide environment directorate. Organisation for Economic Co-operation and Development, Paris
Peters GP (2010) Carbon footprints and embodied carbon at multiple scales. Curr Opin Environ Sustain 2:245–250
Pfister S, Koehler A, Hellweg S (2009) Assessing the environmental impacts of freshwater consumption in LCA. Environ Sci Technol 43:4098–4104
Radermacher W, (1992) Konzept fuer eine umweltoekonomische Gesamtrechnung des Statistischen Bundesamtes. National German Statitical Agency
Ritthof M, Rohn H, Liedtke C (2002) Calculating MIPS. Resource productivity of products and services. Wuppertal Institut, Wuppertal
Ruini L, Marino M, Pignatelli S, Laio F, Ridolfi L (2013) Water footprint of a large-sized food company: the case of Barilla pasta production. Water Resour Ind 1:7–24
Saurat M, Ritthoff M (2013) Calculating MIPS 2.0. Resources 2:581–607
Schandl H, West J (2010) Resource use and resource efficiency in the Asia–Pacific region. Glob Environ Chang 20:636–647
Schandl H, Grünbühel C, Haberl H, Weisz H (2002) Handbook of physical accounting – measuring bio-physical dimensions of socio-economic activities MFA – EFA – HANPP. Institute for Social Ecology, Vienna
Schmidt-Bleek F (1992) MIPS – a universal ecological measure. Fresenius Environ Bull 2:407–412
Schmidt-Bleek F, Bringezu S, Hinterberger F, Liedtke C, Spangenberg J, Stiller H, Welfens M (1998) MAIA. Einführung in die Materialintensitätsanalyse nach dem MIPS-Konzept. Birkhauser, Berlin, Basel
Seppelt R, Manceur AM, Liu J, Fenichel EP, Klotz S (2014) Synchronized peak-rate years of global resources use. Ecol Soc 19:50
Solomon S, Qin D, Manning M, Chen Z, Marquis M, Averyt K, Tignor M, Miller H (2007) Climate change 2007: the physical science basis. Contribution of working group I to the fourth IPCC assessment report. Intergovernmental Panel on Climate Change
Spangenberg J (1995) Sustainable Europe. The study. Friends of the Earth, Brussels
Steen-Olsen K, Weinzettel J, Cranston G, Ercin AE, Hertwich EG (2012) Carbon, land, and water footprint accounts for the European Union: consumption, production, and displacements through international trade. Environ Sci Technol 46:10883–10891

Tukker A, Bulavskaya T, Giljum S, de Koning A, Lutter S, Simas M, Stadler K, Wood R (2014) The global resource footprint of nations. Carbon, water, land and materials embodied in trade and final consumption, Leiden/Delft/Vienna/Trondheim
UNEP (2003) Evaluation of environmental impacts in life cycle assessment. United Nations Environment Programme
UNEP (2009) Towards sustainable production and use of resources: assessing biofuels. International Panel for Sustainable Resource Management
UNEP (2012) Global environmental outlook 5. United Nations Environment Programme, Nairobi
United Nations (2012a) SEEA-W. System of Environmental-Economic Accounting for Water. United Nations Statistics Division
United Nations (2012b). System of environmental-economic accounting. Central Framework. United Nations Statistical Division, New York
van der Voet E, van Oers L, Moll S, Schütz H, Bringezu S, de Bruyn S, Sevenster M, Warringa G (2005) Policy review on decoupling: development of indicators to assess decoupling of economic development and environmental pressure in the EU-25 and AC-3 countries. EU Commission, DG Environment, Brussels
van der Voet E, van Oers L, de Bruyn S, de Jong F, Tukker A (2009) Environmental impact of the use of natural resources and products. CML, Leiden
Weinzettel J, Hertwich EG, Peters GP, Steen-Olsen K, Galli A (2013) Affluence drives the global displacement of land use. Glob Environ Chang 23:433–438
West J, Schandl H (2013) Material use and material efficiency in Latin America and the Caribbean. Ecol Econ 94:19–27
Wiedmann T, Minx J (2007) A definition of 'Carbon Footprint'. ISA/UK Research & Consulting, Durham
Wiedmann T, Schandl H, Lenzen M, Moran D, Suh S, West J, Kanemoto K (2015) The material footprint of nations. Proc Natl Acad Sci 112:6271–6276
WU (2015) Global material flow database. Material extraction data. Technical report, Version 2015.1. Vienna University of Economics and Business (WU), Vienna. www.materialflows.net
WWF, Global Footprint Network, Water Footprint Network, Zoological Society of London (2014) Living planet report 2014. Species and spaces, people and places. WWF, Gland, Switzerland
Yu Y, Feng K, Hubacek K (2013) Tele-connecting local consumption to global land use. Glob Environ Chang 23:1178–1186

Chapter 9
Regenerative Cities

Herbert Girardet

Abstract The concept of regenerative cities is seeking to address the relationship between cities and their hinterland, and beyond that with the more distant territories that supply them with water, food, timber and other vital resources. We need to re-enrich the landscapes on which cities depend, and this includes measures to increase their capacity to absorb carbon emissions. Creating a restorative relationship between cities, their local hinterland and the world beyond, means harnessing new opportunities in financial, technology, policy and business practice. This text argues that the established horizon of urban ecology should be expanded to include all the territories involved in sustaining urban systems. Urban regeneration thus takes on the meaning of eco-regeneration.

Keywords Urban • Sustainability • Regeneration • City • Region

9.1 Introduction

At the start of the twenty first century, humanity is becoming a predominantly urban species and this historic development represents a fundamental, systemic change in the relationship between humans and nature. Urban-based economic activities account for 55 % of GNP in the least developed countries, 73 % in middle income countries and 85 % in the most developed countries (UN Habitat 2006).

Modern cities, then, are defined by the concentration of economic activities and intense human interaction. This is reflected in high average levels of personal consumption and the efficient supply of a great variety of services at comparatively low per-capita costs. But the environmental impacts of an urbanising humanity are a great cause for concern. Apart from a near monopoly on the use of fossil fuels, metals and concrete, an urbanizing humanity now consumes nearly *half* of nature's annual photosynthetic capacity as well.

H. Girardet (✉)
The Club of Rome, Winterthur, Switzerland

University of the West of England, Bristol, UK
e-mail: herbie@easynet.co.uk

S. Shmelev (ed.), *Green Economy Reader*, Studies in Ecological Economics 6,
DOI 10.1007/978-3-319-38919-6_9

Since the industrial revolution the process of urbanisation has become ever more resource-intensive, and it significantly contributes to climate change, loss of soil carbon, natural fertility of farmland, and the loss of biodiversity all over the world. The ravenous appetite of our fossil-fuel powered lifestyles for resources from the world's ecosystems has severe consequences for all life on Earth, including human life.

Cities have developed resource consumption and waste disposal habits that show little concern for the consequences. Addressing this issue is the primary task of this paper.

The larger and the richer the city, the more it tends to draw on nature's bounty from across the world rather than its own local hinterland. Human impacts on the world's ecosystems and landscapes are dominated by the ecological footprints of cities which now stretch across much of the Earth. They can be hundreds of times larger than the cities themselves. In an urbanising world, cities need to rapidly switch to renewable energy and to actively help restore damaged ecosystems.

The WWF states in its Living Planet Reports that in the last 30 years a third of the natural world has been obliterated (WWF 2010). 40–50 % of Earth's ice-free land surface has been heavily transformed or degraded by human activities, 66 % of marine fisheries are either overexploited or at their limit and atmospheric CO_2 has increased more than 30 % since the advent of industrialisation (Vitousek et al. 1997). Helping to reverse this collision course between humans and nature is a new challenge for most national politicians, but even more for urban politicians, planners and managers, and for architects, civil engineers and city dwellers.

The challenge today is no longer just to create sustainable cities but truly *regenerative* cities: to assure that they do not just become resource-efficient and low carbon emitting, but that they positively enhance rather than undermine the ecosystem services they receive from beyond their boundaries. A wide range of technical and management solutions towards this end are already available, but so far implementation has been too slow and too little.

Most importantly, the transformative changes that are required to make cities regenerative call for far-reaching strategic choices and long-term planning as compared to the short-term compromises and patchwork solutions that characterise most of our political decision making systems at all spheres of government.

In recent years there has been a proliferation of *urban regeneration* initiatives focussed on the health and wellbeing of urban citizens and the urban fabric – the 'inner-urban environment' – particularly in rich countries such as Britain, Germany and the USA. Such initiatives have received much funding and media attention, and they have improved the lives of millions of people. In various countries *Urban Regeneration Associations* have been established to address problems such as deindustrialisation, depopulation, congestion, aging infrastructure, run-down sink estates and associated matters.

But the concept of *regenerative cities* goes further – seeking to address the relationship between cities and their hinterland, and beyond that with the more distant territories that supply them with water, food, timber and other vital resources. We need to re-enrich the landscapes on which cities depend, and this includes measures to increase their capacity to absorb carbon emissions. Creating a *restorative relationship* between cities, their local hinterland and the world beyond, means harnessing new opportunities in financial, technology, policy and business practice.

This text argues that the established horizon of *urban ecology* should be expanded to include all the territories involved in sustaining urban systems. *Urban regeneration* thus takes on the meaning of *eco-regeneration.*

Creating regenerative cities thus primarily means one thing: *Initiating comprehensive political, financial and technological strategies for an environmentally enhancing, restorative relationship between cities and the ecosystems from which they draw resources for their sustenance.*

9.2 Cities as Ecological and Economic Systems

Towns and cities need sustenance for their people and this requires elaborate ecological and economic systems (Fig. 9.1).

In his book 'The Isolated State' the prominent nineteenth century economist Johann Heinrich von Thünen described the way in which human settlements, in the absence of major transport systems, are systemically tied into the landscape surrounding them through various logically arranged modes of cultivation.[1] In fact, they have an active, symbiotic relationship with it: they also assure its continuing

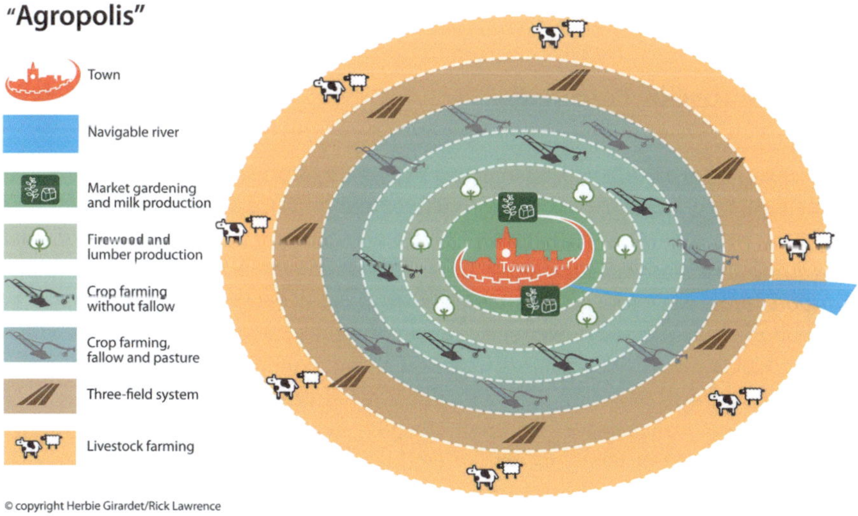

Fig. 9.1 Agropolis

[1] en.wikipedia.org/wiki/Johann_Heinrich_von_Thünen

productivity and fertility by returning appropriate amounts of organic waste to it. In this text I have chosen to use the term 'Agropolis' for this traditional type of settlement system.

Von Thünen pioneered the view that the way cultivated land in close proximity to towns and cities is utilised is a logical function of two interconnected variables – the cost of transporting produce to market, and the land rent a farmer can afford to pay. He describes how isolated communities are surrounded by concentric rings of varying land uses. Market gardens and milk production are located closest to the town since vegetables, fruit and dairy products must get to market quickly. Timber and firewood, which are heavy to transport but essential for urban living, would be produced in the second ring. The third zone consists of extensive fields for producing grain which can be stored longer and can be transported more easily than dairy products, and can thus be located further from the city. Ranching is located in the fourth zone since animals can be raised further away from the city because they are 'self-transporting' on their own legs. Beyond these zones lies uncultivated land of less economic relevance to urban living.

In many parts of the world traditional towns and cities, in the absence of efficient transport systems, had these kind of symbiotic relationships to the landscapes from which they emerged, depending on nearby market gardens, orchards, forests, arable and grazing land and local water supplies for their sustenance. Until very recently, many Asian cities were still largely self-sufficient in food as well as fertiliser, using human and animal wastes to sustain the fertility of local farms (King 1911). Can we learn from these traditional systems in the future whilst utilizing more up-to-date methodologies and technologies?

9.3 The Rise of Petropolis

The industrial revolution caused a virtual explosion of urban growth that continues to this day. Steam engine technology enabled the unprecedented concentration of industrial activities in urban centres. Cities increasingly cut the umbilical cord between themselves and their local hinterland and became global economic and transport hubs. This process has undermined local economies, as new modes of transportation have made it ever easier to supply food, raw materials and manufactured products from ever greater distances. Cities are no longer centres of *civilisation* but of *mobilisation*, with access to global resources as never before.

The phenomenal changes in human lifestyles made possible by the *Age of Fire* were also reflected in new concepts of land use planning, particularly for accommodating the road space needed for motor cars. The vast, low density urban landscapes that appeared in the USA, Australia and elsewhere are defined by the ubiquitous use of cars or *petromobiles* – the word *automobile* implies that they are self-powered which clearly they are not.

The modern city could be described as 'Petropolis': all its key functions – production, consumption and transport – are powered by massive injections of petroleum and other fossil fuels. But there is ever growing evidence that the resulting dependencies are ecologically, economically and geopolitically untenable, particularly because the fossil fuel supplies on which modern cities depend are, most definitely, finite.

Even though we know that we live on a finite planet, infinite economic and urban growth is still taken for granted. While the world's population has grown fourfold in the twentieth century, urban populations and global resource consumption have increased 16 fold and are still rising. It took around 300 million years for oil, gas and coal to accumulate in the earth's crust and we are on track to burn much of it in just 300 years – now at a rate of well over a million years per year. Cities are particularly responsible for this: despite taking up only 3–4 % of the world's surface area they use approximately 80 % of its resources and also discharge similar proportions of waste. These figures are still increasing.

The highly problematic patterns of fossil-fuel dependent urbanisation are still expanding across the world. Today urbanisation and economic and financial globalisation are closely connected. Cities have become globalised centres of production as well as consumption, with throughputs of unprecedented quantities of resources and industrial products being the norm in the wealthier countries. In emerging countries, too, urbanisation is closely associated with ever increasing per-capita use of fossil fuels and with impacts on ever more distant ecosystems. The rapid growth of cities such as Dubai with its vast airport, world record skyscrapers, artificial islands and low-density desert suburbs, is the latest and most astonishing example of this.

We are seeing ever more extraordinary contraptions appear across the face of the Earth to extract fossil fuels from the Earth's crust, to refine them and to deliver them into our cities and homes. With most of the 'easy' coal, oil and gas now used up, new kinds of highly problematic extraction methods have come to underpin the existence of our urban systems. Mountain top removal in places such as West Virginia has become the basis for ever larger scale open-cast coal mining operations. In Alberta, tar sand mining pollutes vast amounts of water that is used to melt the tar contained within the sands. Off-shore oil platform operators are now drilling as much as 10 km down into the E arth's crust in ever more hostile waters. Is this foolhardiness or the epitome of human ingenuity?

Modern cities have often been established on former forest and farmland. City people rely on a steady supply of natural resources from across the planet and consumers are often oblivious to the environmental consequences. Yet there is much evidence that urban resource consumption is fundamentally undermining ecosystems across the world on whose integrity cities ultimately depend.

And much of what goes in must come out again. Contemporary urban systems discharge vast quantities of solid, liquid and gaseous wastes. Where do they end up? We all have a vague idea that the *solid waste* we throw away is buried in landfills in the urban vicinity or may be trucked away to distant locations. But few of us know what is contained in the *liquid waste* we discharge from our homes and what ultimately happens to it.

And what about air pollution? In mega-cities such as Mexico City or Beijing people are still being forced to breathe horrendously polluted air. As long as people experience pollution directly as a local health problem they demand efforts to clean it up. But the detrimental effects of acid fumes such as sulphur and nitrogen oxides on forests and farm crops downwind from cities and power stations is outside most people's everyday experience.

And greenhouse gas emissions affecting the global climate imply a shift of concern from impacts on *human health* to impacts on *planetary health* which is much more difficult for us to face up to. And the global *ecological footprints* of our cities are an even more abstract concept, well beyond the personal experience of most citizens.

The challenge now is to insure that we will face up to the environmental impacts of urban living before they start to hit home in the form of health problems, higher food or energy prices, storms and sea level rises. Communicating the dangers of such *boomerang effects*, which could soon undermine the very existence of our modern cities, is a huge challenge for educators and policy makers.

9.4 Petropolis and Planetary Boundaries

The 'planetary boundaries' that are becoming evident in the face of global industrialisation, urbanisation and population growth have major implications for urban planning and governance. We must face up to the fact that cities are *dependent systems* whose reliance on external inputs for their sustenance is likely to become ever more precarious. The process of *entropication* – of combining resources into products and producing wastes faster than they can be converted back into useful resources – has to be dealt with by deliberate measures of policy and management. Our living planet cannot cope with the ever increasing accumulation and degradation of natural resources in our cities without appropriate measures being taken to replenish the global biosphere and to reduce our impacts on the atmosphere (Fig. 9.2).

A large part of the increase of carbon dioxide in the atmosphere is attributable to combustion in and on behalf of the world's cities. 200 years ago atmospheric CO2 concentrations were around 280 parts per million, but since then they have risen to 390 ppm. Until recently it was widely assumed that we could get away with doubling pre-industrial concentrations. But gradually it has become clear that this could cause the planet to overheat, with dire consequences for all life. Climatologists then gradually brought the target figures down from 550 to 450 ppm, particularly as they discovered the extent of warming that has already occurred in the Arctic Circle. Whilst global temperatures have increased by an average of 0.8 °C, in the Arctic they have gone up much more.

The Arctic regions appear to be exceedingly sensitive to anthropogenic CO_2 emissions. According to the Intergovernmental Panel on Climate Change (IPCC) "Arctic temperatures have increased at almost twice the global average rate in the last 100 years (…) Temperatures at the top of the permafrost layer have generally increased since the 1980s (…) by up to 3 °C"(IPCC 2008a). An increase in arctic

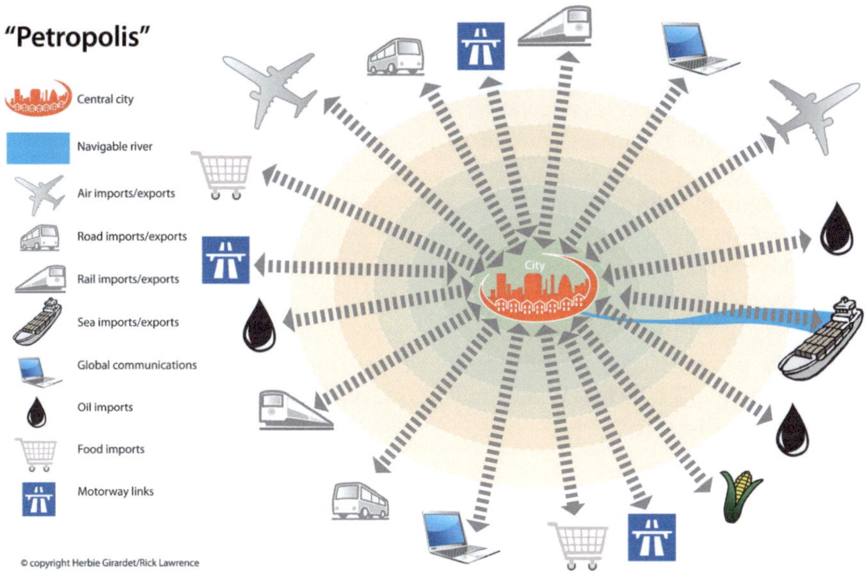

Fig. 9.2 Petropolis

temperatures could further accelerate greenhouse gas discharges into the atmosphere, particularly due to methane release from melting permafrost. This positive feedback loop could fuel global warming even more (IPCC 2008b).

In the Arctic, the rapid collapse of Greenland glaciers has become a particular focus of concern (WWF 2009). This is a major reason why many climatologists are now calling for an actual *reduction* of CO_2 concentrations from 390 to 350 parts per million.[2] This, in turn, has huge implications for the way we design and manage our cities, how we power them, where we locate them and how they relate to the world's ecosystems.

In recent years the most dramatic population growth has occurred in giant coastal cities, particularly those in Asia and Africa. In fact, with expansion of global trade, coastal populations and economies have exploded on every continent. Of the 17 megacities of over 10 million people around the globe, 14 are located in coastal areas. 40 % of the world's cities of 1–10 million people are also located near coastlines. Careless development practices have caused important habitats such as wetlands, coral reefs, sea grasses, and estuaries to be degraded or destroyed (Sale et al 2008). And with substantial sea level rises expected by the end of the twenty first century, major northern coastal mega-cities and greenhouse gas emitters such as London, New York and Shanghai, could well become the primary victims of their fossil fuel burning, whilst also affecting southern low-lying mega-cities such as Calcutta, Dhaka and Lagos (The Times 2009).

[2] 350.org, www.350.org/about/science. Accessed 20 Nov 2015

Box 9.1: WWF Living Planet Report 2010

"Since 1970 the global Living Planet Index has fallen by 30 %, which means that, on average, species population sizes were 30 % smaller in 2007 than they were in 1970. Following current trends, by 2030 humanity will need the capacity of two Earths to absorb CO_2 waste and keep up with natural resource consumption. Higher income nations have an average per capita environmental footprint that is around five times larger than that found in poorer nations.

The implications are clear. Rich nations must find ways to live much more lightly on the Earth, to sharply reduce their footprint, in particular their reliance on fossil fuels. World leaders have to deliver an economic system that assigns genuine value to the benefits we get from nature: biodiversity, the natural systems which provide goods and services like water, and ultimately our own well-being."

The concept of Petropolis, the fossil fuel powered city which is the current global 'urban archetype', needs to be challenged fundamentally as its systemic flaws become increasingly evident. These are some of the dominant trends: demand for fossil fuels, energy costs, carbon emissions, climate instability and sea levels are increasing, whilst global reserves of natural resources and the time left for action is steadily decreasing. But, crucially and hopefully, so is the cost of renewable energy!

9.5 Creating the Solar City

Some people simply want large modern cities to go away. But given that for the time being urbanization is a global trend, ways have to be found for cities to minimise their systemic dependence on fossil fuels and their unsustainable use of natural resources. A rapid switch towards powering our cities with renewable energy is a crucially important starting point. The key question to which an urgent answer is needed is: how can cities that are the product of fossil fuel-based technologies be powered by renewable energy instead? We have addressed this issue in some detail on our recent publication, Peter Droege's report '100 % Renewable Energy – and beyond – for Cities' (World Future Council 2010).

Our planet derives its energy supply from the sun and the Earth's core and, ultimately, these two primary energy sources need to be used to power our cities. The good news is that in the last few years rapid strides have been aded with a wide spectrum of renewable energy technologies.

Technology and policy go closely hand-in-hand: Germany, Spain and another 50 countries and regions around the world have chosen to introduce feed-in tariffs which make the installation of renewable energy systems a cost effective proposition. Owners of solar PV roofs in Germany, Spain, Portugal or Greece are entitled to sell the electricity they produce back to the grid at up to four times the price of conventional power stations. The benefits for national economies have been significant, reducing fossil fuel imports, carbon emissions, as well as environmental damage. In Germany the total cost per household to implement these renewable energy schemes is just five Euro per household per month. As a result of feed-in legislation, 18 % of Germany's electricity now comes from hydro power, solar power and wind farms and 300,000 new jobs have been created in 10 years. This approach to energy policy has also led to significant breakthroughs in technology, and in the design of buildings.

Recently constructed building complexes such as the Solarsiedlung in Freiburg, for example, are designed to produce more energy than they actually require.[3] The highly energy efficient 'plus-energy' buildings with south facing solar roofs are a model for intra-urban renewable energy production. Outside Seville 'concentrated solar power' technology has been pioneered which utilises an array of mirrors that focus beams of sunlight onto the top of towers through which liquid is circulated which drives steam turbines and generators. Seville is well on its way to become the world's first large city to power itself with solar energy supplied from its hinterland, as well from installations on roof tops within the city (Inhabitat 2007). A major new technological breakthrough is thin-film solar electric cells. These can be produced in printing machines which apply a photo-sensitive ink onto an aluminium or plastic foil. These new thin film technologies are bringing the cost of solar electricity ever closer to full cost competitiveness with conventional power generation. In Germany arrays of thin-film solar power stations can be found around a growing number of towns and cities.

Solar thermal technology has been used for many years in the Mediterranean. It is also becoming common place in less sunny countries such as Austria and Germany. Now it is also making rapid strides in China. In fact it has become the world leader. Solar hot water systems are now used by 20 % of its households many of whom never had the benefit of hot water before. "Experts project that by 2010 the number of solar water heaters installed in China will equal the thermal equivalent of the electrical capacity of 40 large nuclear power plants. Globally, solar water heaters have the capacity to produce as much energy as more than 140 nukes."[4]

In September 2010 this ground-breaking building hosted the 4th World Solar Cities Congress.[5] The 75,000 square metre 'sun-dial' building includes exhibition centres, scientific research facilities, meeting and training facilities and a hotel. It is a Chinese government sponsored showcase of energy efficient solar design and solar technology that is likely to highly influential in a country so far better known for its rapid expansion of coal fired power station capacity.

Wind power is also a solar technology because the Earth's air currents are driven by sunlight. The technological breakthroughs in this field have been facilitated by government policies. Denmark was the first country to introduce feed-in tariffs for wind energy 25 years ago. The advances in this technology have been astounding.

[3] Solarsiedlung, www.solarsiedlung.de/. Accessed 20 Nov 2015

[4] Environmental Graffiti, www.environmentalgraffiti.com/…solar…water-capacity…/822. Accessed 20 Nov 2015

[5] China Solar City, www.chinasolarcity.cn/Html/dezhou/index.html. Accesssed 20 Nov 2015

In 1985 50 KW wind turbines were the norm, but by 2010 their energy output has risen to as much as 5 megawatts – 100 times greater. In countries with long coastlines such as Britain, large scale wind farm development is now well under way. The Thames Array of 500 large turbines will start construction in the Thames Estuary in early 2011, and its 1000 megawatt capacity will supply some 30 % of London's domestic electricity.[6]

While it is desirable for cities to produce much of their energy from within their own territory or from their immediate hinterland, very large cities may require additional renewable energy supplies from further afield. Networks of interconnected solar, wind, hydropower and geothermal systems are now under development. The Desertec project which is supported by major European companies is intended to link the renewable energy resources of Europe, the Middle East and North Africa, and elsewhere similar projects are proposing to supply electricity across continents like North America and Asia via new direct-current 'smart supergrids'.[7]

However, none of these efforts will be sufficient without simultaneously introducing comprehensive energy demand management systems for our cities. For more efficient energy use new insulation materials will enable the retrofit of buildings from within without a significant loss of interior space. Three centimetres of 'vacuum insulation panels', for instance, have much the same performance as 30 cm of conventional insulation materials.

Meanwhile the argument for increasing urban density has been gaining much credence. We can enhance transport energy efficiency through designing for *proximity*. We need to get people walking and cycling rather than driving their cars wherever possible and in this we have much to learn from the compact layout of traditional cities. Meanwhile transport technologies are also changing. Just 10 years ago car manufacturers could barely imagine making cars that did not run on petrol or diesel. Today, all mainstream manufacturers are working on hybrid or electric or fuel cell-powered cars which are promising to become the norm in a matter of years.

All of these measures, taken together, can dramatically change the energy production and consumption patterns of our cities whilst also creating major new economic sectors in our urban regions.

9.6 The Metabolism of Cities: From Linear to Circular

Similar to nature's organisms, cities as 'eco-technical super-organisms' (Girardet 2004) have a definable metabolism – the transformation of resources into vital functions. Nature essentially has a circular zero-waste metabolism: every output by an organism is also an input which replenishes and sustains the whole living environment. In contrast, the metabolism of many modern cities is essentially linear, with resources flowing through the urban system without much concern about their

[6] London Array, www.londonarray.com. Accessed 20 Nov 2015

[7] Desertec, www.desertec.com. Accessed 20 Nov 2015

origin, and about the destination of wastes. Inputs and outputs are considered as largely unrelated. Fossil fuels are extracted from rock strata, refined and burned, and the waste gases are dischargedinto the atmosphere. Raw materials are extracted, combined and processed into consumer goods that ultimately end up as rubbish which cannot be beneficially reabsorbed into living nature. In distant forests, trees are felled for their timber or pulp, but all too often forests are not replenished.

Similar processes apply to food: nutrients and carbon are taken from farmland as food is harvested, processed and eaten. The resulting sewage, with or without treatment, is then discharged into rivers and coastal waters downstream from population centres, and usually not returned to farmland. Rivers and coastal waters all over the world are 'enriched' both with sewage and toxic effluents, as well as with the run-off of mineral fertilizer applied to the farmland used for feeding cities.

This linear, open-loop approach is utterly unsustainable. In an urbanising world aiming for long-term viability it cannot continue. The environmental externalities of urban resources use can no longer be ignored. Unless we learn from nature how to create *circular systems*, an urbanising world will continue to be an agent of global environmental decline.

Planners seeking to design resilient urban systems should start by studying the ecology of natural systems. On a predominantly urban planet, cities will need to adopt circular metabolic systems to assure their own long term viability as well as that of the rural environments on which they depend. Outputs will need to become inputs into the local and regional production system. Whilst in recent years a very substantial increase in recycling of paper, metals, plastic and glass has occurred, much more needs to be done. Most importantly, it is crucial to convert organic waste into compost, and to return plant nutrients and carbon to farmland feeding cities, to assure its long-term fertility.

The *local* effects of urban resource use also need to be better understood. Cities accumulate large amounts of materials within them. Vienna with some 1.6 million inhabitants, every day increases its actual weight by some 25,000 tonnes.[8] Much of this is relatively inert materials, such as steel, concrete and tarmac. Other materials, such as heavy metals, have discernible environmental effects as they gradually leach from the roofs of buildings and from water pipes and accumulate in the local environment. Nitrates, phosphates or chlorinated hydrocarbons build up in soils and water courses, with potentially negative impacts for the health of future inhabitants.

Creating a circular urban metabolism can create resilient cities and create many new local businesses and jobs. A critical issue today, as cities become the primary human habitat, is whether urban living standards can be maintained whilst the local and global environmental impacts of cities are brought down to a minimum. To get a clearer picture of the 'performance' of cities, it helps to draw up balance sheets comparing urban resource flows across the world. It is becoming apparent that similar-sized cities supply their needs with a greatly varying throughput of resources (Fig. 9.3).

---

[8] Prof. Paul Brunner, Technical University, Vienna, personal communication

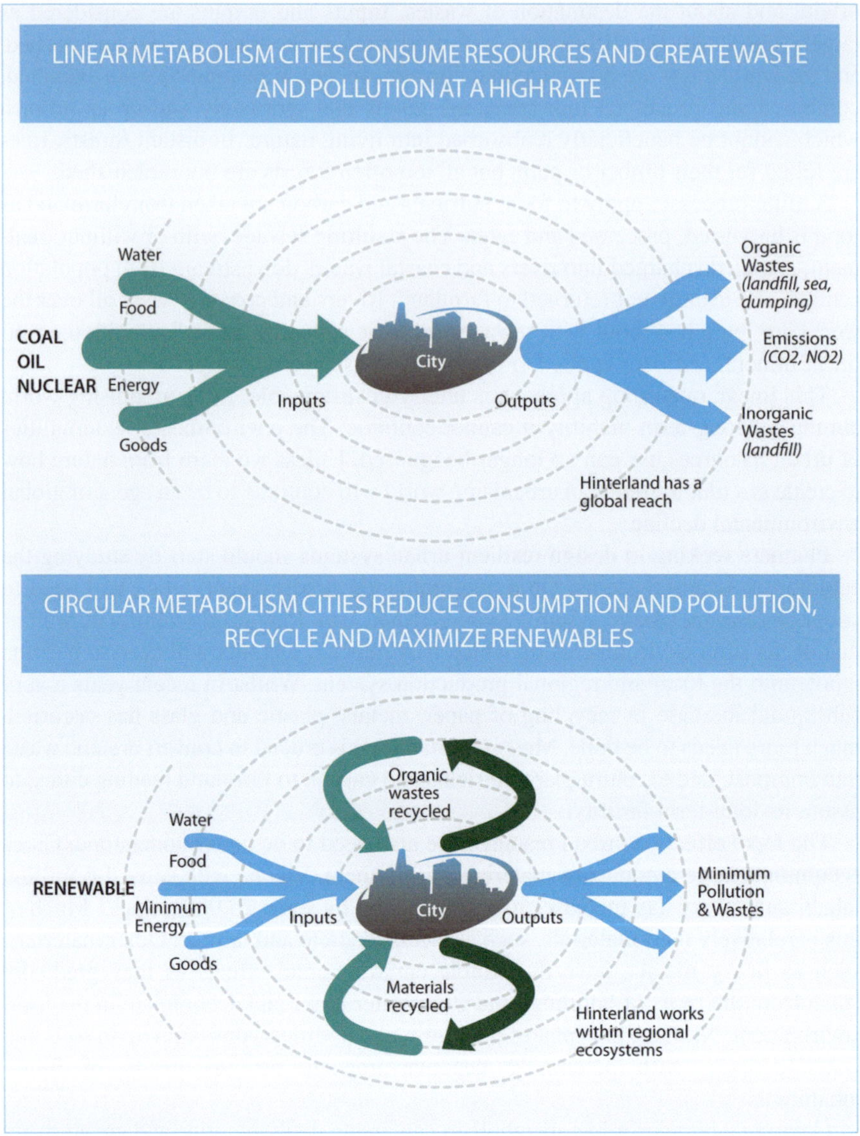

Credit: Herbert Girardet.

Fig. 9.3 Urban Metabolism

One estimate suggests that a North American city with 650,000 people requires some 30,000 km^2 of land to meet domestic needs, without even including the environmental demands of its industries. In comparison, an Indian city of this size would require just 2800 km^2, or less than ten per cent of an American city.[9]

[9] The International Institute for Sustainable Development, Urban and Ecological Footprints, www.gdrc.org/uem/footprints/

Most large cities across the world have been studied in considerable detail and usually it won't be very difficult to compare their use of resources. In developed country cities, disposability and built-in obsolescence still permeates collective behaviour. In contrast, in developing countries large cities have a much lower per capita resource throughput and much higher recycling rates, since recycling and re-use are an essential part of local economies.

9.7 Food for Cities

In many parts of the world, urban growth has been directly linked with mechanization of farming and rural depopulation. Food is supplied to cities by ever more energy intensive production systems. For example, in the United States one farmer, with his complex array of fossil fuelled equipment, typically feeds 100 urban people.

But ten times more fossil fuel energy goes into this type of food production system than the calories that are actually contained in the food we get to eat. We need to find much more efficient ways of supplying food for our cities. This includes a new emphasis on local food production. It is well documented that in Cuba, 'intra-urban' organic agriculture now supplies large amounts of food to cities such as Havana. China has a national policy of surrounding its cities with belts of cultivated land. Such 'peri-urban' food growing systems are also reappearing in the US where farmers' markets supplied by local growers are becoming popular again.

In the United States significant 'intra-urban' agriculture initiatives are also under way. Detroit, once a city of two million people, has contracted to less than 900,000 people, with vast areas of land now lying derelict. Its 139-square-mile surface area is larger than San Francisco, Boston, and Manhattan combined. After studying the city's options of reusing derelict land within Detroit at the request of civic leaders, the American Institute of Architects came to this conclusion in a recent report: "Detroit is particularly well suited to become a pioneer in urban agriculture at a commercial scale." Similar options are now being considered for New Orleans, St. Louis, Cleveland and Newark.[10]

In Denver the Living City Block project goes beyond urban agriculture. It is aiming to create an example of a replicable, scalable and economically viable framework for the resource efficient redevelopment of existing cities. "Starting with a block and a half of Denver's historic Lower Downtown district, Living City Block will create a demonstration of a regenerative urban centre. LCB will draw on selected partners from around Denver, the U.S. and the world to develop and implement a working model of how one block within an existing city can be transformed into a paradigm for the new urban landscape."[11]

---

[10] Smart Planet, www.smartplanet.com/business/blog/…detroit…urban…/4232/. Accessed 20 Nov 2015

[11] Living City Block, www.livingcityblock.org/. Accessed 20 Nov 2015

Box 9.2: Urban Food: The Case of Cuba

According to Cuba,s Ministry of Agriculture, some 150,000 acres of land is being cultivated in urban and suburban settings, in thousands of community farms, ranging from modest courtyards to production sites that fill entire city blocks. Organoponicos, as they are called, show how a combination of grassroots effort and official support can result in sweeping change, and how neighbours can come together and feed themselves. When the food crisis hit in 1989, the organoponicos were an ad hoc response by local communities to increase the amount of available food. But as the power of the community farming movement became obvious, the Cuban government stepped in to provide key infrastructure support and to assist with information dissemination and skills sharing. Most organoponicos are built on land unsuitable for cultivation. They rely on raised planter beds. Once the rganoponicos are laid out, the work remains labour-intensive. All planting and weeding is done by hand, as is harvesting. Soil fertility is maintained by worm composting. Farms feed their excess biomass, along with manure from nearby rural farms to worms that produce a nutrient-rich fertiliser. Crews spread about two pound of compost per square yard on the bed tops before each new planting.

www.i-sis.org.uk/OrganicCubawithoutFossilFuels.php

Even very large cities can source substantial amounts of the vegetables and fruit they require from the urban territory and the surrounding countryside. However, grain supplies require much larger areas of land and most will have to be supplied from farmland further afield.

9.8 The Ecosystems Beyond

But renewable energy, urban agriculture and resource efficient redevelopment are only part of the story of creating truly regenerative cities. Above all else we need to address the relationship between cities and the ecosystems beyond their boundaries on which they will continue to depend even if major redevelopment initiatives are taken within cities. This brings us back to the ecological footprint concept. Calculating the ecological footprint of densely populated areas, such as a city or small country with a comparatively large population – such as New York, Singapore or Hong Kong – invariably leads to the perception of these cities as 'parasitic' because they have little intrinsic bio-capacity, and instead must rely upon large territories elsewhere.[12]

The ecological footprint of a city is a measure of its demand on the Earth's biologically productive land and sea area. It compares that demand with the entire

[12] Wikipedia, en.wikipedia.org/wiki/Ecological_footprint. Accesssed 20 Nov 2015

planet's ecological capacity to regenerate, and to absorb critical waste outputs such as carbon dioxide in the Earth's living fabric. The footprint methodology enables us to estimate how many Earths would be needed to support humanity if everybody lived a particular lifestyle. In 2006, the biologically productive area per person worldwide was 1.8 global hectares (gha). But since the per capita footprint of a large European city such as London was 5.6 gha per person at that time, three planet Earths would be required if all the Earth's inhabitants lived like Londoners. Since it is not so easy to make new planets, reducing London's per capita footprint is obviously a rather important undertaking.

Box 9.3: Sanitation for Soil

Fertile soil is the most crucial factor in sustaining huge populations on Earth. Not only food supply, but also water renewal, regional and global climate as well as drought and flood prevention depend directly on rich living soil. Political support of short term profits of powerful multinational companies, earning from the ultimately destructive use of synthetic fertilisers and pesticides, have destroyed millions of farms that could otherwise have produced food in a sustainable way with keeping their soil fertile. Industrial agriculture has led to a global depletion of soil quality at an alarming extent. Another major global problem is lack of sustainable sanitation. The dominant flush toilet system transports plant nutrients from our stomachs to the seas and is not capable of recovering them. Even the sludge from sewage treatment plants, that are in place for only around 10 % of the wastewater worldwide, is always very low in major nutrients. Phosphate is no longer available to plants after precipitation. Sludge is still high in pollutant concentrations, too. In addition, flush sanitation is causing millions of child deaths though contamination of surface waters with faecal matter.

Both problems – loss of soil fertility and sanitation/water pollution – can be solved together in a rather simple way. This was demonstrated by ancient civilisations in the Amazon: All they left was the best soils in world, as well as very beautiful ceramics.

The 'terra preta' soils where made from organic waste including excreta. A number of very feasible high- and low-tech options for sanitation producing rich fertile soils are now available. In turn this can result in zero sewage discharges into water bodies whilst helping farmers to work with organic agriculture the natural way. Their soils will create healthy food. Societies should not allow the agro-chemical industry to turn ever more farmers across the world into total dependency any longer. If we want a future for People and Planet we need to make every effort to take re-establish the connection between urban organic waste and soil fertility and to shift globally towards Terra Preta Sanitation for our cities.

Prof. Ralf Otterpohl, Technical University, Hamburg-Harburg

The largest section of an ecological footprint is the area required for food production. A key problem with the farming systems supplying the bulk of food, and particularly grain, to urban populations is that both carbon and plant nutrients are removed from farmland as food is harvested and these are not returned back to the land. Agricultural land is kept productive by applications of artificial fertilisers which have been shown to have negative effects both on soil structure and soil organisms. Meanwhile the plant nutrients contained in urban sewage are flushed into rivers and coastal waters, or intercepted in sewage systems, never to be returned to the land. In a regenerative city, new ways have to be found to intercept these nutrients, as well as the carbon content of food waste and sewage.

Shifting from urban systems that damage and degenerate ecosystems to ones that renew and sustain the health of ecosystems on which they depend requires a fundamental rethink of urban systems design. The regenerative development of cities is a comprehensive approach that goes beyond established concepts of sustainable development. Cities need to proactively contribute to the replenishment of the run-down ecosystems – including farm soils, forests and marine ecosystems – from which they draw resources for their survival. And while cities continue to burn fossil fuels, they also need to find ways of assuring that their carbon dioxide emissions are reabsorbed through 'bio-sequestration' in soils and forests (Girardet and Mendonca 2009).

The CO_2 output of cities is far too large for trees within their territories to be able to absorb. Every year we are now discharging nearly 10 billion tonnes of carbon per year (Science Daily 2008) of which four to five billion tonnes are not being reabsorbed into the world's ecosystems but which are accumulating in the atmosphere. This is the primary cause of the climate change problem that we are faced with. Can this issue be addressed by cities? Well, some have made a start: In Adelaide, South Australia, large-scale reforestation has been initiated to assure that the surrounding countryside can absorb a substantial proportion of its carbon emissions. Some two million trees have been planted in the last 7 years for carbon sequestration, erosion control and general environmental improvement.[13] Another million will be planted by 2014. A few other cities are now involved in similar initiatives. Internationally, the 'Billion Trees Campaign' initiated by UNEP in 2007 recorded that over 10 billion trees had been planted by 2010, a quarter of these by urban community groups, NGOs and local or national governments (UNEP 2010).

9.9 From Petropolis to Ecopolis

One of the primary tasks at the start of the twenty first century is to try and map out what is *necessary* in order to try and expand the boundaries of what becomes politically *possible*. The challenge is to find ways of making cities function differently from the way they do today without increasing the costs to financially challenged city administrations.

---

[13] Million Trees, www.milliontrees.com.au/. Accessed 20 Nov 2015

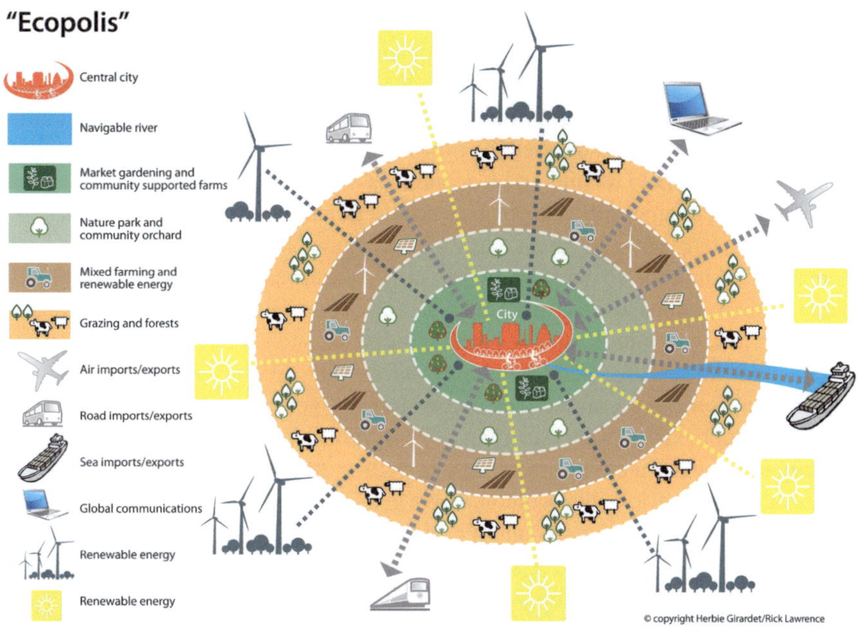

Fig. 9.4 "Ecopolis"

The new task facing of urban planners, civil engineers and managers, in close cooperation with the general public, is to create spatial structures that satisfy the needs of city people whilst also assuring their ecological and economic resilience. We need to provide secure habitats that allow us to move about our cities efficiently, and we want them to provide pleasant spaces for work, recreation and human interaction. We want urban environments that are free from pollution and waste accumulation. But we also need to get to grips with the impacts of cities beyond their boundaries.

It is often said by urban analysts that cities should be seen as the places where solutions to the world's environmental and climate problems can most easily be implemented because as places where most people live closely together they have the potential to make efficient use of resources. It is also in cities where people interact most strongly and where key decisions, and particularly financial decisions, are being made all the time.

This is where the concept of 'Ecopolis' – the ecologically as well as an economically restorative city – needs to assert itself, drawing together the various themes discussed in this text into one comprehensive concept (Fig. 9.4).

Of course, modern cities tend to be much larger than traditional human settlements and this makes reintegration into their local hinterland much more difficult. The reality is that far more people have to be accommodated in cities today than a couple of 100 years ago and this needs to be taken account of in developing concepts for creating resilient human settlements fit for the twenty first century.

In recent years there has been much talk about *peak oil*. Are we also heading for *peak globalisation*? Many cities have a problem of job scarcity due to the relocation of manufacturing jobs to other parts of the world as a result of economic globalisation. In addition, vast amounts of money are still spent on importing fuels to our cities from distant places. Could the creating of resource efficient cities, largely powered by renewable energy, help rebuild urban economies and bring jobs back to our cities?

Creating environmentally regenerative cities is a challenge that urban administrators and educators have not really had to deal with until now. This challenge has been made more difficult since the privatisation of services in recent years has reduced the capacity of city administrations to create integrated urban systems. But the awareness is growing that integrated, restorative planning and management of cities presents major new opportunities for reviving urban economies and creating new businesses and jobs.

Policy makers, the commercial sector and the general public need to jointly develop a much clearer understanding of how cities can develop a restorative relationship to the natural environment on which they ultimately depend. The underlying incentive is that positive outcomes are likely to be beneficial for both global ecology as well as the urban economy. Many reports indicate that a wide range of new businesses and many new job opportunities could be created from a steady move towards efficient use of resources.

To initiate projects for restoring the health of forests, soils and aquatic ecosystems that have been damaged by urban resource demands certainly goes beyond strictly urban policy initiatives. Creating parameters for appropriate action will involve both political and business decisions – with a spectrum ranging from transnational, to national and to urban levels of decision making. It involves drawing up novel legal frameworks and addressing the profit logic of companies involved in natural resource extraction.

Box 9.4: The Value of Ecosystems Services

We cannot manage what we do not measure and we are not measuring either the value of nature's benefits or the costs of their loss. We seem to be navigating the new and unfamiliar waters of ecological scarcities and climate risks with faulty instruments. Replacing our obsolete economic compass could help economics become part of the solution to reverse our declining ecosystems and biodiversity loss. We need a new compass to set different policy directions, change incentive structures, reduce or phase out perverse subsidies, and engage business leaders in a vision for a new economy. Holistic economics – or economics that recognise the value of nature's services and the costs of their loss – is needed to set the stage for a new "green economy".

www.guardian.co.uk/commentisfree/cif-green/2010/feb/10/pavansukhdev-natures-economic-model

Pavan Sukhdev, drawing on his report 'The Economics of Ecoystems and Biodiversity'

The national policies needed to set parameters for regenerative urbanisation include both 'sticks' such as waste disposal taxation and carbon taxes, and 'carrots' such as feed-in tariffs for renewable energy and support schemes for local food production. Cities need to take advantage of the opportunities inherent in environmentally restorative development, harnessing the huge variety of talents and experiences present in cities for better decision making.

For instance, the city government of Porto Alegre, Brazil, decided some years ago to involve the general public in the processes of budget-setting. This creative process challenges citizens to actively contribute their views. All citizens can now have a say about what their tax money should be spent on – better schools, better transport, playgrounds, parks, renewable energy installations, and so on. Through this novel participatory process Porto Alegre has become a truly dynamic, participatory city, and the ideas pioneered there are being copied in cities across the world.[14]

The ecological, economic, social and *externalities* of our urban systems need to be addressed in new ways. We need creativity and initiative at the local level, but we also need appropriate national policy frameworks to enable useful things to happen locally. Without national policy initiatives, enhanced by lively public debate, the necessary changes won't happen fast enough, if at all. For example, feed-in tariffs for renewable energy in Denmark and Germany came out of vigorous public demand that was turned into national policy which was then implemented primarily at the local level.

It is important to emphasise, then, that the creating of regenerative cities as described in this text will require not just changes in approaches to land use and resource use planning at the local level but that national and trans-national policies have to be initiated.

Cities take resources from nature. The new challenge is for cities to find ways to continuously help regenerate natural systems from which they draw resources.

Internationally, cities need to work closely together to develop and implement policies for regenerating regions across the world that have been damaged and depleted by urban consumption patterns. One or two organisations, such as the Climate Alliance of European Cities,[15] which brings together 1500 towns and cities across Europe, have made a tentative start at helping cities to take responsibility for their global climatic and environmental impacts. Much, much more needs to be done.

9.10 Policies for Creating Regenerative Cities

Enshrining regenerative urbanism as an organising principle for urban development practice seems compelling because it offers a number of major new opportunities for local social and economic well-being:

[14] Sustainable Cities, www.sustainablecities.dk. Accessed 20 Nov 2015

[15] Klimabuendnis, www.klimabuendnis.org. Accesssed 20. Nov 2015

Key Principles

- National policy: Frameworks for enabling regenerative urban development
- Urban policy: Integrated, regenerative urban planning as key organising principle
- Green Savings: reducing waste, recycling materials and cutting costs
- Green Economy: new businesses and jobs by environmental protection and restoration
- Green Talent: investing in technical, entrepreneurial and workforce skills

Energy Sufficiency

- Use the 2000 Watt Society concept as an operative policy principle
- Modify building codes to make resource efficient building practice the norm

'Solar City' Development

- Mandate solar city development as national policy priority
- Prioritise feed-in legislation for renewable energy systems, allowing owners to sell electricity at advantageous rates
- Support renewable energy as an important new manufacturing industry
- Create enabling policies for renewable energy development in the urban hinterland

Water Security

- Balance urban, agricultural and commercial uses of water, and their relative social, economic and environmental benefits
- 'Waterproof' cities by encouraging water efficiency and rainwater collection in households and businesses
- Make waste water recycling and storm water reuse a central plank of water policy

Implementing Zero Waste

- Develop new enterprises for processing organic wastes into soil enhancing materials
- Make sewage reprocessing and nutrient capture a central plank of waste management
- Implement policies for the cost effective reprocessing of all technical wastes
- Use zero waste policy to create new green businesses and jobs

Local Food

- Encourage local peri-urban food production for local markets
- Encourage the development of community supported agriculture and farmers markets
- Ensure the use of composted, city-derived bio-waste for urban farming

Sustainable Transport

- Create new pedestrian zones wherever possible
- Create a comprehensive network of dedicated cycle lanes across cities
- Encourage public transport by improving its attractiveness, frequency and flexibility
- Stimulate development of new electric and fuel cell vehicle technology
- Encourage car sharing as a key feature or urban transport

Nature and the City

- Encourage tree planting for biodiversity and soil erosion control in and around the city
- Make carbon sequestration a key aspect of tree planting initiatives
- Develop initiatives to help restore forests and wetlands in remoter areas

Green Business

- Boost the creation of green business by effective use of government procurement
- Encourage resource efficiency in all businesses
- Create 'green business incubators' across the city
- Make environmental resilience the basis for new businesses and jobs

A Culture of Restorative Urbanisation

- Utilise both global networks and local expertise in developing restorative urbanisation
- Ensure that it is addressed in the education system, and through meetings and events
- Encourage imaginative reporting on urban restoration measures by the media
- Produce regular reports on implementation of eco-restoration policies and practices
- Ensure that all citizens take a stake in restorative development

The challenge now is to initiate a mutual learning process in which cities across the world can exchange experiences and information about best policies and practices of regenerative urbanism. The World Future Council intends to make a major contribution to this vitally important process.

References

Girardet H (2004) Cities, people, planet – urban development and climate change. Wiley, London

Girardet H, Mendonca M (2009) A renewable world. Green Books, Dartington, chapter 2

Inhabitat (2007) www.inhabitat.com/2007/05/21/sevilles-solar-power-tower. Accessed 20 Nov 2015

IPCC (2008a) Fourth assessment report. Summary for policy makers. www.ipcc.ch/pdf/assessment-report/ar4/wg1/ar4-wg1-spm.pdf. Accessed 20 Nov 2015

IPCC (2008b) Fourth assessment report. Climate change 2007: Synthesis report https://www.ipcc.ch/publications_and_data/ar4/syr/en/main.html. Accessed 20 Nov 2015

King FH (1911) Farmers of forty centuries: organic farming in China, Korea, and Japan, Courier Dover Publications London Array, www.londonarray.com. Accessed 20 Nov 2015

Sale PF, Butler IV MJ, Hooten AJ, Kritzer JP, Lindeman KC, Sadovy de Mitcheson YJ, Steneck RS, & van Lavieren H (2008) Stemming decline of the coastal ocean: rethinking environmental management, UNU-INWEH, Hamilton, Canada, http://inweh.unu.edu/wp-content/uploads/2013/05/Stemmingdecline.pdf, Accessed 20 Nov 2015

Science Daily (2008) Global carbon emissions speed up, Beyond IPCC Projections www.sciencedaily.com/releases/2008/09/080925072440.htm. Accesssed 20 Nov 2015

The Times (2009) Major cities at risk from rising sea level threat – CAPCOA www.timesonline.co.uk/tol/news/environment/article6938356ce.e, Tuesday, December 1st, 2009. Accessed 20 Nov 2015

UNEP (2010) Billion tree campaign. http://www.unep.org/pdf/BTC_Booklet.pdf

UN Habitat (2006) The state of the world's cities 2006/2007, The millennium development goals and urban sustainability: 30 years of shaping the habitat agenda. United Nations Human Settlements Programme (UN-HABITAT), New York

Vitousek PM, Lubchenco J, Mooney HA, Melillo J (1997) Human domination of Earth's ecosystems. Science 277:494–499

World Future Council (2010) 100% Renewable Energy and Beyond – For Cities, HafenCity University Hamburg and World Future Council Foundation, Hamburg, Germany www.worldfuturecouncil.org/fileadmin/user_upload/PDF/100__renewable_energy_for_citys-for_web.pdf. Accessed 20 Nov 2015

WWF (2010) Living planet report. wwf.panda.org/about_our_earth/…/living_planet_report/2010_lpr2. Accessed 20 Nov 2015

WWF (2009) Arctic climate feedbacks: global implications, executive summary. www.worldwildlife.org/…/WWFBinaryitem15234.pdf

Chapter 10
Multidimensional Sustainability Assessment for Megacities

Stanislav Shmelev

Abstract Urban sustainability assessment is required for the purposes of establishing strategic directions for 'greening' our cities to reduce the environmental impact of their performance, improve employment and economic viability and enhance the quality of life. This chapter considers large world cities: London, New York, Hong Kong, Los Angeles, Sao Paolo, Rio de Janeiro, Paris, Berlin, Singapore, Shanghai, Sydney and Tokyo. To assess urban sustainability performance, we applied multi-criteria decision aid tools to compare the cities on the range of dimensions. The tools chosen for this assessment are ELECTRE III, NAIADE and APIS. The results have shown that Singapore dominates the sustainability rankings in most multi-criteria applications, showing particular strength in economic and environmental dimensions and a slightly less strong performance in the social dimension according to the APIS results. The chapter explores innovative sustainability strategy and new governance structures in Singapore and discusses the reasons for such success.

Keywords Megacity • Sustainability • Policy • Benchmarking • Multi-criteria decision aid (MCDA) • Indicators

10.1 Introduction

Urban sustainability is understood as a multi-dimensional capacity of a city to operate successfully in economic, social and environmental domains simultaneously. This topic receives a lot of attention in the EU, USA and increasingly China and Latin America since the Rio Summit of 1992. UNEP Green Economy Report highlights urban sustainability as one of its important dimensions (UNEP 2011). Major city governments formed a C40 partnership to make important steps forward in climate change mitigation at the urban scale. A review of international sustainable urban policy developments have been offered in a plethora of works (Hall 2014),

S. Shmelev (✉)
Environment Europe Ltd, 38 Butler Close, Woodstock Road, Oxford OX2 6JG, Oxfordshire, UK
e-mail: director@environmenteurope.org

S. Shmelev (ed.), *Green Economy Reader*, Studies in Ecological Economics 6,
DOI 10.1007/978-3-319-38919-6_10

(Hall et al. 2010), (Hall and Pfeiffer 2000), Girardet (2014). A new interdisciplinary perspective on a sustainable city, representing it as a system of interacting dimensions: urban planning, energy, transport, material flows, green space, etc. has been proposed in (Shmelev and Shmeleva 2009). Such multidimensional nature of an urban system determined the choice of a main analytical approach for sustainability assessment employed in the present article, namely the methodology of Multi-Criteria Decision Aid (Roy 1996).

Urban economies depend on the natural world and the functioning of ecosystems on the territory of a much larger region than the city itself, making them important systems for ecological economics research (Girardet and Mendonca 2009). Cities rely on a wide array of ecosystem processes and functions, sometimes called ecosystem services (Bolund and Hunhammar 1999), (Gomez-Baggethun and Barton 2013), (Shmelev and Anger 2013a, b), which are broadly defined as *economic* (provisioning: water, food, fibre, energy), *ecological* (regulation and maintenance: biogeochemical cycling, soil formation, photosynthesis, pollination, air quality regulation) and *social* (cultural: cultural diversity, educational values, inspiration, aesthetic values) (Shmelev 2012). The interaction of two approaches, namely a systemic description of the city and the analysis of city-ecosystem interactions, could define the science of urban sustainability.

This paper aims to consider 12 major megacities to identify the sustainability leaders as well as cities experiencing the strongest sustainability challenges. The analytical tools chosen for the analysis are Principal Component Analysis and three different Multi-Criteria Decision Aid (MCDA) tools, namely, ELimination Et Choix Traduisant la REalité (ELECTRE III), Novel Approach to Imprecise Assessment and Decision Environment (NAIADE) and Aggregated Preference Indices System (APIS). Additionally, it aims to test several MCDA tools for sustainability assessment to provide guidance on how they could be more used for sustainability assessment. First, we analyse a large set of 20 indicators describing the 12 leading megacities in Europe, North and South America, Asia and Oceania. We perform the Principal Component Analysis and identify major relationships between social, economic and environmental factors. We then reduce the set of indicators and perform Multi-Criteria Analysis using different tools. MCDA analysis is used here to identify best practices and explain the success of urban policies. We conclude with a description of sustainability strategies and policies adopted in the leading city of our pool, which could help us to understand its success.

The article is organized as follows. Section 10.1 offers an introduction to the topic. Section 10.2 provides a review of ecological-economic applications for urban sustainability. Section 10.3 discusses data and indicators used in the present paper. Section 10.4 presents the results of the Principle Component Analysis. Section 10.5 explains the Multi-Criteria Decision Making methodologies used in this paper. Section 10.6 presents the results of ELECTRE, NAIADE and APIS methods. Section 10.7 discusses the sustainability strategies and policies in the city identified as a sustainability leader. Section 10.8 concludes.

10.2 Urban Sustainability Analysis Methods

There is a well-developed set of tools and methods that have been used in ecological economics, industrial ecology and operations research to support problem solving at various scales, which could ultimately lead to the development of a new economy. Such methods, identified as most relevant for ecological economics by (Shmelev 2012), include material flows analysis, input–output analysis, optimization and multi-criteria decision aid. In recent years there has been a surge in research and publications applying these tools to urban systems. Our first task was to review applications of ecological-economic methods mentioned above to urban systems. The highlights of this new research are presented in Table 10.1 as an illustration of how the chosen methods were applied to several sustainability issues in urban systems: water, resources, waste and CO_2 emissions.

Material flows analysis (MFA) is a tool, accounting for the weight of resources that are being extracted domestically, are imported and then accumulated, processed or recycled in the national economy, and then emitted into nature in the form of gaseous, liquid or solid residues or exported (Eurostat 2001). The additional element of MFA is in the accounting for the unused fraction associated with resource extraction – amount of overburden and soil that needs to be transferred to extract a resource. The research in material flows accounting in relation to cities was started with a publication focused on Hong Kong (Newcombe et al. 1978). More recently, studies focused on Hamburg, Vienna and Leipzig (Hammer and Giljum 2006), Singapore (Schulz 2007), Beijing (Zhang et al. 2009), Paris (Barles 2009) and Lisbon (Rosado et al. 2014) enhanced our understanding of the material flows at the city scale and contributed towards filling the gap in data availability. It should be mentioned that national MFA datasets have been the first ones to appear and urban and regional datasets are still rare.

Water Footprint emerged as a detailed concept through the efforts of the Water Footprint Network[1] and comprises both production-based and consumption-based water use, the latter illustrating dependence of the economy on water (predominantly used in agriculture) to feed the city. Essentially, it is a partial case of MFA. Very few urban water footprints exist, among them the study on Milan (Vanham and Bidoglio 2014) and previous comparative work on US and Chinese cities (Jenerette et al. 2006), Oslo (Venkatesh and Brattebø 2012) and Berlin, Delhi and Lagos (Hoff et al. 2013).

Carbon Footprint calculation or *Carbon Accounting* has been influenced largely by the Kyoto agreements. The situation at the urban level has been characterised by the C40 partnership initiated by the Mayor of London, Mr Ken Livingstone, which united the leading megacities in the efforts to mitigate climate change, often despite the national position on the Kyoto protocol. Carbon Disclosure project has published an annual cities report since and reports on CO_2 emission inventories and action plans for such cities as San Francisco (San Francisco Department for the

[1] (http://waterfootprint.org/)

Table 10.1 Studies focused on application of ecological-economic methods to water, resource, waste and CO_2 emissions reduction issues in various cities[1]

Dimensions methods	Water	Resources	Waste	Emissions CO_2
MFA	Milan (2014)[2]	Lisbon (2014)[9]	London (2011)	San Francisco (2004)[17] London (2009)[18]
	Berlin (2013)[3]	Paris (2009)[10]		
	Delhi (2013)[4]	Singapore (2008)[11]		Paris (2011)[19]
	Lagos (2013)[5]	Hambourg (2006)[12]		New York (2012)[20]
	Oslo (2011)[6]	Leipzig (2006)[13]		Rio de Janeiro (2011)[21]
	New York (2006)[7]	Vienna (2006)[14]		
	Beijing (2009)[8]	Beijing (2009)[15]		
		Hong Kong (1978)[16]		
Input-output	Beijing (2012)[22]	Beijing (2010)[24]	Suzhou (2012)[25]	Helsinki (2013)[28]
	Chongquing (2006)[23]		Seattle (2012)[26]	Beijing (2013)[29]
			Chongquing (2006)[27]	Copenhagen (2011)[30]
				Sydney (2004)[31]
Optimization	Tabriz (2010)[32]		Mexico (2013)[35]	
	Shanghai (2012)[33]		Taichung (2012)[36]	
	Sydney (2012)[34]		Beijing (2011)[37]	
MCDA	Granada (2012)[38]		Beijing (2010)[41]	Paris (1982)[48]
	Athens (2012)[39]		Kampala (2013)[42]	
	Berlin (2004)[40]		New York (2006)[43]	
			Amsterdam (2006)[44]	
			Moscow (2006)[45]	
			Budapest (2006)[46]	
			Barcelona (2006)[47]	

[1]The sources of relevant case studies are mentioned in subsequent sections further in the text
[2]Vanham and Bidoglio (2014)
[3]Hoff et al. (2013)
[4]Hoff et al. (2013)
[5]Hoff et al. (2013)
[6]Venkatesh and Brattebø (2012)
[7]Jenerette et al. (2006)
[8]Zhang et al. (2009)
[9]Rosado et al. (2014)
[10]Barles (2009)

(continued)

Table 10.1 (continued)

[11]Schulz (2007)
[12]Hammer and Giljum (2006)
[13]Hammer and Giljum (2006)
[14]Hammer and Giljum (2006)
[15]Zhang et al (2009)
[16]Newcombe et al. (1978)
[17]San Francisco Department for the Environment (2004)
[18]City of London (2009)
[19]Marie de Paris (2011)
[20]City of New York (2012)
[21]City of Rio de Janeiro (2011)
[22]Zhang et al. (2012)
[23]Okadera et al. (2006)
[24]Zhou et al. (2010)
[25]Liang and Zhang (2012)
[26]Leigh et al (2012)
[27]Hui et al. (2006)
[28]Ala-Mantila et al. (2013)
[29]Chen et al. (2013)
[30]Hallegatte et al. (2011)
[31]Lenzen et al. (2004)
[32]Zarghami (2010)
[33]Lü et al. (2012)
[34]Mortazavi et al. (2012)
[35]Santibañez-Aguilar et al. (2013)
[36]Chang et al. (2012)
[37]Dai, Li and Huang (2011)
[38]Ruiz-Villaverde et al. (2012)
[39]Kandilioti and Makropoulos (2012)
[40]Simon et al. (2004)
[41]Xi et al. (2010)
[42]Oyoo et al. (2013)
[43]Munda (2006)
[44]Munda (2006)
[45]Munda (2006)
[46]Munda (2006)
[47]Bautista and Pereira (2006)
[48]Roy and Hugonnard (1982)

Environment 2004), London (City of London 2009), Paris (Marie de Paris 2011), Rio de Janeiro (City of Rio de Janeiro 2011) and New York (City of New York 2012) followed.

Input–output analysis is an economic tool designed by Wassily Leontief (Leontief 1936, 1970), which presents the economic system as a web of interconnected types of economic activity or sectors, namely agriculture, energy generation, oil and gas extraction, computer manufacturing, education, health care, etc. There has been an increased interest in using input–output analysis for environmental applications (Shmelev 2010), (Shmelev and Shmeleva 2012). especially in relation to CO_2 emissions (Peters and Hertwich 2006) and water use (Dietzenbacher and Velázquez 2007) at the macro scale, but the urban scale remains a promising new area of research.

Very recently urban input–output models have been built to assess the water use of such Chinese cities like Chongquing (Okadera et al. 2006) and Beijing (Zhang et al. 2012), urban metabolism with a focus on resources in Beijing (Zhou et al. 2010) and waste for Shenzhen (Ni et al. 2001), Chongquing (Hui et al. 2006) and Suzhou (Liang and Zhang 2012). A paper focused on the application of input–output analysis to E-waste recycling in Seattle (Leigh et al. 2012) characterises the situation in a city in an OECD country. Research on input–output analysis of CO_2 emissions for urban systems has been quite abundant: Sydney (Lenzen et al. 2004), Vienna (Ornetzeder et al. 2008), Copenhagen (Hallegatte et al. 2011), Suzhou (Liang et al. 2012), Beijing (Chen et al. 2013) and Helsinki (Ala-Mantila et al. 2013).

Optimization, a mathematical method which allows one to find the solution providing a minimum or a maximum of a certain goal function on a constrained set of possible alternatives, is another tool which is quite widely applied in urban sustainability research. Several varieties of optimization techniques are known depending on the structure of the problem being solved and the type of data available: mixed integer programming, multiobjective optimization, linear programming. In the urban sphere such tools have been most widely applied in water management with cases in Sydney (Mortazavi et al. 2012), Tabriz (Zarghami 2010), Shanghai (Lü et al. 2012); waste management in Genova (Minciardi et al. 2008), Palermo (Galante et al. 2010), Beijing (Xi et al. 2010) and (Dai et al. 2011), Taichung (Chang et al. 2012), and Mexico city (Santibañez-Aguilar et al. 2013).

Multi-criteria decision aid has been applied widely for urban sustainability assessments and decision support due to its ability to find compromise between conflicting goals and priorities. The method has been developed by Bernard Roy (Roy 1985, 1991), who has applied the ELECTRE tool to assess possible locations for new underground stations in Paris (Roy and Hugonnard 1982), which ultimately reduces CO_2 emissions by providing easier access to public transport for residents. MCDA tools have been applied for solving the urban water management issues in the case of Berlin (Simon et al. 2004), Granada (Ruiz-Villaverde et al. 2012) and Athens (Kandilioti and Makropoulos 2012). Waste management issues have been studied with MCDA in the context of Beijing (Xi et al. 2010), Kampala (Oyoo et al. 2013), Dakar (Kapepula et al. 2007) and Barcelona (Bautista and Pereira 2006).

In the context of urban sustainability assessment (Munda 2006, 2005a, 2005b) has been one of the first to suggest using MCDA tools to compare cities on their sustainable development performance.

10.3 Urban Sustainability Indicators

The following cities were selected for our analysis: New York, Los Angeles, San Francisco, Rio de Janeiro, Sao Paulo, London, Paris, Berlin, Tokyo, Hong Kong, Shanghai, Singapore and Sydney. The reasons for selecting the cities were economic importance (all cities feature in world top 30 cities by GDP) and environmental impacts (all cities are part of the C40 network focused on greenhouse gas emissions mitigation). These world capitals combined comprise 8 % of global GDP and 2.8 % of global CO_2 emissions, employ 105 mln people, use 13.8 bln tonnes of water per year and generate 50 mln tonnes of municipal solid waste per year combined and are equivalent to the whole economy of Japan by GDP and Germany by CO_2 emissions.

The panel of indicators on sustainable urban development has been assembled based on the United Nations Guidelines and Methodologies on Sustainable Development Indicators (UN 2007), a Sustainable Development Goals framework (UN 2015), a Sustainable Development Indicators Frameworks (UNECE 2013), (EC 2009), a wide range of sources from Eurostat, city governments (City of London 2009; City of New York 2012; City of Rio de Janeiro 2011; Marie de Paris 2011; San Francisco Department for the Environment 2004; Singapore 2009), Siemens European Green City Index (Siemens 2009), World Cities Culture Forum (Mayor of London 2014), UN Habitat (UN HABITAT 2013) and World Bank publications (World Bank 2013), LSE Going Green Report (LSE 2013) as well as considerations on availability of data. The indicators following the International Urban Sustainability Indicators List proposed in (Shen et al. 2011) include the economic characteristics, such as income per capita; social and cultural dimensions, such as unemployment rate, income differentiation rate in the form of a Gini coefficient and higher education level, and, finally, a wide range of ecological-economic or environmental dimensions, including the share of green space, CO_2 emissions, average PM_{10} concentrations, water use per capita per day, waste generation per capita per day and recycling rates. Selection of individual indicators for cities, chosen for the present paper, was based on our dynamic sustainability assessment carried out for countries (Shmelev 2011; 2016) and adapted for the urban scale.

The process of indicator selection for the study was performed in two parts. First, a large set of 20 criteria was analysed, including economic indicators (income per capita at PPP, number of large companies headquartered in the city, creative industries employment), environmental indicators (CO_2 emissions per capita, share of nuclear energy, PM_{10} emissions, water use per capita, waste generation per capita, recycling rates) and socio-cultural indicators (unemployment rate, Gini Index of income inequality, art galleries, number of firms released per year, visits to top 5 museums per year, number of bookshops, etc.). After performing a Principal Component Analysis, identifying redundant variables and focusing on socio-economic and socio-environmental variables, the set of criteria took its final shape numbering 11 criteria as a result of several iterations.

Figure 10.1 illustrates the multidimensional complexity of the analysis by presenting three indicators for all cities in their standardized form (with means subtracted from the raw figures and the results divided by standard deviations).

Comparing the performance of cities on income per capita at PPP, unemployment and CO_2 emissions, we can observe the strong position of Singapore and Tokyo which have relatively low unemployment figures and CO_2 emissions and relatively high income levels. At the same time, a relatively weaker position of Los Angeles on unemployment and CO_2 emissions and Berlin on unemployment and low incomes can be clearly seen.

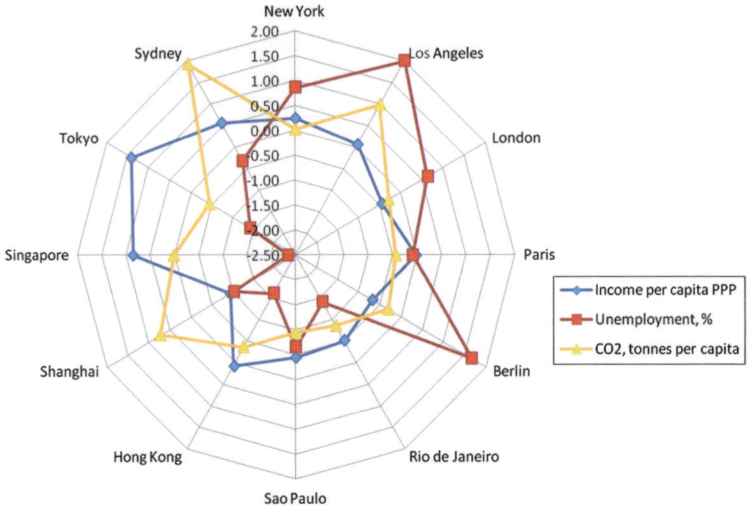

Fig. 10.1 Comparative performance of top megacities in economic, social and environmental dimensions: income per capita, unemployment and CO_2 emissions per capita, standardized data, 2013

Incomes per capita at PPP have been highest in Tokyo and Singapore and lowest in Shanghai, Rio de Janeiro, Sao Paolo and Berlin. The highest number of international companies was headquartered in Tokyo (613), followed by New York (217), London (193) and Paris (168). Creative economies were largest in Sao Paolo (17.4 %), Rio de Janeiro (14.9 %), London (12 %), Tokyo (11.2 %) and Berlin (10 %). Unemployment has been highest in Los Angeles (13.9 %), followed by Berlin (11.7 %) and New York (9.6 %); the lowest unemployment among the cities in the sample was in Singapore (1.8 %), Hong Kong (3.1 %), and Tokyo (3.6 %). The largest shares of citizens with higher education have been recorded in Shanghai (42.9 %), London (41.9 %), Berlin (39 %), followed by Paris (35.8 %); the lowest figures were observed in Rio de Janeiro (14.5 %) and Sao Paolo (19 %). The highest levels of income differentiation measured in the form of Gini coefficient were recorded in Rio de Janeiro (0.53), Hong Kong (0.53), New York (0.504); the lowest – in Tokyo (0.25), Berlin (0.28) and Shanghai (0.32).

The environmental dimension in our assessment was represented by CO_2, PM_{10}, water use, waste generation, and recycling rates. The largest per capita CO_2 emission values were recorded for Sydney (20.3 t/person/year), Los Angeles (13,3 t/person/year) and Shanghai (12,9 t/person/year), which can be explained by the share of the coal energy in the energy mix. On the other hand, CO_2 emissions were lowest in Sao Paolo (1.4 t/person/year) and Rio de Janeiro (2.1 t/person/year) due to the dominance of hydro energy and bio-fuel based car fleet. PM_{10} emissions were highest in Shanghai (81 μG/m^3), Rio de Janeiro (64 μG/m^3), and Hong Kong (55 μG/m^3) and lowest in Sydney (12 μG/m^3), New York (21 μG/m^3), Tokyo (23 μG/m^3) and Los Angeles (25 G/m^3), which is largely related to the fuel standards and quality and efficiency of the car fleet.

Water use has been highest in Los Angeles (707.8 l per person per day), Shanghai (411.1 l per person per day) and Hong Kong (371.2 l per person per day) and lowest in Berlin (152 l per person per day), London (167 l per person per day) and Sydney (200 l per person per day). Cities with the largest share of green public space were Singapore (47 %), Sydney (46 %), Sao Paolo, Hong Kong (41 %) and London (38.6 %). Much less green were Shanghai (2.6 %) and Tokyo (3.4 %).

In terms of waste generation per capita, the most polluting cities were: Rio de Janeiro (1.6 kg per person per day), New York (1.47 kg per person per day) and Paris (1.46 kg per person per day). Less polluting were Tokyo (0.77 kg per person per day), Singapore (0.85 kg per person per day) and Shanghai (0.90 kg per person per day). Recycling rates were phenomenally high in Los Angeles (76.4 %), Sydney (66 %) and Singapore (61 %) and were very low in Rio de Janeiro (1 %), Sao Paolo (1 %) and Shanghai (1 %).

10.4 Principle Component Analysis

The pattern of social, economic and environmental dimensions for the world megacities needs to be explored with statistical tools to reveal hidden relationships and trends. We have identified Principal Component Analysis as a means to cross-check the criteria set for consistency, exclude highly redundant variables and identify the main factors according to which the megacities could be grouped and classified. First, we calculate a correlation matrix of urban sustainability indicators (Table 10.2). The indicators of UNESCO cites, other museums, other historic sites, visits to top five museums, area, population and Gross Regional Product have been excluded from the analysis due to high correlation with other indicators.

Table 10.2 presents the matrix of the Pearce coefficients of correlation between various criteria. The correlations, which were significantly different from 0 with a significance level a = 0.5 are denoted in bold. It is most interesting to acknowledge the negative and significant correlation between the % of people employed in creative industries and overall CO_2 emissions. This fact reinforces the value of a creative sector, which includes art, music, cinema, theatre, museums and galleries for sustainable urban development.

Using Principal Component Analysis (PCA) and limiting the set of principal components to three, we have obtained the following factor loadings (Table 10.3). The nature of the indicators selected under Factor 1, Factor 2 and Factor 3 allows us to use ‘Social’, ‘Economic’ and ‘Environmental’ labels for them respectively. Social dimension included higher education level, unemployment, art galleries, cinema screens, number of films released and visits to top five museums per year. Incidentally, the share of nuclear energy was chosen to be part of the social dimension by the PCA method. Economic factors included income per capita at PPP, number of large companies headquartered in the city, and green space, the latter pointing towards a degree of prosperity. Environmental factors included emissions of CO_2, water use, waste generation and recycling rate.

Table 10.2 Correlation matrix of social, economic and environmental indicators of urban sustainability for megacities

Variables	Income per capita PPP	Large Companies (McKinsey)	Higher Education Level, %	Unemploy-ment, %	GINI	CO2, tonnes per capita	Share of Nuclear	PM_{10} Annual mean mG/m^3	Green space, %	Water use per capita, l per day	Waste generation per capita, kg per day	Recycling rate	National Museums	Art Galleries	Cinema Screens	Number of films released a year	Bookshops per 100000	Visits to top 5 museums per	Interna-tional tourists per year	Creative industries employ-ment, %
Income per capita PPP	**1**	**0.675**	-0.319	-0.270	-0.095	0.049	0.115	-0.563	0.100	0.069	-0.328	0.562	-0.511	0.207	-0.165	0.525	0.234	-0.071	0.163	-0.389
Large Companies (McKinsey)	**0.675**	**1**	0.009	-0.144	-0.388	-0.152	0.333	-0.329	-0.378	-0.011	-0.376	0.084	-0.284	0.534	0.045	**0.811**	0.341	0.218	-0.070	0.105
Higher Education Level, %	-0.319	0.009	**1**	0.459	**-0.600**	0.481	0.261	-0.146	-0.380	-0.158	-0.160	0.107	0.561	0.507	0.474	0.247	-0.070	0.560	-0.216	-0.193
Unemploy-ment, %	-0.270	-0.144	0.459	**1**	-0.097	0.236	0.210	-0.450	-0.403	0.249	0.575	0.361	-0.071	0.348	0.527	0.330	-0.184	0.342	-0.264	0.007
GINI	-0.095	-0.388	**-0.600**	-0.097	**1**	-0.342	0.042	0.235	0.272	0.291	**0.620**	-0.184	-0.200	-0.170	0.090	-0.413	0.000	-0.146	0.317	0.006
CO2, tonnes per capita	0.049	-0.152	0.481	0.236	-0.342	**1**	-0.267	-0.246	-0.015	0.267	0.010	0.568	0.117	-0.148	0.267	-0.161	-0.054	-0.162	-0.136	**-0.618**
Share of Nuclear	0.115	0.333	0.261	0.210	0.042	-0.267	**1**	-0.164	-0.455	0.005	0.163	0.008	0.404	**0.781**	0.534	**0.607**	0.273	0.506	0.108	0.022
PM10 Annual mean mG/m3	-0.563	-0.329	-0.146	-0.450	0.235	-0.246	-0.164	**1**	-0.133	0.206	-0.088	**-0.678**	0.509	-0.368	-0.032	-0.570	0.004	-0.335	0.140	0.157
Green space, %	0.100	-0.378	-0.380	-0.403	0.272	-0.015	-0.455	-0.133	**1**	-0.425	-0.025	0.139	-0.249	-0.473	**-0.589**	-0.506	0.106	-0.041	0.412	-0.067
Water use per capita, l per day	0.069	-0.011	-0.158	0.249	0.291	0.267	0.005	0.206	-0.425	**1**	0.389	0.308	-0.080	-0.088	**0.587**	-0.018	-0.055	-0.439	0.038	-0.365
Waste generation per capita, kg per day	-0.328	-0.376	-0.160	0.575	**0.620**	0.010	0.163	-0.088	-0.025	0.389	**1**	0.086	-0.190	0.052	0.433	-0.104	-0.046	0.108	-0.014	0.156
Recycling rate	0.562	0.084	0.107	0.361	-0.184	0.568	0.008	**-0.678**	0.139	0.308	0.086	**1**	-0.252	0.072	0.144	0.282	0.129	-0.010	0.215	**-0.701**
National Museums	-0.511	-0.284	0.561	-0.071	-0.200	0.117	0.404	0.509	-0.249	-0.080	-0.190	-0.252	**1**	0.144	0.267	-0.196	0.219	0.182	0.243	-0.193
Art Galleries	0.207	0.534	0.507	0.348	-0.170	-0.148	**0.781**	-0.368	-0.473	-0.088	0.052	0.072	0.144	**1**	0.572	**0.799**	0.078	**0.806**	-0.108	0.031
Cinema Screens	-0.165	0.045	0.474	0.527	0.090	0.267	0.534	-0.032	**-0.589**	**0.587**	0.433	0.144	0.267	0.572	**1**	0.302	-0.206	0.266	-0.235	-0.161
Number of films released a year	0.525	**0.811**	0.247	0.330	-0.413	-0.161	**0.607**	-0.570	-0.506	-0.018	-0.104	0.282	-0.196	**0.799**	0.302	**1**	0.156	0.498	-0.182	0.098
Bookshops per 100000	0.234	0.341	-0.070	-0.184	0.000	-0.054	0.273	0.004	0.106	-0.055	-0.046	0.129	0.219	0.078	-0.206	0.156	**1**	0.165	**0.782**	-0.176
Visits to top 5 museums per capita	-0.071	0.218	0.560	0.342	-0.146	-0.162	0.506	-0.335	-0.041	-0.439	0.108	-0.010	0.182	**0.806**	0.266	0.498	0.165	**1**	0.085	0.135
International tourists per year	0.163	-0.070	-0.216	-0.264	0.317	-0.136	0.108	0.140	0.412	0.038	-0.014	0.215	0.243	-0.108	-0.235	-0.182	**0.782**	0.085	**1**	-0.422
Creative industries employment, %	-0.389	0.105	-0.193	0.007	0.006	**-0.618**	0.022	0.157	-0.067	-0.365	0.156	**-0.701**	-0.193	0.031	-0.161	0.098	-0.176	0.135	-0.422	**1**

Values in bold are different from 0 with a significance level alpha=0.05, the colours of the heat map indicate positive (red) or negative (blue) correlation values

Table 10.3 Correlations between variables and social, economic and environmental factors for Urban Sustainability Indicators

	SOCIAL	ECONOMIC	ENVIRONMENTAL
Income per capita PPP	0.271	0.887	0.130
Large Companies (McKinsey)	0.573	0.549	-0.336
Higher Education Level, %	0.615	-0.391	0.072
Unemployment, %	0.547	-0.399	0.408
GINI	-0.415	-0.172	0.119
CO2, tonnes per capita	0.135	-0.007	0.753
Share of Nuclear	0.694	-0.145	-0.262
PM10 Annual mean mG/m3	-0.538	-0.510	-0.269
Green space, %	-0.591	0.407	0.038
Water use per capita, l per day	0.057	-0.203	0.642
Waste generation per capita, kg per day	0.035	-0.443	0.361
Recycling rate	0.337	0.482	0.750
National Museums	0.098	-0.516	-0.152
Art Galleries	0.903	-0.061	-0.260
Cinema Screens	0.612	-0.547	0.346
Number of films released a year	0.876	0.316	-0.214
Bookshops per 100000	0.085	0.393	-0.142
Visits to top 5 museums per capita	0.643	-0.097	-0.348
International tourists per year	-0.205	0.336	0.044
Creative industries employment, %	-0.094	-0.279	-0.699

The Principle Component Analysis is aimed at reducing complexity of multidimensional data sets by identifying key factors or components, which are linear combinations of the original variables. It allows one to simultaneously represent both observations and variables for two selected components in the factor space (Fig. 10.2). Proximity of the variable vectors in this chart (e.g. number of large companies and income per capita) signifies a high correlation between the variables in question. The length of a variable vector denotes the variance of a variable, and the angle between a vector and an axis signifies the importance of the contribution of the corresponding variable to a factor.

Figure 10.2 presents the space of Social vs Economic dimension. It can be seen from the figure that Tokyo has a high income per capita, many large companies, high recycling rates, a large number of films released per year, relative low Gini index of income inequality and low values of PM_{10} emissions. London, Paris, Berlin and Los Angeles have significantly higher education levels, significant levels of unemployment, a lot of art galleries and visitors to their museums, but less green space than Singapore, Hong Kong and Sydney. Sao Paolo, Rio de Janeiro and Shanghai have high levels of income inequality, PM_{10} pollution, relatively large creative industries but low average incomes and low recycling rates.

As can be seen from Fig 10.3, Paris, London, New York, Tokyo and Berlin are located close to each other, which denotes similarity in the Socio-Environmental space. Los Angeles exhibits high levels of waste generation, water use and CO_2

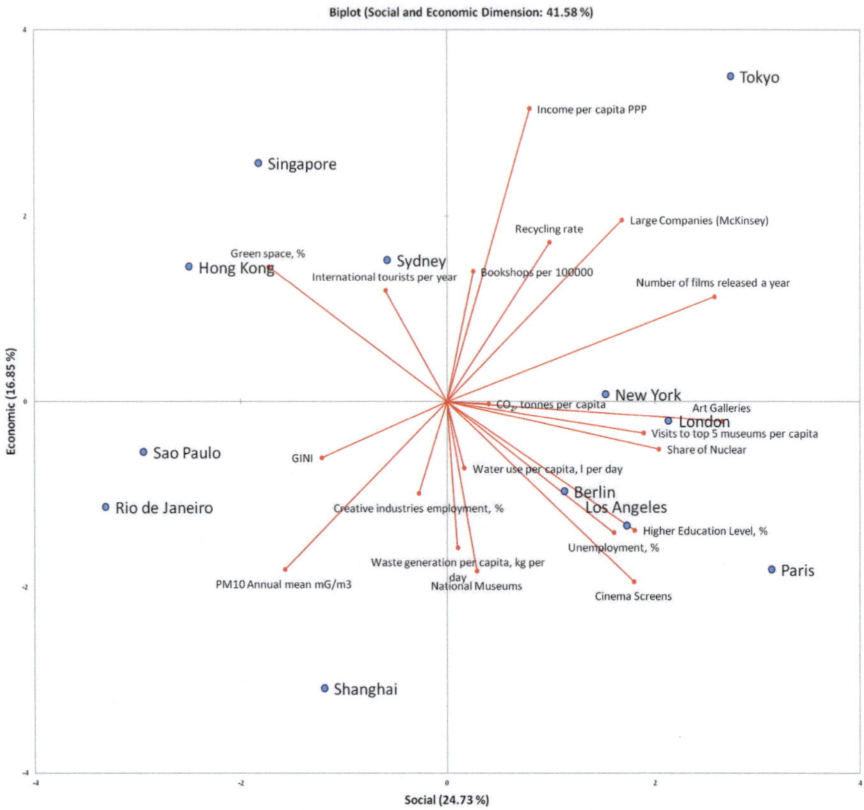

Fig. 10.2 Comparative assessment of mega cities with principal component analysis: social vs economic factors

emissions per capita and low levels of PM_{10} emissions. London, Paris and Tokyo are relatively close to each other in the socio-environmental space and show a significant number of large companies, art galleries and low levels of income inequality.

Focusing on the Economic vs Environmental factor space (Fig. 10.4) we could note that Singapore and Sydney exhibit high income levels and high recycling rates and low emissions of PM_{10}. Tokyo has very low levels of unemployment and income inequality and is quite different from Paris, London, Rio de Janeiro and Sao Paolo, which are clustered very closely together. Los Angeles is shown here as a clear outlier with high use of water and waste generation and high CO_2 emissions as well as high unemployment. Singapore performs well on the income level, green space and recycling but slightly worse on national museums and creative industries. Berlin, Paris, London, Sao Paolo and Rio de Janeiro are clustered together exhibiting a high

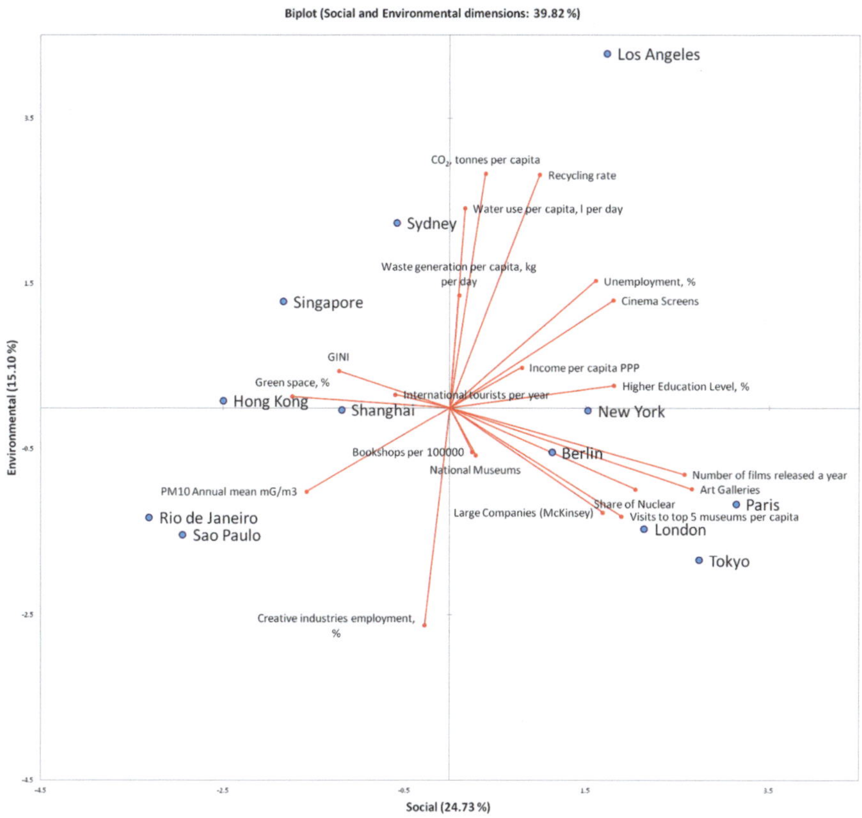

Fig. 10.3 Comparative assessment of mega cities with principal component analysis: social vs environmental factors

share of creative industries, relatively low CO_2 emissions, but lower levels of recycling or incomes than in Singapore or Sydney.

Based on the results of the Principle Component Analysis and excluding highly correlated indicators, the final set of indicators for the Multi-Criteria Decision Aid (Table 10.4) was assembled, focusing mostly on socio-economic and socio-environmental dimensions of sustainability. We gave slightly more attention to the economic and environmental measures, because it was particularly important for us to see how megacities are achieving success in finding the balance between economic and environmental performance. Social variables like unemployment, income differentiation level and higher education level were included but cultural variables focusing on the number of museums, art galleries and bookshops were omitted in MCDA analysis presented in this paper.

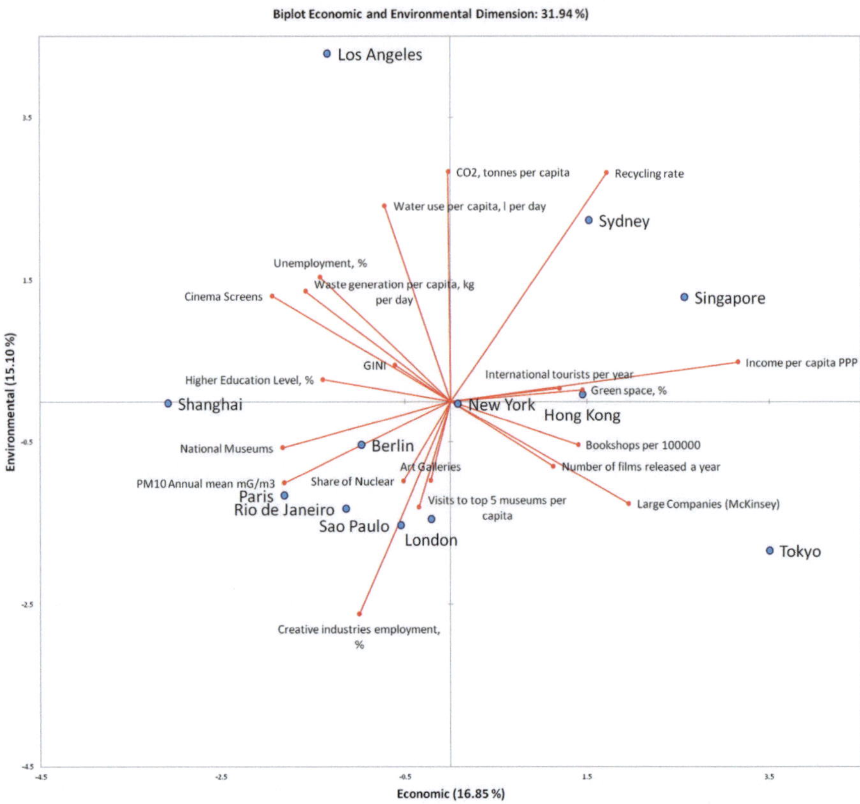

Fig. 10.4 Comparative assessment of mega cities with principal component analysis of megacities: economic vs environmental factors

10.5 Multi-Criteria Decision Aid Methodology

The method of Multi-Criteria Decision Aid (MCDA) is usually understood in terms of alternatives or courses of actions that need to be compared; criteria that are used to assess performance of these actions; a multi-criteria aggregation procedure (MCAP) and recommendations that could be given. Prof. Roy suggested using several types of 'problematique' – classes of problems that could be solved with the help of multicriteria methods. In the present study the MCDA methodology was chosen for its ability to treat several dimensions of data simultaneously and its capacity to integrate such information via multi-criteria aggregation procedure (MCAP) with or without converting the data of different nature into a single composite index. Currently there are various MCDA methods available in the form of standalone software packages or libraries of tools.

Table 10.4 Performance of top world cities on sustainability criteria

	Area, sq km	Population	GRP, mln	Income per capita PPP	Higher education level, %	Unemploy-ment, %	GINI	CO2, tonnes per capita	PM10 Annual mean μG/m3	Green space, %	Water use per capita, l per day	Waste generation per capita, kg per day	Recycling rate
New York	1214.4	8,135,133	1,406,000	30,498	33.3 %	9.6	0.504	7.90	21	14.0	262.3	1.47	27.0
Los Angeles	10,510.0	9,818,605	747,306	27,915	29.2 %	13.9	0.479	13.30	25	6.7	707.8	1.90	76.4
London	1572.0	7,825,200	565,000	20,182	41.9 %	8.2	0.340	6.20	29	38.4	167.0	1.30	34.0
Paris	12,012.0	11,797,021	505,985	26,497	35.8 %	6.1	0.490	5.20	38	9.4	300.0	1.46	23.0
Berlin	892.0	3,460,725	30,000	14,544	39.0 %	11.7	0.280	5.90	26	14.4	152.0	1.17	42.2
Rio de Janeiro	1200.0	6,320,646	194,900	16,971	14.5 %	3.6	0.530	2.10	64	29.0	301.3	1.60	1.0
Sao Paolo	1500.0	11,253,503	370,000	17,708	19.0 %	5.0	0.500	1.40	38	41.0	220.5	1.33	1.0
Hong Kong	1104.0	7,071,600	357,475	28,766	19.4 %	3.1	0.530	5.31	55	41.0	371.2	1.36	39.0
Shanghai	6340.5	23,474,600	213,000	5472	42.9 %	4.2	0.320	12.9	81	2.6	411.1	0.90	1.0
Singapore	710.0	5,183,700	311,566	47,180	22.8 %	1.8	0.463	7.86	32	47.0	308.5	0.85	61.0
Tokyo	2130.0	13,159,388	743,826	55,766	25.5 %	3.6	0.250	4.89	23	3.4	320.2	0.77	40.0
Sydney	12,144.5	4,575,532	91,618	33,285	35.0 %	5.8	0.321	20.30	12	46.0	200.0	1.27	66.0

It was decided to apply Multi-Criteria Decision Aid for the sustainability assessment of megacities for the following three reasons. First, MCDA tools, namely outranking methods based on pair-wise comparisons of alternatives, allow simultaneous consideration of several criteria without giving them explicit weights. This could be helpful given the fact that policy makers or the public often do not know which weights to use and sometimes prefer to focus on strong sustainability or weak sustainability, rather than to give a higher weight to a particular dimension. Strong sustainability is understood here as a setting where less compensation is allowed among criteria and weak sustainability– where more compensation is accepted (Martinez-Alier et al. 1998). Elimination and Choice Expressing Reality (ELECTRE III) method and Novel Approach to Imprecise Assessment and Decision Environment (NAIADE) method are well suited for this task. Second, both these methods address *problematique* γ (Roy 1996), which is oriented towards a creation of a complete or partial preorder on the set of alternatives, making them appropriate instruments for comparing the megacities. Third, for modelling various policy priorities and the explicit treatment of weights we have chosen to use a powerful Aggregated Preference Indices System (APIS) method, which performs a sensitivity analysis using a Monte-Carlo technique. Using three MCDA tools in parallel allows us to cross-check the results and better understand the reasons for a particular ranking.

ELimination Et Choix Traduisant la REalité (ELECTRE, ELimination and Choice Expressing REality) is a family of discrete multicriteria outranking methods originally proposed by Bernard Roy in the 1960s (Roy 1985). ELECTRE is composed of two major parts: first, the construction of one or several outranking relations, which aims to comprehensively compare each pair of actions; second, an aggregation procedure that combines the results into a web of domination relationships. The nature of the recommendation depends on the *problematique* being addressed: choosing, ranking or sorting.

Novel Approach to Imprecise Assessment and Decision Environment (NAIADE) is a discrete multi-criteria method whose impact (or evaluation) matrix may include either crisp, stochastic or fuzzy measurements of the performance of an alternative with respect to a judgement criterion (Munda 1995). No traditional weighting of criteria is used in this method. The whole procedure can be divided in three main steps: (1) pairwise comparison of alternatives, (2) aggregation of all criteria, (3) evaluation of alternatives. The method is based on the concept of the fuzzy preference relation. This tool has been used in dynamic multidimensional sustainability assessment in (Shmelev and Rodriguez-Labajos 2009).

The Aggregated Preference Indices System (APIS) method, developed by Nikolai Hovanov (Hovanov 1996), is based on the Bayesian model of uncertainty randomization. An extensive description of the method is contained in the recent publications: Hovanov et al. (2009), Afgan et al. (2000) and Polydecision (2008). The ASPID method is designed to compare complex objects, given a range of criteria describing their performance using an additive aggregated preference index and a measure of dominance reliability. The weights in APIS are randomized by using a

Monte-Carlo method, which can be interpreted as a kind of sensitivity analysis. The policy priorities that we defined imply relationships between weights, which translate into constraints in the weights optimization problem. An optimization problem was run to derive all those weights that satisfy the pre-set policy priority constraints. The MCDA results can then be presented as distributions of the performance scores taking into account uncertainty in weights coefficients rather than treating them as accurate point estimates.

In the present study we have chosen a synthetic approach which combines the use of three MCDA methods, namely ELECTRE III, NAIADE and APIS.

10.6 Multi-Criteria Decision Aid Applications

10.6.1 ELECTRE

The application of ELECTRE III method to the comparative sustainability analysis of largest world cities reveals the following tendencies. Figs. 10.5, 10.6 and 10.7 depict the webs of domination relationships among the cities obtained through the pair-wise comparisons within the ELECTRE III tool. The arrow between two cities denotes a relationship of domination in the sense of the criteria chosen. Lack of such an arrow points to incomparability. According to the set of 11 criteria selected for this assessment, ELECTRE III, under the assumption that a 1 % difference in the value of each criterion is sufficient for preference or domination (q) and less than 1 % presents an indifference (p) coupled with the assumption of equal weights (Fig. 10.5), reveals the following pattern. Singapore is the most sustainable city in the set; followed by Sydney, then Tokyo, followed by London and Berlin, which are incomparable with each other. London is, in turn, dominating New York, Paris and Hong Kong, followed by Sao Paolo and Rio de Janeiro. Berlin, in turn, is dominating Rio de Janeiro, Los Angeles and Shanghai, with Paris and Hong Kong being incomparable, and it is difficult to differentiate between Los Angeles and Shanghai in the present sustainability assessment.

At the same time, under the assumption that a 10 % difference in the value of each criterion is sufficient for domination and less than 5 % presents an indifference, Singapore and Tokyo are featured at the top of the web of domination relationships, followed by Sydney and New York, which are incomparable with each other which, in turn, are followed by London, Berlin and Sao Paolo (Fig. 10.6).

Finally, under the assumption that a 50 % difference in the value of each criterion is sufficient for domination and less than 10 % presents an indifference, the web of domination relationships looks like that presented in Fig. 10.7. Singapore and London dominate Tokyo, which in turn dominates Sydney and New York, which both dominate Berlin, seen as more sustainable than Paris, Hong Kong and Sao Paolo which, in turn, dominate Los Angeles, Shanghai and Rio de Janeiro.

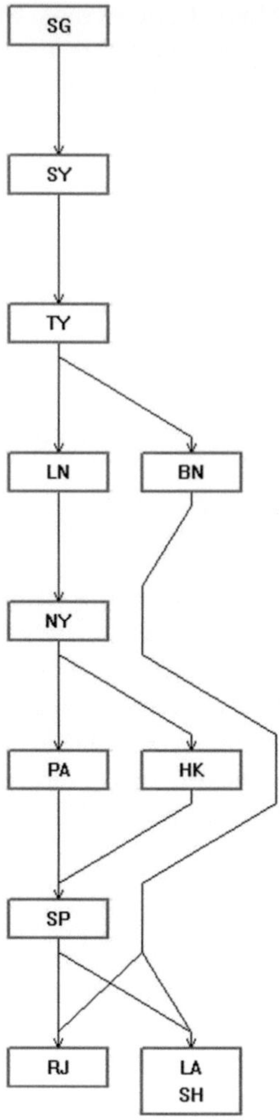

Fig. 10.5 Multi-criteria sustainability performance assessment web of domination relationships among megacities, ELECTRE III: A=0.01 indifference A=0.01 for preference

10.6.2 NAIADE

The results obtained with the NAIADE method present a somewhat similar picture. Under the assumption that the minimum requirement for fuzzy relation is 0.5, the web of domination relationships looks like the one presented in Fig. 10.8. Here Singapore is seen as most sustainable followed by New York, Sydney and London (which are incomparable with each other) in turn followed by Hong Kong, Sao Paolo, Tokyo, Paris, Berlin, Los Angeles and featuring Rio de Janeiro and Shanghai at the bottom of the sustainability hierarchy.

Under the assumption that the minimum requirement for fuzzy relation is 0.1 (weak sustainability) the results appear in the form presented in Fig. 10.9. Here Singapore is also occupying the top of the sustainability web and is followed by

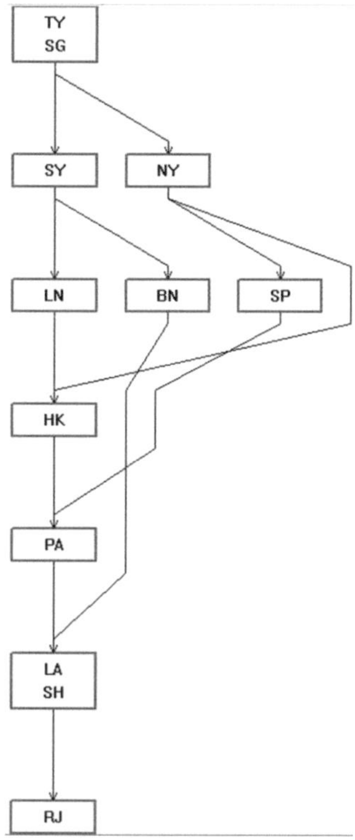

Fig. 10.6 Multi-criteria sustainability performance assessment web of domination relationships among megacities, ELECTRE III: A=0.05 for indifference A=0.1 for preference

New York, Sydney and London, which are incomparable with each other, in turn followed by Sao Paolo, Hong Kong, Tokyo, Paris, Berlin, Rio de Janeiro, Shanghai and Los Angeles; the latter being incomparable and situated at the bottom of the sustainability hierarchy.

Finally, under the assumption that the minimum requirement for fuzzy relation is 0.9 (strong sustainability) the results are shown in Fig 10.10. In this assessment, Singapore is at the top of the sustainability web followed by New York, Sydney and London which are incomparable with each other, in turn followed by Hong Kong, Tokyo and so on with Shanghai seen as least sustainable of the set of 12 world cities.

10.6.3 APIS

The multicriteria assessment tool APIS allows the analyst a slightly different degree of freedom. Using a special feature of this method, which allows the user to test different policy priorities, we will look at three different cases: Economic criteria dominant over Environmental, Environmental dominant over Economic and Social dominant over Economic. The results are presented as charts indicating the

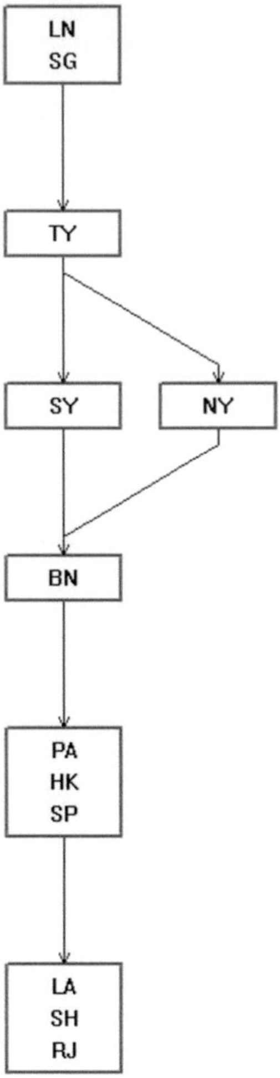

Fig. 10.7 Multi-criteria sustainability performance assessment web of domination relationships among megacities, ELECTRE III: A=0.1 indifference A=0.5 for preference

performance scores (mid-points of the red bars on the vertical axis) of each city accompanied by the measures of uncertainty (distribution) of each performance score (red bars). Additional blue bars denote the probability of domination of one alternative over the other if there is an overlap between distributions of performance scores. The results presented in n indicate that under the dominance of Economic over Environmental criteria the most sustainable cities are Tokyo, Singapore and New York with the least sustainable being Berlin, Rio de Janeiro and Shanghai. Under the assumption of the dominance of Environmental over Economic the dominant top sustainability leaders are Singapore, Sydney and London with the least sustainable being Shanghai, Los Angeles and Rio de Janeiro. Under the assumption of the priority of Social over Economic criteria, we see a different result altogether: Sydney, Tokyo, London and Singapore are seen as most sustainable with Los Angeles, Rio de Janeiro, New York and Sao Paolo seen as least sustainable.

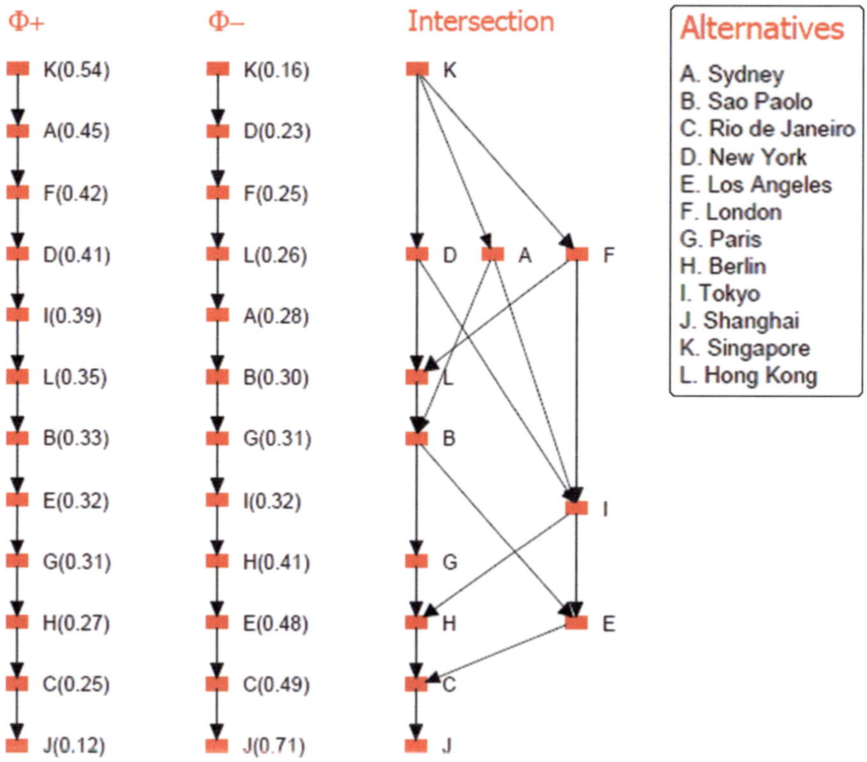

Fig. 10.8 Multi-criteria sustainability performance assessment web of domination relationships among megacities: NAIADE, A = 0.5

Overall, New York has been leading in the set of 12 megacities on Gross Regional Product; Los Angeles, on recycling; London and Shanghai, on higher education; Paris, on the share of nuclear energy; Berlin, on water use; Rio de Janeiro and Sao Paolo, on CO_2 emissions; Singapore, on unemployment and green space; Tokyo on income per capita, the number of large companies, Gini index of income inequality and waste generation per capita and Sydney, on PM_{10} emissions(Figs. 10.11, 10.12 and 10.13).

10.7 Discussion

In the light of the information presented above using three different sustainability assessment tools with multicriteria procedures (ELECTRE III, NAIADE and APIS) there is a considerable degree of convergence in the results pointing towards Singapore as definitely the most sustainable of all 12 cities in the pool, with perhaps only the social dimension performing slightly worse than economic and environmental. APIS tool has proven to be a useful assessment tool for weak sustainability.

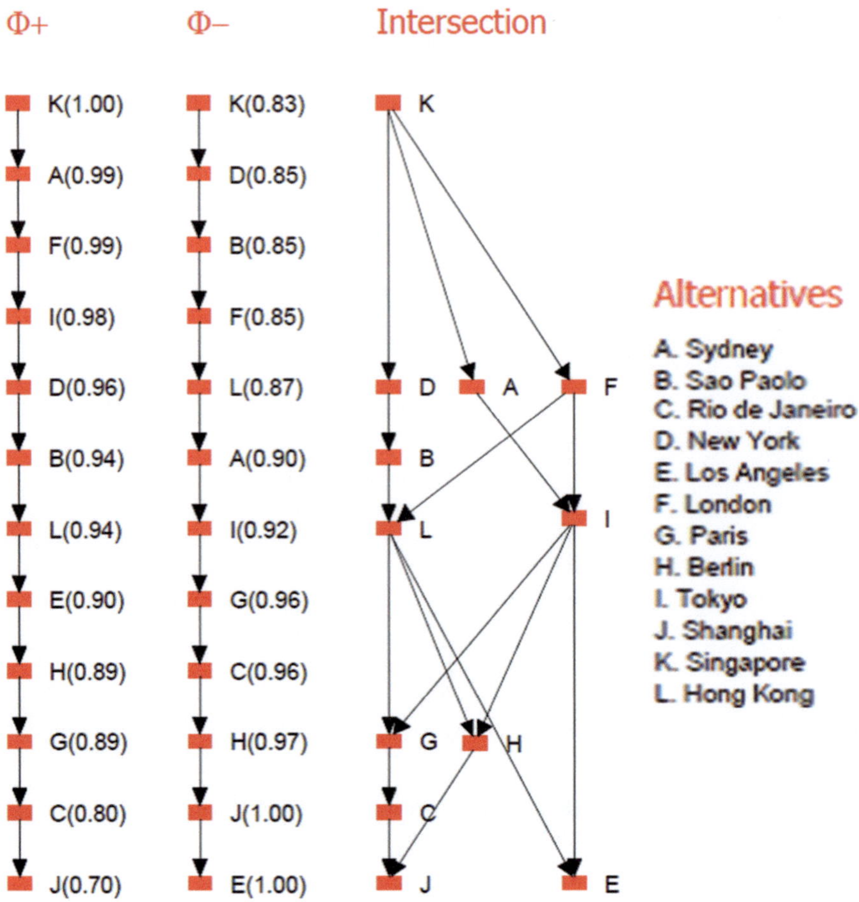

Fig. 10.9 Multi-criteria sustainability performance assessment web of domination relationships among megacities: NAIADE, A = 0.1

It provided additional information through sensitivity analysis, highlighting the leaders in sustainability focusing on economic, social or environmental dimensions. ELECTRE and NAIADE tools, covering a stronger sustainability end of spectrum, can be helpful if the goal is to accept incomparabilities without projecting the overall sustainability performance index into one dimension. At the same time, acknowledging the understanding of the sustainability continuum from weak to strong, it should be said that the differentiation between weak and strong sustainability is not operationalized well enough even in NAIADE, although the method has been designed to explicitly address these issues. It is hard to differentiate between various degrees of the strength of sustainability or degrees of compensation among criteria. A more transparent theoretical approach is required to clarify the strong-weak sustainability relations within the MCDA field.

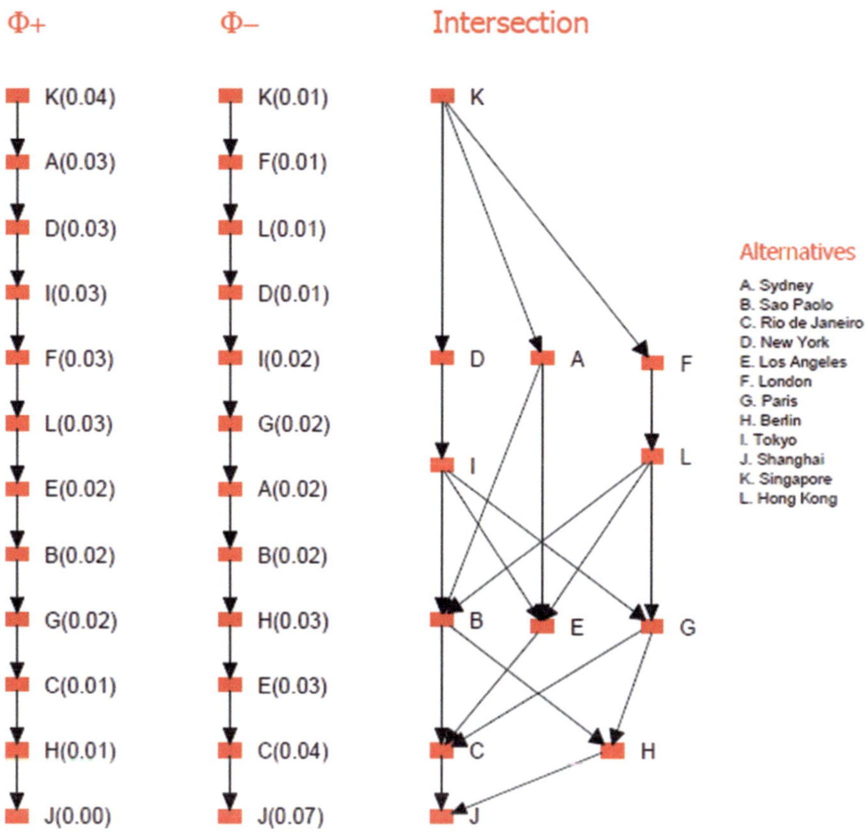

Fig. 10.10 Multi-criteria sustainability performance assessment web of domination relationships among megacities, NAIADE, A = 0.9

In this regard it would be highly beneficial to turn to the experience of Singapore in implementing a sustainability strategy and using policy instruments to achieve sustainability at the city-state level, which is more significant than that of other world capitals.

Singapore has created an Inter-Ministerial Committee on Sustainable Development (IMCSD), formed in January 2008 (Singapore 2009). This body was set up to formulate a national strategy for Singapore's sustainable development. The IMCSD was co-chaired by the Minister for National Development, Mr Mah Bow Tan, and the Minister for the Environment and Water Resources, Dr Yaacob Ibrahim. The members included the Minister for Finance, Mr Tharman Shanmugaratnam, the Minister for Transport, Mr Raymond Lim, and the Senior Minister of State for Trade and Industry, Mr S Iswaran. Setting very high aims of reaching a 70 % recycling rate by 2030, achieving a 35 % improvement in energy efficiency from 2005 levels by 2030 and reaching a level of domestic water consumption of 140 l per person per day by 2030, the Strategy for Sustainable Growth formulated in 2009 presented a road map to the situation we observe today. The aim of the strategy for Singapore was to

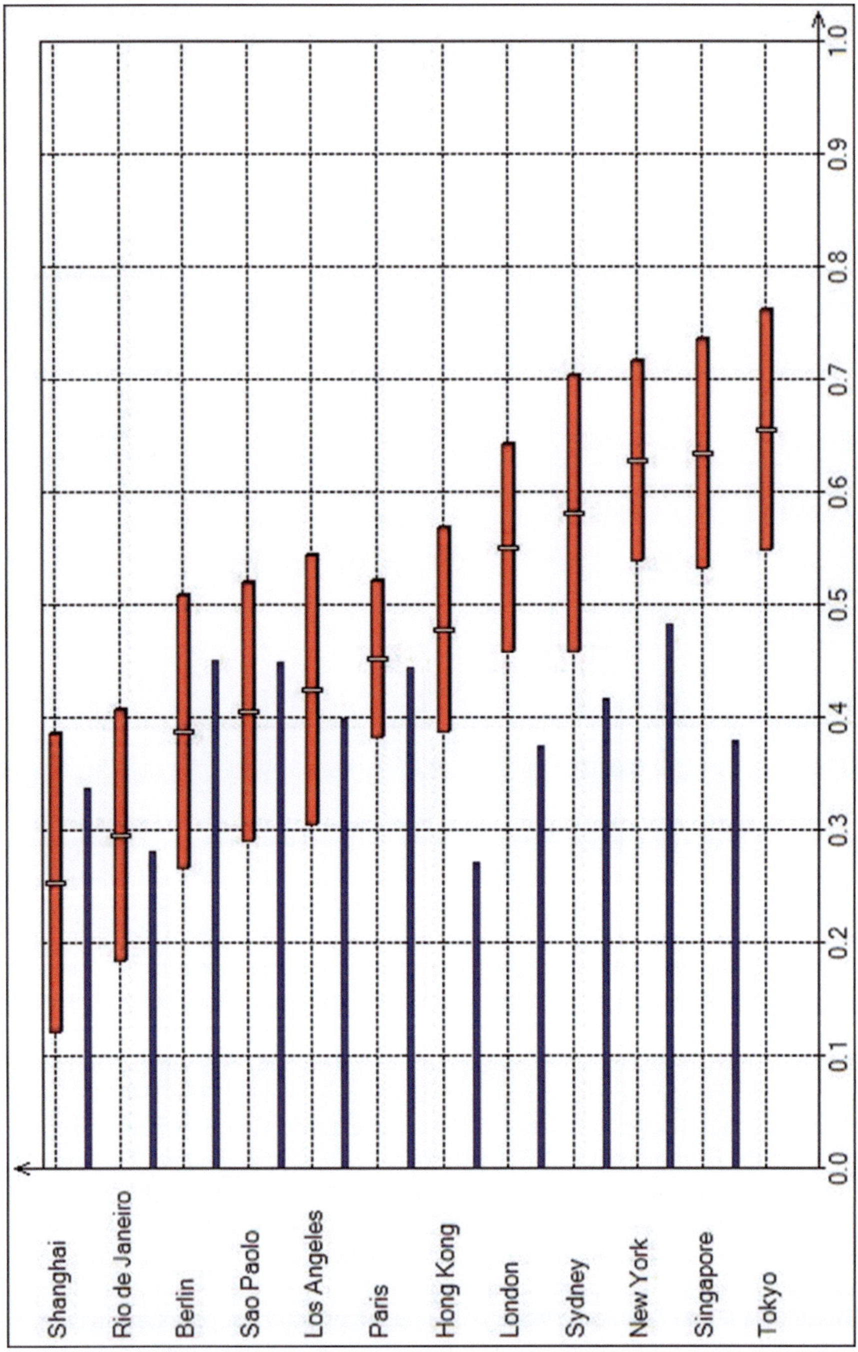

Fig. 10.11 Multi-criteria comparative sustainability performance assessment of megacities: APIS, priority of economic over environmental criteria

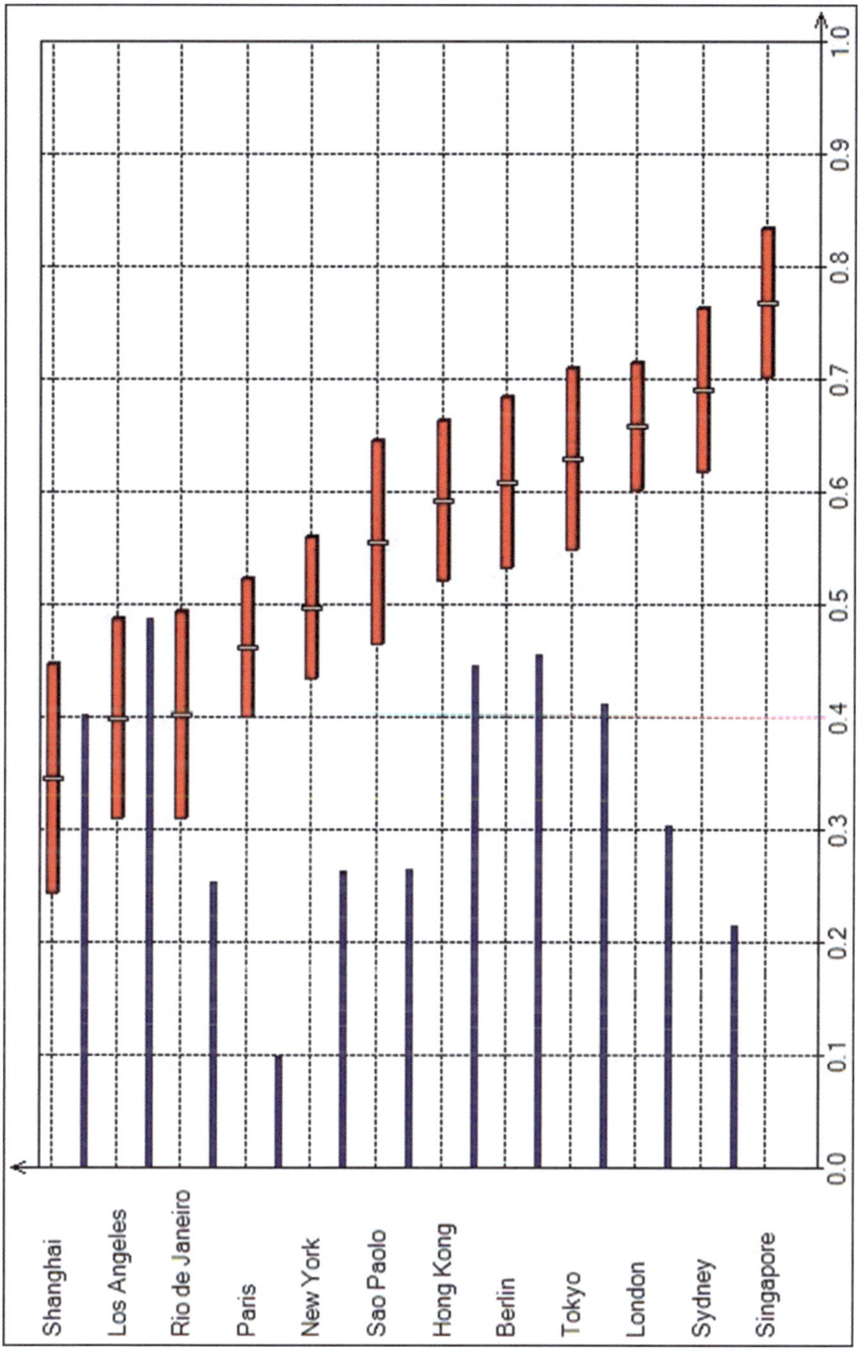

Fig. 10.12 Multi-criteria comparative sustainability performance assessment of megacities: APIS, priority of environmental over economic criteria

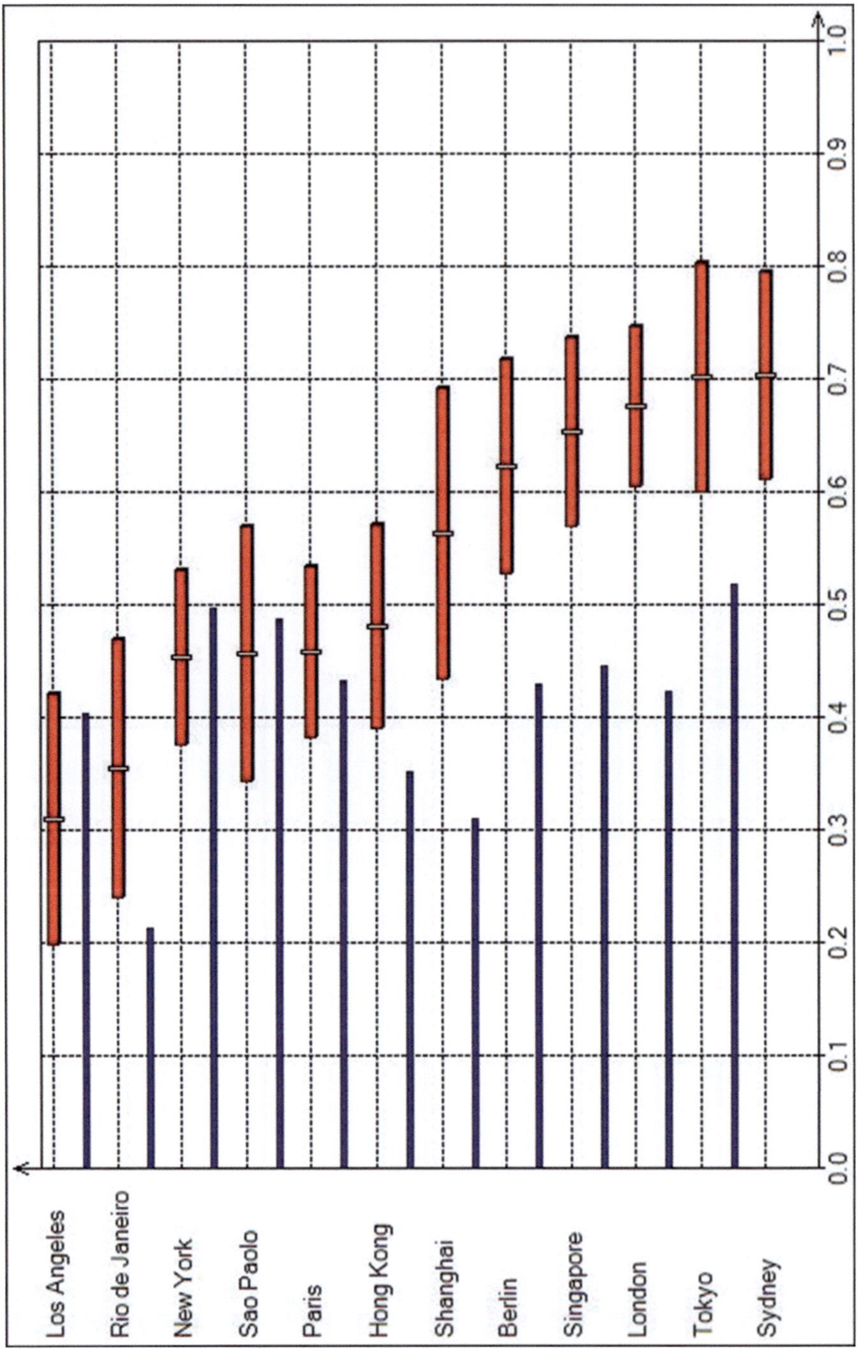

Fig. 10.13 Multi-criteria comparative sustainability performance assessment of megacities: APIS priority of social over economic criteria

become the top city in Asia in terms of quality of life and to develop as a sustainable, high density city that is clean and green, with excellent connectivity and a sense of space. The strategy set the aims to reduce the levels of PM2.5 to 12 mg/m^3 by 2020, to have 0.8 ha of green space for every 1000 residents and at the same time ensure that70% of all journeys in the city are made by public transport. As of 2011, Singapore achieved a recycling level of 61 %. In the process of designing the Blueprint 700, people from the non-governmental organisations, businesses, grassroots organisations, academia, media and mayors were consulted and over 1300 suggestions were received from the public. Knowing that Singapore does not have viable resources of renewable energy (wind, geothermal or hydropower), the strategy focused on: (1) *raising energy efficiency* by pricing energy appropriately to reduce environmental impacts, providing information for better decisions, boosting energy-efficient industry designs, processes and technologies, building capabilities in renewable energy, promoting resource-efficient buildings, promoting public transport, expanding water supply, improving water efficiency, and minimizing waste. At the same time the decision was taken to to stimulate facilitation of household recycling and enhance land use planning. The further priority of (2) *enhancing urban environment* was aimed at reviewing emission standards, adopting new technologies, pricing pollution, improving water quality, making the city cleaner, improving transport links, enhancing the city's greenery and conserving urban biodiversity. The third priority of (3) *building capabilities* implied investment in R&D and facilitation of international sharing of knowledge and, finally, the fourth goal of (4) *fostering community action* was focused on promoting community efforts, promoting industrial efficiency and stimulating development of the the public sector (Singapore 2009).

In the past 10 years, Singapore estimated the potential damages from congestion to be in the range of $2–3 bln per annum and introduced a smart-card innovation for public transport, designed by IBM, which covered road tolls, bus travel, taxis, metro and even shopping and was capable of registering 20 mln transactions per day and collected extensive traffic data, allowing city administration to constantly change routes to minimize congestion. The National Water Agency developed the Newater initative, through which a new Siemens-designed desalination plant and a water recycling scheme provide up to 30 % of Singapore's water needs, and two thirds of the Singapore's land surface became a water catchment area. Between 2000 and 2007 the share of electricity produced by natural gas increased from 19 % to 79 % in Singapore, thereby reducing the harmful CO_2 emissions. Since 2005, over 1650 buildings in Singapore were made environmentally friendly. At the moment around 80 % of residents of Singapore are living in public housing provided by the Housing and Development Board. In a short space of time using a highly focused and strategic approach, Singapore achieved a great deal in the *economic* sphere by attracting 7000 international companies and securing one of the highest per capita income levels in the world, in the *social* sphere by keeping very low unemployment at 1.8 % and achieving success in the *environmental* sphere by reducing the amount of waste generated per person and increasing recycling levels to 61 %, keeping PM_{10} pollution at a relatively low level of 32 μG/m^3 through developing public transport and increasing the green space to 47 % of its territory.

It is particularly reassuring to see such tremendous success achieved in Singapore through intensive interdepartmental and interdisciplinary collaboration (Singapore 2009); the case for which was outlined in our earlier paper (Shmelev and Shmeleva 2009).

The leading position of Singapore on unemployment is explained by the innovative government policies in this field: generous subsidies are offered for the upgrading of skills through the Continuing Education and Training (CET) system. Within the system, jobseekers receive subsidies in the range of 70–90 % of training course fees. The government also offers Place-and-Train (PnT) programmes to help jobseekers find work and pick up the skills they need in their new jobs. At the same time, the government recently improved the Work-Life Grant under the WorkPro scheme to provide support to companies introducing and maintaining good work-life practices, such as flexible work arrangements.

10.8 Conclusion

In this paper we focused on megacities; which are global centres for economic activity. Megacities are responsible for a considerable share of global emissions of greenhouse gases, require considerable amounts of water and produce substantial volumes of waste. We applied Principal Component Analysis to a large set of indicators comprising economic, environmental and social dimensions. This helped to highlight the relative positions of various cities on aggregated dimensions and projections of individual variables on the same plane. Computation of the correlation matrix helped to eliminate highly redundant indicators from the set. The application of multi-criteria decision aid allowed us to produce a multidimensional web of domination relationships among the top 12 world cities on 11 sustainability criteria with the help of ELECTRE and NAIDE. At the same time, application of APIS produced aggregate performance scores of megacities under three policy priorities: priority of economic over environmental criteria, priority of environmental over economic and priority of social over economic criteria. Uncertainty was taken into account via randomisation of the weights using a Monte Carlo technique. The assessment carried out using three multi-criteria tools identified sustainability leaders (Singapore, London, Sydney) and those that are lagging behind (Shanghai, Rio de Janeiro, Los Angeles). Application of APIS confirmed the results of the Principal Component Analysis and identified Singapore as the leader under a priority of environmental over economic criteria, Tokyo as the leader under a priority of economic over environmental criteria and Sydney as a leader under a priority of social over economic criteria. The result has put the performance of individual cities within the global context and presented the indicator- based sustainable development performance of individual cities within a coherent framework of multi-criteria decision aid. Learning from best practices and worst cases in this context provides an invaluable insight for policy reform to create smarter, greener, more compact, socially diverse, economically strong and less polluting cities around the world.

References

Afgan NH, Carvalho MG, Hovanov NV (2000) Energy system assessment with sustainability indicators. Energ Policy 28:603–612
Ala-Mantila S, Heinonen J, Junnila S (2013) Greenhouse gas implications of urban sprawl in the Helsinki metropolitan area. Sustainability 5(10):4461–4478
Barles S (2009) Urban metabolism of Paris and its region. J Ind Ecol 13(6):898–913
Bautista J, Pereira J (2006) Modeling the problem of locating collection areas for urban waste management. An application to the metropolitan area of Barcelona. Reverse Prod Syst 34(6):617–629
Bolund P, Hunhammar S (1999) Ecosystem services in urban areas. Ecol Econ 29:293–301
Chang YJ, Chu CW, Lin MD (2012) An economic evaluation and assessment of environmental impact of the municipal solid waste management system for Taichung City in Taiwan. J Air Waste Manag Assoc 62(5):527–540
Chen GQ et al (2013) Three-scale input–output modeling for urban economy: carbon emission by Beijing 2007. Commun Nonlinear Sci Numer Simul 18(9):2493–2506
City of London (2009) City of London carbon footprint
City of New York (2012) Inventory of New York City greenhouse gas emissions
City of Rio de Janeiro (2011) Greenhouse gas inventory and emissions scenarios of Rio de Janeiro, Brazil
Dai C, Li YP, Huang GH (2011) A two-stage support-vector-regression optimization model for municipal solid waste management – a case study of Beijing, China. J Environ Manage 92(12):3023–3037
Dietzenbacher E, Velázquez E (2007) Analysing Andalusian virtual water trade in an input–output framework. Reg Stud 41(2):185–196
EC (2009) Sustainable development indicators an overview of relevant framework programme funded research and identification of further needs in view of EU and international activities. 127 pp
Eurostat (2001) Economy-wide material flow accounts and derived indicators. A methodological guide. Office for Official Publications of the European Communities, Luxembourg
Galante G et al (2010) A multi-objective approach to solid waste management. Waste Manag 30(8–9):1720–1728
Girardet H (2014) Creating regenerative cities. Abingdon, Routledge, 216 pp
Girardet H, Mendonca M (2009) A renewable world. Energy, ecology, equality. A report to the world future council. Green Books, Totnes
Gomez-Baggethun E, Barton DN (2013) Classifying and valuing ecosystem services for urban planning. Ecol Econ 86:235–245
Hall P (2014) Good cities, better lives: how Europe discovered the lost art of urbanism, Planning, history and environment series. Routledge, London, 356pp
Hall P, Pfeiffer U (2000) Urban future 21: a global agenda for twenty-first century cities. Routledge, London, 384pp
Hall P, Buijs S, Tan W, Tunas D (2010) Megacities. Exploring a sustainable future. 384
Hallegatte S et al (2011) Assessing climate change impacts, sea level rise and storm surge risk in port cities: a case study on Copenhagen. Clim Chang 104(1):113–137
Hammer M, Giljum S (2006) Materialflussanalysen der Regionen Hamburg, Wien und Leipzig
Hoff H et al (2013) Water footprints of cities – indicators for sustainable consumption and production. Hydrol Earth Syst Sci Discuss 10(2):2601–2639
Hovanov NV (1996) Analiz i sintez pokazatelej pri informazionnom defizite. St Petersburg State University Press, St Petersburg, 196 pp
Hovanov NV, Yudaeva M, Hovanov K (2009) Multicriteria estimation of probabilities on basis of expert non-numeric, non-exact and non-complete knowledge. Eur J Oper Res 195(3):857–863
Hui Y et al (2006) Urban solid waste management in Chongqing: challenges and opportunities. Waste Manag 26(9):1052–1062

Jenerette GD et al (2006) Contrasting water footprints of cities in China and the United States. Ecol Econ 57(3):346–358
Kandilioti G, Makropoulos C (2012) Preliminary flood risk assessment: the case of Athens. Nat Hazard 61(2):441–468
Kapepula KM et al (2007) A multiple criteria analysis for household solid waste management in the urban community of Dakar. Waste Manag 27(11):1690–1705
Leigh NG, Choi T, Hoelzel NZ (2012) New insights into electronic waste recycling in metropolitan areas. J Ind Ecol 16(6):940–950
Lenzen M, Dey C, Foran B (2004) Energy requirements of Sydney households. Ecol Econ 49(3):375–399
Leontief W (1936) Quantitative input and output relations in the economic systems of the United States. Rev Econ Stat 18(3):105–125
Leontief W (1970) Environmental repercussions and the economic structure: an input–output approach. Rev Econ Stat 52(3):262–271
Liang S, Zhang T (2012) Comparing urban solid waste recycling from the viewpoint of urban metabolism based on physical input–output model: a case of Suzhou in China. Waste Manag 32(1):220–225
Liang S, et al. (2012) Sustainable urban materials management for air pollutants mitigation based on urban physical input–output model.In: 8th world energy system conference, WESC 2010, 42(1), pp 387–392
LSE (2013) How cities are leading the next economy going green. A global survey and case studies of cities building the green economy, final report. LSE Library Services, London, 115 pp
Lü YP et al (2012) Cost-effectiveness-based multi-criteria optimization for sustainable rainwater utilization: a case study in Shanghai. Urban Water J 10(2):127–143
Marie de Paris (2011) Le Bilan Carbone de Paris. Bilan des émissions de gaz à effet de serre
Martinez-Alier J, Munda G, O'Neill J (1998) Weak comparability of values as a foundation for ecological economics. Ecol Econ 26:277–286
Mayor of London (2014) World cities culture report. 132 pp
Minciardi R et al (2008) Multi-objective optimization of solid waste flows: environmentally sustainable strategies for municipalities. Waste Manag 28(11):2202–2212
Mortazavi M, Kuczera G, Cui L (2012) Multiobjective optimization of urban water resources: moving toward more practical solutions. Water Resour Res 48(3):W03514
Munda G (1995) Multicriteria evaluation in a fuzzy environment. Theory and applications in ecological economics, Contributions to Economics Series. Physica-Verlag, Heidelberg, 255 pp
Munda G (2005a) "Measuring sustainability": a multi-criterion framework. Environ Dev Sustain 7(1):117–134
Munda G (2005b) Multiple criteria decision analysis and sustainable development. In: Figueira J, Greco S, Ehrgott M (eds) Multiple-criteria decision analysis. State of the art surveys, International series in operations research and management science. Springer, New York, pp 953–986
Munda G (2006) Social multi-criteria evaluation for urban sustainability policies. Resolving environmental conflicts:combining participation and muli-criteria analysis. 23(1):86–94
Newcombe K, Kalma JD, Aston AR (1978) The metabolism of a city: the case of Hong Kong. Ambio 7(1):3–15
Ni JR et al (2001) Total waste-load control and allocation based on input–output analysis for Shenzhen, South China. J Environ Manag 61(1):37–49
Okadera T, Watanabe M, Xu K (2006) Analysis of water demand and water pollutant discharge using a regional input–output table: an application to the city of Chongqing, upstream of the three Gorges Dam in China. Ecol Econ 58(2):221–237
Ornetzeder M et al (2008) The environmental effect of car-free housing: a case in Vienna. Ecol Econ 65(3):516–530
Oyoo R, Leemans R, Mol AP (2013) The determination of an optimal waste management scenario for Kampala, Uganda. Waste Manag Res 31(12):1203–1216

Peters GP, Hertwich EG (2006) Pollution embodied in trade: the Norwegian case. Glob Environ Chang 16(4):379–387
Polydecision (2008) Decision support system APIS for MEAD. Advanced user guide.http://poly-decision.com/blog/. 63 pp
Rosado L, Niza S, Ferrão P (2014) A material flow accounting case study of the Lisbon metropolitan area using the urban metabolism analyst model. J Ind Ecol 18(1):84–101
Roy B (1985) Méthodologie Multicritère d'Aide à la Décision. Economica, Paris
Roy B (1991) The outranking approach and the foundations of electre methods. Theor Decis 31(1):49–73
Roy B (1996) Multicriteria methodology for decision aiding. Kluwer Academic Publishers, Dordrecht
Roy B, Hugonnard JC (1982) Ranking of suburban line extension projects on the Paris metro system by a multicriteria method. Transp Res A Gen 16(4):301–312
Ruiz-Villaverde A, González-Gómez F, Picazo-Tadeo AJ (2012) Public choice of urban water service management: a multi-criteria approach. Int J Water Resour Dev 29(3):385–399
San Francisco Department for the Environment (2004) Climate action plan for San Francisco. Local actions to reduce greenhouse gas emissions
Santibañez-Aguilar JE et al (2013) Optimal planning for the sustainable utilization of municipal solid waste. Waste Manag 33(12):2607–2622
Schulz NB (2007) The direct material inputs into Singapore's development. J Ind Ecol 11(2):117–131
Shen LY, Ochoa JJ, Shah MN, Zhang X (2011) The application of urban sustainability indicators: a comparison between various practices. Habitat Int 35:17–29.
Shmelev SE (2010) Environmentally extended input–output analysis of the UK economy: key sector analysis. University of Oxford, Oxford, QEH Working Paper Series – QEHWPS183, 33 pp.
Shmelev SE (2011) Dynamic sustainability assessment: the case of Russia in the period of transition (1985–2008). Ecol Econ 70(11):2039–2049
Shmelev S (2012) Ecological economics: sustainability in practice. Springer, Dordrecht/New York
Shmelev S (ed) (2016) Green economy reader. Lectures in ecological economics and sustainability. Springer (in press)
Shmelev SE, Anger A, (2013a) Macroeconomic implications of ecosystem service change and management. WP2.1 literature review. Report for the UK national ecosystem assessment follow-up phase
Shmelev SE, Anger A (2013b) Macroeconomic implications of ecosystem service change and management. WP2.3 methods and models. Report for the UK national ecosystem assessment follow-up phase
Shmelev SE, Rodríguez-Labajos B (2009) Dynamic multidimensional assessment of sustainability at the macro level: the case of Austria. Ecol Econ 68(10):2560–2573
Shmelev SE, Shmeleva IA (2009) Sustainable cities: problems of integrated interdisciplinary research. Int J Sustain Dev 12(1):4–23
Shmelev SE, Shmeleva IA (2012) Sustainability analysis: an interdisciplinary approach. Palgrave, New York
Siemens (2009) European green city index. Assessing the environmental impact of Europe's major cities. 51 pp
Simon U, Brüggemann R, Pudenz S (2004) Aspects of decision support in water management—example Berlin and Potsdam (Germany) I—spatially differentiated evaluation. Water Res 38(7):1809–1816
Singapore (2009) A lively and liveable Singapore: strategies for sustainable growth. 130 pp
UN (2007). Indicators of sustainable development: guidelines and methodologies. New York
UN (2015) Technical report by the Bureau of the United Nations Statistical Commission (UNSC) on the process of the development of an indicator framework for the goals and targets of the post-2015 development agenda, Working draft. 44 pp

UNECE (2013) Framework and suggested indicators to measure sustainable development. Prepared by the Joint UNECE/Eurostat/OECD Task Force on measuring sustainable development, 27 May 2013, 179 pp
UNEP (2011) Towards a green economy. Pathways to sustainable development and poverty eradication. UNEP, Nairobi
UN HABITAT (2013) Planning and design for sustainable urban mobility: global report on human settlements. 348 pp
Vanham D, Bidoglio G (2014) The water footprint of Milan. Water Sci Technol 69(4):789–795
Venkatesh G, Brattebø H (2012) Assessment of environmental impacts of an aging and stagnating water supply pipeline network. J Ind Ecol 16(5):722–734
World Bank (2013) Building sustainability in an urbanizing world. A partnership report
Xi BD et al (2010) An integrated optimization approach and multi-criteria decision analysis for supporting the waste-management system of the City of Beijing, China. Eng Appl Artif Intel 23(4):620–631
Zarghami M (2010) Urban water management using fuzzy-probabilistic multi-objective programming with dynamic efficiency. Water Resour Manag 24(15):4491–4504
Zhang Y, Yang Z, Yu X (2009) Evaluation of urban metabolism based on energy synthesis: a case study for Beijing (China). Ecol Model 220(13–14):1690–1696
Zhang Z, Shi M, Yang H (2012) Understanding Beijing's water challenge: a decomposition analysis of changes in Beijing's water footprint between 1997 and 2007. Environ Sci Technol 46(22):12373–12380
Zhou SY, Chen H, Li SC (2010) Resources use and greenhouse gas emissions in urban economy: ecological input–output modeling for Beijing 2002. Commun Nonlinear Sci Numer Simul 15(10):3201–3231

Chapter 11
The Economics of Avoiding Dangerous Climate Change

Terry Barker

Abstract The problem of avoiding dangerous climate change requires analysis from many disciplines. Mainstream economic thinking about the problem has shifted after the Stern Review from a single-discipline focus on cost-benefit analysis to a more inter-disciplinary and multi-disciplinary risk analysis, already evident in the IPCC's Assessment Reports. This shift is more evidence of the failure of the traditional, equilibrium approach in general to provide an adequate understanding of observed behaviour, either at the micro or macro scale. The economics of the Stern Review has been accepted by governments and the public as mainstream economic thinking on climate change, when in some critical respects it represents a radical departure from the traditional treatment. The conclusions regarding economic policy for climate change have shifted from "do little, later" to "take strong action urgently, before it is too late". This chapter sets out four issues of critical importance to the new conclusions about avoiding dangerous climate change, each of which have been either ignored by the traditional literature or treated in a misleading way that discounts the insights from other disciplines: the complexity of the global energy-economy system (including the poverty and sustainability aspects of development), the ethics of intergenerational equity, the understanding from engineering and history about path dependence and induced technological change, and finally the politics of climate policy.

Keywords Climate change • IPCC • Cost-benefit analysis • Multi-criteria analysis • Macroeconomics • Uncertainty

---

This chapter is a revised and updated version of an earlier paper discussing the topic in the context of the Stern Review. See Barker (2008).

T. Barker (✉)
Department of Land Economy, University of Cambridge, Cambridge, UK

School of Environmental Sciences, University of East Anglia, Norwich, UK
e-mail: tsb1@cam.ac.uk

S. Shmelev (ed.), *Green Economy Reader*, Studies in Ecological Economics 6,
DOI 10.1007/978-3-319-38919-6_11

11.1 Introduction

Since the early 1990s, it has been internationally recognized that one consequence of economic activity has been the accumulation of greenhouse gases and that this may lead to climate change. This already threatens development in poor countries that are most vulnerable to climate variability. If unchecked, it will threaten future generations with unknown but potentially catastrophic climate events, given the availability of fossil carbon at current prices relative to carbon-free alternatives, which could raise concentrations to levels not seen for millions of years. At the same time the costs of reducing the emissions have been agreed as negligible in relation to expected growth in incomes. Yet, after over two decades, global action has been limited and emissions have continued to grow. Indeed, the political recognition of the urgency of the problem has only become evident at a global scale since the publication of the Stern Review of the economics of climate change in 2006.

One reason for this delay has been the response by traditional economists, dating from the publication of the first IPCC Report in 1990, to the concern and calls for action by the climate scientists. The response came in the form of the mis-application of a tool developed by equilibrium economists for prescriptive public policy: the cost-benefit analysis. This chapter addresses the question of how and why the focus of the *economics* of climate change has shifted from the single-discipline cost-benefit analysis, as in the IPCC Second Assessment Report in 1995, to the multi-disciplinary uncertainty analyses in the subsequent IPCC Reports of 2001, 2007 and 2014 and the radically different policy prescriptions of the 2006 Stern Review.

The application of traditional cost-benefit analysis has yielded, with some exceptions, policy prescriptions of very low carbon taxes and delayed action until more information is available on the problem and more R&D has been done to lower the costs of any response. The new uncertainty analysis, in contrast, suggests that a political global decision needs to be taken urgently on targets to avoid dangerous climate change and that cost-effective and equitable policies and measures should be implemented strongly without delay to accelerate progress towards a complete decarbonisation of the world economy.

Defining dangerous in this context is a social and political task. Governments have, over the years, repeatedly committed themselves to a target of a maximum rise of 2 °C for average long-term global temperatures by 2100 above pre-industrial levels, most recently in the UN Summit in New York, 2014.[1] This target has implications for greenhouse gas (GHG) concentrations over the period to 2100, and the pathways of GHG emissions to 2100. The IPCC's Fifth Assessment Report (AR5) on mitigation (Working Group 3) (2014, p.12) has summarized the implications.[2] For the likely achievement of limiting the temperature increase to 2 °C by 2100, GHG concentrations by that year should be in the range 430–480 ppm CO_2-eq or less, cumulated GHG emissions should be in the range 550–1300 $GtCO_2$ over the

[1] http://www.un.org/climatechange/summit/2014/09/2014-climate-change-summary-chairs-summary/

[2] http://report.mitigation2014.org/spm/ipcc_wg3_ar5_summary-for-policymakers_approved.pdf

period 2011–2050, and the reduction in GHG emissions compared to 2010 should be 41–72 % by 2050. The scenarios suggest that global CO_2 emissions will have to become negative (through sequestration and storage) by 2080 and beyond. Therefore to be reasonably sure of avoiding dangerous climate change defined as a 2 °C rise or less, the world should be aiming for almost complete decarbonisation by 2050 or earlier, to be safer. All sectors in all countries should be aiming to stop emitting GHG into the atmosphere as soon as possible without excessive cost and at reasonable carbon prices.[3] Without urgent action, the risks of losing coral reefs and pristine tropical rainforests appear significant, e.g. for a rise above 1–3 °C, the AR5 WG2 Summary has "very high confidence" of widespread mortality of coral reefs.

In this chapter I argue that the shift in economic thinking to urgent and strong action is more evidence of the failure of equilibrium economics in general to provide an adequate and coherent explanation of observed behaviour, from the personal (micro) to the global (macro) scales, with the aim of guiding policy. Equilibrium economics underlies most textbooks of economics and journals of economic theory. It is theory largely unsupported by formal scientific observation and empirical data. Over the past 50 years, it has been increasingly recognized as dependent on false assumptions about human behaviour and physical systems, and as based on a rigid and ill-informed interpretation of utilitarian ethics. The continued use of the assumptions, most pertinently in the application of cost-benefit analysis and computable general equilibrium models for climate policy, strongly suggests that their only justification is that they are required for the mathematics and computation to be tractable. Any empirical support for the theory has been generally incoherent, *ad hoc* and rhetorical, with the most outstanding example the fact that the multi-sectoral equilibrium modelling of GHG mitigation policy, which dominates the literature, is based typically on 1 year's data (and this is simply to calibrate the model to yield results of the right magnitude, rather than to provide empirical validation of the results).

In contrast, the Stern Review considers cost-benefit analysis of climate change as a marginal analysis inappropriately applied to a non-marginal multi-disciplinary systemic problem (p. 50). Both Stern (p. 163) and the IPCC Reports after 1995 take a multi-criteria approach rather than a narrowly monetary one and question cost-benefit analysis. This is one reason for the intemperate response from some traditional economists to the Stern Review.[4] The effect of the more general approach is

[3] The carbon price associated with this target is not the topic of this paper, but it is of great interest. With limited literature to assess such a price, AR5 suggests that it is in the range $US50 to $US100/tCO2 by 2030, rising after 2030 and in year 2010 prices. See p. 31 http://report.mitigation2014.org/drafts/final-draft-postplenary/ipcc_wg3_ar5_final-draft_postplenary_technical-summary.pdf

[4] Tol and Yohe (2007, p. 233–4) accuse Stern of "substandard analysis" and "dubious economics"; Nordhaus (2007a, p. 5) claims that Stern is "political" and the discount rates used are "extreme" (p. 6); Dasgupta (2007, 2008, p. 7) accuses Stern of choosing parameter values so "that they yield desired answers." Such charges are perhaps more appropriately made against neoclassical economics as in (Mirowski 1989, 2002; Barker 1996; Nelson 2001; DeCanio 2003; Ackerman and Nadal 2004; Foley 2006; Nelson 2006; Beinhocker 2006; Ackerman and Stanton 2014). The charges also contrast with the praise by leading mainstream economists and Nobel Laureates quoted in the

to criticise, qualify, and generally undermine such equilibrium thinking in the assessment of the costs and benefits of climate change and its mitigation. This is not new: government rejection of the cost-benefit valuation of human lives goes back to the IPCC Second Assessment Report. What is new is that the economics of the Stern Review has been accepted by governments and the public as mainstream and consensus economic thinking, when in critical respects it represents a radical departure from a traditional deterministic treatment and its messages for economic policy.

This chapter takes an interdisciplinary approach to the economics of dangerous climate change, contrasting the cost-benefit analysis (CBA) with the new economics of risk that acknowledges and respects the insights and analysis from other disciplines, namely the other social sciences, climate and geographical sciences, ethics, history, engineering as well as complexity and evolutionary theory.[5] This chapter sets out four issues of critical importance to climate policy, each of which have been either ignored by traditional economic modelling of the problem, or treated in a misleading way that discounts the insights from heterodox economics and other disciplines.

1. The global economy, with its use of energy and emissions of GHGs, is a complex, nonlinear dynamic system with technological change inherent in economic growth. Climate effects and responses to climate policy are potentially non-marginal changes to the system in the context of strong uncertainty.
2. The problem of intergenerational equity from climate change and mitigation is primarily an ethical one, and should be informed by moral philosophy rather than economics in isolation. Traditional economic models adopt an extreme form of utilitarianism, with a questionable choice and use of discount rates, ignoring the philosophical literature and the concept of justice.
3. Engineering and history inform GHG mitigation policy through studies of the production processes involving the supply and demand of energy, in particular the technical possibilities of accelerated decarbonisation of the energy system. Economic history is critical in understanding the relationship between mitigation policies, economic development and technological change because the time scales are long in relation to the life-times of most energy-using capital, so that the technologies involved can respond to carbon-price signals over the policy period. Traditional models assume continuity and path independence.
4. The politics of mitigation implies unstable alliances and trade-offs between governments and political parties. By the use of the social welfare function (required for the calculus), traditional economists simplify social choices and pre-empt political negotiation, claiming an optimality for their subjective assumptions and market interpretations.

Review, such as Mirrlees, Sen, Stiglitz, and Solow, and support after publication from Arrow, DeLong and Deaton among others. There are far too many references to include here, so I have only included examples of the mainstream literature and the new literature.

[5] Other disciplines are just as important, including the humanities, but I am restricting the discussion to a few issues. In particular, I do not consider in any depth the analysis of development, which is crucial for adaptation to climate change

This is an ambitious agenda and the issues can only be summarized, relying on the underlying literature. The implications for new economic thinking are often unclear, provisional and tentative in the face of the certainties of some cherished traditional beliefs. This chapter covers the issues one by one, but first the economics needs to be further explained and the Stern Review and IPCC Reports interpreted in the light of the shift in economic thinking.

11.2 "Traditional" and "New" Economics Approach to Climate Change

Understanding and solving the climate problem requires the synthesis and co-development of many disciplines with different traditions and links between them. Economics is relevant because it explains why human behaviour might lead to climate change (via economic choices and the use of the atmosphere for free waste disposal) and it provides systematic methods for assessing (and monetizing) costs and benefits of different activities and policies. However, it is important to distinguish between a general definition of economics[6] and the particular approach used (pre-Stern) in most of the literature on the economics of climate change. Following Beinhocker (2006) and Maréchal (2007) I shall call this particular approach "traditional" economics defined as

> …the set of concepts and theories articulated in… textbooks. It also includes concepts and theories that peer-reviewed surveys claim, or assume, that the field generally agrees on. (Beinhocker 2006, p. 24)[7]

---

[6] Economics is the study of social activity undertaken with its primary purpose the expectation of reward, which usually involves money, the motivations of such activity and its consequences both good and bad, e.g. for equity and the environment. In contrast the neoclassical economist Robbins (1932) defined economics as "the science which studies human behaviour as a relationship between ends and scarce means which have alternative uses" (p. 16), asserting that economics is a value-free science.

[7] It has become debatable whether neoclassical equilibrium economics is mainstream anymore (Colander et al. 2004; Dequech 2007). This paper argues that Stern has shifted the mainstream away from the traditional neoclassical approach to climate-change economics. See (Maréchal 2007) for a supporting view. For a more general view see (Hodgson 2007). Prominent economists are acknowledging that for macroeconomic growth "The right way to think about this complex set of issues is not clear, but it is clear that the competitive paradigm cannot be fully appropriate." (DeLong and Summers 2001). Arrow (2007) accepts the Stern estimates of costs and benefits, quoting the range of GDP mitigation costs (3.4 % to −3.9 %) from the meta-analysis in (Barker et al. 2006a, b). Akerlof (2007), the 2007 President of the American Economic Association, exposes the weaknesses of neoclassical macroeconomics and suggests that early Keynesian macroeconomic theory has more explanatory power. Deaton (2007) supports the Stern treatment of discount rates: "Whatever it is that is generating market behaviour, it is not the outcome of an infinitely lived and infinitely far-sighted representative agent whose market and moral behaviours are perfectly aligned, and who we can use as some sort of infallible guide to our own decisions and policies." (p. 4)

This traditional economics is epitomized by Samuelson's *Economics*, later co-authored by Nordhaus,[8] and based on the neoclassical mathematical synthesis promoted by Samuelson that came to dominate mainstream economics thinking in the late twentieth century. I shall contrast the traditional economics with a "new" economics, as in the title of Boulding's 1992 book, including complexity, evolutionary and Post Keynesian theory and emphasizing institutions, non-linear dynamics, and deep uncertainty.

Neoclassical economics is defined as being characterized by an emphasis on rationality, via the use of utility maximization, an emphasis on equilibrium, and by neglect of strong kinds of uncertainty, particularly of fundamental uncertainty (Dequech 2007, p. 290). The traditional economic approach to modelling climate change has been almost exclusively and narrowly neoclassical, adopting a version of expected utility theory with human welfare usually translating into private market consumption per head in the applied models. The theory is applied to utility across countries with huge differences in consumption and over 100 years or more into the future, when consumption can rise perhaps 20 or 30 times over. The formal approach to the problem has been cost-benefit analysis beginning in the 1990s (e.g. Cline 1991, 1992; Nordhaus 1991a, b, 1994, 2007b; Nordhaus and Boyer 2000). In simplified terms, the costs of climate change are set against the benefits of taking action in a way that allows comparison with other potential uses of public revenues. The price of carbon (the CBA literature is focused on the specific "social cost of carbon") is calculated in a supply-demand framework, such that the costs of climate change arising from the marginal addition of CO_2 into the atmosphere are matched in equilibrium with the benefits of not making the addition, i.e. the mitigation or marginal abatement, giving the "maximum benefit" for humanity, discounted over all future generations. The treatment of uncertainty and risk has been to ignore deep uncertainty and convert the asymmetric risks of long-term irreversible damages into certainty-equivalent damages, which are then discounted when compared to short-term costs of mitigation. The outcome is an "optimal" price of carbon, an indefinite "optimal" rise in global temperatures, and very modest prescriptions for action.

This method rests on the idea that individual preferences are fixed and utilities can be aggregated and converted into well-behaved mathematical equations in a "social welfare function", and differentiated to give stable marginal properties, as

[8] Samuelson and Nordhaus (2001) is the 17th edition of a textbook originally published by Samuelson in 1948. Yohe (1989), another contributor to the traditional literature, has published a study guide to this textbook. Nordhaus is taken as the exemplar in this chapter (rather than a straw neoclassical man) because of his economics, methods and distinction in the field. He provides full details of equations, exposing all his assumptions. His approach is followed by many others and his models are widely used for academic climate policy analyses. It is characteristic of his work, and of his school, that qualifications to the economics or results (e.g. extreme events, recycling of carbon tax revenues or induced technological change) are introduced independently, but subsequently there is usually a reversion to the core model even when the qualifications produce radical effects on the results (Nordhaus 1993, 2007b). Other problematic features one would expect in a CBA, e.g. environmental co-benefits of GHG mitigation, are ignored. Eventually, if the qualification becomes mainstream, it is accommodated in the model or analysis (Mirowski 1989, 2002).

the basis for climate policy. It also crucially assumes that all natural services can be converted to money and back again at any time, i.e. that there are no irreversible effects of climate change. These and other assumptions render CBA inadequate and misleading for climate policy analysis (Azar and Lindgren 2003; van den Bergh 2004; Ackerman and Heinzerling 2004). In summary, CBA "does not yield transparent or objective evaluations of benefits; rather, it renders the discussion of benefits obscurely technical, excluding all but specialists from participation. At the same time, political debate continues behind the veil of technicalities, as rival experts battle over esoteric valuation problems." (Ackerman 2004). The approach (Arrow et al. 1996; Pearce et al. 1996) became highly controversial in the IPCC Second Assessment Report, when the governments rejected an assumption used in the analysis that the costs of human life should be different for different countries, depending on their income levels. Since the CBA literature is voluminous, it is covered in subsequent IPCC reports, but any implications for policy are heavily qualified and extensive lists are given of damages that are not or cannot be monetised (IPCC 2007, Table SPM3).

However CBA continues to be the theoretical basis for those advocating the traditional climate policies. A leading example of the post-Stern CBA approach to the climate problem is Nordhaus (2007b), who concludes that "the Gore and Stern proposals…are more costly than [doing] nothing."(p. 177) A peer-reviewed survey of the costs of "doing nothing" in the CBA is to be found in (Tol 2005), which covers 80 peer-reviewed estimates, i.e. fulfilling Beinhocker's definition of traditional economics. Tol concludes "One can therefore safely say that, for all practical purposes, climate change impacts may be very uncertain but [it] is unlikely that the marginal damage costs of carbon dioxide emissions *exceed* [\$14/tCO_2][9] and [they] are likely to be substantially smaller than that." (p. 2073, italics added). An update (Tol 2007) covering 125 such studies confirms the median estimate of \$4/tCO_2. For comparison, a carbon tax of \$5/tCO_2 in 2005 dollars (arguably close to the starting rates of the "optimal" tax from this mainly 1990s literature before any extra allowance for risks of dangerous climate change was introduced in response to the issues raised by the Stern Review) converts to an increase of \$2.5/bbl in US crude oil prices in 2005, or about 2 % in US gasoline prices. Such numbers from this established body of literature appear unbelievably small if they are taken to be, as intended, the estimated "optimal" carbon price for preventing climate catastrophe.[10]

[9] To avoid confusion, all estimates have been converted to \$/tCO_2 in this chapter. The price base is given where known. Despite their apparent precision, Tol's estimates cover different base years and price bases, because he does not appear to have converted the estimates to a consistent basis. They are all in real terms in his original sources, but typically adopt different base years for the discounting and prices. This lack of comparability in the underlying data renders his averages undetermined, a problem not mentioned by Tol.

[10] They are also five times smaller than the carbon prices revealed by the market before the Great Recession starting in 2009 led to over-supply of allowances. Phase 2 of the EU emission trading scheme has a carbon price over \$20/tCO2 (March 2008), arguably to achieve a much weaker internal EU target than that implied as optimal by the CBA literature. Weitzman (2009) has shown how a more considered treatment of catastrophe radically changes the traditional conclusion.

In summary, the key policy messages, extremely stylized,[11] from traditional economics have been: (1) introduce a small carbon tax, rising indefinitely in real terms and, if technological change is considered, (2) wait rather than act now because the costs of mitigation will fall as a result of technological change. These policy prescriptions have come from an approach relying on deterministic cost-benefit analysis and high discount rates, assuming that technological change is exogenous to the economic system or not affected by the carbon price, and ignoring deep uncertainty.

Contrast the tone and implications of these statements with Stern's conclusions: "the benefits of strong early action far outweigh the economic costs of not acting" and "even at moderate levels of warming, all the evidence…shows that climate change will have serious impacts on world output, on human life and on the environment." (pp. xv and xv1). Stern acknowledges that technological change is partly driven by economic factors, such as the price of carbon, implying that the benefits from waiting are replaced by benefits from acting so as to induce the change and reduce the future costs (Köhler et al. 2006). Stern commissioned an application of the PAGE model (Hope 2003), which computes a probabilistic risk-based CBA, and considered the CBA approach as one of three lines of evidence from the literature.[12] Stern asserts that the economics of climate change are more appropriately concerned with risk rather than return, and with the development of technologies for mitigation, both features of the problem that has been evident from the early 1990s, when the scientific assessments began in earnest. This in turn implies that the economic problem is one of achieving political targets and lowest costs compatible with equity and effectiveness, rather than with the political and scientific problem of choosing the targets themselves.

[11] Dasgupta (2007, p. 5) summarises the traditional view (incorrectly but reflecting academic and political perceptions) as: "Nordhaus…has been studying the economics of climate change for over three decades. The most remarkable conclusion of his studies—conducted on his Dynamic Integrated Model of Climate and the Economy (DICE)—has been that, despite the serious threats to the global economy posed by climate change, little should be done to reduce carbon emissions in the near future; that controls on carbon should be put into effect in an increasing, but gradual manner, starting several decades from now. This conclusion has withstood the many modifications Nordhaus and others have made to the climate science embodied in DICE. Their idea is not that climate change should not be taken seriously, but that it would be more equitable (and efficient) to invest in physical and human capital now, so as to build up the productive base of economies (including, especially, poor countries), and divert funds to meet the problems of climate change at a later date." Nordhaus (2007b p. 237) in fact has carbon taxes starting in 2005. Later work on the RICE model yields strongly escalating carbon prices, with a median of $14 $/t$CO_2$ in 2015 (2005 prices) and much higher prices if the analysis uses the Stern report discount rates (Nordhaus 2011). Others using DICE (see note 14), but altering the assumptions, come up with Stern-like conclusions (Ackerman et al. 2010).

[12] It is significant that his critics have mainly criticized technical assumptions about discounting in a CBA and have asserted that he has based his conclusions on this CBA. In fact the results of the CBA are to be viewed as "indicative only" and to be "interpreted with great caution" (p. 174) and their discussion takes up only 16 out of nearly 700 pages of the Review.

Contrast also the pre-Stern traditional conclusion that "optimal" temperatures should rise indefinitely (Nordhaus 1994, p. 89; 1997, p. 324)[13] with the potential impacts of such warming from the IPCC: "Partial loss of ice sheets on polar land could imply metres of sea level rise projected to occur over millennial time scales, but more rapid sea level rise on century time scales cannot be excluded." (IPCC 2007, p. 13). Baer and Mastrandrea (2006) present a risk analysis of such dangers: "a precautionary approach requires near immediate efforts to 'bend the curve' of global emissions, and much steeper reductions [80% by 2050] than are currently contemplated" (p. 8). The traditional approach is unsuitable mainly because it simplifies or just ignores such deep uncertainties of long-term climate projections. It requires knowledge about the far distant future under climate conditions radically different from those of today (van den Bergh 2004), as well as many other assumptions about human welfare and behaviour needed to make the mathematics tractable (Nelson 2006 p. 93).

However, the literature coming to the conclusions from traditional economists does not exist in isolation from that of other disciplines addressing the same problem. Climate scientists address the likelihood and risks of extreme events, and draw conclusions about what "one can safely say, for all practical purposes". Ethics considers issues of inter-generational equity, when climate damages are uncertain and far in the future. Engineering and architecture give insights into how the capital stock can be designed to reduce greenhouse gas emissions. Economic geography and history provides understanding as to how economies grow and how technologies diffuse and evolve. Political science considers how societies make decisions regarding public policy. Furthermore, economics is not confined to the study of equilibrium in various guises, assuming groups of identical representative agents, with entirely self-interested consumers and known, quantifiable social welfare functions. All these economic assumptions and more are standard in the traditional cost-benefit analysis that lies behind the conclusions of Nordhaus and Tol[14] (DeCanio 2003). The weakening of the neoclassical analysis of climate change has been accompanied by a more general undermining of the ideology and prescriptions of traditional economics by deconstruction of the origins of the theory in physics and cybernetics (Mirowski 1989, 2002). Behavioural economics going back to (Kahneman and Tversky 1989) has revealed key relevant empirical findings for risk aversion and utility maximisation that are inconsistent with traditional treatments. Complexity theory and agent-based modelling has developed a theory of economic evolution (Arthur 1994; Beinhocker 2006).

Traditional economics has developed an approach to climate change, which has persistently ignored the conclusions and insights of other disciplines. The new

[13] "...the optimal climate change policy reduces long-run global warming from 6.6 to 6.2 °C." (Nordhaus 1997, p. 324). His more recent "optimal" rate is closer to 3 °C, see below.

[14] The literature covered by Tol requires the monetization of the costs of climate change and almost without exception adopts the standard assumptions. Alternatives approaches are provided by Barker 1996; Barker et al. 2006a, b; Ackerman and Heinzerling 2004; Padilla 2004; Maréchal 2007.

economics, as it has developed in the literature covered by Stern, is more pluralistic and respectful of other disciplines. The CBA approach is formally replaced by a Multi-Criteria Analysis (MCA) developed in management science and applied to sustainable development (Munda 2005) in which socio-economic, ecological, and ethical perspectives are taken into account. Most of the insights and techniques of CBA can be incorporated into the market (monetised) criterion in MCA, but as in the Stern Review and IPCC Reports, intrinsic values (non-monetised) of human suffering, damages to nature, and risks and uncertainties are also taken into account as criteria for social choice.

11.3 Uncertainty in Economic Systems: Equilibrium Versus Complexity

A critical issue in the understanding of climate change and its mitigation is the treatment of uncertainty. The use of the word "dangerous" in the UNFCCC objective[15] is interpreted by the IPCC (IPCC 2007, p.19) in terms of risks of climate change being balanced against the risks of threatening economic sustainability by the response measures. The problem is clearly one entailing unknown risks (uncertainties) of the climate system responding to the anthropogenic pulse of additional greenhouse gas, primarily CO_2, into the atmosphere being compared to the largely known risks associated with mitigation policies and the even lower risks of co-benefits of mitigation, such as improved air quality. The treatment of these risks and uncertainties distinguishes the traditional and the new economic analysis.

The classic text (Knight 1921) defines risk as the property of outcomes with quantifiable probabilities and uncertainty as that with unknown probabilities. Keynes made a similar distinction: 'By uncertain knowledge I do not mean merely to distinguish what is known for certain from what is only probable. About these matters there is no scientific basis to form any calculable probability whatever. We simply do not know.' [1973, pp. 113–14 quoted by Holt (2007)]. In the traditional CBA, the form of the probability density function is simply assumed, deep uncertainties are set aside despite the complexity of the scientific understanding, the risks are assumed to be symmetric despite the key feature of disproportionately higher risks from higher temperatures, and average discount rates are used for both costs and benefits, despite the differences in risks (Frederick et al. 2002). Many CBAs are simply deterministic, neither converting uncertainty into "certainty equivalence" nor subjecting the final model to a sensitivity analysis.[16] However, the shape of the damage function is uncertain and that the climate science suggests a significant (i.e. more than 1 %) chance of catastrophe, defined by Weitzman (2009) as average

[15] UNFCCC, 1992, states as its objective: "to achieve stabilization of greenhouse gas concentrations in the atmosphere at a low enough level to prevent dangerous anthropogenic interference with the climate system."

[16] Dietz, et al. (2007a) show how risk analysis can change the conclusions of the CBA.

global temperatures rising by more that 5 °C above pre-industrial. Weitzman has extended the traditional CBA analysis to convert the uncertainty into a second-order risk and his interpretation of the results is to question the validity of CBA: "Perhaps in the end the climate-change economist can help most by *not* presenting a cost-benefit estimate for what is inherently a fat-tailed situation with potentially unlimited downside exposure as if it is accurate and objective." Instead, he argues, economists should be "explaining better to policy makers that the artificial crispness conveyed by conventional IAM-based CBAs is *especially and unusually* misleading". And that "in rare situations like climate change, we may be deluding ourselves and others with misplaced concreteness if we think that we are able to deliver anything much more precise than this with even the biggest and most-detailed climate-change IAMs as currently constructed and deployed." (p.42)

The CBA estimates prevalent in the literature on the economics of climate change are highly misleading because the studies set aside or ignore deep uncertainty in costs and benefits.[17] A critical example is that the global long-run growth rates are almost entirely exogenous in the models (coming from labour supply and technology), so that the uncertainties of the effects of climate change on labour and technologies and then on growth are ignored. The estimated costs of climate change and the optimal carbon tax rates from the CBAs are essentially the subjective views of Nordhaus, Tol[18] and others presented as scientifically precise results. Tol in 2005 presented quantified ranges although he repeatedly acknowledges the uncertainty of the results; in contrast in 2007, responding to Weitzman's analysis, he acknowledges that the upper bounds to the costs may be unquantifiable.[19] Another example is the conclusion from Nordhaus' DICE study (July 2007) that "The total discounted economic damages with no abatement are in the order of $23 trillion" (p. 181). Such an estimate is much too precise because the uncertainties in the projections include possible worlds with temperatures >5 °C above pre-industrial averages. This matters because the economists who have made the calculations have developed a "scientific" consensus in defending the approach. They have argued that the "social cost of carbon" can be precisely estimated and ranges given with sufficient precision to form climate policy. Wietzman's point is that the upper boundaries in such ranges

[17] I exclude from this critique the integrated assessment modelling of Hope (2013), whose PAGE model emphasises uncertainties.

[18] Tol ignores the value judgment underlying all cost-benefit analysis, namely that the equilibrium outcome of rational self interest is in some sense optimal for society. He presents his survey of a set of essentially subjective estimates, often done by the same clique of authors, as observations of probability density functions, when they are basically undetermined (see note above), not properly sampled, or even independent.

[19] "… the uncertainty is so large that a considerable risk premium is warranted.… More importantly, there is a 1 % probability that the social cost of carbon is greater than $78/tC. This number rapidly increases if we use a lower discount rate – as may well be appropriate for a problem with such a long time horizon – and if we allow for the possibility that there is some truth in the scare-mongering of the gray literature." Tol 2007 (conclusion). Tol (2005) does not mention a risk premium or that the damages may be unbounded, although the fat tail of damages is emphasized as a feature of the analysis.

e.g. a top 95 percentile of about \$100/tCO_2 in Tol's 2005 study, are subjective and misleading. With notable exceptions[20] the literature on the social cost of carbon, by ignoring some uncertain damages, truncating probabilities and discounting risky outcomes, has been promoting an over-optimistic picture of the uncertainties of climate damage for the last 20 years.

There are also uncertainties associated with the costs of mitigation. In traditional economics, the Marginal Abatement Cost (MAC) is the cost of abating the marginal emission of GHGs and the Marginal Abatement Benefit (MAB) is the benefit, in terms of the reduction in the damages caused by the emissions and converted into money. By assumption, as the level of abatement rises, costs go up and benefits go down, so that there is an equilibrium solution for this level in which the marginal cost equals the marginal benefit, utility is maximized and the optimal carbon tax rate can be calculated. When used correctly, *marginal* means 'vanishingly small' since it is calculated by differentiating cost curves that are required by the theory to be continuous, but in equilibrium models of climate change 'marginal' is used, loosely, to refer to discrete, system-wide changes, e.g. a shift from fossil fuel to hydrogen in transportation. The *benefits* in the MAB are the avoided costs of "doing nothing" as discussed above, including the monetary estimates of risking long-term catastrophe. The *costs* in MACs can be a large range of disparate, but largely shorter-term, costs both private and public (or social) with and without market-based valuations, but all associated with the abatement. It is clear that these costs do not normally include any political costs of introducing abatement policies and measures. These economic costs may be offset by ancillary environmental benefits or improvements in efficiency from the use of tax revenues, but the CBA literature often either ignores these associated benefits or sets them aside as being too uncertain or assumed to be managed by non-climate policies.

There are serious problems with the MAC concept and the total costs derived from the models that use it.[21] The first problem lies in the treatment of uncertainty and technology. The reasoning assumes that the future schedules of costs are known in advance and independent of policy. In fact they are uncertain and evidence shows that they are likely to respond to policy: low-carbon technologies can be expected to develop in response to higher real prices of carbon, just as energy-saving technologies have responded to energy prices (Popp 2002). If so, the MAC schedule is not independent of the cost of carbon, so the schedule is both uncertain and unstable. As noted above, in general the risks and uncertainties of abatement are much less than those of the damages. If decision-makers are risk averse, or wish to follow the Precautionary Principle, the fact that abatement costs are less uncertain is likely to justify action involving higher costs than what otherwise would be the case. Additionally mitigation reduces the risks of the damages *and* of future adaptation,

[20] Examples, using Nordhaus' DICE model are the studies by Mastrandrea and Schneider (2001, 2004), Azar and Lindgren (2003) and Ackerman and Findlayson (2007).

[21] This is not to undermine the usefulness of the static incremental abatement schedule, showing abatement options as a set of non-linear steps rising as the carbon price rises, and usually referred to as the MAC.

i.e. it reduces the cascade of risks from the emissions. A deterministic equalisation of estimates of costs and benefits without taking into account these uncertainties ignores fundamental differences between them.

A second problem is the one Stern identifies as the non-marginal nature of the economics. As a result of discontinuities and path dependence in the economic system, the placing of system changes within the apparatus of continuous cost schedules is misleading because different mixes of old and new technological systems (e.g. a mix of oil-based and hydrogen-based systems) appear to be highly unlikely because of economies of scale and specialization and lock-in effects. The complex system effects may be large enough to achieve significant reductions in costs under new technologies. If there are indivisibilities, e.g. a global electricity grid for low-cost renewable generation, there is no longer a unique solution for the equilibrium carbon tax. There is evidence for such system properties from the investigation of future costs of energy systems undertaken by researchers at IIASA (Gritsevskyi and Nakicenovic 2000). It is also obvious at the micro scale that the technologies and costs of mitigation are not continuous. This appears to be the case at the macro scale because of network economies and technological lock-in. Not only are there significant discontinuities in the abatement cost schedule, the costs are likely to go up or down for different levels of abatement depending on the technological system under study.

The assessments of the Stern Review (p. 338) and the IPCC's AR5 (WG3, Technical Summary, p. 31)[22] of escalating macroeconomic costs of mitigation as targets become more stringent (450 ppm CO_2-eq and below) are also open to question, since the underlying literature largely assumes continuity and limited technological and institutional options for mitigation. Macroeconomic costs may not escalate when policies lead to decarbonisation, although carbon prices, energy investments and the policy "effort" are all likely to rise, perhaps disproportionately and the macroeconomic costs do become more uncertain at higher carbon prices.[23] All the available mitigation options have not been investigated and included in the models in terms of the speed at which they need to be implemented or their eventual scale. Extrapolation of the costs in the literature (Barker and Jenkins 2007) suggests that, depending on policies, macroeconomic costs for more stringent mitigation will remain negligible,[24] but risks of policy mistakes increase.

The modelling of economic risks in the context of climate change has been taken forward with the post-Stern work of Dietz et al. (2007a, b), in which the effects of adding risk to the CBA are shown to increase the costs of climate change significantly.

---

[22] http://report.mitigation2014.org/drafts/final-draft-postplenary/ipcc_wg3_ar5_final-draft_post-plenary_technical-summary.pdf

[23] Tol (2007) asserts that high carbon taxes would bankrupt some countries. He seems to be confusing tax revenues with tax payments. The tax revenues accrue to governments and benefit their finances; they benefit the population if used wisely. One such use is to ensure that all home owners adopt low-GHG technologies. The tax payments may not be a problem if safeguards to protect vulnerable social groups predate or accompany the introduction of a carbon tax.

[24] See Barker and Crawford-Brown (2013) for a discussion of the use of meta-analysis to suggest that macroeconomic costs should be higher.

The key features of a risk analysis of the problem are that the risks and uncertainties associated with the climate damages are much greater, because of systemic irreversibilities and non-linearities, compare to those of mitigation, which are largely known from past experience, and that the air quality benefits of mitigation are even less risky, more immediate, and well-documented (Barker et al. 2007).

A more flexible "new economics" alternative modelling approach to equilibrium modelling (Barker et al. 2006a, b, 2012) is based on the economic history of institutional structures. It emphasizes the importance of accounting and economies of specialization and allows for increasing returns to scale in the factor demand equations. In critical sectors, technology is modelled to allow for reductions in unit costs as the scale of production increases and the markets develop. Scenarios incorporating system-wide changes in technology, e.g. those involving the hydrogen economy, can be developed consistently. This approach does not impose costs of mitigation by assumption, unlike general equilibrium, so that an alternative low-carbon world economy may be less costly than business as usual, depending on the reductions in costs that emerge when new technologies come into widespread use.

11.4 Economic Ethics, Intergenerational Equity and the Discount Rate

Neoclassical economists claim that their work is value-free (Robbins 1932), scientific (Nordhaus 2007b) or purely descriptive (Pearce et al. 1996; Nordhaus 2007b). In doing so, it has been plausibly argued that they are drawing on nineteenth century science to promote a secular, rationalist religion (Nelson 2001, p. 133). Their faith is in the path-independence of consumer preferences and producer technologies, a faith shown to be empirically false in psychology, physics and history. Their thinking, apparently logical, is based on the fallacy that "the pursuit of self interest is guided by objective laws to a socially beneficent outcome" when instead this pursuit involves moral choices, at both personal and social levels (Foley, p. xiii).

Nordhaus (2007b, p. 140–1) characterizes economics as scientific in being peer-reviewed and reproducible; he derives the discount rate from a pure description of the market rather than from a consideration of ethics and moral philosophy. He contrasts his approach with that of the Stern Review, which he finds unscientific.[25] The many critiques of the Stern Review have been dominated by the discussion of

[25] There is a literature devoted to the issue of whether economics is a science or not. See (Mirowski 1989, 2002; Weintraub 2002; Katzner 2003). It is a science in that theory and observation are considered together when and where possible or in that mathematics is a science (Samuelson's position). However, neoclassical path-independent economics as a mathematical science is strictly a branch of mathematics rather than economics, since it violates a basic law of physics, the Second Law of Thermodynamics. Nordhaus (2007b) himself is ambiguous about whether economics is a science or not, since he repeatedly distinguishes economics from science.

the discount rate[26] (e.g. Nordhaus 2007a; Dasgupta 2007, 2008; Tol and Yohe 2007). The "pure rate of time preference" is one component of the discount rate used in calculating the costs of doing nothing in relation to climate change. Stern adopts a pure rate of time preference of near zero, drawing on moral arguments, compared with the rate adopted by the traditional literature, e.g. 1.5%pa in (Nordhaus 2007b) down from 3%pa in earlier applications of his DICE model. The difference in discount rates between Stern and the traditional literature is one of the reasons (cited by several of the critics as the main reason) for the much higher costs of 5–20 % of global GDP "now and forever" estimated by Stern (p. xv)[27] for business-as-usual of 2–5 °C warming or more, compared with the cost of 6 % of GDP for a 10 °C warming from Cline (1992), who also used a near-zero discount rate as quoted in the IPCC Second Assessment Report (Pearce et al. 1996, p. 208), but had a different approach to risk, yielding lower costs.

The detailed deconstruction of this difference in the "costs of doing nothing" is covered by Quiggin (2008), so the discussion here can be brief. The first point is that moral philosophers have long debated the relative weighting to be given in utility theory between social groups living at different times. The Stern Review commissioned a review of the ethics of climate change from Broome (2006), who had written earlier on the issue (1992). He makes uncomfortable reading for traditional economists, partly because he insists, rightly, that economics is not ethics-free, that basing economics on the ethics of individuals assumed to be entirely self interested can go badly wrong, and that "willingness to pay" is invalid as a means of valuation (Broome 2008). This is in direct contradiction to the analysis of Pearce et al. (1996, p. 196 197) and Nordhaus,[28] when they contrast prescriptive with descriptive valuations of human life. In considering the ethics of climate change, Broome positions *justice* centre stage, arguing that those who cause climate change should cease to do

---

[26] See Quiggin (2008) for an explanation. The use of a zero discount rate specifically for climate damages in a cost-benefit analysis of climate change (Hasselmann et al. 1997) anticipated in some respects Stern's use of low discount rates and also set off a fierce debate with those supporting an aggregate discount rate for all types of damage (including loss of human life) or sectors, which they justified by the traditional neoclassical treatment (Heal 1997; Nordhaus 1997) relying on the assumption of social groups being identical representative agents having full information and foresight. This traditional approach also denies any significance to the empirical finding of differences between sectors in the discount rates actually used (e.g. private rates being several times public discount rates). Hasselmann's reply (1999) also anticipates the emerging resolution of the post-Stern debate discussed here, specifically the conclusions in (Hoel and Sterner 2007) from a two-sector model.

[27] It is not the only reason. Dietz et al. (2007a) provide four reasons for the higher costs in the Stern Report. Previous studies, with important exceptions, have "(i)…mostly omitted to adequately employ the probabilistic results of recent science; (ii)…tended to consider a narrow range of impacts, a product of focusing largely on 2–3 °C warming, whereas we now know that there is a possibility of far higher temperatures; (iii)…not used the economics of risk to the extent appropriate; (iv)…not paid adequate attention to the underlying ethics." The overall effect has been to give "on average, strong downward bias on damage estimates in the previous literature". (pp. 156–157)

[28] Nordhaus (2007b) claims that his 1.5%pa pure rate of time preference is "designed to provide the most accurate projections rather than to be normative in nature." (p. 40).

so because it is unjust, and if they cannot cease, then they should compensate those who suffer.

Justice as a theory of ethics (Rawls 1971) deserves serious attention as an alternative to utilitarianism in climate-change analysis. Consider two population groups: a well-off urban majority, burning fossil fuels, and a subsistence rural minority, dependent on the weather for food and water. Assume that the costs of mitigation are negligible as the literature suggests. Assume also that the rural minority do not share in average global growth; they can be said, in Rawls' words, to be the 'least advantaged group'. In his theory, the standard of living of the most advantaged would be justified only if their privileges maximised the welfare of the least advantaged group, for example through the general effect of incentives on the economy. Let us assume there is no such Rawlsian theory of justice in place. If policy were to be formulated according to the traditional assumptions with the results described above, the outcome would be a triple injustice:

1. The rural minority have not been responsible for today's GHG concentrations causing climate change, and have not benefited from the comfort and power given by the fossil energy services, yet being dependent on the weather suffer most of the consequences.
2. The minority will suffer much more from future climate change because droughts and floods threaten their subsistence income, and they cannot buy their way out of the problems.
3. The minority's future income is discounted by an average dominated by the well-off majority's income growth, so their future utility counts for much less in the CBA of global climate-change policy. This outcome is a direct consequence of the discounting of average consumption net of climate damages (including mortality) by Nordhaus and others, supported by Dasgupta (2007).[29]

Since global inequalities over the last century have been increasing (Piketty 2014) and a subsistence minority of countries (and social groups within countries) may continue into the far future, the assumptions may well be more realistic than those of the traditional model. Rawlsian ethics would focus social policy on preventing the climate change and caring for the subsistence minority. Instead, traditional models have been used to justify weak policies and inaction.

Broome (2006) also considers expected-utility theory alongside justice as a guide to social policy. Importantly he distinguishes "value" from "utility" and allows for intrinsic value in human life and nature. He considers the utilitarian view of climate change, arguing that (1) lives should not be valued by the method of willingness-to-pay, which makes the value of people's lives depend on how much they can afford to spend on prolonging them and (2) future lives should not be

[29] Like Broome, Dasgupta (2005) is an authority on economics and ethics, but he argues that traditional economics has solved the ethical problems. However, Dasgupta does not mention the ethical problems involved with the averaging done in CBA, when the assumed monetary estimates of health and lives are discounted.

discounted in value relative to present lives of similar quality.[30] The argument that because people in the future are expected to be better off in real money terms, so that we can then discount a monetary value of their lives (or their health) runs into serious logical and moral problems, which are not solved by recourse to the term "statistical lives". Nordhaus (2007b, p.47) is discounting the quantity and quality of human life when he includes valuations of mortality and morbidity in the damages from catastrophe (which are stated to include health damages).[31] Implicitly, those who discount such damages at 1.5%pa and higher are valuing the next generation's lives and health at a fraction of their own. An equal valuation would transform the policy prescriptions towards urgent action and high carbon tax rates.

Nordhaus and others rely on the market to provide an estimate of the social discount rate. The preferences underlying the rate are assumed to be fixed and to take into account far-future climate damages. Such assumptions are not empirically valid and the procedure short-circuits the political process, in which for example democratically elected politicians aim to lead and change preferences (see below). The preferences are also assumed to take a particular form, in which no *ethical* preferences are allowed, although in fact people might prefer that natural resources be preserved as a matter of principle, even though they have no utility for them. Finally these authors are assuming fungibility of natural and man-made assets, i.e. that they all have monetary values and can be exchanged. Irreversible changes, e.g. warming of the oceans leading to loss of coral reefs for the indefinite future, means that such exchange is impossible. Hoel and Sterner (2007) have explored an extension to the traditional model allowing for human and natural services and the likelihood that as the natural services become scarcer, their prices will rise. They conclude that under reasonable assumptions the discount rate could become negative.

It is the implicit assumption on the part of traditional economists of a 'moral' superiority of the market that is at the heart of this debate. Utilitarian philosophers will have none of it. Traditional economists evade this implication of their analysis, claiming that they are being descriptive rather than prescriptive, but their logic does not stand up to scrutiny.[32] This is economics as a religion (Nelson 2001), in which society is composed of self-interested individuals, whose behaviour is to be assumed rational, then to be interpreted and described by economics as a mathematical science, e.g. in finding and using the pure rate of time preference, or the value of

[30] Broome's view on discounting is supported by the utilitarian philosopher R M Hare, who likewise argued that a discount rate above zero cannot be justified ethically (Hare 1981, p.100–101).

[31] Dasgupta (2007) supported Nordhaus's approach, but not his adopted pure rate of time preference. In contrast, Stern argues that human lives and environmental quality should be treated separately (p. 165), although the PAGE model (used to calculate the 5–20 % range on costs) appears to include valuations of human life and health implicitly in its damage functions. Dasgupta (2008) no longer supports Nordhaus, and concludes that "an optimum policy may not exist", and (implicitly) that CBA is "an overly formal analysis" leading to "misplaced concreteness" in its conclusions (since 1991) on climate change.

[32] See Broome (2008). Nelson (2008) reveals other hidden value judgments that may underlie the traditional neoclassical approach to climate change, leading to the tendency to rigidity and blindness to errors evident in the critical responses to the Stern Review.

human life. The underlying fallacy is that market forces lead by themselves to intrinsically good outcomes (Foley 2006). A "new economics" approach is to acknowledge that there are ethical, aesthetic and other values, and that all life should not be converted into money, with the exchangeability that money permits (Ackerman and Heinzerling 2004; Gowdy 2005). The use of the discount rate to account for time preference and risk should be re-thought to allow for subjective time preference and a risk analysis independent of the return (Price 2005). Climate change economics should learn from organizational sciences applied to management of high-risk activities (Nelson 2008). The distribution of rights consistent with sustainable development should be considered (Padilla 2004). The anti-utilitarian moral philosopher Bernard Williams has criticized the reductionism of "utilitarian thought" and "the device of regarding all interests, ideas, aspirations and desires as on the same level, and all representable as preferences of different degrees of intensity, perhaps, but otherwise to be treated alike. The assimilation does not give our convictions enough weight in our own calculations. At the same time, it can give other people's convictions too much weight" (Williams 1985, p86). The utilitarian approach has much to offer, but its claims should be qualified and limited by considerations other than that of utility, such as those of justice, well-being of future lives, and benevolence.

11.5 Engineering and History: Induced Technological Change and the Costs of GHG Mitigation

The costs of GHG mitigation in traditional economics are derived from the production function, a concept basic to the determination of the allocation and growth of economic output, conventionally measured as marketed output, i.e. GDP in national accounts. In the models this function takes special forms, typically Cobb–Douglas or Constant Elasticity of Substitution (CES) with tractable properties: they are continuous, typically with constant returns to scale, and reversible in that outputs can be expressed in terms of their inputs of labour, capital, materials, and *vice versa*, a feature that contradicts path dependence, i.e. the second law of thermodynamics.[33] This economics has been derived by analogy with physical process of the first law of thermodynamics by Walras drawing on nineteenth Century textbooks of physics (Mirowski 1989; Beinhocker 2006) without an adequate treatment of time or the later second law with the underlying physical requirement that all processes involve increasing entropy. In the case of the burning of fossil fuels, this means the return into the atmosphere of the original CO_2 captured by plants and fossilized over millions of years as fuels.

[33] The second law of thermodynamics is an expression of the universal law of increasing entropy, stating that the entropy of an isolated system which is not in equilibrium will tend to increase over time, approaching a maximum value at equilibrium. (Wikipedia, 15.01.08)

More striking still, technological change has been assumed in the traditional multi-industry treatment to be independent of production change, implying no learning by doing or by researching (Barker et al. 2007, 11.5). If the general equilibrium models are to include such endogenous technological change it is usually grafted on to the neoclassical production function by linking it with an engineering model, typically for the energy supply and demand sectors. The outcome is inconsistent in that endogenous technological change is allowed for energy output but not other sectors (as well as carbon prices, many other relative prices will change as an effect of climate policies) nor for other economic variables, such as exports, employment or even consumption. It is also incomplete, in that it ignores the potential interaction between the information economy, energy and new low-GHG technologies, which accelerates their adoption and diffusion throughout the world economy.

The aggregate production functions, used in the equilibrium economic models, have been subject to detailed and severe criticism over many years, both of the underlying theory (Mirowski 1989; Felipe and Fisher 2003; DeCanio 2003) and of the validity of the empirical estimates (Felipe and McCombie 2005). Theoretically, the use of an aggregate production function requires two (heroic) assumptions: (1) that it is a meaningful exercise to combine the processes of e.g. furniture-making, oil-refining, and food-retailing, and (2) that *all* markets are perfectly competitive. Empirically, the use of National Accounts data on flows in current prices to estimate production functions is usually flawed, because the dataset imposes an accounting identity on the monetary value of production and the combined values of the inputs to production (namely materials, labour and capital), when capital services are measured as residual profits. The estimates in the literature are often based on accounting identities, not causal relationships, and hence the very good fits obtained are entirely artifacts of the data.[34] The implication of the production function in the traditional models, both the one-sector models of Nordhaus and others and the multi-industry general-equilibrium models, is that because the functional form assumes that the economy is at full employment and maximal efficiency, *any* climate policy leads to costs in the form of loss of potential output. It is this feature that leads to the contrast between the energy-engineering, bottom-up models, providing estimates of some 6 $GtCO_2$-eq mitigation potential by 2030 at net negative costs, i.e. “no regret options”, compared with no such unrealized potential estimated by the top-down equilibrium models (IPCC 2007, p. 14–16). The potential for energy saving assessed by countless engineering studies is simply ignored in the equilibrium models by assuming full information, maximum efficiency and full employment, now and forever, in violation of the facts. New evolutionary economics can provide insights into the non-economic barriers to energy efficiency and how they may be overcome (Maréchal 2007, pp. 5183–5184).

[34] See (Felipe and McCombie 2005). The empirical basis of the functions actually included in the climate-policy models is even more compromised, being no more than a collection of guess-estimates from an inconsistent literature (DeCanio 2003).

The traditional treatment of production also normally rules out of court any modelling outcome that increases the growth rate of the economy as an outcome of climate policy. There are many conditions under which GDP may increase, e.g. use of carbon tax revenues to reduce distortionary taxes, the effect of policy in reducing the widespread underemployment in developing countries, and the possibilities of more productive technological pathways. Although documented in the theoretical and empirical literature (Barker et al. 2007, 11.4 and 11.5), these conditions are routinely and implicitly set aside by assumption in the traditional treatment.

However, complexity economists (Arthur 1994) strongly argue for path dependency and increasing returns and economic historians have long argued that technological change and economic growth are intimately related (Maddison 2001) and path dependent (David 2001). The scientific requirement to decarbonise the global energy system is in effect suggesting the need for another technological revolution, implying major structural shifts in the energy industries and requiring the diffusion of low-carbon technologies, particularly across the developing world, which holds the greatest potential for adoption, radical changes and impacts. In contrast to the eighteenth and nineteenth century changes, the context is now one of the global spread of almost free information, instantaneously. The potential for global, induced technological change to reduce costs and even increase GDP is recognized in the Stern Review, as is the modelling that relaxes the assumption of constant returns to scale. In contrast to the equilibrium approach, such modelling has the great advantage that it aims to explore technological and institutional options that give rise to opportunities rather than costs, making the problem for international negotiations much more tractable (sharing out the benefits of a technological revolution, to put it crudely, but a revolution that will only happen if countries co-operate). At the same time it should be recognized that badly designed policies and regulatory frameworks, as in banking, could lead to potential inflation or financial collapse of investments and programmes.

11.6 Social Choice

The switch in policy required to address the climate change problem is an issue of social choice. Traditional economics approaches this problem by the use of the social welfare function, which is a mathematical equation, or a set of equations, in an economic model, intended to represent the social good. However, the concept is fundamentally flawed. When national governments act, it is much more likely to be 'in the national interest' than in any formal manner capable of being represented as a 'criterion function', an 'objective function' or a 'social welfare function' as some key concepts are known in equilibrium modelling of the economy and the environment. As Arrow (1967, p. 736) remarks about Samuelson's neoclassical treatment, 'Whose behaviour or whose judgment is referred to in the social welfare function is never clarified.'

In theory, the concept depends on the validity of adding up the welfare of households or people such that the aggregate social welfare function is stable and predictable over time. Arrow (1950) showed that for a set of reasonable assumptions (*inter alia*: a heterogeneous population, universality, "independence of irrelevant alternatives") such aggregation is impossible except under extremely restrictive conditions. Traditional economics has resorted to assuming that members of the population, or social groups such as households or firms, are in fact identical "representative agents", whose welfare and behaviour can be aggregated. This assumption, required for the macroeconomic equilibrium models to be theoretically valid in relation to microeconomic behaviour, is 'both unjustified, and leads to conclusions that are usually misleading and often wrong.' (Kirman 1992, p.117). The aggregate approach also often ignores specific issues of equity and the distribution of wealth, which are especially important for climate change economics because the costs tend to be met disproportionately by those who cannot afford insurance, relocation or adaptation investments.

In addition, the social welfare function is not politically viable. The idea that there is a stable relationship between different policy objectives such as reduction of greenhouse gases, economic development, growth in consumption, reduction in unemployment or in the rate of inflation, does not make sense when the actual political process of policy formation is considered, whatever the political complexion of the government or the prevailing consensus about sound policy promoted by international organizations such as the OECD, the IMF or the World Bank. Institutional decision-making (e.g. that by national governments) is normally characterized by the achievement of consensus between people and groups with potential conflicts of interest. If this is so, it is quite easy to envisage the simultaneous pursuit of conflicting goals, and the sudden alteration of policies as different interest groups gain precedence. There is no escaping the fact that the goals of national, economic and social policy are different for different interest groups, and that the national interest cannot be restricted to a fixed formula. In the face of these difficulties, traditional economists have resorted to another counter-factual assumption (usually implicit, but required for an optimal solution), that of a global planner, i.e. a policy dictatorship for good or bad.[35]

Social choice regarding the climate policy involves social groups, "stakeholders", such as government, industry, NGOs, and political parties, in a process of consensus (Ostrom 1990). But it also involves information and the law (Heinzerling and Ackerman 2007). A real choice requires the equal and simultaneous presentation of feasible alternatives. When a policy is the subject of political debate and possible

[35] The traditional cost-benefit analysis has been taken up by Lomborg in the "Copenhagen consensus" to promote the idea that global problems other than climate change (e.g. HIV) are more worthy of funding. It is inherently unlikely that the national interest can be identified with the functions for aggregate utility in the equilibrium models and then solved to obtain 'optimal' set of policies. If an attempt were made to elicit the function by asking a series of hypothetical questions of governments, rather than Lomborg's selection of economists, it would fail because the answers would most likely be inconsistent in terms of the equation. In any case, politicians would usually refuse to commit themselves on hypothetical questions.

implementation by government, policy advisors consider the benefit that such implementation would produce in each of various mutually exclusive 'states of nature' that might follow it, the good being considered for each group affected over space and time.

The actual process of developing such information for the global community has been chaotic. Different governments have produced their own analysis, sometimes as in the case of the US Administration in 2001, selectively choosing scenario results to meet obvious political requirements. Hasselmann and Barker (2008) discusses a way forward to improve the information basis in developing international climate policy under the UNFCCC and the IPCC. These bodies have arisen out of the international political process and are in keeping with decentralised and varied political structures. This process has brought questions of equity to the fore as witnessed by the crisis in the IPCC's adoption of the Second Assessment Report in 1995, with the neoclassical economists' insistence on valuing human life on an insurance basis. The use of values of "statistical lives" came into conflict with a perception that human life at present and in the future should be valued equally irrespective of income or circumstance, for the purpose of agreeing international policy. The governments of the developing countries arguing their case for equality prevailed over the expert IPCC economists advising them. However, it is perfectly feasible that a consensus approach in international negotiation can help to establish policies and social values in difficult and controversial areas, such as abatement of climate change, where the interests of different countries and future generations are to be taken into account. For example, the IPCC's summaries for policymakers are agreed by all governments explicitly at international meetings.

11.7 Towards a New Economics of Climate Change

The IPCC's skeptical approach to the use of cost-benefit analysis (CBA) as the sole basis for the economics of climate change has been supported by Stern. The CBA of climate change after Stern has been developed by Weitzman (2009) to the point of destruction. Just as a central bank, faced with the risk of the collapse of the banking system, will act on perceived risks rather than a monetary CBA, so governments have eschewed CBAs in which the "optimal" solution involves risks of dangerous climate change. The intemperate and rushed reaction by a clique of neoclassical economists criticizing the economics of the Stern Review illustrates the sensitivity to the implied criticism of their conclusions.

The subsequent development of the literature in supporting Stern's conclusions illustrates how radical the shift in mainstream economic thinking has been. It is now acknowledged that the economics of climate change is now more appropriately concerned with uncertainty rather than return, a feature of the problem that has been evident from the early 1990s, when the scientific assessments began in earnest. It can also reasonably be argued that CBA is useless for the climate problem because of the uncertainty and risks of catastrophe. The discounting of costs and benefits in which risks are converted into certainty equivalents and discounted at market rates has been shown to be misleading and biased. This in turn implies that the economic problem is one of achieving political targets, based on scientific evidence, at lowest costs compatible with equity and effectiveness, rather than with the economics of choosing the targets themselves.

The new information and evidence on the increasing risks of climate change has reinforced earlier perceptions about the dangers, and raised scientific and political alarm, but the general message has been to strengthen the evidence, and bring it home to the public through debate on weather-related catastrophes. In summary, the problem is clear and the solution appears to be almost costless in macroeconomic terms, but requires the long-term transformation of the global energy system. Decarbonisation of the global energy system by 2050 seems to be feasible at reasonable carbon prices (Barker and Crawford-Brown 2014) with benefits to health and negligible effects on economic growth, but it will require unprecedented global co-ordinated action.

Acknowledgements I would like to thank Frank Ackerman, Eric Beinhocker, John Broome, Simon Dietz, Klaus Hasselmann, Richard Lewney, John McCombie, John Quiggin, Serban Scrieciu, David Taylor and Rachel Warren for providing helpful comments on earlier versions of this paper. I also acknowledge the support of the UK Research Councils in funding my work as part of the Tyndall Centre for Climate Change Research.

References

Ackerman F (2004) Priceless benefits, costly mistakes: what's wrong with cost–benefit analysis? Real-World Econ Rev 25:2–7

Ackerman F, Finlayson IJ (2007) The economics of inaction on climate change: a sensitivity analysis. Clim Pol 6(5):509–526

Ackerman F, Heinzerling L (2004) Priceless: on knowing the price of everything and the value of nothing. The New Press, New York

Ackerman F, Nadal A (eds) (2004) The flawed foundations of general equilibrium: critical essays on economic theory. Routledge, London

Ackerman F, Stanton EA (2014) Climate economics: the state of the Art. Routledge, London

Ackerman F, Stanton EA, Bueno R (2010) Fat tails, exponents, extreme uncertainty: simulating catastrophe in DICE. Ecol Econ 69(8):1657–1665

Akerlof G (2007) The missing motivation in macroeconomics. AEA presidential address, January 2007

Arrow KJ (1950) A difficulty in the concept of social welfare. J Polit Econ 58(4):328–346

Arrow KJ (1967) Samuelson collected. J Polit Econ 75(5):730–737

Arrow KJ (2007) Global climate change: a challenge to policy economists, economists' voice. http://www.bepress.com/ev. Cited June

Arrow KJ, Cline W, Maler KJ, Munasinghe M, Squitieri R, Stiglitz J (1996) Intertemporal equity, discounting and economic efficiency. In: Bruce J, Lee H, Haites E (eds) Climate change 1995 – economic and social dimensions of climate change. Cambridge University Press, Cambridge, pp 125–144

Arthur WB (1994) Increasing returns and path dependence in the economy. University of Michigan Press, Ann Arbor

Azar C, Lindgren K (2003) Catastrophic events and stochastic cost–benefit analysis of climate change. Clim Chang 56:245–255

Baer P, Mastrandrea M (2006) High stakes designing emissions pathways to reduce the risk of dangerous climate change. IPPR, London

Barker T (1996) Space–time economics. Cambridge Econometrics, Cambridge

Barker T (2008) The economics of avoiding dangerous climate change. An editorial essay on the Stern Review. Clim Chang 89:173–194. doi:10.1007/s10584-008-9433-x

Barker T, Crawford-Brown, D (2013) Are estimated costs of stringent mitigation biased? a commentary on "Counting only the hits? The risk of underestimating the costs of stringent climate policy" by Massimo Tavoni and Richard S. J. Tol' Clim Change. doi:10.1007/s10584-010-9867-9

Barker T, Crawford-Brown, D (2014) (eds) Decarbonising the world economy. Imperial College Press, London

Barker T, Jenkins K (2007) The costs of avoiding dangerous climate change: estimates derived from a meta-analysis of the literature. A briefing paper for the United Nations Human Development Report 2007

Barker T, Pan H, Köhler J, Warren R, Winne S (2006a) Decarbonizing the global economy with induced technological change: scenarios to 2100 using E3MG. Energy J 27:241–258

Barker T, Qureshi MS, Koehler J (2006b) The costs of greenhouse-gas mitigation with induced technological change: a meta-analysis of estimates in the literature. 4CMR, Cambridge Centre for Climate Change Mitigation Research, University of Cambridge, Cambridge, UK

Barker T, Bashmakov A, Alharthi M, Amann L, Cifuentes J, Drexhage M, Duan O, Edenhofer B, Flannery M, Grubb M, Hoogwijk FI, Ibitoye CJ, Jepma WA, Pizer K, Yamaji 'Mitigation from a cross-sectoral perspective' Chapter 11 in Climate Change 2007: Mitigation. Contribution of Working Group III to the Fourth Assessment Report of the Intergovernmental Panel on Climate Change [Metz,B Davidson, OR Bosch, PR Dave, R Meyer LA (eds)], Cambridge University Press, Cambridge, United Kingdom and New York, NY, USA

Barker T, Anger A, Chewpreecha U, Pollitt H (2012) A new economics approach to modelling policies to achieve global 2020 targets for climate stabilisation', special issue on 'economic policies of the New thinking in economics'. Int Rev Appl Econ 26(2):205–211

Beinhocker E (2006) The origin of wealth: evolution, complexity and the radical remaking of economics. Random House Business Books, London

Boulding K (1992) Towards a new economics. Edward Elgar, Brookfield

Broome J (1992) Counting the cost of global warming. White House Press, Cambridge

Broome J (2006) Valuing policies in response to climate change: some ethical issues. A contribution to the work of the Stern Review of the economics of climate change

Broome J (2008) Why economics needs ethical theory. In: Basu K, Kanbur R (eds) Welfare, development, philosophy and social science: essays for Amartya Sen's 75th birthday, vol 3. Oxford University Press, Oxford

Cline WR (1991) The economics of the greenhouse effect. Econ J 101(407):920–937

Cline WR (1992) The economics of global warming. Institute for International Economics, Washington, DC

Colander D, Holt R, Rosser JB Jr (2004) The changing face of mainstream economics. Rev Polit Econ 6(4):485–499

Dasgupta P (2005) What do economists analyze and why: values or facts? Econ Philos 21:221–278

Dasgupta P (2007) Comments on the Stern Review's economics of climate change. Natl Inst Econ Rev 199:4–7

Dasgupta P (2008) Discounting climate change. J Risk Uncertain 37:141–169

David P (2001) Path dependence, its critics, and the quest for 'historical economics'. In: Garrouste P, Ioannides S (eds) Evolution and path dependence in economic ideas: past and present. Edward Elgar, London

Deaton, A (2007) Letter from America—On transatlantic vices, or Stern in America. Royal economic society, New letter, October
DeCanio S (2003) Economic models of climate change a critique. Macmillan, New York
DeLong JB, Summers LH (2001) The new economy: background, historical perspective, questions, and speculations. Proceedings, Federal Reserve Bank of Kansas City, pp 11–43
Dequech D (2007) Neoclassical, mainstream, orthodox, and heterodox economics. J Post Keynes Econ 30(2):279–302
Dietz S, Hope C, Stern N, Zenghelis D (2007a) Reflections on the Stern Review (1): a robust case for strong action to reduce the risks of climate change. World Econ 8(1):121–168
Dietz S, Hope C, Patmore N (2007b) Some economics of 'dangerous' climate change: reflections on the Stern Review. Glob Environ Chang 17:311–325
Felipe J, Fisher FM (2003) Aggregation in production functions: what applied economists should know. Metroeconomica 54:208–262
Felipe J, McCombie JSL (2005) How sound are the foundations of the aggregate production function (in symposium on the aggregate production function). East Econ J 31(3):467–488
Foley D (2006) Adam's fallacy: a guide to economic theology. Harvard University Press, Cambridge, MA
Frederick S, Loeenstein G, O'Donohue T (2002) Time discounting and time preference: a critical review. J Econ Lit XL 40(2):351–401
Gowdy J (2005) Towards a new welfare economics for sustainability. Ecol Econ 53:211–222
Gritsevskyi A, Nakicenovic N (2000) Modelling uncertainty of induced technological change. Energy Policy 28(13):907–922
Hare RM (1981) Moral thinking. Oxford University Press, Oxford
Hasselmann K (1999) Intertemporal accounting of climate change: harmonizing economic efficiency and climate stewardship. Clim Change 41:333–350
Hasselmann K, Barker T (2008) The Stern Review and the IPCC fourth assessment report: implications for interaction between policymakers and climate experts. Clim Chang 89:219–229
Hasselmann KS, Hasselmann R, Giering OV, Hv S (1997) Sensitivity study of optimal CO_2 emission paths using a simplified Structural Integrated Assessment Model (SIAM). Clim Change 37:345–386
Heal G (1997) Discounting and climate change. Clim Change 37:335–343
Heinzerling L, Ackerman F (2007) Law and economics for a changing world. Harv Law Policy Rev 1(2):331–362
Hodgson GM (2007) Evolutionary and institutional economics as the new mainstream? Evol Inst Econ Rev 4(1):7–25
Hoel M, Sterner T (2007) Discounting and relative prices. Clim Change 84:265–280
Holt R (2007) What is post Keynesian economics? In: Forstater M, Mongiovi G, Pressman S (eds) Post Keynesian macroeconomics. Routledge, London
Hope C (2003) The marginal impacts of CO_2, CH_4 and SF_6 emissions. Judge institute of management research paper 2003/10, University of Cambridge, UK
Hope C (2013) Critical issues for the calculation of the social cost of CO_2: why the estimates from PAGE09 are higher than those from PAGE2002. Clim Chang 117(3):531–543
IPCC (2007) Climate change 2007: synthesis report. IPCC, Geneva
IPCC (2014) Climate change mitigation summary for policymakers. IPCC, Geneva
Kahneman D, Tversky A (1989) Prospect theory: an analysis of decisions under risk. Econometrica 47(2):263–291
Katzner D (2003) Why mathematics in economics? J Post Keynes Econ 25(4):561–573
Keynes JM (1973) The general theory and after, collected writings, vol XIV. Macmillan, London
Kirman AP (1992) Whom or what does the representative individual represent? J of Econ Perspect 6(2):117–136
Knight FH (1921) Risk, uncertainty, and profit. Hart, Schaffner & Marx; Houghton Mifflin Company, Boston, MA

Köhler J, Grubb M, Popp D, Edenhofer O (2006) The transition to endogenous technical change in climate-economy models. Energy J 27:17–55
Maddison A (2001) The world economy: a millennial perspective. OECD, Paris
Maréchal K (2007) The economics of climate change and the change of climate in economics. Energy Policy 35:5181–5194
Mastrandrea MD, Schneider SH (2001) Integrated assessment of abrupt climatic changes. Clim Pol 1:433–449
Mastrandrea MD, Schneider SH (2004) Probabilistic integrated assessment of "dangerous" climate change. Science 304:571–575
Mirowski P (1989) More heat than light economics as social physics: physics as nature's economics. Cambridge University Press, Cambridge
Mirowski P (2002) Machine dreams economics becomes a cyborg science: light economics as social physics: physics as nature's economics. Cambridge University Press, Cambridge
Munda G (2005) Multiple criteria decision analysis and sustainable development. In: Figueira J, Greco S, Ehrogott M (eds) Multiple criteria decision analysis: state of the art surveys, vol 78, International series in operations research & management science. Springer, Berlin, pp 953–986
Nelson RH (2001) Economics as religion from Samuelson to Chicago and Beyond. State University Press, Pennsylvania
Nelson JA (2006) Economics for humans. University of Chicago Press, Chicago
Nelson JA (2008) Economists, value judgments, and climate change: a view from feminist economics. Ecol Econ 65:441–447
Nordhaus WD (1991a) A sketch of the economics of the greenhouse effect. Am Econ Rev 81(2):146–150
Nordhaus WD (1991b) To slow or not to slow: the economics of the greenhouse effect. Econ J 101(7):920–937
Nordhaus WD (1993) Optimal greenhouse-gas reductions and tax policy in the "DICE" Model. Am Econ Rev 83(2):313–317
Nordhaus WD (1994) Managing the global commons: the economics of climate change. MIT, Cambridge, MA
Nordhaus WD (1997) Discounting in economics and climate change. Clim Change 37:315–328
Nordhaus WD (2007a) A review of the Stern Review on the economics of global warming. J Econ Lit 45(3):686–702
Nordhaus WD (2007b) The challenge of global warming: economic models and environmental policy, 24, July 2007, Yale
Nordhaus WD (2011) Estimates of the social costs of carbon: background and results from the RICE-2011 model, October 2011, Cowles Foundation Discussion Paper number 1826, Yale
Nordhaus WD, Boyer J (2000) Warming the world: economic models of global warming. MIT, Cambridge, MA
Ostrom E (1990) Governing the commons: the evolution of institutions for collective action. Cambridge University Press, Cambridge
Padilla E (2004) Climate change, economic analysis and sustainable development. Environ Values 13:523–544
Pearce DW, Cline WR, Achanta AN, Fankhauser S, Pachauri RK, Tol RSJ, Vellinga P (1996) The social costs of climate change: greenhouse damages and the benefits of control. In: Bruce J, Lee H, Haites E (eds) Climate change 1995—economic and social dimensions of climate change. Cambridge University Press, Cambridge, pp 125–144
Piketty T (2014) Capital in the twenty-first century. Harvard, Cambridge
Popp D (2002) Induced innovation and energy prices. Am Econ Rev 92(1):160–180
Price C (2005) An intergenerational perspective on effects of environmental changes: discounting the future's viewpoint. In: Innes JL, Hickey GM, Hoen HF (eds) Forestry and environmental change: socioeconomic and political dimensions. Report 5 IUFRO Task Force on environmental change

Quiggin J (2008) Stern and his critics on discounting and climate change: an editorial essay. Clim Chang 89:195–205

Rawls J (1971) A theory of justice. Harvard University Press, Cambridge, MA

Robbins L (1932) An essay on the nature and significance of economic science. Macmillan, London

Samuelson PA, Nordhaus WD (2001) Economics, 17th edn. McGraw-Hill, New York

Stern N (2006/2007) The economics of climate change: the Stern Review. Cambridge University Press, Cambridge (References are to the book, published in 2007)

Tol RSJ (2005) The marginal damage costs of carbon dioxide emissions: an assessment of the uncertainties. Energy Policy 33(16):2064–2074

Tol RSJ (2007) The social cost of carbon: trends, cutliers and catastrophes. Economics discussion papers, 2007–44. http://www.economics-ejournal.org/economics/discussionpapers/2007-44

Tol RSJ, Yohe GW (2007) A review of the Stern Review. World Econ 7(4):233–250

van den Bergh JCJM (2004) Optimal climate policy is a utopia: from quantitative to qualitative cost-benefit analysis. Ecol Econ 48:385–393

Weintraub ER (2002) How economics became a mathematical science. Duke University Press, Durham

Weitzman ML (2009) On modeling and interpreting the economics of catastrophic climate change. Rev Econ Stat 91(1):1–19

Williams BAO (1985) Ethics and the limits of philosophy. Routledge, London

Yohe GW (1989) A study guide to accompany Samuelson & Nordhaus. McGraw-Hill, New York

Chapter 12
A Precauctionary Strategy to Avoid Dangerous Climate Change is Affordable: 12 Reasons

Jeroen C.J.M. van den Bergh

Abstract There is a widespread sense that a sufficiently stringent climate mitigation policy, that is, a considerable reduction of greenhouse gas emissions to avoid extreme climate change, will come with very high economic costs for society. This is supported by many cost–benefit analyses (CBA) and policy cost assessments of climate policy. All of these, nevertheless, are based on debatable assumptions. This paper will argue instead that safe climate policy is not excessively expensive and is indeed cheaper than suggested by most current studies. To this end, climate CBA and policy cost assessments are critically evaluated, and as a replacement 12 complementary perspectives on the cost of climate policy are offered.

Keywords Climate change • Policy • CBA • Integrated assessment models • Social cost of carbon • Solar energy • Happiness

12.1 Introduction

It is generally felt that a climate policy which stabilizes atmospheric concentrations of greenhouse gases (GHGs) at a 'safe' level will be extremely expensive, whether measured in terms of monetary costs, reduced GDP growth or forgone welfare. This

---

This article is a considerably shortened (about 50 %) version of an article previously published under the terms of the Creative Commons Attribution Noncommercial License ("open access") as: J.C.J.M. van den Bergh (2010). Safe climate policy is affordable – 12 reasons. *Climatic Change* (2010) 101:339–385. Original section 3 and many footnotes have been deleted. In addition, Section 2 and 5 (here 4) have been shortened. In addition, references have been updated and text has been rewritten here and there.

J.C.J.M. van den Bergh (✉)
ICREA, Barcelona, Spain

Institute for Environmental Science and Technology, Universitat Autònoma de Barcelona, Barcelona, Bellaterra, Spain

Faculty of Economics and Business Administration and Institute for Environmental Studies, VU University Amsterdam, Amsterdam, The Netherlands
e-mail: jeroen.bergh@uab.es

S. Shmelev (ed.), *Green Economy Reader*, Studies in Ecological Economics 6,
DOI 10.1007/978-3-319-38919-6_12

is supported by a number of influential economic cost–benefit analyses of climate policy as reviewed in Kelly and Kolstad (1999) and Tol (2008a, b). In this paper it will be argued that the application of cost–benefit analysis (CBA) to climate change and policy should be judged as being overly ambitious. To avoid the many fundamental and practical problems associated with CBA and the associated notion of 'optimal' climate policy, it will be argued that a better option is to adopt a more modest and practical approach, namely examining the cost of a safe climate policy. This reflects a policy aimed at a stable and safe level of atmospheric GHG concentrations—thus focusing on mitigation, not adaptation. The combination of risk aversion, pervasive uncertainty, and extreme climate change and events motivates such a safe or precautionary approach as a rational alternative to an optimal climate policy. In fact, (avoiding) extreme climate change may be regarded as the ultimate reason for us to worry about and respond to climate change. Even two strong advocates of using CBA to analyze climate change, Tol and Yohe (2007, pp. 153–154), state: "A cost–benefit analysis cannot be the whole argument for abatement. Uncertainty, equity, and responsibility are other, perhaps better reasons to act."

It will be argued here that the cost of climate policy has so far been approached from too narrow a perspective. This will involve a discussion of fundamental problems associated with applying CBA to climate change and policy. Spash (2007) concludes that cost-effectiveness studies are not much better than CBA's. Indeed, studies attempting to assess the monetary cost of climate policy make many debatable assumptions as well. Nevertheless, the shortcomings are less serious than in the case of climate CBA studies because the monetization of climate damage is avoided. Since some of the shortcomings of CBA's and cost assessments of climate policy cannot be resolved, one cannot hope for a single model analysis of climate policy to provide the definite insight about its cost let alone its optimality.

This paper will therefore offer an alternative approach consisting of assessments of the cost of climate policy from a range of complementary perspectives. Together, these aim to avoid or surpass the limits of existing CBA and policy cost studies. The alternative approach can be seen as trying to determine the economic and social costs of a safe or reasonably safe—given all sorts of uncertainties involved—climate policy by considering a range of perspectives to somehow bound the "cost space". The focus on a safe or precautionary climate mitigation policy can be regarded as the outcome of a qualitative risk analysis, as will be discussed in Sect. 12.3. Twelve perspectives on the cost of climate policy are offered. Together they deliver quite an optimistic conclusion, namely that climate policy is not excessively expensive and is certainly cheaper than suggested by most current studies. In other words, our global society can afford to invest in a safe climate policy. This should serve as relevant information for all politicians who fear severe economic consequences from stringent regulation of GHG emissions.

The remainder of this article is organized as follows. Section 12.2 briefly argues the failure of cost–benefit analysis of climate policy. Section 12.3 presents the main arguments in favor of a safe, precautionary approach to climate policy. Given that the current economic approaches to assessing the (net) costs of climate policy have severe limitations, they are prone to generating inaccurate estimates. This means there is a need for an alternative approach, as offered in Sect. 12.4. It presents the new approach consisting of 12 perspectives on, and interpretations of, climate policy costs that move beyond current model assumptions and limitations. Section 12.5 provides conclusions.

12.2 The Failure of Cost–Benefit Analyses of Climate Policy

The history of climate CBA shows enormous variation in estimates. For example, whereas early studies (e.g., Nordhaus 1991) excluded adaptation to and benefits of climate change, later studies did take them into account and arrived at lower climate damage costs. Despite variation, most climate CBA studies share many basic assumptions. These have received considerable criticism, much of which is difficult to resolve (e.g., Ayres and Walters 1991; Broome 1992; Barker 1996; Azar 1998; Neumayer 1999; Spash 2002; DeCanio 2003; van den Bergh 2004; Padilla 2004; Gowdy 2008; Tol 2008b; Ackerman et al. 2009; and various responses to the Stern Review). Criticism has been directed, among others, at the assumed behavior of economic agents, the social welfare objective used, the treatment of small-probability-high-impact scenarios, discounting and social discount rate values, monetary valuation of a human life, and the neglect or incomplete treatment of certain cost categories.

A main criticism is that the analysis of climate policy should not be conceptualized as a problem suitable for quantitative cost–benefit analysis but as one of risk analysis, since the cost of climate damage cannot be assessed with any acceptable degree of certainty. Weitzman (2007, p. 703) says about this: "The basic issue here is that spending money to slow global warming should perhaps not be conceptualized primarily as being about consumption smoothing as much as being about how much insurance to buy to offset the small chance of a ruinous catastrophe that is difficult to compensate by ordinary savings." The latter means that social welfare losses due to extreme climate change cannot be reversed or undone through adaptation. Especially the treatment of extreme climate change and climate events characterized by a combination of small probabilities and large impacts has been argued to not go together well with an expected value approach to cost–benefit analysis. This view is the motivation for the approach adopted in this paper, namely an assessment of the cost of a (reasonably) safe climate policy. This specific, fundamental criticism is addressed in more detail in Sect. 12.3, which will result in an extended argument in favor of a precautionary approach to climate mitigation policy.

Perhaps the most important shortcoming of current economic studies of climate policy relying on CBA is that they incompletely account for extreme and irreversible climate scenarios, such as: extreme low or high temperatures; a slow-down or halting of the global thermohaline circulation, of which the Gulf Stream is a part; an extreme increase of the world's mean sea-level over centuries due to the collapse of the ice sheets on Greenland and West Antarctica; 'runaway dynamics' caused by positive feedback mechanisms in the biosphere, such as substantial emissions of methane (with a much higher warming potential than CO2) from permafrost regions; changes in climate subsystems such as the 'El Niño Southern Oscillation'; acidification of the oceans due to high atmospheric CO2 concentrations, meaning a deterioration in the living conditions for marine organisms with yet unforeseen effects; and extreme weather events, notably extreme rainfall, an increased probability of heat waves and droughts, and an increased intensity of hurricanes due to warmer seas. If, moreover, such changes take place rapidly, then insufficient time for adaptation will contribute to higher damage costs. The omission of these extremities from CBAs is incomprehensible given that the ultimate reason for studying climate change is—or in any case should be—a concern for extreme events which will fundamentally alter the environmental conditions for humans and the rest of the biosphere. In fact, studies that have incompletely taken into account extreme events should not be taken too seriously—they really involve nothing more than toy models—and the respective authors should be modest about the policy implications of their analyses (see also Azar and Lindgren 2003). In particular, studies omitting extreme events will underestimate the cost of climate change, or the benefits of climate policy, and therefore be biased against safe climate policy. The problem is, of course, that most worst-case climate change scenarios cannot be accurately quantified.

The differential treatment of extreme climate events offers one explanation for the wide range of damage cost estimates of GHG emissions that one can find in the literature (Tol 2005; Fisher and Morgenstern 2006). Tol (2008a) performs a meta-analysis of them, suggesting that the most reliable estimate cannot be the outliers,thus explicitly questioning the high damage estimates used in the Stern Review. However, a meta-analysis assumes that all studies are equally valuable unless one weights studies, for instance, by giving a relatively high weight to more recent studies using updated information. But since Tol does not apply such a weighting scheme, the outcome of his analysis is dominated by the large share of (older) studies which neglect or incompletely address extreme climate change scenarios and events. The meta-analysis thus hides the fundamental shortcomings of the primary studies, even though it gives the impression of being an objective aggregation. An aggregation based on accounting for four shortcomings of previous studies leads to a lower bound to the social cost of CO_2 emissions equal to \$125 (van den Bergh and Botzen 2014).

Other limitations and weaknesses of CBAs of climate policy have been well documented. Tol (2008b) lists the many imperfections in a refreshingly critical and honest account of climate damage cost studies.[1] In particular, he notes the neglect in existing studies of the impact of climate change on human conflict, large-scale biodiversity loss, economic development, and human population/demography. Most models take immediate adaptation for granted by assuming rational behavior by

[1] What is disappointing, though, is that after listing an impressive number of uncertainties andmissing elements in existing cost studies and presenting a range of marginal carbon cost estimates as wide as \$20–669/tC (Tol 2008b, Table 2), Tol proposes to use a carbon tax in the lower range of \$26–50/tC.

economic agents. A general shortcoming is the neglect of any impacts beyond 2100 in many studies.

Next, over long-term horizons, such as in climate change analysis, CBA is extremely sensitive to discounting and particularly the choice of (social) discount rate. A large part of the variation in results of studies that have undertaken a quantitative CBA of climate policy is due to this discount rate sensitivity. The debate on intergenerational discount rates was revived by the Stern Review (Stern et al. 2006). There are several fundamental objections to be made against discounting as formalized in the famous Ramsey formula, as well as objections against the choice of parameter values in applying this formula in climate change studies. For overviews of the various arguments see Dasgupta (2007), Quiggin (2008), Ackerman et al. (2009), and van den Bergh (2010, Section 2). A concise review and evaluation is in van den Bergh and Botzen (2014) who conclude that the arguments in favor of low discount rates outnumber and are more convincing than those in favor of high rates. They also argue that using a high discount rate effectively means giving little attention to low-probability, high-impact scenarios as these tend to involve extreme events far away in time. Most importantly, perhaps, as noted by Arrow (2007), even with a much higher social discount rate than the one resulting from the Stern Review's assumptions, and well above the value range accepted by most economists (3–6 %), the cost–benefit argument for stringent climate policy remains valid. Dominant researchers in the field as Nordhaus (DICE model) and Tol (FUND model) do not give much credit to fundamental objections against social time preference discounting and instead harshly judge the Stern Review as representing a "decidedly-minority paternalistic view", "lowest bound of just about any economist's best-guess range" and "nonconventional assumptions that go so strongly against mainstream economics". However, speaking of mainstream economics in relation to climate policy analysis does not do justice to the fundamental criticism of the suitability of CBA as a method to evaluate climate policies, as summarized above. One can indeed interpret fierce attacks on the Stern Review as a "historical accident", to use a term from the literature on path-dependence: if Cline (1992) and Stern had been the dominant players in the field, those arriving late on the scene and wanting to use high discount rates would have likely received fierce criticism for making unorthodox assumptions.

CBA is an attractive and reasonable evaluation method for well-bounded problems (local, sectoral) with limited time horizons, non-extreme and manageable uncertainties, reversible scenarios, and limited income inequality. But its application to global, long-term climate change and policy questions runs into severe problems.[2] Here CBA is not merely stretched to its extreme but breaks down. This does not mean that one has to reject qualitative-type of CBA thinking. Indeed, it is diffi-

[2] Notice that application of CBA to acid rain and related SO2 and NOx emissions reduction policies has not received so much attention, even though this problem is more limited in scope than climate change and policy. The economic research on acid rain has been dominated by cost-effectiveness analysis, with RAINS developed at IIASA probably being the best-known study of this type (Alcamo et al. 1990).

cult to escape thinking in terms of trade-offs between qualitative costs or the disadvantages and benefits or advantages of any choice. Such a qualitative, conceptual approach is in fact needed to support a precautionary approach to climate policy (van den Bergh 2004). But unlike the quantitative CBA approach, its qualitative counterpart expresses clearly that specific, detailed statements about the social optimality of choices in the context of climate policy are very, and possibly overly ambitious.

12.3 Arguments for a Safe, Precautionary Approach to Climate Policy

If it can be argued that a safe climate policy means considerably lower net costs than the absence of such a policy, it is rational to be in favor of such a policy. This represents a kind of cost-effectiveness combined with precaution, given the uncertainties involved, aimed at avoiding extreme damage costs due to climate change. As a guide we can take Nordhaus and Boyer (2000) estimate of 10 % and the Stern Review's estimate of almost 20 % potential GDP damage cost of extreme climate change (Stern et al. 2006). As noted in Sect. 12.2, considerably lower damage costs require the omission of relevant extreme climate events and scenarios. If we compare these figures with climate policy cost estimates by IPCC (2007), which are in the range of 1–4 % of global GDP, then safe climate policy is clearly seen to be socially efficient. The slogan used by some environmental NGOs is surprisingly appropriate: 'the most expensive climate policy is doing nothing'.

The combination of small probabilities and large impacts associated with extreme climate change and climate events does not go together well with an expected value approach to cost–benefit analysis, and moreover does not reflect the way humans generally tend to evaluate such problems (Botzen and van den Bergh 2009; Quiggin 2008). This can partly be understood through different treatments of risk aversion in expected and non-expected utility approaches. Low-probability, high-impact scenarios have a small expected value compared to more certain changes associated with less extreme costs, and as a consequence receive a relatively low weight in CBA analysis. This effectively means a risk-neutral or riskloving approach. Nevertheless, one may perceive such costs as very undesirable and hence place a considerable value on preventing low-probability, high-impact events from occurring, especially when such events are irreversible and involve the loss of non-substitutable goods or services, as is the case with climate change.

In line with this view, Loulou and Kanudia (1999) and van den Bergh (2004) have proposed studying climate change using a precautionary principle formalized via a minmax regret goal. This represents more risk aversion than an expected value approach and less risk aversion than, for example, maximin net benefits.

Tol (2008b, p. 10), a fervent believer in climate CBA, supports the precautionary approach to climate policy evaluation implicitly by stating that in view of the strongly right-skewed distribution of climate change damage costs (median \$14/tC, mean \$93/Tc, 95 percentile \$350/tC; Tol 2005): "The policy implication is that emission reduction should err on the ambitious side". Dietz et al. (2007, p. 250) make a convincing plea for precaution in climate policy as well: "Those who deny the importance of strong and early action should explicitly propose at least one of three arguments: (1) there are no serious risks; (2) we can adapt successfully to whatever comes our way, however big the changes; (3) the future is of little importance. The first is absurd, the second reckless, and the third unethical."

Environmental economists have long thought about uncertainty, irreversibility and precaution, which has given rise to option value theory. But surprisingly they have refrained from systematically applying it to the most relevant case of irreversible environmental change, namely climate change (an exception is Schimmelpfennig 1995). In brief, this would mean that the foregone benefits of a certain 'preservation scenario' (i.e. safe climate policy) are included as a cost category of the 'development scenario' (i.e. no policy, leading to climate change). The resulting option value can be interpreted as the value of flexibility to either accept climate change at a later date or not, where the flexibility is due to investing in GHG emissions reduction to avoid the irreversible build-up of greenhouse gases in the atmosphere. Ha-Duong (1998) applies the notion of quasi-option to climate policy, which states that precaution allows for learning about climate change in terms of risks, costs, and adaptation opportunities. Admittedly, a main weakness of applying (quasi-)option value theory to climate change policy is that it takes expected utility theory as a basis, which, as argued above, is problematic in view of the low-probability, high-impact scenarios associated with climate change.[3]

Gollier et al. (2000) have shown the precautionary principle to result from a rational decision formalized as dynamic optimization under uncertainty and irreversibility involving Bayesian updating/learning. The conditions for precautionary action turn out to depend on risk aversion and "prudence". The latter is captured by the third derivative of the utility function and reflects the degree to which an individual increases his savings in response to an increase in uncertainty about future revenues (Kimball 1990). Other approaches than expected utility maximization and minimax regret to support a precautionary policy are maximin utility and nonlinear methods like prospect theory or rank-dependent utility theory, which one can characterize either as rational or boundedly rational (but not irrational) approaches. Although experts seem not to entirely agree on the best theoretical approach to address decisions in the face of low-probability, high-impact scenarios, a defensible

[3] Several authors have theoretically studied climate policy given economic (investment) irreversibilities. They conclude that there is then a risk of overinvestment in economic capital (manufactured and human) and that current emissions reduction policy should be slightly laxer than without learning (Kolstad 1996; Ulph and Ulph 1997). However, these findings do not suggest a move away from precaution, since climate irreversibility is characterized by much more extended time scales than economic irreversibility, while for climate capital, unlike for economic capital, no substitutes are available. These studies can also be criticized for employing an expected utility approach.

approach seems to be to give relatively more attention or weight to extreme case scenarios, which comes down to a kind of minimax regret approach.

In the face of extreme uncertainty a quantitative analysis will not necessarily be able to offer more informative insight than a mere qualitative analysis. The reason is that the extreme uncertainty does not disappear by adding more quantitative sophistication to the method of analysis or by reducing uncertainty to (subjective) risk. All existing models that include uncertainty somehow apply arbitrary probability distributions to extreme climate events and changes (surveyed by van den Bergh 2004). These models regard investments in emissions reduction as a decision on risky investments, but they insufficiently reflect the irreversibility of climate change, the extreme uncertainty (content and likelihood) associated with certain scenarios and events, and the non-insurability against extreme climate change and events due to risks being highly correlated for all regions in the world.

A somewhat different way to understand the rationale behind a precautionary approach to climate policy is based on comparing the likelihood and features of climate and economic instability. This represents a kind of risk management view, which conceptualizes climate policy as the outcome of a trade-off between the risks and costs associated with natural and economic instabilities. However, these two risks are neither on equal par nor symmetric. One may even go as far as to say they are of a different order and thus simply incomparable. This can be reasoned as follows. With a given global environment under a stringent climate policy, humans cannot predict economic changes with certainty, but they can guide and control them within boundaries. Economic stability can then be maintained. For example, if a stringent climate policy turns out to create too high economic costs and too much instability, the policy may be altered or adapted. However, under extreme climate change—due to a lax or lacking climate policy—one has to reckon with macro-scale risks, with catastrophic and irreversible changes in the coupled climate-biosphere system which cannot be controlled by any public policy, even though impacts may in some cases be ameliorated by climate adaptation policies. Governments will then be unable to avoid extreme impacts on the world economy, and economic policy will have a very hard time stabilizing economic responses to extreme climate change. In fact, a severe climate crisis may very well stimulate an unprecedented economic crisis. All in all, economic adaptation and policy under stable natural, climate conditions, enhanced by a stringent climate policy, are easier and safer than responding to unstable natural conditions resulting from a lax climate policy. This is consistent with the view of Azar and Schneider (2003, p. 331): "Thus, we do not see costs and benefits in a symmetrical cost–benefit logic, but rather as an equity problem and a risk management dilemma." The Stern Review also shares this standpoint, and many other observers have made similar statements.

The extensive literature on resilience and ecosystem functioning also suggests that we should be extremely careful in tinkering with the biosphere through humaninduced climate change, as this may cause discrete, structural changes in all kinds of ecosystems (freshwater, marine, rangeland, wetland, forest, arctic) when certain critical thresholds of GHG concentration in the atmosphere are surpassed (Holling 1986). The risk of extreme events or disasters, as documented in Sect. 12.2, is relevant here, as many of them will considerably affect basic conditions for many ecosystems. In addition, the uncertain synergy between biodiversity loss and climate change is relevant. Biodiversity supports the stability of ecosystem functions and related services to humans, while biodiversity loss is being enhanced by climate change. Against this background, some have even denied the relevance of normal scientific analyses of complex issues like climate change and climate policy on the basis of the climate system being complex and able to show catastrophic behavior (Rind 1999). Add to this the other dimensions of global change that may interact with climate change in nonlinear and unknown ways, such as land use, deforestation, water use, destruction of wetlands, acid rain, acidification of the oceans, and human control over a sizeable portion of primary production. Complexity implies that causal connections between a multitude of potential factors and effects cannot be identified, let alone be quantified. Against this background, a 'post-normal science' has been pleaded for, characterized by "uncertain facts, values in dispute, high stakes and urgent decisions" (Funtowicz and Ravetz 1993). The climate problem meets all four characteristics.

The foregoing set of considerations suggests that the implementation of a precautionary principle in climate policy emerges as a rational strategy. Neither decisionmaking based on quantitative CBA nor waiting until more information is available are convincing strategies. An often-heard argument against the precautionary principle is that climate policy means that alternative public goals have to be sacrificed. But whereas, for instance, less health care and education can indeed reduce growth and welfare, they are unlikely to cause extreme and discrete changes at a global scale. For this reason, climate policy needs to be treated as fundamentally different from many other areas of public policy.

Finally, Van den Bergh (2010) discusses the more modest cost (so no full CBA) assessment studies of a safe climate policy and reviews the methods and assumptions that have been used to produce the main cost estimates. Because of lack of space, we refer here to the original article (Sect. 12.4).

12.4 Twelve Reasons Why a Safe Climate Policy Is Affordable

The section below presents 12 new, complementary perspectives on the cost of climate policy.

12.4.1 Perspective 1: Extrapolating Learning Curves for Renewable Energy

The easiest way to reason about the cost of climate policy is by considering a most likely definite solution to the core problem, that is, the emission of greenhouse gases, notably carbon dioxide. Renewable energy really offers the only definite solution, as it can in principle support the supply of electricity and other types of energy carriers in a carbon-free way. Moreover, from the perspective of rebound risks, indirect energy use due to energy conservation or efficiency improvements (Sorrell 2007), renewable energy has a major advantage over energy conservation. Van den Bergh (2010) argues why within this category solar photovoltaics (PV) is a main candidates for future dominance.

van der Zwaan and Rabl (2003, 2004) have analyzed scenarios of the price and cost of solar PV on the basis of experience or learning curves. Such curves convey that overall production costs tend to decline with an increase in cumulative production. It is true that overall costs not only capture learning and innovation (R&D) effects but also change in market prices of inputs (notably material inputs). The latter may sometime increase which can (temporarily) reverse the normal, negative relation between cumulative production and costs. Nevertheless, generally speaking learning curves are seen as quite a robust tool to examine the long-run cost behavior of technologies. For solar photovoltaic (PV) energy, a most likely or middle scenario delivers an estimate on an order of magnitude equal to US$60 billion associated with a cumulative production of about 150 GWp (note: in 2004 cumulative production was about 1 GWp). This amount of money represents an extra expenditure over the investment in fossil fuel electricity, which is needed to make solar PV competitive with electricity produced from fossil fuels (van der Zwaan and Rabl 2004, Table 2, progress ratio 0.8). If learning is favorable, then US$30 billion (at 50 GWp) is a better estimate, while if learning is slow the cost may rise to US$300 billion (at 1000 GWp).

12.4.2 Perspective 2: Global Climate Policy Cost Normalized by OECD GDP

Here the cost of worldwide climate policy will be normalized by the GDP of OECD countries. This can be justified on the basis of their historical contribution to climate change (Botzen et al. 2008) as well as their currently high incomes relative to the rest of the world, i.e. historical and intra-generational fairness. We can then take the range of 1–4 % suggested by a survey of studies by IPCC (2007) as one basis for a climate policy cost estimate. The second estimate can be drawn from the previous section, where the cost of public support to make solar PV competitive was estimated to be in the range of US$30 billion to US$300 billion with a best, middle estimate of US$60 billion. These costs result in only 0.17 % (with an uncertainty range of 0.08–1.65 %) of the joint GDP of the 30 OECD countries in 2007 (which was US$ 36,316 billion; OECD 2008). An equal distribution would simply come down to 60/30 = US$2 billion per country, which is not a shocking figure. If the investment were spread over the course of 10 years, then it would amount to only US$200 million per country per year (over 10 years) or on average 0.017 % of GDP (with an uncertainty range of 0.008–0.17 %). In the worst case scenario, this would imply a cost to a family with a net income of €25,000 about €40; in the most likely case this would be €4, and in the most favorable case €2, over a 10-year period.

In 2007, OECD income was about 55 % of world GDP (about US$66 trillion). If OECD would carry all the cost of climate policy, and taking the climate policy cost range identified by IPCC (1–4 %), this would lead to an average cost for OECD countries equal to 1.8–7 % of GDP. This is significantly higher than the estimates based on public support of solar PV. Why is that so? First, the 4 % is quite a high estimate, and it is likely that the 1 % estimate is a more reasonable order of magnitude, yielding 1.8 % for the OECD countries. This is, however, still about 100 times larger than the yearly middle estimate and ten times the yearly upper endestimate (assuming a 10-year investment period to make solar PV competitive) of the cost of public support of solar PV. One important reason is that climate policy initially will indeed be more expensive as solar PV is still maturing, meaning that it can not make a significant contribution to reducing GHG emissions. However, according to the scenario sketched under perspective 1 in Sect. 12.5, after a 10-year period solar PV should fairly quickly take over the market and provide the major means of reducing GHG.

Therefore, during the first 10 years one should expect a relatively high cost of 1.8 % and subsequently a rapid drop in the cost of climate policy to 0.017 % (with an uncertainty range of 0.008–0.17 %). This pattern should not come as a surprise, as it simply reflects an initial investment in R&DDD and then enjoying the returns on this investment. This is consistent with the suggestion by Sandén and Azar (2005) that we need to enter a decade of experimentation with low carbon technologies.

12.4.3 Perspective 3: Delayed GDP Growth

If it is true that climate policy will cost about 1 % of GDP per year, then given that economic growth in many countries has historically been around 2 % on average, and in some countries higher, this would mean that net growth, after discounting the cost of climate policy, would still be positive, and that one would reach a certain level of income with a delay.

A related perspective on the cost of climate policy was proposed by Azar and Schneider (2002). They take as a starting point studies suggesting that the absolute cost of reaching what is regarded by the IPCC as "safe" concentrations of CO2 is in the range of 1–20 trillion US$. Although this may seem impressive, it turns out to imply only a few, namely 1–3, years' delay in achieving a specific level of income in the distant future. The delay evidently depends on income growth. Global income during the twenty-first century is expected to increase about tenfold (on average 2.35 % per annum). Azar and Schneider (2002, p. 77) calculate that "if the cost by the year 2001 is as high as 6 % of global GDP and income growth is 2 % per year, then the delay time is 3 years. . .". This 3-year delay is moreover easily dominated by random noise given the uncertainties involved in GDP movements over a period of one century. That is, uncertainty over such a long time horizon might translate in a variation of the final GDP level (i.e. after one century) which exceeds the 6 % figure. This all means there is little reason to worry about the long-term negative effects of climate policy on the economy. In other words, seen in a long-term perspective, the costs of a stringent climate policy are marginal in economic terms. Aznar and Schneider further note that "…the global economy is expected to be an order of magnitude larger by the end of this century…we would still be expected to be some five times richer on a per capita basis than at present, almost regardless of the stabilization target."

12.4.4 Perspective 4: Happiness Instead of GDP

Economic evaluation of climate policy is often cast in terms of lost GDP. This seems attractive, as the economic and welfare impact is captured in a simple, aggregate number. However, it neglects that implicit assumptions and judgments about the relationship between wellbeing, happiness, and GDP have been strongly criticized (van den Bergh 2009), from the angles of inequity, lexicographic needs, informal activities and environmental degradation. This has given rise to questioning the use of indicators like income and GDP as proxies for social welfare and progress. There is much support for the view that beyond a certain threshold, which has been passed by most rich countries, average income increases do not translate in significant rises in well-being. In particular, this research indicates that somewhere between 1950 and 1970, the increase in welfare stagnated or even reversed into a negative trend in most industrials (OECD) countries, in spite of steady GDP growth, the so-called "Easterlin Paradox" (Easterlin 1974). This supported by the 'Eurobarometer surveys', the half-yearly opinion polls of the inhabitants of the EU member states, as well as by aggregate indicators of sustainable income based on GDP corrections, notably the ISEW and (derived) GPI indicators (Lawn and Clarke 2008). Of course, one should not expect a rigid threshold to apply generally for all countries, cultures, and times. A country comparison clarifies that happiness is characterized by diminishing returns on increases in GDP per capita. This means, not surprisingly, that for poor, developing countries the correlation of income and well-being is higher than for rich countries.

Three stylized facts assessed by happiness research can explain the observed delinking of income and happiness (van Praag and Ferrer-i-Carbonell 2004). First, income and income growth contribute considerably to happiness if people are poor or countries are in a low development phase, as extra income will be mainly spent on basic needs. Second, although people may enjoy short-term or transitory increased happiness effects, ultimately they will adapt or get used to a higher income and changed circumstances in various other dimensions. One explanation for this is that our senses can only handle a limited amount of stimuli, and ultimately satisfaction or boredom ensues. Since most people are not aware of the phenomenon of adaptation, they continue striving for 'more'. This is reflected by a range of terms used by different researchers: 'addiction', 'hedonic adaptation', 'hedonic treadmill', and 'preference drift'. Third, people compare their situation with that of others in a peer group, so their welfare has a relative component. This is associated with status-seeking and rivalry in consumption. In addition, studies have consistently found that income-independent factors greatly influence individual welfare or happiness, the most important ones being health, having a stable family (partner, children), personal freedom (political system), and being employed. Certain studies reported below also point out the relevance of environmental and climate factors

The implication of the foregoing stylized facts is that absolute individual income at best imperfectly, and beyond a certain threshold hardly, correlates with individual welfare (Clark et al. 2008). Relative income turns out to be critical. But at the societal level, relative income changes are largely a zero-sum game: what one wins another loses.

Therefore, using effects on happiness instead of GDP as a criterion for judging climate policy is likely to provide quite different conclusions. Three considerations are relevant here. First, although climate policy may lead to a slower pace of economic growth, the foregoing discussion suggests that this translates into a smaller or even insignificant loss in happiness terms, depending on which country or group of people is considered. Secondly, climate policy aimed at preventing extreme events implies avoidance of serious reductions in happiness, given that happiness directly depends on climate, i.e. it involves direct non-market effects on individuals and households. This means that the economic and welfare effects of climate change measured in GDP terms may underestimate the real impact on happiness. Especially extreme climate events are not easily captured by GDP or other monetary cost terms, as argued in Sect. 12.2. Extreme climate change will have a profound impact on local and regional sea levels, temperatures, and weather patterns. This can in turn cause extreme effects on resource availability (notably clean water), human health, human security, vulnerability of poor people in regions with low productivity (Sahel countries), migration, and violent conflicts. It is virtually impossible to cost-account for these, even though it is clear that human happiness and basic needs are then seriously at stake. Third, although climate change may not affect the happiness of people in Western countries much, for people in poor countries it may mean that their basic needs will come under threat, which is likely to create severe and structural losses in happiness. In addition, richer people and richer countries can more easily adapt to climate change so that they can restore or approximate their old happiness levels. This is because rich countries are characterized by high levels of wealth (financial reserves), high average education, good access to modern technologies, and a generally high capacity for collective action.

Although no serious climate policy study has employed a happiness type of criterion or goal, a few studies have examined the impact of climate conditions on happiness. For example, Rehdanz and Maddison (2005) and Frijters and van Praag (1998) econometrically examines the relationship between temperature and happiness and find significant effects. The shortcoming of these an many other partial analyses is that they consider small temperature changes or differences and give no attention to large changes or even extreme climate change or events. As a result, these studies may deliver an overly optimistic and insufficiently representative general picture of how people's happiness responds to climate change.

Cohen and Vandenbergh (2008) consider the lessons that can be learned from happiness research for climate policy, focusing on consumers. Taxes on pollutive consumption with a positional good character has two benefits: it reduces the status externality due to reduced consumption of such goods (Ireland 2001), and it reduces the total pollution associated with the consumption. Layard (2005) suggests taxing income to stimulate leisure and temper "status games" with respect to income and consumption. This may reduce status effects and pollution related to goods consumption equally, although this will depend on the shift in consumption (e.g., more holidays to distant countries will give rise to increased air traffic with associated GHG emissions).[4]

A provision to the above arguments is that people may adapt to a changed climate in the sense of being initially (negatively) affected in their happiness, while later slowly recovering their old happiness level. However, such adaptation is difficult to imagine for extreme climate change and events. Finally, note that adopting a happiness approach may also affect the discount rate debate. The reason is that one would then be less inclined to discount as this would mean that the happiness of a person in the future would be valued less than that of a person living now. When more general, abstract notions like costs and benefits are employed instead, as in CBA studies, specific people and their happiness disappear from the picture, making the case for discounting easier to defend.

12.4.5 Perspective 5: Comparison With Large Public Investments: Iraq War, Financial Crisis, Military R&D and Sectoral Subsidies

The cost of climate policy or more particularly of making solar PV a competitive technology might be seen as a large public project. This suggests a comparison with other public projects. Two large 'projects' will be considered here, namely the Iraq war and combating the financial crisis. Van den Bergh (2010) also considered R&D investment in the military sector, and expenditures on subsidies to economic sectors.

Stiglitz and Bilmes (2008) have estimated the cost of the Iraq war to the United States to be at least US$3 trillion (3000 billion). Hartley (2006) notes that the economic costs of war receive far less attention than political, moral, legal and military

[4]The happiness perspective also affects the evaluation of other types of policies. Frank (1985), Ireland (2001) and Layard (2005) illustrate specific findings of happiness research as applied to economic policy: (extra) taxation of working overtime, (extra) taxes on status goods, limiting commercial advertising, and restricting flexible labor contracts. Although from a traditional economic growth perspective these look like bad measures, they are positively evaluated from a real welfare or happiness perspective.

considerations. He suggests that the US could have bribed Saddam Hussein by offering him and his family US$20 billion to leave Iraq, giving the Iraqi people US$50 billion, and on top of that save US$30 billion given that the cost of the war was ex ante (grossly under-) estimated at US$100 billion.

Another interesting comparison is with the financial crisis in 2008/2009. The USA decided overnight to reserve US$700 billion to stabilize the US banking system. Governments in Europe are likely to have reserved a similar amount. For example, The Netherlands created a €20 billion fund to stabilize the financial sector and he UK spent about €44 billion to take a majority share in four large British banks to rescue them. In total, OECD countries may have invested more than US$2 trillion (2000 billion) to stabilize the financial system. One may argue that some of the guarantees offered by countries in response to the financial crisis are in fact only creating reserves or represent investments in (shares of) banks rather than being effective spending, but nevertheless the countries or at least their governments were willing to set aside so much money in response to a threat without the support of any cost–benefit analysis

So governments worldwide have invested roughly US$5 trillion in the Iraq war and countering the financial crisis jointly. We can compare this with the range of climate policy cost estimates, i.e. 1–4 % of world GDP (US$66 trillion in 2007), or 0.7–2.7 trillion US$, which is only 14–54 % of the aforementioned public investments. If one focuses on the cost range of making solar PV competitive, i.e. US$30 billion to US$300 billion with a middle scenario estimate of US$60 billion (Perspective 1 in this section), then as a proportion of the current investments in Iraq and the financial crisis this comes down to a central estimate of about 1 % and a range of 0.6–6 %. In other words, if these percentages of current public investments would be diverted to renewable energy, we would very likely solve the problems of energy scarcity and climate change. If the cost of making solar PV competitive is compared only to the cost of the Iraq war, then the assessed central estimate of US$60 billion and the higher end estimate of US$300 billion result in only 2 % and a uncertainty range of 1–10 % of the expenditures on the Iraq war.

12.4.6 Perspective 6: The Current Cost of Energy Is Fairly Low

Here it is argued that current fossil fuel-based energy (gasoline and electricity) is cheap, too cheap in view of associated negative externalities. The latter is especially true if the cost of CO2 reflects extreme climate events and scenarios (van den Bergh and Botzen 2014).

The falling cost of energy in various areas can be observed by considering the share of energy cost in total national income. The ratio of (all) energy expenditures to GDP since the 1970s shows a pattern that starts at around 8 %, increases to about 14 % in the early 1980s and then drops again to levels below those of 1970 and recently increases again (EIA 2008). This illustrates that—in any case, until recently—the cost of energy can be judged as fairly low. Even though energy is the fundamental input to all human economic activity, roughly 90 % of income is spent on things other than energy. Moreover, continuous GDP growth and an almost constant share of energy costs in it suggest that the disposable income after energy expenditures has increased over time.

A disadvantage of the aggregate approach to measuring energy expenditures as a share of GDP is that it hides income inequality. Generally, low income families spend a larger part of their income on energy, and they will also see a relatively rapid increase in the cost share when energy prices rise. The shares can differ between low, middle, and high incomes from 15 %, 5 % and 2 %, respectively. This suggests that for some people, energy use may represent a considerable expenditure, while for many it does not. Roberts (2008) regards households as "undoubtedly fuel poor" when they are spending more than 10 % of their income on energy just to meet basic requirements. This 10 % threshold may reflect, however, that we take a very low share of energy cost in income for granted simply because this is a historical fact. Income inequality does suggest, though, that a serious climate policy raising energy prices might need to be complemented by an income redistribution policy (e.g., as part of shifting taxes from labor to energy).

Another indication that the cost of energy is not very high or even low is that the long-term average oil price (US crude oil prices adjusted for inflation in 2006 US$), if calculated from 1869 to 2007, equals $21.66 per barrel for world oil prices, and for the post-1970 period, $32.23 (http://www.wtrg.com/prices.htm). In addition, the sharp increase in the oil price in 2007–2008 did not give rise to serious, sustained social unrest. This all means that there is room for safe climate policy, which will undoubtedly increase the price of energy.

12.4.7 Perspective 7: Stimulating a Fundamental Social–Technical Transition

Climate change policy is not a simple, one-dimensional policy or an instrument with a clear cost, rather a complex process of multilevel and multi-dimensional change involving the unlocking of a dominant, undesirable system of fossil fuel technologies and infrastructures, and changing institutions, incentives, knowledge bases, and international cooperation. This is hoped to stimulate a "social-technical transition to sustainability", involving structural changes in the economy, including technological innovations and alterations in sector structure, demand side patterns, products types and designs, and institutional arrangements. Such qualitative changes are not well captured in one-dimensional monetary indicators, be it cost measures or foregone GDP growth.

Against this background, Prins and Rayner (2007) argue in favor of "placing investment in energy R&Don a wartime footing". Earlier, former US Vice-President and Nobel Peace laureate Al Gore made a similar call for a "global Marshall Plan". Various others have referred to the Manhattan Project and New Green Deal in this context. Sufficient R&D on de-carbonized energy technologies and a transition to sustainable energy technologies are indeed not guaranteed by environmental regulation alone. One important reason is the lock-in features of fossil fuel energy and related technologies like vehicles with combustion engines. Case studies of historical transitions show that a number of conditions need to be met for a transition to occur (Geels 2005). One of these is public investment in infrastructure and basic (fundamental) research. The history of nuclear fission shows this clearly; it received strong support through direct subsidies and military R&D (in the USA). Several other technologies have benefited greatly from public R&D, particularly investments in military R&D. Notable in this respect are information and communication technologies (ICT), supporting technologies like solid state electronics, semiconductors, transistors, integrated circuits, data transmission networks, and of course basic software codes. All these have received massive funding from the (American) military complex, usually with the motivation of the Cold War.

In many countries, agriculture also has received a great deal of public support, both to maintain the status quo (protection) and to foster certain transitions (Green revolution). For example, the post-war transition in Dutch agriculture was extensively funded by the government through investment subsidies, financial compensation for taking out land, public investment in land consolidation, and the creation and maintenance of drainage systems. This was motivated by a strong urge to achieve food security and self-sufficiency. Similarly, if one recognizes a stable climate as a basic condition for human life and activity, one needs to seriously invest in it.[5]

12.4.8 Perspective 8: Behavior, Learning and Substitution

Closely related to the previous transition perspective is a behavioral perspective. Many substitution opportunities at the level of inputs, sectors, and demand are insufficiently recognized by existing models because of aggregation and limits of empirical data. Notably, stringent climate policy will move prices outside ranges historically observed, so that, for instance, the empirical price elasticities of demand may underestimate potential responses. The more substitution opportunities exist, the easier it is for systems to adapt in a way so as to reach a similar performance level without much additional cost. Moreover, models often do not reflect the fact that in the long run people can change fundamental choices that affect their energy use, or the very many ways in which individuals can adapt to a higher energy cost. For instance, car users can adopt the following strategies: changing the time they drive (outside peak hours), carpooling, using other means of transport (walking,

[5] I am grateful to Frank Geels for suggesting these examples.

biking, public transport), traveling less, being more efficient in combining trips, and in the longer run changing jobs or houses to reduce commuting distances.

A particular aspect of the behavior of firms and individuals is learning and innovation. Sagar and van der Zwaan (2006) examine learning-by-doing in relation to renewable energy and note various learning mechanisms: at the individual worker level (education, learning-by-operating so as to develop tacit skills), within a firm (learning-by-manufacturing), within the industry (learning by copying), across different industries, and within supply-demand interactions (learning-by-implementing, such as integrating PV systems into buildings, on roofs, which involves institutional structures such as for financing and equipment maintenance). Feedback from users to producers and from products to processes, along with systemic improvements (adjustment of all elements, such as institutions, markets, integrated building components, production chain) lead to falling overall costs of the renewable energy technology. Generally, the literature shows that adding endogeneity of growth, i.e. R&D or learning instead of exogenous technological change, reduces policy cost estimates (Söderholm 2007).

It is fair to add that some types of bounded rationality may lead to higher estimates for certain policy cost categories than the rational agent assumption. The energy gap literature illustrates this. Firms do not always invest in profitable energy conservation opportunities for various reasons. One is that agents do not have full information; another is that they do not minimize overall costs but instead focus on what they regard as main activities or investments, which does not include energy conservation; and habitual behavior has also been suggested as an explanation. Information provision and other strategies to stimulate more rational responses as part of climate policy may increase energy conservation (rebound effects not considered) and thus reduce the cost of effective policy. A good translation of insights from behavioral to environmental, energy, and climate economics is currently lacking and would be needed to shed more light on these issues (Brekke and Johansson-Stenman 2008).

12.4.9 Perspective 9: Ancillary Benefits

As discussed in Sect. 12.2, CBA studies of climate policy have omitted many benefits or avoided cost categories. The euphemistic term employed for some of these is ancillary benefits or co-benefits of policy. One that has received ample attention is that the reduction of GHGs generated by fossil fuel combustion will sometimes go along with reductions in other emissions, notably acidifying substances (nitrogen oxides and sulfur dioxide). For example, HEAL (2008) estimates that if the European Union raised its GHG emission target from the current 20–30 % (in line with IPCC recommendations), then additional co-benefits in the range of €6.5–25 billion per year would result from health savings arising from an associated reduction in emissions of fine particles, nitrogen oxide, and sulfur dioxide. All avoided cost categories in CBA studies of climate policy can be regarded as ancillary benefits. Van den Bergh and Botzen (2014) try to quantify these and arrive at a lower bound to the social cost of carbon (or CO_2 to be more precise).

The strong connection between scarce fossil fuel resources and greenhouse gas emissions from combusting fossil fuels also creates a relevant co-benefit. Notably, solving emissions problems by creating new sources of energy (renewable) will mean reducing problems of energy resource scarcity, avoiding potential fierce oil peak shocks, enhancing energy security, and avoiding conflicts over scarce energy resources. For example, a study assessing the social cost of the OPEC oil cartel to the US identified four cost categories, namely wealth transfer to OPEC, cost of strategic petroleum reserve, total GNP loss due to price shocks and shortages, and military costs. This resulted in an estimated cost ranging from about US$150 to 400 billion per year (1990$) during the period 1974–1985 (Green and Leiby 1993).

12.4.10 Perspective 10: Upward Bias in Ex Ante Estimates of Regulation Cost

Various studies indicate that there is often a gap and sometimes even a large gap between ex ante and ex post estimates of the costs of environmental regulation, including both private and public-administrative costs (Harrington et al. 1999). MacLeod et al. (2009) find this for a wide range of environmental policies in European countries, including policies aimed at water and air pollution, health, food safety, fuel standards, directives on combustion plants, and animal welfare. There are two important reasons why ex ante cost assessments may deliver overestimates. First, information on actual costs is often provided by firms having an interest or stake. As a result, those being regulated may provide overly high estimates of individual abatement costs. This can be due to strategic behavior to resist implementation of stringent regulations, or simply to individual uncertainty about (future) abatement costs. Standard environmental economics somehow recognizes these problems, regarding price regulation as having the advantage that it decentralizes the problem of environmental regulation, and not requiring governments to have full information about pollution abatement technologies and associated costs (Baumol and Oates 1988). A second reason for ex ante overestimates is that they may neglect or underrate the potential for reduction of abatement costs through polluters' innovation, learning, and adaptation (see van den Bergh 2010, Section 4).

12.4.11 Perspective 11: International Cooperation and Agreements

An additional important factor influencing cost estimates of climate policy is the presence (or absence) of international agreements, or more generally international cooperation between countries on climate policy and related technological diffusion. If international agreements are absent or weakly constrain individual countries, vast differences in policy may exist between countries. As a result, the costs of stringent climate policy for industries or consumers may be high since it will mean a loss in the international competitive position of industries as well as leakage of emissions from countries with stringent to those with less stringent policies. Instead, a stringent climate policy agreed upon by all countries in the world would mean a level playing field that reduces the policy cost, as competitive disadvantages and emission spillover is avoided. The relationship between policy cost and international cooperation is like a vicious circle. As long as governments think that the cost of safe climate policy is high, they will refrain from committing themselves to a stringent international climate agreement. However, as long as such an agreement is lacking, the cost of unilaterally stringent climate policy will be excessively high because of the loss of competitive position.

12.4.12 Perspective 12: Lack of Insurance Against Climate Change

Currently, private insurance with premiums that reflect the risk of extreme events like those possibly caused by climate change, such as flooding and hurricanes, is largely lacking in most countries (Botzen and van den Bergh 2008). This has three consequences for judging the cost of climate policy. First, it means that there is no efficient sharing of climate-related risks which would reduce the overall costs of the consequences of both climate change and climate policy. Second, the absence of insurance means that appropriate incentives for adequate adaptation to climate risks and changes is lacking. Third, it also means disoptimal incentives for stimulating producers, consumers, (re)insurance companies, and even governments to efficiently reduce greenhouse gas emissions. At present, insurers are already actively involved in promoting reductions in greenhouse gas emissions (Botzen et al. 2009). Such efforts are likely to become stronger if more climate change risks were covered through private insurance. Both insured and insurers have incentives to limit climate risk in case increases in the frequency and severity of natural hazards are reflected in a higher cost of offering insurance and higher premiums. Moreover, with insurance, adaptation at the individual and social level will be more adequate so that climate mitigation policy may need to be less stringent and thus less expensive. In other words, with adequate insurance arrangements in the face of climate-related risks, safe climate mitigation policies will turn out to be more efficient, i.e. less expensive. This is especially true since climate insurance would imply many indirect economic effects because insurance affects the direct and indirect costs of economic activities and therefore works as a price signal of risk. If climate policy is undertaken in the presence of adequate insurance arrangements for risks related to climate change, or if such a policy includes incentives for insurance companies to undertake these arrangements, then the cost of climate policy will be lower than without such arrangements.

12.5 Conclusions

This paper has argued that both cost–benefit analysis and cost assessment or accounting of climate policy using quantitative models are overly ambitious, despite the fact that we can evidently learn much from them. The multi-perspective approach to evaluating the cost of a safe, precautionary climate policy as presented here can be regarded as a way out of the never-ending debate on the usefulness and feasibility of cost–benefit analyses of climate policy. Indeed, if climate policy is seen as a precautionary strategy to avoid unpredictable and irreversible natural as well as economic catastrophes rather than as a way to optimize social welfare (or GDP growth) in the face of GHG emission–climate–economic damage feedback, then a focus on qualitative risk analysis and cost assessment of climate policy makes more sense than a quantitative cost–benefit analysis. This is true both for methodological reasons—CBA possibly represents an overly risk-loving decision-maker—and for practical reasons—quantification of extreme events with small probabilities simply is not feasible.

The paper has tried to credibly defend, using various arguments, that a safe or precautionary approach to climate policy is indeed rational. If one does not accept one argument: there are 11 others waiting in line. The set of 12 perspectives together provide a strong case for the view that a safe climate policy is likely to be affordable and cheaper than most previous studies have suggested.

The happiness or subjective well-being perspective on the cost of climate policy emerges as possibly the most important new view. It is pertinent to introduce it into the debate on climate policy to arrive at a correct picture of what we really gain and sacrifice if we undertake a stringent, safe climate policy worldwide. In terms of happiness or real welfare, climate policy looks much less expensive than in terms of lost GDP, while climate change was evaluated as much more expensive in terms of happiness than in terms of GDP.

Finally, on the basis of various quantitative indicators it was argued that energy is currently not very expensive, so there is considerable leeway for increasing its price through climate policy. Indeed, an effective and safe climate policy cannot avoid raising energy prices considerably, certainly if one wants to simultaneously minimize the rebound effects of energy conservation and efficiency improvements, restructure demand and supply in the economy in a sustainable direction, and stimulate a transition to renewable energy sources. In addition, one will need countervailing distributional measures to avoid energy poverty (e.g., recycling carbon tax revenues to low incomes, or block-pricing for carbon or energy). To keep promising but expensive energy technology paths open, technological subsidies (notably for R&D) will be needed as well.

Of course, while the costs of a safe climate policy may be manageable at global and national levels, as argued here, such a policy will pose serious challenges for particular economic sectors. But this is entirely logical and acceptable, since higher energy costs will regulate and restructure the economy and affect energy-intensive products, processes, firms, and industries relatively severely. Higher energy prices and costs will thus set into motion a process of creative destruction, which is an inevitable component in the transition to a low-carbon economy. Postponing such a transition will only make it more expensive, while safe levels of atmospheric GHG concentration will get out of reach.

Acknowledgements Wouter Botzen, Laurens Bouwer, Frank Geels, John Gowdy, Stijn van den Heuvel, Richard Tol and two anonymous reviewers provided useful suggestions on the original, longer article.

References

Ackerman F, DeCanio SJ, Howarth RB, Sheeran K (2009) Limitations of integrated assessment models of climate change. Clim Change 95:297–315. http://www.springerlink.com/content/c85v5581x7n74571/fulltext.pdf
Alcamo J, Shaw R, Hordijk L (eds) (1990) The RAINS model of acidification. Science and strategies in Europe. Kluwer, Dordrecht
Arrow KJ (2007) Global climate change: a challenge to policy. The Economists' voice 4(3):Article 2. Available at: http://www.bepress.com/ev/vol4/iss3/art2
Ayres RU, Walters J (1991) Greenhouse effects: damages, costs and abatement. Environ Resour Econ 1:237–270
Azar C (1998) Are optimal CO2 emissions really optimal? Four critical issues for economists in the greenhouse. Environ Resour Econ 11(3–4):301–315
Azar C, Lindgren K (2003) Catastrophic events and stochastic cost–benefit analysis of climate change. Clim Change 56(3):245–255
Azar C, Schneider SH (2002) Are the economic costs of stabilising the atmosphere prohibitive? Ecol Econ 42:73–80
Azar C, Schneider SH (2003) Are the economic costs of stabilising the atmosphere prohibitive? A response to Gerlagh and Papyrakis. Ecol Econ 46:329–332
Barker T (1996) A review of managing the global commons: the economics of global change, by WD Nordhaus. Energy Environ 7(1):85–88
Baumol WJ, Oates WE (1988) The theory of environmental policy, 2nd edn. Cambridge University Press, Cambridge
Botzen W, van den Bergh JCJM (2008) Insurance against climate change and flooding in The Netherlands: present, future and comparison with other countries. Risk Anal 28(2):413–426
Botzen W, van den Bergh JCJM (2009) Bounded rationality, climate risks and insurance: is there a market for natural disasters? Land Econ 85(2):266–279
Botzen WJW, Gowdy JM, van den Bergh JCJM (2008) Cumulative CO2 emissions: shifting international responsibilities for climate debt. Clim Pol 8:569–576
Botzen WJW, van den Bergh JCJM, Bouwer LM (2009) Climate change and increased risk for the insurance sector: a global perspective and an assessment for The Netherlands. Nat Hazards 52(3):577–598
Brekke KA, Johansson-Stenman O (2008) The behavioural economics of climate change. Oxf Rev Econ Policy 24(2):280–297
Broome J (1992) Counting the cost of global warming. White House Press, Cambridge
Clark AE, Frijters P, Shields MA (2008) Relative income, happiness, and utility. J Econ Lit 46(1):95–144
Cline WR (1992) The economics of global warming. Institute for International Economics, Washington, DC

Cohen MA, Vandenbergh MP (2008) Consumption, happiness, and climate change. RFF discussion paper 08–39, Oct 2008. Resources for the Future, Washington, DC
Dasgupta P (2007) Commentary: the Stern Review's economics of climate change. Natl Inst Econ Rev 199:4–7
DeCanio SJ (2003) Economic models of climate change: a critique. Palgrave-Macmillan, New York
Dietz S, Anderson D, Stern N, Taylor C, Zenghelis D (2007) Right for the right reasons: a final rejoinder on the Stern Review. World Econ 8(2):229–258
Easterlin RA (1974) Does economic growth improve the human lot? Some empirical evidence. In: David PA, Reder MW (eds) Nations and households in economic growth: essays in honour of Moses Abramowitz. Academic, New York
EIA (2008) Annual energy outlook 2008. Energy Information Administration, Washington, DC. Available at http://www.eia.doe.gov/oiaf/aeo/
Fisher C, Morgenstern RD (2006) Carbon abatement costs: why the wide range of estimates? Energy J 27(2):73–86
Frank RH (1985) Choosing the right pond: human behavior and the quest for status. Oxford University Press, New York
Frijters P, van Praag BMS (1998) The effects of climate on welfare and wellbeing in Russia. Clim Change 39:61–81
Funtowicz S, Ravetz JR (1993) Science for the postnormal age. Futures 25:739–755
Geels FW (2005) Technological transitions and system innovations: a co-evolutionary and socio-technical analysis. Edward Elgar, Cheltenham
Gollier C, Jullien B, Treich N (2000) Scientific progress and irreversibility: an economic interpretation of the precautionary principle. J Public Econ 75:229–253
Gowdy JM (2008) Behavioral economics and climate change policy. J Econ Behav Organiz 68:632–644
Green DL, Leiby PN (1993) The social costs to the US of monopolization of the world oil market 1972–1991. Oak Ridge National Laboratory, Report 6744, Oak Ridge
Ha-Duong M (1998) Quasi-option value and climate policy choices. Energy Econ 20(5–6):599–620
Harrington W, Morgenstern RD, Nelson P (1999) On the accuracy of regulatory cost estimates. Resources for the Future, Washington, DC
Hartley K (2006) The costs of war. Centre for Defence Economics, The University of York. http://www.york.ac.uk/depts/econ/documents/research/iraq_conflict_vox_june2006.pdf
HEAL (2008) The co-benefits to health of a strong EU climate change policy. Health and Environment Alliance (HEAL), Climate Action Network Europe (CAN-E) and WWF. Available at http://www.climnet.org/Co-benefits%20to%20health%20report%20-september%202008.pdf
Holling CS (1986) The resilience of terrestrial ecosystems: local surprise and global change. In: Clark WC, Munn RE (eds) Sustainable development of the biosphere. Cambridge University Press, Cambridge
IPCC (2007) Fourth assessment report. Intergovernmental Panel on Climate Change, Geneva
Ireland NJ (2001) Optimal income tax in the presence of status effects. J Public Econ 81:193–212
Kelly DL, Kolstad CD (1999) Integrated assessment models for climate change control. In: Folmer H, Tietenberg T (eds) The international yearbook of environmental and resource economics 1999/2000. Edward Elgar, Cheltenham
Kimball MS (1990) Precautionary saving in the small and in the large. Econometrica 58:53–73
Kolstad C (1996) Learning and stock effects in environmental regulations: the case of greenhouse gas emissions. J Environ Econ Manage 31:1–18
Lawn P, Clarke M (2008) Sustainable welfare in the Asia-Pacific: studies using the genuine progress indicator. Edward Elgar, Cheltenham
Layard R (2005) Happiness: lessons from a new science. Penguin, London

Loulou R, Kanudia A (1999) Minimax regret strategies for greenhouse gas abatement: methodology and application. Oper Res Lett 25:219–230
MacLeod M, Ekins P, Vanner R, Moran D (eds) (2009) Understanding the costs of environmental regulation in Europe. Edward Elgar, Cheltenham
Neumayer E (1999) Global warming: discounting is not the issue, but substitutability is. Energy Policy 27:33–43
Nordhaus WD (1991) To slow or not to slow? The economics of the greenhouse effect. Econ J 101:920–937
Nordhaus WD, Boyer J (2000) Warming the world: economic models of global warming. The MIT Press, Boston
OECD (2008) OECD in figures 2007. OECD, Paris. www.oecd.org/infigures
Padilla E (2004) Climate change, economic analysis and sustainable development. Environ Values 13:523–544
Prins G, Rayner S (2007) Time to ditch Kyoto. Nature 449:973–975
Quiggin J (2008) Stern and his critics on discounting and climate change: an editorial essay. Clim Change 89(3–4):195–205
Rehdanz K, Maddison DJ (2005) Climate and happiness. Ecol Econ 52:111–125
Rind D (1999) Complexity and climate. Science 284(5411):105–107
Roberts S (2008) Energy, equity and the future of the fuel poor. Energy Policy 36(12):4471–4474
Sagar AD, van der Zwaan B (2006) Technological innovation in the energy sector: R&D, deployment and learning-by-doing. Energy Policy 34:2601–2608
Sandén BA, Azar C (2005) Near-term technology policies for long-term climate targets—economy wide versus technology specific approaches. Energy Policy 33:1557–1576
Schimmelpfennig D (1995) The option value of renewable energy: the case of climate change. Energy Econ 17(4):311–317
Söderholm P (2007) Modelling the economic costs of climate policy. Research report 2007–17, Department of Business Administration and Social Sciences, Luleå University of Technology
Sorrell S (2007) The rebound effect: an assessment of the evidence for economy-wide energy savings from improved energy efficiency. UK Energy Research Centre, London. http://www.ukerc.ac.uk/Downloads/PDF/07/0710ReboundEffect/0710ReboundEffectReport.pdf
Spash CL (2002) Greenhouse economics. Routledge, London
Spash CL (2007) Climate change: need for new economic thought. Economic and Political Weekly, 10th Feb 2007, 483–490
Stern N, Peters S, Bakhshi V, Bowen A, Cameron C, Catovsky S, Crane D, Cruickshank S, Dietz S, Edmonson N, Garbett S-L, Hamid L, Hoffman G, Ingram D, Jones B, Patmore N, Radcliffe H, Sathiyarajah R, Stock M, Taylor C, Vernon T, Wanjie H, Zenghelis D (2006) Stern Review: the economics of climate change. HM Treasury, London. http://www.hm-treasury.gov.uk/stern_review_climate_change.htm. Later published as Stern N (2007) The economics of climate change: the Stern Review. Cambridge University Press, Cambridge
Stiglitz JE, Bilmes L (2008) The three trillion dollar war: the true cost of the Iraq conflict. Norton, New York
Tol RSJ (2005) The marginal damage costs of carbon dioxide emissions: an assessment of the uncertainties. Energy Policy 33(16):2064–2074
Tol RSJ (2008a) The social cost of carbon: trends, outliers, and catastrophes. Economics, the openaccess, open-assessment. E-Journal 2(25):1–24
Tol RSJ (2008b) The economic impact of climate change, Working paper no 255. Economic and Social Research Institute, Dublin
Tol RSJ, Yohe GW (2007) A stern reply to the reply of the review of the Stern Review. World Econ 8(2):153–159
Ulph A, Ulph D (1997) Global warming, irreversibility and learning. Econ J 107:636–650
van den Bergh JCJM (2004) Optimal climate policy is a utopia: from quantitative to qualitative cost-benefit analysis. Ecol Econ 48:385–393

van den Bergh JCJM (2009) The GDP paradox. J Econ Psychol 30(2):117–135
van den Bergh JCJM (2010) Safe climate policy is affordable – 12 reasons. Clim Change 101:339–385
van den Bergh JCJM, Botzen WJW (2014) A lower bound to the social cost of CO2 emissions. Nat Clim Chang 4(April):253–258
van der Zwaan B, Rabl A (2003) Prospects for PV: a learning curve experience. Sol Energy 74(1):19–31
van der Zwaan B, Rabl A (2004) The learning potential of photovoltaics: implications for energy policy. Energy Policy 32:1545–1554
van Praag B, Ferrer-i-Carbonell A (2004) Happiness quantified: a satisfaction calculus approach. Oxford University Press, Oxford
Weitzman ML (2007) A review of the Stern Review on the economics of climate change. J Econ Lit 45(3):703–724

Chapter 13
Renewable Energy in the UK: A Slow Transition

David Elliott

Abstract As a case study in the global technological, political and economic transition to the use of sustainable energy sources, this chapter looks at the way renewable energy technology has been developed in the United Kingdom (UK) in the context of its overall response to climate change. In particular it highlights the impacts that differing views on the role of market competition have played. It argues that the market-oriented approach to the support of renewable energy adopted by the UK has been a key reason why it has, with some exceptions, been relatively slow to develop its very large renewable energy resource compared to most other EU countries. It suggests that, under present policies, this relatively poor showing may not improve, especially given the UK's strong commitment to expanding nuclear power.

Keywords Renewable energy • UK energy policy • Nuclear • Subsidies • Feed-in tariffs • Support mechanisms

13.1 Introduction

A global transition to the use of renewable energy sources, in parallel with a commitment to energy saving, is widely seen as a key response to the threat of climate change (IPCC 2014). However this transition is occurring at different rates, and under different political and economic conditions, around the world. While hydro electric power is already widely used in both developed and undeveloped countries, supplying about 16 % of global electricity, in terms of the 'new' renewables (e.g. the use of wind and solar energy), the industrialised nations have led the way, although there are significant differences in approach and success.

This chapter looks at the case of the United Kingdom (UK) in the context of the European Union's overall energy and climate policies. It reviews the approach to

---

D. Elliott (✉)
The Open University, Milton Keynes, UK
e-mail: d.a.elliott@open.ac.uk

S. Shmelev (ed.), *Green Economy Reader*, Studies in Ecological Economics 6,
DOI 10.1007/978-3-319-38919-6_13

developing and deploying renewable energy technologies adopted in the UK, in terms of the targets set and the support mechanisms used. In particular it highlights the impacts that differing views on the role that market competition can play in the selection and promotion of new energy technologies like renewables.

There are of course other energy options, including nuclear power and the continued use of fossil fuels, but with carbon emissions captured and stored. Some countries, including the UK, have followed these options up, along with renewables and energy efficiency, with varying degrees of commitment. Although starting from a low level, the emphasis in terms of investment in new energy technology in recent years in most countries has however mainly been on renewables, and they seem likely to remain the fastest growing new energy supply options globally (WEC 2013).

Most governments have sought to stimulate the development of new low carbon energy technologies by introducing market adjustment mechanisms to take some account of environmental and climate change concerns, such as energy and carbon taxes or trading systems. However, the most successful approaches, at least in terms of boosting the take-up of renewables, have involved direct subsidies to help new entrants compete with the well- established energy options. The design of these support systems, and in particular the extent of their market orientation, has varied, with the outcomes varying (Lesser and Su 2008). This chapter looks at the approach adopted in the UK, where a market-based approach has been adopted.

13.2 Renewable Energy Globally and in the UK

The use of renewable energy is expanding rapidly around the world, in the EU, the USA and China especially, with wind energy playing a major role. By the end of 2014 China had around 115 GW of wind generation capacity installed, the USA about 66 GW, Germany 40 GW, and the world as whole 370 GW (WWEA 2015). To put that in perspective, total live global nuclear capacity in 2015 was around 340 GW. Photovoltaic solar has also expanded rapidly, reaching around 180 GW globally by the end of 2014, while solar thermal (for space and water heating) is at around 326 GW (REN21 2014). The use of biomass for heat and/or power, as well as for transport fuel, is also expanding, although its growth may be constrained by land-use issues, but these do not apply to the newly emerging offshore renewables, including wave and tidal energy, and only to a very limited extent to geothermal energy (Elliott 2013a).

In all, renewable now supply over 22 % of the electricity used globally, from around 1560 GW of generation plant, and 19 % of global primary energy consumption, compared with 2.6 % of global energy from nuclear (REN21 2014). The International Energy Agency says that wind, solar, bio-energy and geothermal use may grow 40 % by 2018, twice the 20 % rate in 2011, supplying 25 % of global electricity by 2018 (IEA 2013a).

Much of this expansion, and subsequent growth, could be in Asia, in China especially, where the aim is to have 200GW of wind capacity in place by 2020 and to obtain around 15 % of primary energy (not just electricity) from non-fossil fuels, mostly renewables, by 2020. Renewables already supply over 17 % of China's electricity, compared to around 2 % from nuclear. The wind output alone is now larger than that from nuclear. The USA is, in effect, trying to compete with China by investing in key renewables, wind especially. Renewables supply around 13 % of its electricity at present, and although there are no formal national targets, there are some quite high state level targets, e.g. California's 33 % by 2020 electricity target, excluding hydro. Japan's energy policy is still in flux, following the Fukushima nuclear disaster, but interim plans have suggested obtaining up to 24 % of electricity from renewables by 2030. Targets for expansion in Africa are generally less ambitious, but hydro already supplies a large proportion of power in some African countries, and that is even more the case in South America, where there are also some quite ambitious expansion programmes for new renewables (Elliott 2013a)

The current EU target is to obtain 20 % of all its energy, not just electricity, from renewable sources by 2020 and 27 % by 2030. Targets for the longer term are also being negotiated, with a proposal that the EU 2050 Roadmap should include a renewable energy target of between 55 % (in the lowest scenario) and 75 % (in the highest scenario). In the latter case, 97 % of all electricity would by then be supplied by renewables (EC 2011). Certainly a range of independent studies have suggested that it could be feasible to get near to 100 % of total EU electricity from renewables by 2050 at reasonable costs, given proper support and attention to energy saving (EREC 2010; ECF 2010; PWC 2010).

In parallel, some independent/academic studies looking globally have reached similar conclusions. Indeed some say renewables could possibly supply nearly all energy globally by 2050 (WWF 2011; Jacobson and Delucchi 2011). The usually conservative International Energy Agency has published a scenario in which renewables supply 75 % of global electricity by 2050 (IEA 2010), while the Intergovernmental Panel on Climate Change has claimed that renewables could supply up to 77 % of total global energy by 2050 (IPCC 2011).

Clearly the resource is large, and the technologies for exploiting it are developing rapidly (Boyle 2012). In a 2011 report on 'Deploying Renewables' the International Energy Agency said that 'a portfolio of renewable energy technologies was becoming cost- competitive in an increasingly broad range of circumstances, in some cases providing investment opportunities without the need for specific economic support.' (IEA 2011). That process in now well underway.

Within the EU, several countries are already approaching their 2020 renewable energy targets, with Austria, Denmark, Finland, Latvia, Portugal and Sweden in the lead in percentage terms (Eclareon 2011), and three have already surpassed their 2020 targets (Bulgaria, Estonia and Sweden). Some have ambitious follow-up targets. For example, Denmark is aiming to obtain all its power and heat from renewables by 2030, and aims to be 'zero carbon' by 2050 (Richardson et al. 2011).

Germany has now set interim target corridors of obtaining 40–45 % of its electricity from renewables by 2025 and 55–60 % by 2035, and then plans to expand that in stages to at least 80 % by 2050 (Maue 2012; GEB 2014). Moreover, the German Advisory Council on the Environment (SRU) has claimed that a transition to 100 % renewable electricity by 2050, rather than just 80 %, is possible (SRU 2011).

By contrast, the UK has been relatively cautious. Although it has made a commitment to cut its greenhouse gas emissions with respect to a 1990 baseline by at least 80 % by 2050, it has only made a commitment to obtaining 15 % of its energy from renewables by 2020- the target agreed with the EU.

Progress towards meeting that relatively low EU-agreed target has been relatively slow compared to what has been achieved by many other EU members, some of whom have agreed much higher 2020 targets with the EU. For example Austria's 2020 renewable energy target is 34 %, Denmark's is 30 %, Finland's 38 %, Latvia's 40 %, Portugal's 32 % and Sweden's is 49 %

In terms of technology deployment, the UK trails well behind in most areas. By 2015 the UK had only managed to install around 12 GW of wind generation, including offshore, compared with over 40 GW in much less windy Germany. The UK's record with PV solar has also been also poor. Although it has improved of late, it had still only reached around 5 GW by 2015, compared to Germany at 36 GW. See Table 13.1 for a cross-EU comparison.

As this indicates, the UK at that point fell third from the bottom on each measure, only beating Malta and Luxembourg. To be fair, these comparisons include some countries that, unlike the UK, have large existing hydro capacities, as well as significant biomass production. But that mainly involves the leaders e.g. Sweden, Latvia, Finland and Austria. Even if these capacities are removed, the ranking is basically unchanged and UK still comes out near the bottom.

13.3 Why so Slow?

Given that the UK has some of the best renewable resources in the EU, and possibly the world, it is perhaps surprising that it has only developed them to a limited extent so far. Offshore wind is now being exploited somewhat more successfully, with the UK leading the world, at around 5 GW by 2015, but it seems clear that, so far, overall, the UK has not done well. One explanation could be the financial support systems it adopted- first the Non Fossil Fuel Obligation (NFFO) and then the Renewables Obligation (RO).

Competitive market systems like the NFFO and RO have been much less successful at building renewables capacity than guaranteed-price Feed-In Tariffs (FiTs), as pioneered by Germany and widely used in the rest of the EU. The reason appears to be that the systems are radically different not so much in terms of the total amount of money allocated, but in terms of how the allocation was organised.

Under the NFFO, which ran in a series of rounds from 1990 to 1998, there were competitive capacity auctions, set against targets for renewables. There were some winning bids at low prices but some of the developers were unable to complete the projects – they had underbid. Not much capacity was installed as a result, especially in the later rounds (Mitchell 2000)

Table 13.1 EU renewable energy supply and targets. Renewable energy in final gross energy consumption %

Country	Supply in 2010 (ranked)	Supply in 2012	EU agreed 2020 target
Sweden	47.6	52.4	49
Latvia	33.1	33.0	40
Finland	33.0	36.4	38
Austria	30.9	32.2	34
Portugal	26.8	24.7	32
Estonia	25.6	27.8	25
Romania	24.1	22.1	24
Denmark	23.5	24.2	30
Slovenia	18.5	20.2	25
Lithuania	18.3	20.8	23
Spain	15.1	16.7	20
France	13.3	13.7	23
Bulgaria	12.8	17.7	16
Germany	12.3	12.4	18
Greece	11.2	12.5	18
Italy	11.2	13.8	17
Poland	10.6	12.4	15
Czech Republic	10.4	11.3	13
Slovakia	9.5	10.6	14
Hungary	8.2	9.8	13
Ireland	6.1	7.5	16
Cyprus	6.0	7.0	13
Belgium	5.6	5.3	13
Netherlands	4.4	4.5	14
United Kingdom	3.8	4.1	15
Luxembourg	2.8	3.1	11
Malta	0.4	0.3	10

Source: Eurobserver 2012/2013. http://www.eurobserv-er.org/pdf/press/year_2012/RES/English.pdf and http://www.eurobserv-er.org/pdf/press/year_2013/res/english.pdf

The subsequently introduced RO system, which is still in place, has arguably not been much more successful, facing developers with economic risks and uncertainties (Toke 2005, 2007; Mitchell et al 2006). The value of the tradable Renewable Obligation Certificates given to companies for each MWh of green power supplied is determined by market trading and cannot be predicted: so it is hard for companies to get good interest rates from banks when they seek to borrow money to invest in new capacity. The result is that they have had to charge consumers more. By contrast, under the Feed-In Tariff system, prices are guaranteed for years ahead at known levels, making the investment climate much more stable, leading to lower consumer prices.

For example, in 2005/2006 the UK's RO system cost consumers 3.2 pence/kilowatt hour, whereas, in 2006 the German FiT only cost consumers 2.6 pence/kilowatt hour, despite Germany having a much larger wind capacity in areas with generally much less wind than in the UK, and also supporting the installation of increasing amounts of much more expensive PV solar capacity (Ernst and Young 2008). France which, started to develop renewables even later than Germany, quickly matched the UK's wind capacity, using a FiT. So did Italy. Most EU countries now use FiTs.

After campaigning by pressure groups, in 2010 the UK government adopted a small FiT scheme, but only for micro-generation projects, under 5 MW, and with an allocation limit. The scheme was rapidly oversubscribed by consumers wishing to install PV solar. However the RO continued to be promoted as the main support system, especially for wind and other larger renewables.

FiTs may not have been ideal for supporting the take-up of initially higher cost technologies like PV solar, since the cost pass-through to consumers was high. So, although PV prices have fallen (in part due to the success of FiTs across the word), as take-up accelerated, the extra consumer cost proved provocative in the UK, and also across the EU, with caps being imposed to limit the impact on consumers. However, as the German experience showed, FiTs were well suited to supporting more economically competitive on-land wind projects, with relatively low cost pass-through to consumers. By contrast, that was clearly not the case for the RO.

One of the additional reasons why the UK did not install as much wind capacity as elsewhere, could be that the RO's competitive market trading approach led companies to locate wind projects at the most profitable sites, which are usually in environmentally sensitive, high wind speed, upland areas. This has led to a backlash against many wind projects – many have been opposed by local people and have been turned down by local planning authorities, usually on the basis of visual intrusion. By contrast, there is much less opposition elsewhere in the EU, where FiTs have made it viable to locate projects in low-land sites, and also for many projects to be initiated and run locally. For example in Denmark, which now gets around 39 % of it electricity from wind, the FiT that was used at one stage led to about 80 % of the wind projects being owned by local community-based wind co-ops or by farmers. They often quoted the old Danish proverb 'your own pigs don't smell'. In the UK, the RO makes it hard for co-operatives to get investment capital for smaller local projects. So far there are only three wind co-operatives in existence.

The UK government had been strongly committed to the RO system, but eventually it had to admit that it was an expensive approach to supporting wind and other renewables. In 2011 the government decided to go ahead with a variable price 'Contracts for Difference' (CfD) scheme for renewables, but also extended it to cover nuclear power and Carbon Capture and Storage (DECC 2011, 2012). Although rather oddly labeled by the UK government as a Feed -In system, like the RO, the CfD is a market orientated approach, and project contract auctions have been included to further enhance competition, harking back to the NFFO (DECC 2014a).

Some critics argued, why not adopt a guaranteed-price FiT instead? After all, by 2015, the small UK FiT, despite being marginalised, had led PV solar to expand from almost nothing to around 5GW. However, the CfD is set to become the main support mechanism, replacing the RO for new projects entirely by 2017 (DECC

2014b). Whether it will be any more successful at stimulating the deployment of renewables than the RO remains to be seen. Its operation has be set within an overall annual finance cap, fixed at £300 m for the first auction round starting in 2014, so that competition for funding was fierce. Moreover, in 2015, the government blocked further support for solar farms above 5 MW and proposed cutting support for all on-shore wind projects, including CfD support, an arguably odd policy given that they were the cheapest of the main renewable options, needing only a little more support to become fully competitive.

The competitive market approach adopted in the RO, and retained and enhanced in the CfD, has not been the only problem facing renewables in the UK. Planning issues certainly have had very significant impacts. While, as argued above, some of the planning problems do seem be the result of the RO's competitive approach, some are the result of other primary factors, such as lack of full local consultation and poor planning procedures, without sufficient transparency, leading to increased local opposition. Problems with making grid connections have also clearly been a key issue, one that is unconnected to the RO. The high costs and long delays in getting grid connections agreed leads to increased overall costs and risk for developers, who therefore have to raise charges to consumers, thus giving further ammunition to those who claim that wind is expensive. There have also been problems in the context of providing marine cable links to offshore wind projects, with, at one stage, an emphasis on maintaining a competitive approach, so that each wind farm project would have its own very expensive link Fortunately this approach now seems to have been abandoned in favour of a more integrated approach (Elliott 2012a).

13.4 Nuclear Power

While the RO, planning conflicts and grid link issues have been amongst the main practical problems facing renewables in the UK, it could be argued that there was a wider issue- a fundamental lack of commitment to them compared with support for nuclear power.

The UK had developed nuclear power in the 1960s and at one point it was supplying around 25 % of UK electricity. However it had fallen out of favour, due to the relatively high costs compared with gas-fired plants.

In 1995, after a major review, the then Conservative government had concluded that 'providing public sector funds now for the construction of new nuclear power stations could not be justified on the grounds of wider economic benefits and would not therefore be in the best interest of either electricity consumers or tax payers' (Nuclear Review 1995).

In 1998, the incoming Labour government confirmed this view: '*at present nuclear power is too expensive to be economic for new capacity and in current circumstances it is unlikely that new proposals for building nuclear plants will come forward from commercial promoters.*' (Trade and Industry Select Committee 1998).

In 2003 the Labour government produced a White Paper on Energy which said that the current economics of nuclear power *'make it an unattractive option and there are also important issues of nuclear waste to be resolved'* (DTI 2003). Instead it supported renewables and energy conservation.

However, after a new energy review in 2006, it changed its mind, and commented in a new White Paper on Energy 'new nuclear power stations would make a significant contribution to meeting our energy policy goals' (BERR 2007). Nuclear and renewables would both be supported. This policy was also adopted by incoming Conservative-Lib Dem coalition Government. It its revised 2011 National Policy Statements on Energy it said that, by 2025, the UK would need 113 GW of electricity generating capacity, of which at least 59 GW would have to be new capacity, with renewables at around 33 GW, but 26 GW being 'for industry to determine', although it said that it believed that, 'in principle, new nuclear power should be free to contribute as much as possible towards meeting the need for around 18 GW of new non-renewable capacity by 2025' (NPS 2011).

The result is that nuclear, renewables and also Carbon Capture and Storage, are all being supported in parallel. There are problems with this approach. It can lead to conflicts. The nuclear and renewables lobby tend to be wary of each other, reflecting the previous era when nuclear power got the lions share of government funding, and nuclear proponents were often very dismissive of renewables (Elliott 2010). They still are sometimes, and conflicts over funding have continued.

The nuclear industry professed not to need subsidies, but it did want the market structure to be supportive. However, nuclear critics have generally seen the governments proposed radical Electricity Market Reforms as designed primary to help sustain nuclear, by creating the new 'Contacts for Difference' (CfD) support system (covering nuclear, Carbon Capture and storage and, from 2017, some renewables) and also by introducing a unilateral UK carbon price support system to buttress the EU Emission Trading System (Mitchell 2011). These changes may also help some of the larger renewables, but the nuclear industry was likely to benefit most, with, for example, its existing plants being likely to get a large windfall carbon credit bonus (Toke 2010).

Although the Department of Energy and Climate Change (DECC) claimed that renewables would be given similar treatment to nuclear, the first of the proposed new UK nuclear plants, at Hinkley, seems to have been given privileged status. No attempt was made to open the Hinkley project to competitive tendering, whereas, as noted earlier, a competitive project auction system has been introduced for renewables seeking to get CfD support (DECC 2014b). The CfD strike price offered to French company EdF for Hinkley, gives it guaranteed support for 35 years, as opposed to 15 years for CfD contracts for renewables (HM Government 2013, HM Treasury 2013). Moreover, the CfD support level for Hinkley seems to be higher that what is likely to be available or needed by most renewables (on-land wind and PV in particular) by the time the 3.2 GW Hinkley plant might start up (2023), if it goes ahead. In addition, as a final extra level of support, a £10bn loan guarantee has been offered to EdF for the Hinkley project, under the UK's infrastructure support system (HM Treasury 2012).

Overall, it does seem that the approach to nuclear/renewables support, so far, is asymmetrical, and initially it seemed likely that the funding arrangements for the Hinkley project would fall foul of European Commission's state aid rules. However, after some adjustments, with some profit claw-back arrangements added, the EC gave it a go ahead (Europa 2014).

Clearly the UK government is keen to press ahead with nuclear. Several other nuclear projects are in the pipeline, from EdF and other overseas companies, 16GW in all, with some expected construction start dates in the 2020s. By contrast, DECC has published a scenario in which, after expansion up to 2020 (to meet the EU 15 % renewable energy target), the growth of renewables falls off dramatically (DECC 2013).

In addition, the nuclear industry and its supporters clearly have ambitions for even larger scale subsequent expansion, with, for example, following on from the initial 16 GW currently expected, a second expansion phase, after 2030, leading to up to 25 GW of new nuclear capacity in the UK (Grimes and Nuttall 2010). Under the previous Labour administration there had been talk of nuclear providing 35–40 % of UK electricity 'beyond 2030' (Wicks 2009). A 2012 Energy Research Partnership report, produced by the UK National Nuclear Laboratory in consultation with the nuclear industry, suggested that, after the first 16 GW of new plants were built, more could be added bringing the total to over 40 GW by 2050 (ERP 2012). Moreover, a study by the Smith School at Oxford University even looked at a scenario with up to 90 GW of nuclear by 2050 (Smith School 2012). Subsequently the UK government nuclear strategy review suggested that 75 GW might be possible by 2050, supplying 86 % of UK electricity, up from 17 % now (BIS 2013).

This is not the place for a recital of the case for and against nuclear power. Suffice it to say that, so far, it has exhibited a 'negative learning curve', with costs continually rising (Gruber 2010). A major nuclear programme based on new technology might produce a different result, but that is speculative, while it is already clear that the costs of renewables are falling, in some cases rapidly (IRENA 2012). The economic basis for backing nuclear strongly is thus debatable.

13.5 UK Strategic Issues: Diversity

There are more general strategic problems with the approach being adopted in the UK. The UK government says it favours a mix of options, to balance risks, but a 86 % nuclear contribution would hardly represent 'diversity'. Even if a lower nuclear percentage is adopted, leaving more room for renewables and possibly CCS, there are risks in diluting efforts. Diversity is a good strategic principle, but the risk is that spreading resources so widely over three large areas will mean that none are developed well. Moreover, if it is diversity that is required, then renewables are in fact not one option but a wide range of very different technologies, at different stages of development. By contrast, at present at least, nuclear is based on one relatively mature technology (the PWR and its upgrades), while CCS technology is as yet undeveloped.

There are also longer-term advantages from focusing on renewables, not least the fact that the energy sources are indigenous and will never be exhausted, while the conversion technology is likely to continue to become more economic, without the risk of fuel price rises due to external events.

It is sometimes argued that, while that may be true, the UK should wait until the technology has developed (presumably mostly elsewhere) before deploying it widely. But that view ignores 'first mover' commercial and technological advantages. It might be wise for the UK to focus on it strengths, which currently are mostly in the marine renewables field, but more generally, as Barack Obama put it eloquently in 2009 'the country that harnesses the power of the clean, renewable energy will lead the twenty-first century' (Obama 2009).

The UK may not be able to lead in all areas, but it can be a major player in some areas of renewable development and deployment, whereas it stands little chance of leading technologically in nuclear or CCS. Given that the UK probably has the world's best renewable resources (in terms of wind, wave and tidal energy especially), as well as established technological expertise, particularly in offshore engineering and marine technology, there is a strong strategic case for focusing on renewables

Finally, there is the often repeated claim that there will be a need for all the new capacity that can be developed, since demand will double by around 2050. That has been used to justify the nuclear expansion. This claim was challenged in a critical 'Corruption of Governance' report, which argued that the government had been poorly informed on this issue, and that, in fact, that there was no evidence for it. And certainly no need to accept it: Germany was planning to cut electricity demand by 25 % by 2050 and overall energy use by 50 % (Bailey and Blair 2012). The EU as a whole has a target of reducing energy demand by 20 % by 2020 and is looking to more substantial reductions later- 30 % by 2030.

So far this demand-side argument has not had much impact, with the UK government evidently being convinced that demand will grow and that there could be problems meeting it, especially in a situation where oil and gas imports may be constrained. So the governments commitment to nuclear remains strong. The result could be that, within the Western EU, the UK (Scotland apart) could become almost unique in having a major nuclear expansion programme, at a time when many other countries in this region have either reconfirmed their opposition to nuclear (e.g. Austria, Denmark, Ireland, Norway, Portugal, Greece) or are phasing it out (Germany, Italy, Belgium, Switzerland and possibly even, in part, France) and are focusing instead on renewables (Sovacool and Valetine 2012).

In theory, funding for new energy technologies need not be a fixed sum game, but in practice, given budgetary limits, as Scotland's then First Minister Alex Salmond, said, a pound spent on nuclear is a pound not spent on renewables. Certainly countries with major nuclear programmes have so far tended to have small renewables programmes and vice versa.

It is interesting in this context that the devolved government of Scotland has opposed new nuclear projects and has promoted renewables strongly, with nearly 50 % of Scotland's electricity needs being met from renewables by 2015 and plans to expand that to 100 % by 2020 (Scottish Government 2013). This difference in approach was one of the issues driving the move to full Scottish independence. Given that Scotland's wind and hydro account for around 36 % of UK green power at present, independence would have a major impact (Goodall 2014)

13.6 UK Energy Choices

The UK faces some choices. It has been trying to back a mixed approach to energy, using market-determined support mechanisms, like the RO and its replacement, the CfD system, with the balance amongst the technology options essentially being left up to the market. However the government has indicated what it thinks the options should be, on the basis of its overall policies, and has set up the support structure on this basis. It is backing Carbon Capture and Storage (CCS) in the hope of being able to continue to use fossil fuels, and in terms of non-fossil energy supply options, it is supporting both centralised nuclear and decentralised renewables to varying degrees.

While it is sometimes claimed that there is no alternative to supporting this wide range of options, and that for example, renewables could not sustain the UK, several studies have demonstrated the opposite. For example a report from the UK Pugwash group included a scenario in which renewables supplied 80 % of UK energy by 2050, with no nuclear and little CCS, at a cost similar to, or slightly less than, strategies relying mostly on nuclear and CCS, and it also looked at options for phasing out the remaining fossil fuel element, so as to get to 100 % renewables (Pugwash 2013).

The UK governments multiple-option approach may in any case not be viable in the longer term. Firstly it is far from clear that a single mechanism, the CfD system, will be able to support such varied technologies effectively- they are at different stages of development. Secondly, it is far from clear if a market-based system is what is needed for rapid development. That certainly was the lesson from the RO. Focusing on low cost at an early stage of development is perhaps not sensible: most new technologies need support to get down their learning curves. That has been shown clearly by PV: prices have fallen rapidly under FiTs, albeit at some cost to consumers. Grant aid for demonstration projects may be better for some newly emerging technologies: for example only one tidal stream project was able to use the RO and it seems unlikely that many wave or tidal stream projects will prosper under the CfD competition system. Thirdly, it is not clear if it is wise strategically to support all three major areas. As argued above, there is a case for focusing on your strengths, and renewables look like the most appropriate set of options for the UK, offing a range of choices and possibilities.

On the basis of that view, it can be argued that a new strategy is needed, not just for the individual technologies, but for the energy system as a whole. At present most power grid systems are built on the basis of having a few large power plants feeding electricity down the grid to a large number of remote consumers. Some of these plants, often nuclear plants, are kept running continuously to meet 'baseload' demand i.e. the minimum level of demand, while other plants (mostly gas plants) are kept ready to ramp up to full power to meet the daily peaks in demand. For the moment, renewables have simply been added on to this centralised grid system. But they do not fit very well. They are often variable, smaller scale and distributed in and around the country. They need a different more decentralised and flexible grid system (Elliott 2013a).

The newly emerging view is that baseload, far from being vital, is actually, in the new decentralised flexible energy supply and demand system, an inflexible hindrance. In the new system, variable loads and variable supply (from renewables) are balanced via a smart grid with demand-side measures, load peak shaving/delay, energy storage, and increasingly, green power backup sources, along with an EU-wide supergrid network, linking up decentral generation (Elliott 2012b; Greenpeace 2014)

It has been argued that we need to decide which energy model we want to use in future. German Federal Minister of the Environment Norbert Röttgen said in 2010: 'It is economically nonsensical to pursue two strategies at the same time, for both a centralized and a decentralized energy supply system, since both strategies would involve enormous investment requirements. I am convinced that the investment in renewable energies is the economically more promising project. But we will have to make up our minds. We can't go down both paths at the same time'.

In 2011, after the Fukushima nuclear disaster in Japan, Germany clearly did make up its mind. The UK is trying to introduce elements of this new approach, but the commitment to nuclear, buttressed by the revamped electricity market, may make it hard. To be fair, some elements of the market revamp may help some renewables, for example the proposed capacity payment system, designed to reward suppliers who can help balance variable inputs. However the nuclear ambitions may lead to distorted patterns of development and operation, given the technical incompatibility of the nuclear and renewables options, with, for example, wind output having to increasingly be curtailed to avoid shutting down nuclear output when overall demand is low. Certainly nuclear does not offer much help with the grid balancing and curtailment problems associated with using variable renewables, but, as noted above, there are many options for resolving that issue, including energy storage, smart grid demand management and supergrid interconnectors (RAe 2014).

13.7 Conclusions

Most of the issues and problems discussed in this chapter are not limited to the UK. Although the UK seems to have followed a somewhat unique path, around the world, there are a wide range of views and policies on renewable energy, with some being similar to those that have emerged in the UK. One can speculate on what shapes them, and the success or otherwise of renewables. Locational and geographical factors are obviously important, although the UK has near ideal conditions for renewables, so that cannot explain its low level of achievement so far.

Political orientation seems much more relevant. A pan-European study suggested that there was a strong correlation between left leaning governments and commitments to renewable energy and also a correlation between right wing governments and support for nuclear power (Biresselioglu and Karaibrahimoglu 2012). The situation in practice may be more complex and fluid than that. For example, following the Fukushima nuclear disaster, we have seen right leaning governments in Germany and Italy reject nuclear in favour of renewables. However, this seems to have mainly been due to popular pressure, rather than political conviction (Elliott 2013b). In which case, in the UK context, even given a right wing government, change might be possible.

Certainly there are some complex and often changing socio-political factors involved in determining why some countries back nuclear and others do not (Sovacool and Valetine 2012). However, in practice, the choice of technology may come down to simple economic issues. For example, when it pulled out of the UK nuclear programme, Germany utility E.ON explained that 'We have come to the conclusion that investments in renewable energies, decentralised generation and energy efficiency are more attractive – both for us and for our British customers' (Teyssen 2012).

Economic views, often based on political preferences, also influence the type of support system adopted. As noted above, the UK market-orientated RO has not been as successful in terms of building capacity cheaply as the guaranteed price FiT system adopted elsewhere. However, as also noted earlier, the cost of the latter rose as the scale of deployment, of PV especially, grew. In the mid to late 2000s, with political worries about high energy costs mounting, this led to caps being imposed on FiTs across the EU, to reduce consumer surcharges by slowing deployment. More recently there has been pressure from the European Commission to replace the FiT system with a market orientated auction system, something like what has emerged in the UK for renewables under the CfD (Lewis and Chee 2014). In 2014 the German government adopted a similar approach. Clearly then, although there has been strong opposition to these moves (Leidreiter 2014), pointing to the success of FiTs, at the moment, the market approach is winning. Whether that will rebound on nuclear power, with its high costs, remains to be seen.

Views on the likely impacts of an enhanced market approach on renewables will differ, reflecting different ideological positions. Devotees of markets will expect competition to reduce renewable energy technology costs, and thus, in theory, lead to more capacity for the same, or reduced, outlay. This view seems to underlie the critique of FiTs by the German Commission of Experts for Research and Innovation, which claimed that they did not stimulate technological innovation (EFI 2013). An alternative view is that FiTs did stimulate price reduction, for example by building an expanding market for PV solar (reaching 36 GW in Germany by 2014), as well as radical market innovation, enabling many German consumers to invest in self-generation, becoming 'prosumers' (Schleicher-Tappeser 2012; IEA 2014). Along with parallel initiatives by over 900 local energy co-ops in Germany, that has helped renewables to expand rapidly.

While cutting the FiTs may reduce costs to consumers in the short term, 'greens' and renewable energy supporters argue that this will slow the deployment of renewables, and that this is how it will reduce consumer costs. Though, they add, since increasingly expensive fossil fuel will therefore have to be used more, longer-term, cutting FiTs will push prices, and emissions, up. Time will tell who is right. Longer term most renewables looks like being competitive with conventional sources: indeed in some markets some already are. So FiTs, or any other type of subsidy, should not be needed. But in the initial stages, given that they have to challenge entrenched energy systems, if they are to expand rapidly, renewables need support to get established.

As far as the UK is concerned, the UK government strongly backed the 2014 European Commission proposal, in the context of an overall 40 % by 2030 carbon reduction target, to avoid mandatory national renewable energy targets in future, leaving the choice of technology up to each country (EC 2014). Some countries may choose to emphasize nuclear and/or gas/CCS, rather than renewables. The UK's support for this position could be because, having only reached just over a 5 % energy contribution from renewables by 2015, it may find it hard, on the basis of current policies, to achieve even the existing 15 % by 2020 renewable energy target, much less the new 27 % by 2030 indicative renewable energy target the EC has proposed for the EU as whole for 2030. The UK government's recent support cuts and planning blocks for solar farms and on-land wind may make achieving the 2020 target even harder.

Rather than aiding competition, these policies seem to be primarily about cutting expenditure. While as noted above, some argue that market competition will lead to more renewables being developed and deployed at lower costs, the adoption of market-orientated approaches by the UK and some other countries around the world does seem to mainly be a response to recession and economic constraints, and, if so, it may lead to a lower level of renewable deployment. For example, in the World Energy Council's 2050 global energy market-led 'Jazz' scenario, the share of renewables in electricity generation is 31 % and in its more policy-led 'Symphony' scenario, 48 %.

In terms of the role that nuclear power may play, WEC said that while *'the share of renewable energy sources will increase from around 15 % in 2010 [of primary energy] to almost 20 % in Jazz in 2050 and almost 30 % in Symphony in 2050,… nuclear energy will contribute approximately 4 % of total primary energy supply in Jazz in 2050 and 11 % in Symphony globally – compared to 6 % in 2010'* (WEC 2013).

The implication is that, if nuclear is to expand, it will require policy support, otherwise, left to the market, it will decline. Whereas renewables will expand under either approach, although faster with policy support. This is certainly in line with emerging views on the relative economics of nuclear and renewables- renewables are getting cheaper (IRENA 2012; IEA 2013b), while nuclear, at least within the EU and USA, seems to be getting more expensive, and cannot compete (Boccard 2014; Agora 2014)

Whether WECs projections for renewables (at best 48 % by 2050) will prove to be correct, remains to be seen. As indicated above, there are rival projections which suggest that, given the right policies, renewables could supply near 100 % of electricity, and possibly even of energy, by around 2050 in many countries, including the UK. That would clearly be challenging. However, if rapid expansion on that scale is to be attempted, then arguably, as the UK experience seems to confirm, market based approaches may not be the best way forward.

References

Agora (2014) Comparing the cost of low-carbon technologies: what is the cheapest option?. Agora Energiewende, Berlin. http://www.agora-energiewende.org/topics/optimisation-of-the-overall-system/detail-view/article/klimaschutz-wird-mit-erneuerbaren-deutlich-preiswerter-als-mit-atomkraft/

Bailey R, Blair L (2012) 'A Corruption of Governance', unlock democracy and the association for the conservation of energy, London

BERR (2007) Energy white paper: meeting the energy challenge. Department of Business Enterprise & Regulatory Reform, London

Biresselioglu M, Karaibrahimoglu Y (2012) The government orientation and use of renewable energy: case of Europe. Renew Energy 47:29–37. http://www.sciencedirect.com/science/article/pii/S0960148112002479

BIS (2013) Nuclear industry strategy. Department for Business Innovation and Skills, London. https://www.gov.uk/government/organisations/department-for-business-innovation-skills/series/nuclear-industrial-strategy

Boccard N (2014) The cost of nuclear electricity: France after Fukushima. Energy Policy 66(March):450–461. http://www.sciencedirect.com/science/article/pii/S0301421513011440

Boyle G (ed) (2012) Renewable energy. Oxford University Press, Oxford

DECC (2011) Planning our electric future: a white paper for secure, affordable and low carbon electricity. Department of Energy and Climate Change, London. https://www.decc.gov.uk/en/content/cms/legislation/white_papers/emr_wp_2011/emr_wp_2011.aspx

DECC (2012) The energy bill: electricity market reforms. Department of Energy and Climate Change, London. www.decc.gov.uk/en/content/cms/news/pn12_151/pn12_151.aspx

DECC (2013) Energy and emissions update. Department of Energy and Climate Change, London

DECC (2014a) Electricity market reform: allocation of contracts for difference. Department of Energy and Climate Change, London. https://www.gov.uk/government/uploads/system/uploads/attachment_data/file/271919/Competitive_allocation_consultation_formatted.pdf

DECC (2014b) Electricity market reform: contracts for difference. Department of Energy and Climate Change, London. www.gov.uk/government/publications/electricity-market-reform-contracts-for-difference

DTI (2003) Energy white paper 2003: our energy future- creating a low carbon economy. Department of Trade & Industry, London

EC (2011) Energy roadmap 2050: a secure, competitive and low-carbon energy sector is possible. European Commission, Brussels. http://europa.eu/rapid/pressReleasesAction.do?reference=IP/11/1543&type=HTMLIEA

EC (2014) 2030 framework for climate and energy policies. European Commission, Brussels. http://ec.europa.eu/clima/policies/2030/index_en.htm

ECF (2010) Roadmap 2050. European Climate Foundation, Brussels. http://www.roadmap2050.eu
Eclareon (2011) RES-Integration, report to the EC DG ENG, on member states progress. http://www.eclareon.eu/en/res-integration
EFI (2013) Research, innovation and technological performance in Germany. Commission of Experts for Research and Innovation (EFI). http://www.e-fi.de/fileadmin/Gutachten/EFI_Report_2013.pdf
Elliott D (2010) Nuclear power and renewables in the UK: can we have both? In: Elliott D (ed) Nuclear or not? Palgrave Macmillan, Basingstoke
Elliott D (2012a) Windpower: opportunities, limits and challenges. In: Szarka J et al (eds) Learning from windpower: governance and societal perspectives on sustainable energy. Palgrave, Basingstoke
Elliott D (2012b) Emergence of European supergrids. Energy Strateg Rev 1:171–173
Elliott D (2013a) Renewables: a review of sustainable energy supply options. Institute of Physics Publications, Bristol. http://iopscience.iop.org/book/978-0-750-31040-6
Elliott D (2013b) Fukushima: impacts and implications. Palgrave Macmillan, Basingstoke
EREC (2010) Rethinking 2050. European Renewable Energy Council, Brussels. http://www.rethinking2050.eu
Ernst and Young (2008) Renewable energy country attractiveness indices, Ernst and Young report, Quarter 1–2
ERP (2012) 'UK Nuclear Fission Technology Roadmap', produced for the energy research roadmap by the UK National Nuclear Laboratory. http://www.energyresearchpartnership.org.uk/nucleartechnologyroadmap
Europa (2014) Commission concludes modified UK measures for Hinkley Point nuclear power plant are compatible with EU rules. European Commisons press release. http://europa.eu/rapid/press-release_IP-14-1093_en.htm
GEB (2014) German energy blog info service. http://www.germanenergyblog.de/?p=15647#more-15647
Goodall C (2014) Energy and Scottish independence. Carbon Commentary, 22nd April 2014. http://www.carboncommentary.com/2014/04/22/3544
Greenpeace (2014) 'PowE[R]2030,' Energynautics report for greenpeace Europe, and earlier greenpeace 'Energy [R]evolution' reports. http://www.energynautics.com/news/#GP_EU
Grimes R, Nuttall W (2010) Generating the option of a two-Stage nuclear renaissance. Science, 13th August 2010, pp 799–803.
Gruber A (2010) The costs of the French nuclear scale-up: a case of negative learning by doing. Energy Policy 38:5174–5188. http://www.sciencedirect.com/science/article/pii/S0301421510003526
HM Government (2013) Initial agreement reached on new nuclear power station at Hinkley. UK Government press release. https://www.gov.uk/government/news/initial-agreement-reached-on-new-nuclear-power-station-at-hinkley
HM Treasury (2012) Government publishes legislation to help accelerate major infrastructure investment and house building. UK government press release. https://www.gov.uk/government/news/government-publishes-legislation-to-help-accelerate-major-infrastructure-investment-and-house-building
HM Treasury (2013) Investing in Britain's future. HM Treasury, London. http://www.gov.uk/government/publications/investing-in-britains-future
IEA (2010) World energy outlook 2009. International Energy Agency, Paris
IEA (2011) Deploying renewables: best and future policy practice. International Energy Agency, Paris
IEA (2013a) IEA medium term renewable energy market report 2013. International Energy Agency, Paris
IEA (2013b) RE-COST, Prysma report on behalf of IEA-RETD. International Energy Agency, Paris. http://iea-retd.org/archives/publications/re-cost

IEA (2014) RE Prosumer IEA-RETD report. International Energy Agency, Paris. http://iea-retd.org/archives/publications/e-prosumers-report
IPCC (2011) Special report on renewable energy sources and climate change mitigation (SRREN). Intergovernmental Panel on Climate Change, Geneva. http://www.srren.org
IPCC (2014) Fifth Assessment report, Intergovernmental Panel on Climate Change, Geneva http://www.ipcc.ch/
Jacobson M, Delucchi M (2011) Providing all global energy with wind, water, and solar power. Energy Policy 39(3):1154–1190
Leidreiter A (2014) German Renewable Energy Act Reform is not a "Feed-in Tariff 2.0". World Future Council. http://power-to-the-people.net/2014/04/german-renewable-energy-act-reform-is-not-a-feed-in-tariff-2-0/
Lesser J, Su X (2008) Design of an economically efficient feed-in tariff structure for renewable energy development. Energy Policy 36:981–990
Lewis B, Chee FY (2014) New EU rules on energy funding phase out subsidies for renewables. Reuters. http://uk.reuters.com/article/2014/04/09/eu-funding-energy-idUKL6N0N12FP20140409
Maue G (2012) Presentation on the Germany energy programme to a PRASEG seminar in February in London. Parliamentary Renewables and Sustainable Energy Group
Mitchell C (2000) The England and Wales non-fossil fuel obligation: history and lessons. Ann Rev Energy Environ 25:285–312
Mitchell C (2011) Nuclear power is the reason for the new energy regulations. The Guardian, 11th March 2011. http://www.guardian.co.uk/environment/2011/mar/11/nuclear-power-reason-energy-regulations
Mitchell C, Bauknecht D, Connor P (2006) Effectiveness through risk reduction: a comparison of the renewable obligation in England and Wales and the feed-in system in Germany. Energy Policy 34:297–305
NPS (2011) National policy statements for energy, EN-1. Department of Energy and Climate Change, London. http://www.decc.gov.uk/en/content/cms/meeting_energy/consents_planning/nps_en_infra/nps_en_ifra.aspx
Nuclear Review (1995) The prospects for nuclear power in the UK. Nuclear review report Cmnd 2860. HMSO, London
Obama B (2009) Presidential speech to Congress. http://www.whitehouse.gov/the_press_office/Remarks-of-President-Barack-Obama-Address-to-Joint-Session-of-Congress
Pugwash (2013) Pathways to 2050: three possible UK energy strategies. British Pugwash group, London. http://www.britishpugwash.org/recent_pubs.htm
PWC (2010) A roadmap to 2050 for Europe and North Africa. PriceWaterhouse Coopers, London. http://www.pwc.co.uk/eng/publications/100_percent_renewable_electricity.html
RAe (2014) Wind energy: implications of large-scale deployment on the GB electricity system. Royal Academy of Engineering, London. http://www.raeng.org.uk/windreport
REN21 (2014) 'Renewables 2014 Global Status Report', Renewable energy policy network for the 21st century. Paris. http://www.ren21.net/gsr
Richardson K, Dahl-Jensen D, Elmeskov J, Hagem C, Henningsen J, Korstgård J, Kristensen N, Morthorst P, Olesen J, Wier O, Nielsen M, Karlsson K (2011) Denmark's road map for fossil fuel independence.Solut J 2(4). http://www.thesolutionsjournal.com/node/954
Schleicher-Tappeser R (2012) How renewables will change electricity markets in the next five years. Energy Policy 48(September):64–75. http://www.sciencedirect.com/science/article/pii/S0301421512003473
Scottish Government (2013) Electricity Generation Policy Statement – 2013. Devolved government report, Edinburgh. http://www.scotland.gov.uk/Publications/2013/06/5757/0
Smith School (2012) Towards a low carbon pathway for the UK. Smith School of Enterprise and the Environment, University of Oxford
Sovacool B, Valetine S (2012) The national politics of nuclear power. Routledge, London

SRU (2011) Pathways towards a 100 % renewable electricity system. German Advisory Council on the Environment (SRU). http://www.umweltrat.de/SharedDocs/Downloads/EN/02_Special_Reports/2011_10_Special_Report_Pathways_renewables.html

Teyssen J (2012) E.on confirms strategy focused on renewables. Windpower Monthly, quoting Teyssen's comments to German business daily Handelsblatt, 30th March 2012. http://www.windpowermonthly.com/go/windalert/article/1124937/?DCMP=EMC-CONWindpowerWeekly

Toke D (2005) Are green electricity certificates the way forward for renewable energy? An evaluation of the United Kingdom's renewables obligation in the context of international comparisons. Environ Plann C Gov Policy 23:361–374

Toke D (2007) Renewable financial support systems and cost-effectiveness. J Clean Prod 15:280–287

Toke D (2010) Letter to the guardian. 17th June 2010. http://www.guardian.co.uk/environment/2010/jun/17/true-costs-nuclear-power?INTCMP=SRCH

Trade and Industry Select Committee (1998) Conclusions of the review of energy sources for power generation and the government response to the fourth and fifth reports of the Trade and Industry Committee. House of Commons, London, Cmnd 4071 Oct

WEC (2013) World energy scenarios: composing energy futures to 2050. World Energy Council, London. http://www.worldenergy.org/publications/2013/world-energy-scenarios-composing-energy-futures-to-205

Wicks M (2009) Energy security report for the Prime Minister. Department of Energy and Climate Change, London

WWEA (2015) World Wind Energy Association data. http://www.wwindea.org/wp-content/uploads/2015/02/top12.jpg

WWF (2011) The energy report – 100 % renewable energy by 2050. World Wildlife Fund for Nature. http://www.wwf.org.uk/research_centre/research_centre_results.cfm?uNewsID=4565

Part III
Ecological Economics of Social Change

Chapter 14
Social Metabolism and Ecological Distribution Conflicts in India and Latin America

Joan Martinez-Alier, Leah Temper, Mariana Walter, and Federico Demaria

Abstract This chapter draws on results of the project entitled EJOLT (Environmental Justice Organizations, Liabilities and Trade) focused on the analysis of ecological distribution conflicts across the world. We include comparative data on India and Latin America (and also for some variables on Africa and Europe) exploring the links between increases in the social metabolism and the appearance of ecological distribution conflicts. We also analyse the successful resistance movements led by environmental justice organizations and the "valuation languages" deployed by them.

Keywords Environmental justice • Ecological distribution conflicts • HANNP • Material flows analysis

14.1 Introduction

The industrial economy is based on the use of fossil fuels. Therefore there is a need for "fresh" supplies of energy all the time. Materials can be recycled to some extent. When the economy grows, inputs of fossil fuels, biomass (food and feedstuffs, paper pulp, wood, agrofuels), building materials and mineral ores, keep increasing.

The fundamental clash between economy and the environment comes from two facts. First, population growth. In the twentieth century population grew four times. It now seems that "peak population" will be reached at about nine billion by 2050. Second, the social metabolism of industrial economies. The energy from the fossil fuels is used only once, and new supplies must be obtained from the "commodity frontiers" (Moore 2000). Similarly, materials are recycled only in part, and therefore, even an economy that would not grow, would need fresh supplies of bauxite, copper, iron ores. The growth in the number of resource extraction conflicts and also

J. Martinez-Alier (✉) • L. Temper • M. Walter • F. Demaria
ICTA, Institute of Environmental Science and Technology, Autonomous University of Barcelona, Barcelona, Spain
e-mail: Joan.Martinez.Alier@uab.cat; joanmartinezalier@gmail.com

S. Shmelev (ed.), *Green Economy Reader*, Studies in Ecological Economics 6,
DOI 10.1007/978-3-319-38919-6_14

waste disposal conflicts (of which the most noticeable one is that arising from the production of an excessive amount of carbon dioxide), is explained by the social metabolism.

We focus on the trends in the Material Flows of the economies of India and Latin America since 1970, comparing them, and tracing links to ecological distribution conflicts, and therefore to resistance movements of environmental justice.

The EJOLT project (2011–2015) is assembling information on resource extraction conflicts and waste disposal conflicts in many countries, connecting them to general trends in social metabolism and also to their immediate local causes. Conflicts arise because of rapid changes in the structure of the social metabolism, for instance new production and use of some types of biomass (eucalyptus plantations, soybeans monoculture), or a new wave of open cast metal mining or gas fracking, or new plans for waste dumps or incinerators. Some types of materials (sand and gravel, for instance) are not conflictive in some places but become very conflictive in other places.

Questions that we aim to answer in EJOLT include the following. What is the rate of "success" of environmental justice in such conflicts? Does population density help to explain the incidence of socio-environmental conflicts? Is local scarcity of water a relevant factor? How does the presence of indigenous populations affect the rate of successful outcomes? In this chapter, we present only a few results from the EJOLT project focusing on some of the research questions of a new "statistical political ecology".

14.2 The Link from Ecological Economics to Environmental Justice and Political Ecology

Ecological economics studies the social metabolism, using methods developed over the last 20 years. In turn, political ecology studies socio-environmental conflicts (or, equivalently, ecological distribution conflicts). In such conflicts we realize that different valuation languages are used by the social actors, from monetary compensation of damages to ecological values or to the sacredness of mountains or rivers, including also the unalienable rights of indigenous people to their territories. Since ecological economics aims at taking environmental issues into account through not only in money terms, the analysis of such incommensurable valuation languages is of great interest for ecological economists.

The economy is not "dematerializing" in per capita terms. Therefore, there are increasing local and global conflicts over the sharing of the burdens of pollution (including the enhanced greenhouse effect) and over the access to natural resources. Such conflicts show what has been called "the environmentalism of the poor", or also popular environmentalism, livelihood ecology, liberation ecology and the movement for environmental justice (local and global). Such environmental movements may help to move society and economy in the direction of social justice and ecological sustainability (Martinez-Alier 2005).

When did the global environmental justice movement appear? This was earlier than Seattle in 1999 and the World Social Forums of the 2000s which certainly pushed forward the globalization of environmental justice. The alternative "treaties" signed at Rio de Janeiro in 1992 had proved already the many links among environmental groups. Friends of the Earth (born in California) became international, spreading also to Southern countries or rather bringing in organizations which already existed since the 1980s, like CENSAT in Colombia or Acción Ecológica in Ecuador. Nnimmo Bassey of ERA, Nigeria, was president of Friends of the Earth International in the 2010s. But also outside Friends of the Earth, one could find important local organizations in many countries linking the environmentalism of the poor with wider notions of environmental justice and climate justice as the Centre for Science and Environment in Delhi did in 1991.

United States academics and activists such as Robert Bullard (1990) belonging to the EJ movement travelled and became influential in Brazil and South Africa. Academic work was published since the mid-1990s if not before making explicit connections between the EJ movement in the United States born in the early 1980s and other manifestations of EJ in Latin America, Africa and Asia. This connection was obvious after the deaths of Chico Mendes in Brazil in 1988 fighting deforestation and of Ken Saro-Wiwa and his Ogoni comrades in Nigeria in 1995 complaining against the Shell company. By the mid-1990s classic books analyzing environmental justice movements against dams (Mc Cully 1996) and against tree plantations (Carrere and Lohman 1996) had been published (Guha and Martinez-Alier 1997).

Focusing on case studies, the field of Political Ecology (Robbins 2004) studied since the 1980s many environmental conflicts in Southern countries. Going beyond case studies, some researchers generated statistics of conflicts on resource extraction and waste disposal. For instance, Gerber (2011) researched conflicts on industrial tree plantations for wood, palm oil and rubber production which are among the fastest growing monocultures and are currently being promoted as carbon sinks and energy producers. Such plantations cause a large number of conflicts between companies and local populations. Gerber investigated the impacts of the plantations, the social traits of the protesters involved, and the modalities of the conflicts. There are conflicts on rubber trees, eucalyptus, oil palm, gmelina, and acacia trees. He relied on his own case studies in Cameroon and Ecuador and on a literature review corresponding to 58 conflict cases. In EJOLT we followed on Gerber's steps, when doing a large world inventory and Atlas of environmental conflicts, drawing to a large extent on "activist knowledge", not forgetting however that while conflict often signals injustice, many injustices do not immediately produce open conflicts.

14.3 Methods for the Study of Social Metabolism

The primary causes of the ubiquitous movements of resistance are the increase in the social metabolism and the defence of livelihoods against resource extraction. Meanwhile, there are also conflicts on transport and conflicts on urban waste disposal. Which of these types of conflicts existed already long ago, which ones are new? Which are the trends?

The economy may be described in terms of economic indicators such as growth of GDP, savings ratio, budget deficit as percentage of GDP, current account balance in the external sector… Social factors may be taken into account, as in the Human Development Index which nevertheless correlates closely with GDP per capita, and leaves aside environmental and cultural loses.

The economy may also be described in terms of physical indicators. Economic, social, and physical indicators are non-equivalent descriptions. An economy may provide 260 GJ (gigajoules) of energy per person/year, its HANPP (human appropriation of net primary production) is 35 %, material flow amounts to 16 tons per person/year of which fossil fuels account for 5 tons. Of the material flows, 5 tons are imported, 1 ton is exported. Income per capita is 34,000 US$. It ranks 10th in the HDI.

Of another economy, we say that it provides only 35 GJ person/year, its materials flow amounts to only 5 tons person/year, its HANPP is 65 % (a heavily populated country, relying on biomass, with little external trade). Foreign trade is less than 0.3 ton per capita/year of exports or imports. Income per capita is 3 000 US$ (at PPP). It ranks 127th in the HDI. Different regions and different classes of people in such countries could be classified by their metabolic profiles

MEFA -materials and energy flows accounting- is a set of methods for describing and analysing socio-economic metabolism. They examine economies as systems that reproduce themselves not only socially and culturally, but also physically through a continuous exchange of energy and matter with their natural environments and with other socio-economic systems. Material flow accounts draw on methodologies established by the research group led by Marina Fischer-Kowalski (1997) in Vienna and other groups over the last 30 years.

In the Material Flows we calculate first the Domestic Extraction (in tons per year) divided into Biomass, Minerals for Building Materials, Mineral Ores for Metals, and Fossil Fuels. They show different levels and trends in different countries. The Domestic Extraction is denoted as DE. The DMC (Domestic Material Consumption) is equal to Domestic Extraction plus Imports minus Exports. Physical imports and physical exports measure all imported or exported commodities in tonnes. Physical trade balance (PTB) equals physical imports minus physical exports. So countries like Brazil or Russia (among the BRICs) have large Physical Trade Deficits, but not India as a whole.

Such accounts (including carbon or energy "rucksacks", "virtual" water and "embodied HANPP") are relevant to understand current ecological distribution conflicts and also for historical and current debates on ecologically unequal exchange and the ecological debt.

Energy flow accounting (EFA) is an integral part of the analysis of social metabolism. Primary and final energy delivered are usually classified in the statistics according to source. Such energy flows (including hydro-electricity) are also unequally distributed, and in India they are creating not only coal mining conflicts and the new nuclear conflicts but also many conflicts in the Himalayas and in the North East on hydroelectricity. Notice that energy accounts are separate from the Material Flows. The idea of linking economic history to the use of energy goes back to Wilhelm Ostwald, and later to Leslie White and other authors but it was only in the 1980s when several histories of the use of energy in the economy were published. The most interesting EFA indicator is that of Energy Return on Energy Input (EROI). EROI is a useful coefficient for assessing the increasing costs of obtaining energy in developing tar sands or heavy oil in Alberta in Canada or the Orinoco Delta, Venezuela, or when taking oil from the bottom of the sea (an in the Brazil's *pre-sal*) or in the new gas fracking, or for agro-fuels such as those derived from sugar cane or *Jatropha curcas* (with low EROI).

An economic-ecological history would establish the changes in the EROIs over the years noticing an improvement as biomass energy (fuelwood, charcoal) is substituted or supplemented by fossil fuels and, later, indicating perhaps a decline because getting energy while going down the Hubbert curve (after peak oil) will require (it seems) increasing amounts of energy.

14.4 The HANPP

The HANPP (human appropriation of net primary production of biomass) is calculated in three steps. First, the potential net primary production (in the natural ecosystems of a given region or country), NPP, is calculated. Then, the actual NPP (normally, less than potential NPP, because of agricultural simplification and soil sealing) is calculated. The part of actual NPP used by humans and associate beings (cattle, etc.) relative to potential NPP is the HANPP, meant to be an index of pressure on the biodiversity (because the higher the HANPP, less biomass is available for "wild" species). (Haberl et al 2007). So, an increasing HANPP is an indicator of increasing pressure on biodiversity. This should be relevant to do a history of India's conservation areas and threats to its much threatened wildlife, and similarly for Latin America (where the HANPP is lower than in India).

In India, due to high population density and land conversion, and due also to a relatively high use of biomass per capita (which would still be larger if the Indian population ate more meat), the HANPP is very high (as it is also in Bangladesh). Simron Singh et al (2012) put it at 73 % compared to about 40 % in the EU (with comparable population densities), and only 24 % in Japan (that imports much biomass). The HANPP in Latin America is increasing because of deforestation for cattle raising and for domestic or export crops in Brazil, Argentina and other countries. We can also ask the question of who gets the HANPP among groups of humans, as when a commercial tree plantation is planted in a former forest used sparsely by *adivasi* groups.

14.5 Social Metabolism of Latin America

"Social metabolism" refers then to the manner in which human societies organize their growing exchanges of energy and materials with the environment (Fischer-Kowalski 1997). In this chapter we use a socio-metabolic approach to examine the Material Flows (extraction, exports, imports) of Latin America and India and their implications in terms of socio-environmental pressures and conflict. Later, we propose a classification of extractive conflicts based on the commodities at stake.

Material Flow Analysis (MFA) aims to complement the system of national accounts, with a compatible system of biophysical national accounts using *tonnes* per year as the key unit of measurement. Such methodology provides a picture of the physical dimension of the economy, where the total turnover of materials of the socio-economic system can be analyzed historically or cross-section through the accounts of inputs flows (tonnes of biomass, fossil fuels, construction minerals, etc.) or output flows (tonnes of materials exported, waste or pollutant generated). We focus on the input side by taking into account all materials that enter into the national economy and acknowledging also the physical dimension of foreign trade (Hornborg 1998, 2006, 2009). MFA offers a picture of the overall evolution of the pressures exerted by an economy to extract renewable and non-renewable resources.

A socio-metabolic approach acknowledges that inputs into the economy become ultimately outputs from the economy in the form of waste (except for the part that accumulates as a stock, like in buildings). The main output in volume from rich economies (apart from wastewater) is carbon dioxide from the burning of fossil fuels, the excessive production of which is a main source of climate change. Solid wastes produced by the economy are disposed off locally (in landfills or incinerators), or sometimes exported to distant regions or countries. All goods circulate through "commodity chains" (Raikes et al. 2000), i.e. from cradle to grave or from point of extraction to waste disposal. Ecological distribution conflicts occur at different stages as peasant or tribal groups, national or multinational companies, national governments, local or international NGOs, consumer groups, have stakes.

MFAs have been conducted in most OCDE countries and also in Latin America. For instance, MFA for Argentina was studied by Perez Manrique et al. 2013, and by other authors for Colombia, Mexico, Ecuador and Latin America as a whole (Russi et al. 2008; Gonzalez-Martinez and Schandl 2008; Vallejo et al. 2011; West and Schandl 2013). For Argentina, we showed that the recent increase in some forms of biomass extraction (soybeans, in 20 million ha.) and in metal ores, leads to new types of conflicts.

Figure 14.1 shows the trends in extraction of materials in Latin America in the last 40 years and until 2008. By 2008 extraction reached more than 10 tons per

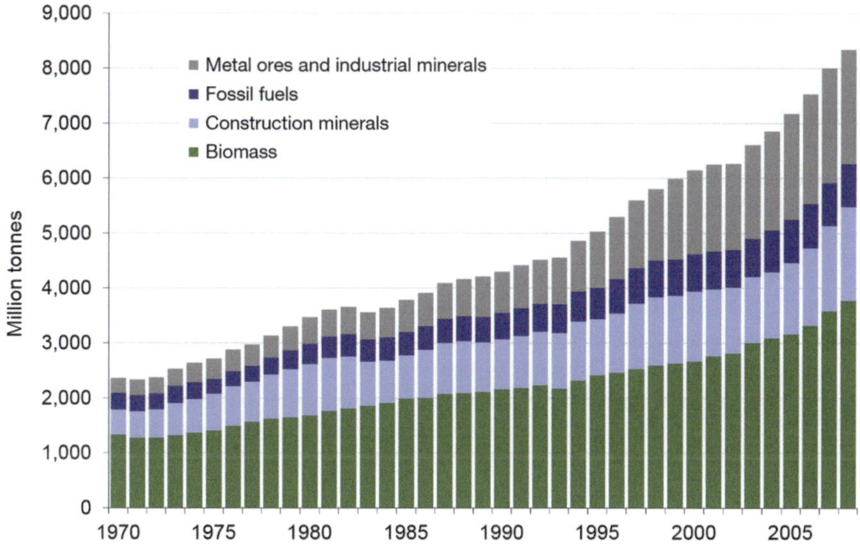

Fig. 14.1 Domestic extraction in Latin America by major category of material for the years 1970–2008, in million tons (Source: West and Schandl 2013)

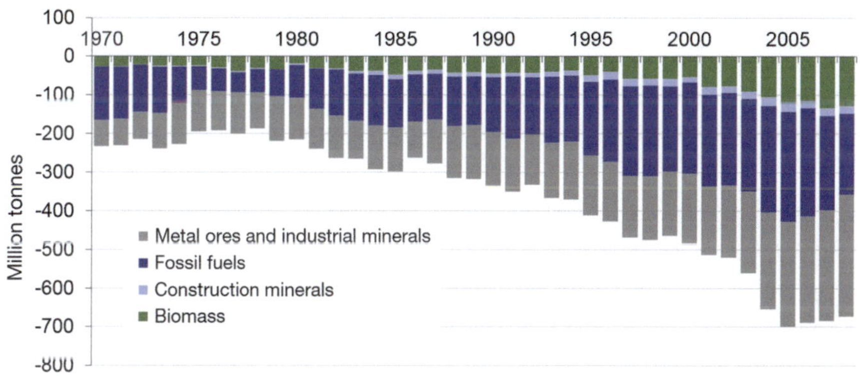

Fig. 14.2 Latin America physical trade deficits (imports – exports) in million tons, 1970–2008 (Source: West and Schandl 2013)

capita, of which over 10 % was exported. Exports exceeded imports (in tons) by a large amount (Fig. 14.2).

Overall there was a fourfold increase in material flows between 1970 and 2008 for domestic consumption and also for exports. The Latin American economy has certainly not become "dematerialized" – one could compare such trends with other continents, such as Europe where the rate of increase of material extraction has been much lower, or with India which per capita has a lower rate of material extraction than Latin America and which is not at all a net exporter in physical terms (Singh et al 2012) (see next section). Such physical indicators are useful to characterize the economic structure of countries or regions.

As remarked above, the Physical Trade Balance (PTB) is the difference between the number of tonnes of materials that are imported from an economy and the number of tons that are exported. In physical terms, if exports are higher than imports, we call this a Physical Trade Deficit, the country is losing "·substance". The Monetary Trade Balance (MTB) is the difference between how much is paid for the imports and how much is earned by exports in monetary terms. Latin American economies and particularly South American economies have a persistent and increasing physical trade deficit (West and Schandl 2013). Exports in tons are larger than imports in tons, resulting in a "deficit" in the same sense that would be applied to a tree plantation that grows less than the harvest rate. Figure 14.2 presents a yearly PTB of the Latin American (which includes Mexico) per type of material from 1970 to 2008. Notice in Fig. 14.2 the increased physical trade deficit for metal ores and industrial minerals, which reflects the growing pressure to extract and export these materials. Latin America as a whole has about 600 million people, so the Physical Trade Deficit is over one ton per capita/year. It is unlikely that this will grow much in the next decades – just maintaining this level implies very heavy pressure at the commodity extraction frontiers.

India as a whole, as we shall see, imports and exports a very small amount per capita (in tons). Latin America as a whole, despite its voluminous exports, can scarcely pay for its imports. In 2014–2015 many South American countries have negative trade balances (in money terms).

We admit that one ton of uranium is environmentally extremely different from one ton of sand and gravel, or one ton of cellulose from one ton of shrimp. However, we are showing here trends within broad material categories rather than focussing on specific commodities.

14.6 India's Social Metabolism

Following the example of Japan for 20 years at least until 2013, economic growth has stopped in many rich countries since 2008 (less by design than by the economic crisis), while in the BRICs but also Peru, Indonesia, Colombia, Turkey and many other countries there has been sustained growth even after 2009. Poverty in terms of income per capita has declined in all such countries, including India.

This growth is achieved at great environmental and social costs. Peasants are squeezed out of the land, tribals in India and elsewhere are being displaced because they happen to live at the commodity extraction frontiers. Biodiversity is being rapidly lost, and the concentration of carbon dioxide in the atmosphere is still increasing at 2 ppm per year.

This is the background to the study of India's Material Flows from 1961 to 2008 carried out by Singh et al and published in *Ecological Economics* in 2012. India per capita consumes less fossil fuels, less building materials and less mineral ores than most other countries. To reach a physical dimension per capita similar to Latin America (which has half the population of India), the total material flows of India (whether locally extracted as at present, or to a relatively small extent imported as coal is from Australia) would have to multiply by not least than a factor of four. One small advantage of India is in some respects its high density of population that diminishes internal transport flows. Another one advantage could have been to go directly to rural electrification through solar energy.

In the 1960s, about three quarters of the total material consumption of India consisted of biomass while construction materials were second in importance. Biomass is for subsistence compared with Latin America, where a lot is for meat consumption or for exports. Fossil fuels and industrial minerals and ores were insignificant in relation to the total flows. In the course of the almost 50 years under study, this changed in quantity and composition. The use of biomass only doubled. Fossil fuel consumption multiplied by a factor of 12.2, industrial minerals and ores by a factor of 8.6, and construction materials by a factor of 9.1 (Singh et al 2012).

Until the 1980s the population grew at a slightly faster pace than material throughput. Throughout the 1960s and 1970s, material use remained at a low and slowly declining level of less than 3 t/cap/year. Only since the early 1980s (10 years before Dr Manmohan Singh became Finance Minister in 1991) a sustained growth in per capita material consumption set in, growing to 4.3 t/cap/year, accelerating in the period since 2004. Taking into account further growth from 2008 to 2013, India is probably at a level of 5 t/cap/year. In comparison, per capita material consumption in EU countries is about 15 tons per person/year.

Notice moreover that in the EU imports are very significant, and they exceed exports in tons by a factor of 4. Contrariwise, in South America exports exceed imports (in tons) by a factor of 3 or 4. (Mexico has a different pattern because of its *maquila* industry). In contrast, India's external trade is more or less in balance in terms of tons, and it is a small part of its physical economy (Singh et al 2012). However, some states in India surely show a considerable Physical Trade Deficit, and these are states (Odisha, Jarkhand), also Goa because of iron ore exports, where many ecological distribution conflicts arise (Martinez-Alier et al. 2014). In Singh et al (2012) we wrote that, in general, construction minerals are abundant, and scarcity (and conflicts) are usually only regional phenomena. But in fact almost all regions of India suffer from the phenomenon of "sand mafias".

The trends in total and per capita material domestic consumption in India (extraction plus imports minus exports), and on physical trade are summarized in Fig. 14.3. Notice than in Fig. 14.1 we gave rates of extraction for Latin America. For India we give rates of Domestic Consumption (extraction + imports – exports) which for India are very similar to rates of extraction because of the physical irrelevance (in tons) of external trade for the economy of India as a whole. We give totals (Fig. 14.3) and per capita results (Fig. 14.4).

Table 14.1 below shows again the Domestic Material Consumption (Domestic Extraction + Imports – Exports) per capita from 1961 to 2008, excluding the Biomass, comparing with growth of GDP and population.

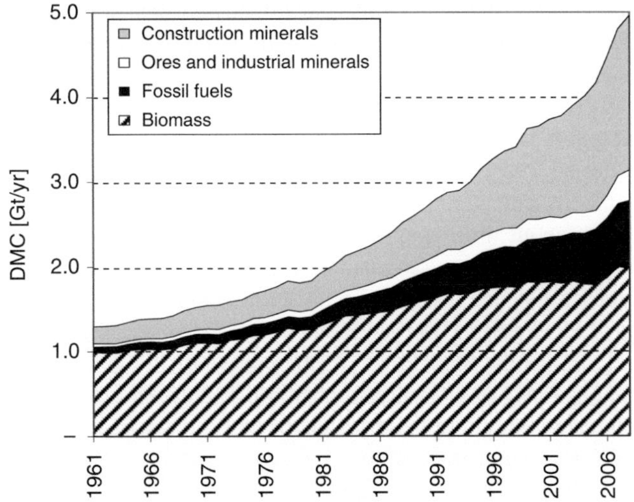

Fig. 14.3 Material flows accounts of India 1961–2008, Domestic Consumption in gigatons per year (1000 million) (Source: Singh et al. 2012)

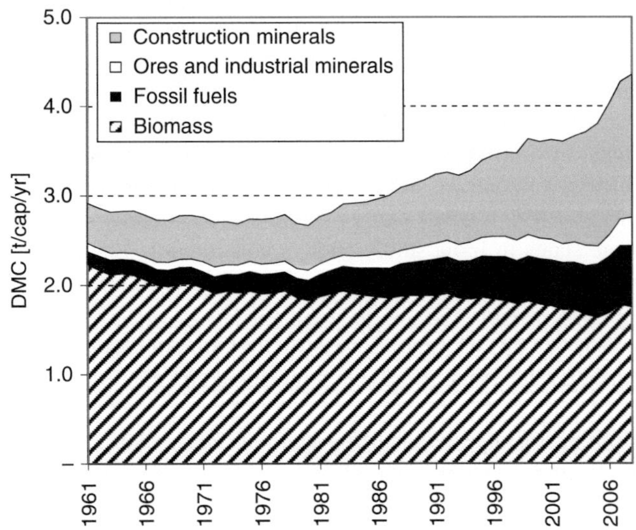

Fig. 14.4 Domestic Material Consumption in India, tons per capita/year (Source: Singh et al. 2012)

Table 14.1 DMC [t/cap/year] of India for the three main groups of mineral and fossil materials and their average annual growth rates (%) in comparison to population and GDP

	1961	1980	2008	1961–1980	1980–2008
GDP [billion USD at const. 2000]	66	156	812	3.5 %	6.0 %
Population [million]	444	687	1140	2.3 %	1.8 %
Fossil energy carriers [DMC t/cap/year]	0.1	0.2	0.6	4.7 %	6.0 %
Ores and industrial minerals [DMC t/cap/year]	0.1	0.1	0.3	3.4 %	5.6 %
Construction minerals [DMC t/cap/year]	0.4	0.5	1.6	2.0 %	6.2 %

Source: Singh et al (2012)

14.7 Ecological Distribution Conflicts

Ecological distribution conflicts are struggles over the burdens of pollution or over the sacrifices made to extract resources, and they arise from inequalities of income and power (Martinez-Alier and O'Connor 1996). The concept of ecological distribution conflicts is born in the intersection between the fields of ecological economics and political ecology that links the emergence of environmental conflicts in the global South with the growth of the metabolism of societies in the global North (which includes parts of China). Political Ecology focuses on the exercise of power in environmental conflicts, in other words, the question is, who has the power to impose decisions on resource extraction, river dams, land use, pollution levels, biodiversity loss, and more importantly, who has the power to determine the procedures to impose such decisions, allowing or excluding some valuation languages (Martinez-Alier 2005, 2009; Robbins 2004).

Ecological distribution conflicts (which, for simplicity, we also call "environmental conflicts" or "socio-environmental conflicts") emerge from the structural asymmetries in the burdens of pollution and in the access to natural resources that are grounded in unequal distributions of power and income, in social inequalities of ethnicity, caste, social class and gender (Martínez-Alier 1997; Martinez-Alier et al 2011). Sometimes the local actors claim redistributions, leading to conflicts, which are often part of, or lead to larger gender, class, caste and ethnic struggles (Robbins 2004). In this line, the concept of "environmental justice" is important. It was born in the United States (Bullard 1990) and it has gained growing acceptance in extractive industries, water use and waste disposal conflicts all over the world.

Not all conflicts are born from immediate metabolic needs. Demand for certain commodities like gold arises in part from the search to have an investment outlet that moreover allows further speculation. Other metals such as copper can also be stored and used as guarantees for speculative loans. The fact remains that energy-carriers (coal, gas, oil, biomass) and metallic minerals are essential inputs for the industrial economy and their use, in total, grows more or less in proportion to the growth of the economy.

Economic change generally occurs for the benefit of some groups and at the expense of others existing or future groups. Externalities can be positive (like the free environmental services provided by a forest) or negative. Negative externalities are not seen here as market failures but rather as (provisional) cost-shifting successes (Kapp 1950) that often give rise to complaints from environmental justice movements or from the "environmentalism of the poor and the indigenous". Optimistic views regarding ecological modernization and "dematerialization" of the economy are confronted with the reality of increased inputs of energy and materials into the world economy, increasing production of waste and ecological distribution conflicts such as those we are mapping in the EJOLT project.

14.8 The Environmentalism of the Poor and the Indigenous

Activists in many countries understand and use the concept of the "environmentalism of the poor". Thus in India (which is its cradle), Sunita Narain wrote on the environmentalism of the poor on 10th January 2011 in an article in *Business Standard* from which I quote:

> The year 2010 was a loud year for the environment. High profile projects – from Vedanta to Posco and Navi Mumbai airport and now Lavasa — hit the headlines for non-compliance with environmental regulations. While 2009 was the 25th anniversary of the Bhopal gas tragedy, it was only last year that we were all outraged by the disaster. The realisation of how every institution – the judiciary, Parliament and government – had miserably failed to provide justice to the victims shocked us deeply.
>
> Then in December, meeting in Cancun, the world took the final step to deny the problem of climate change. It agreed to do nothing to reduce its emissions – at the scale and pace needed.
>
> It would not be wrong to say that virtually all infrastructure and industrial projects – from mining to thermal and hydel and nuclear power to cement or steel – are under attack today from local communities who fear loss of livelihoods. These communities today are at the forefront of India's environmental movement. They are its warriors. But for them environment is not a matter of luxury – fixing the problems of growth, but of survival – fixing growth itself. They know that when the land is mined and trees are cut, their water source dries up or they lose grazing and agricultural fields. They know they are poor. But they are saying, loudly and as clearly as they can, what we call development will only make them poorer. This is what I have called the environmentalism of the poor…The question is where do we go from here? I would argue, we need to keep listening to these voices, not dismiss or stifle them in the name of anti-growth dissent or even Naxalism. This can be done through the strengthening of all the processes of democracy that make us ensure that local people have a voice in development. For instance, the Forest Rights Act demands that the gram sabha (village assembly) in tribal areas must give its written consent to the project before it is cleared… In most cases you will find the concern raised by people is pushed aside as projects are rammed through in the name of industrial development. This must stop…. We must understand that our future lies in being part of the environmentalism of the poor, as this movement will force us all to seek new answers to old problems.

This kind of environmentalism is very different from "the cult of wilderness". The "cult of wilderness" is only one of the varieties of environmentalism. From late 1980s onwards, a different kind of environmentalism was identified as "the environmentalism of the poor", focusing on the global South but (as explained above) closely related to the "environmental justice" movement in the United States (Guha and Martinez-Alier 1997).

The thesis of the "environmentalism of the poor" applies to India and also to Latin America where it has been used since 1991. It does not assert that as a rule poor people or even indigenous peoples feel, think and behave as environmentalists. This is not so. The thesis is that in the many resource extraction and waste disposal conflicts in history and today, the poor are often on the side of the preservation of nature against business firms and the state. This behaviour is consistent with their interests and with their values. The environmentalism of the poor builds on the premise that the fights for human rights and environment are inseparable.

In the environmentalism of the poor as in environmental justice movements in general, it is important to recognise the contribution women make in poor communities both rural and urban (Agarwal 1992). Women more often collect water, gather wood, look for medicinal plants, tend to domestic animals, and grow crops, and therefore they have greater knowledge and awareness of their community's direct dependence on the natural environment. This does not imply that women have an empathy with nature denied to men for biological reasons. The argument is based on social roles. In an urban setting, it is women who often take leading positions in environmental justice conflicts (in contrast to labour union struggles) as regards complaints against waste dumping, or air or water pollution. Women are often the main actors of environmental conflicts.

The environmentalism of the poor relates to actions and concerns in situations where the environment is a source of livelihood. This is reinforced by other values, such as the defence of indigenous territorial rights (appealing to Convention 169 of ILO or *adivasi* rights in India), the claim to the sacredness of particular elements of nature (a mountain, a forest, or even a tree). When livelihood is threatened, those affected will be motivated to act provided that there is a sufficient degree of democracy and they are not suffocated by fear as is often the case. Indeed, a clean and safe environment is a need for all humans rather than a luxury good (Bandyopadhyay and Shiva 1988).

In such conflicts, a variety of "valuation languages" are deployed. Some of them were perhaps more powerful in the past (livelihood, sacredness) than in this era of the generalized market system where even "the fetishism of fictitious commodities" is in the ascendant in schemes for payment for environmental services. One wonders for instance how effective is still in India to oppose the sacredness of a "sacred grove" against a mining project or a dam. Notice in India the attempt at compensation of damages in resource extraction conflicts by ascertaining the so-called Net Present Value (in money terms) of the foregone products and services from forests (Temper and Martinez-Alier 2013).This is a peculiarity of Indian state administration. Other valuation languages apart from money compensation, such as human rights, indigenous territorial rights, environmental justice against "environmental racism", and even the Rights of Nature as in art 71 of the 2008 Constitution of Ecuador, are perhaps gaining in strength across the world.

Table 14.2 classifies environmental conflicts depending on the stage of the commodity chains where they occur, and depending on their geographical reach. One could easily fill in Table 14.2 exclusively with Indian and Latin American examples. Sometimes local conflicts become "glocal", i.e. well known also outside the territory in question, as with the Dongria Kondh vs. Vedanta in Odisha (Padel and Das 2010) or in Latin America, the Chevron-Texaco case in Ecuador.

There is much scope for historical and comparative work on the "valuation languages" deployed in environmental conflicts. In the EJOLT project (www.ejolt.org) (Environmental Justice Organizations, Liabilities and Trade), we hope to collect and map about two thousand by the end of 2015. We classify the conflicts according to the commodity in question, as in the following list.

- **Nuclear**: uranium extraction; nuclear power plants; nuclear waste storage
- **Ore & building materials**: mineral extraction; mineral processing; tailings
- **Waste Management**: e-waste & other waste import zones; ship-breaking; waste privatisation; waste-pickers; incinerators; landfills; uncontrolled dump sites; industrial; municipal waste

Table 14.2 A classification of socio-environmental conflicts

Geographical scope Stage	Local	National and regional	Global
Extraction	Resource conflicts in tribal areas, such as bauxite mining in Odisha, coal or uranium mining in Jharkhand, oil extraction in the Amazon of Ecuador or Peru or the Niger Delta.	Mangrove uprooting.	Worldwide search for minerals and fossil fuels at the "commodity" frontiers.
		Tree plantations for wood or paper pulp.	Bio-piracy.
		Collapses of fisheries.	Attempts at regulation of "corporate accountability"
Transport and trade	Complaints against urban motorways or heavy traffic in rural mining areas because of noise, pollution, landscape loss	Inter-basin water transport.	Oil spills at sea
		Oil/gas pipelines (e.g. Burma to Thailand).	"Ecologically unequal exchange" because of large South to North material flows
		No TAV movement (Italy). TIPNIS road (Bolivia).	
Waste disposal and pollution, post-consumption	Conflicts on incinerators (dioxins) or VOCs, NOx, ozone, particulate matter.	Acid rain from sulphur dioxide.	CO2, CFC: causes of climate change/ozone layer destruction/ocean acidification. POPs even in remote pristine areas.
		Nuclear waste disposal (Yucca Mountain, Nevada, USA; Gastre in Patagonia-Argentina).	
		Ship dismantling (Alang-Sosiya, Chittagong)	Claims for a "carbon debt", Climate Justice.

- **Biomass**: land-grabbing; tree plantations; logging; non-timber products; deforestation; agro-toxics; GMOs; agroofuels; mangroves vs. shrimps; biopiracy & bio-prospection; intensive food production (monoculture & livestock); fisheries
- **Fossil Fuels** & **Climate Justice/Energy**: oil and gas extraction; oil spill; gas flaring; coal extraction; climate change related conflicts (glaciers & small islands);REDD/CDM; windmills; gas fracking
- **Transportation** & **Infrastructure**: megaprojects, high speed trains
- **Water management and Hydric Justice**: dams, water transfers, aquifers, hydroways, desalination
- **Biodiversity**: invasive species, damage to nature, conservation conflicts
- **Industrial & utilities conflicts**: factory emissions, industrial pollution

It is laborious but easy to trace in India and Latin America many hundreds of examples of these types of conflicts, listing the specific commodities in question, the social actors involved, tracing their geographical distribution patterns over time, accounting for the reasons for the occasional successful outcomes (stopping projects).

14.9 Do Indigenous Groups in Latin America and India Take Part Disproportionately in Environmental Conflicts?

This is one issue that we can begin to answer from the results of the EJAtlas. The aswer seems to be "yes" (cf. Pérez-Rincón 2014, for Colombia). This is of relevance in order to trace an intellectual and political connection between global environmental justice and the US environmental justice movement on the 1980s that insisted so much on the struggle against "environmental racism".

Some results are given in Table 14.3. The database form includes a space to list social actors mobilizing in conflicts. Several actors can be listed simultaneously, including "indigenous or traditional communities" and "ethnically/racially discriminated communities".

Where there are indigenous or traditional communities and ethnically discriminated communities (there is a certain overlap of both categories), their participation in the set of conflicts is seemingly larger than their relative number in the population as a whole. Thus in South America, these communities are present in more than half the conflicts. In Europe, the 4 cases reported are in Scandinavia – Sami people.

One question to ask is whether the disproportionate participation of indigenous peoples in environmental conflicts (when they are present in the country in question) is explained by their location at the commodity frontiers or by the fact that trasnational or companies or governments target them because of their presumed low capacity to resist against political pressure. To repeat, interesting comparisons can be made by applying the category of "environmental racism" developed by the US environmental justice movement.

Table 14.3 Environmental conflicts and indigenous groups/traditional communities in India, South America and Europe

	Number of conflicts in EJOLT Atlas	Participation of ethnically/racially discriminated communities *or* indigenous groups or traditional communities	Only ethnically/racially discriminated communities	Only indigenous groups or traditional communities
South America (selected countries)	494	264	87	250
India	184	97	16	95
Europe (selected countries)	117	4	4	0
Africa (selected countries)	116	69	8	67

Source: EJOLT Atlas, www.ejatlas.org (as of 20 November 2014), in India, in selected South American countries (Argentina, Brasil, Colombia, Peru, Ecuador, Chile), in selected European countries (Finland, France, Germany, Italy, Serbia, Spain, Sweden) and in selected African countries (Madagascar, Mozambique, Nigeria, South Africa)

14.10 "Success" in Environmental Conflicts

Other types of analysis can be done. In the inventory of 1259 cases, 215 are reported as "successes" in environmental justice, which often means than the investment project that triggered the conflict, has been stopped. Table 14.4 shows the number of cases in the broad categories of conflicts.

There are differences in the "success" rate (varying between 12 and 25 %) which we cannot explain at present. With more cases, one could try to see whether such "deviations" from the average rate of "success" of 17–18 %, significantly depend for some reason on the commodities in question. In the database forms there is much more detailed information, with over 100 specific commodities.

The average rate of "success" itself might also increase or decrease a little as more cases are brought into the EJAtlas. Perhaps "success" of environmental justice is more easily achieved in some countries than others. If we reach an inventory of 3000 cases worldwide by the end of 2017, I would dare predict that between 450 and 600 will be considered "successes". The database forms (that anybody can read in the EJAtlas webpage) provide space to explain the reasons why a conflict is considered a success, a failure, or a "not sure" case in environmental justice. Inspection of this information freely available in the database forms would also help advance a theory of the meaning of "success" (or failure) in conflicts on environmental injustice.

Table 14.4 Number of conflict cases and rates of success per selected commodity category in EJOLT inventory (as of 20 Nov. 2014)

Category	Total cases	"Success" cases
Mineral ores, building materials	281	58
Fossil fuels/climate change	230	27
Biomass extraction	216	31
Water management	180	28
Industry and Utilities	100	20
Infrastructures and built environment	75	15
Waste management	66	15
Nuclear energy	46	8
Biodiversity conservation	36	8
Tourist recreation	29	5

Finally, is ethnicity a factor in the rate of success? In the total number of conflicts in the inventory of 1259 (at 20 November 2014), in 580 indigenous, traditional and/or ethnically discriminated communities take part, this is in 46 %, with the differences between Europe and other continents that have been remarked upon in the previous section. Of the total number of "successes", 215, only in 91 we have indigenous, traditional and ethically discriminated communities, which is a little over 42 %. Therefore ethnicity/indigeneity/"traditionality" do not seem to be a significant factor in explaining success or failure.

When we take only the social group "ethnically/racially discriminated communities", we have a total of 196 conflict cases (out od 1259), of which 47 are considered as "successes", that is 23%, which is not significantly higher than the overall rate of 17 or 18%. Again, the hypothesis would be that ethnicity is not a factor either in achieving success or in suffering failure.

In general terms, one could ask not only about the factors that explain "success" (the nature of the commodities in question, the social groups mobilizing, the political culture of the states and, importantly, the degree of democracy…) but whether such rate of "success" in environmental conflicts is or is not a factor helping to move the economy in a less unsustainable, more ecological direction. Moreover, in the EJOLT inventory there are many cases classified as "not sure" (414), some of which are or have been very close to "success". So, a rate of "success" (when there is a conflict) of environmental justice of 1/5 or perhaps even 1/4 is a plausible and at the same time an encouraging estimate. Together with the efforts of the eco-efficiency movement and sometimes in alliance with conservationism, the global environmental justice movement can be considered as a very relevant actor in the struggle for sustainability.

We must remember however that there are many environmental injustices that do not produce visible conflicts.

14.11 The Vocabulary of Environmental Justice

In this concluding section, we reiterate a main finding. The conflicts seem to arise not so much because of the overall amount of material flows in the Domestic Extraction of Domestic Consumption (whether it is 5 tons, 10 tons or 15 tons per capita/year) but from the sudden increases in the rates of extraction, transport, and waste disposal. The toxicity of the materials themselves or the perceived dangers of the techniques of extraction or waste disposal, play an important role, but nevertheless we see that in India (and sometimes in Latin America as in the Tunjuelo in Bogota) that there are also conflicts on sand and gravel extraction, in itself a non-toxic material. Another factor in triggering conflicts is possibly the awareness that the materials go to other regions abroad or inside the country itself (in a pattern of ecologically unequal trade) while the risks and the damage impact locally (Muradian et al 2012).

We can show also that one social outcome of so many ecological distribution conflicts is the reinforcement of a transnational movement for environmental justice that operates through networks. Table 14.5 concludes this chapter by describing many of the concepts (of non-academic origin) introduced by activist networks and which are used in their social and political practice.

Table 14.5 The vocabulary of the Global Environmental Justice Movement

Movement	EJOs and main authors	Short description
Environmental justice (EJ)	USA Civil Rights Movement, North Carolina 1982 against environmental injustices (Bullard 1990, 1999).	"People of color" and low-income populations suffer disproportionate harm from waste sites, refineries and incinerators, transport infrastructures.
Environmental racism	Rev Benjamin Chavis, c. 1982	The fight for EJ, against pollution in Black, Hispanic, Indigenous areas, was seen as a fight against *environmental racism.*
Ecological debt	Instituto Ecología Política, Chile, 1992, Acción Ecológica 1997	Rich countries' liability for resource plunder and disproportionate use of space for waste dumping (e.g. GHG).
Popular epidemiology	Brown, P., 1992, 1997	"Lay" local knowledge of illnesses from pollution may be more valid than official knowledge (sometimes absent).
Environmentalism of the poor	A.Agarwal/S. Narain (CSE, Delhi) c. 1989	Struggles by poor/indigenous peoples against deforestation, dams, mining… ; proactive collective projects for water harvesting, and forest conservation
Sacrifice zones	Steve Lerner, book with this title, 2010	Related to the EJ movement in the US, description of several cases.

(continued)

Table 14.5 (continued)

Movement	EJOs and main authors	Short description
Food sovereignty	Via Campesina, c. 1996	People's right to healthy, culturally appropriate, sustainably produced food. Right to define own food and agriculture systems.
Biopiracy	RAFI (Pat Mooney) 1993, popularized by Vandana Shiva	Appropriation of genetic resources (in medicinal or agricultural plants) without recognition of knowledge and property rights of indigenous peoples
Climate justice	CES (Delhi), 1991, Durban Alliance, CorpWatch 1999–2002	Reduce excessive per capita emissions of carbon dioxide and other GHG. "Subsistence emissions vs. luxury emissions". Contraction and convergence.
Water justice, hydric justice	R. Boelens, 2011, EJOs in Latin America (CENSAT).	Water should not run towards money, or towards power. It should go to those needing it for livelihood.
Water as human right	Pablo Solon (Bolivian envoy to UN), Maud Barlow (Council of Canadians).	Human Right to Water recognized at UN level in 2011, as an independent human right.
"Green Deserts"	Brazil, against eucalyptus plantations, *Rede Alerta contra o Deserto Verde*, 1999	Brazilian local term for eucalyptus plantations, used by networked CSO and communities, also by researchers and activists for any tree plantation.
Tree Plantations are not Forests	*Pulping the South*, 1996 by R. Carrere, L. Lohman, World Rainforest Movement	The WRM collects and spreads information on tree plantation conflicts. It proposes a change in the FAO definition of forest, to exclude tree monocultures.
Land grabbing	GRAIN (small pro peasant EJO), 2008	The wave of land acquisitions in Southern countries for plantations for exports, leading to first statistics on land-grabbing
Resource caps	Resource Cap Coalition, RCC Europe, c. 2010	It advocates reduction in global resource use and in poverty. It calls for a *European energy quota scheme* and the ratification of the *Rimini protocol*.
To Ogonize/Yasunize	ERA Nigeria, Acción Ecológica, Oilwatch, 1997–2007	Leave oil in the soil to prevent damage to human rights and biodiversity, and against climate change. Adopted by anti- gas fracking, tar sands and coal mining movements.
Rights of Nature	Ecuador, Constitutional Assembly, 2008	In Constitution of Ecuador 2008, art 71, pushed by Acción Ecológica and Alberto Acosta. Actionable in court.

(continued)

Table 14.5 (continued)

Movement	EJOs and main authors	Short description
Corporate accountability	Friends of the Earth International, 1992–2002	At UN Johannesburg summit, FoE proposed the adoption of a Corporate Accountability Convention, against lukewarm CSR principles.
GPPI (Europe)	*Grands Projets Inutiles Imposés*	Network born c. 2012 against useless public works such as the airport in Nantes, the TAV from Torino to Lyon
"Critical mass", cyclists rights	San Francisco 1992 (Chris Carlsson)	International movement reclaiming the streets with cyclists marching to impose cyclists rights.
Urban waste recyclers movements	c. 2005, GAIA against incineration and "energy valorization" of urban waste.	Unions or cooperatives of urban waste gatherers, with positive environmental impact, including climate change (movements in Delhi, Pune, Bogota).
Urban "guerrilla food gardening"	c. 2000, started by "food justice" networks	Vacant lot food growing, permaculture, community gardening movements in cities around the world.
Toxic colonialism, toxic imperialism	BAN, Basel Action Network, c. 2000	Against long-distance export of waste from rich to poor countries, forbidden by the Basel Treaty, ship-breaking in India, Bangladesh, nuclear, chemical or electronic waste.
Post-extractivism	Latin America E. Gudynas (CLAES), A. Acosta, M. Svampa, 2007.	Against the reprimarization of LA economies. Transition to a sustainable economy based on solar energy. Quotas and taxes on raw materials exports.
Buen Vivir, Sumak Kawsay	Ecuador and Bolivia 2008	Adopted in Constitutions of both countries, inspired by indigenous traditions and by the "post-development" approach.
Indigenous territorial rights, prior consultations	Convention 169 of ILO, 1989; *adivasi* forest rights in India…	In conflicts on mining, oil exploitation, dams… communities ask for applying legislation defending indigenous rights..
"Sand mafias"	Name given c. 2005 by environmental movement, journalists.	The illegal "mining" of sand and gravel in India in rivers or beaches, driven by the growing building and public works industry.
"Cancer villages"	China, popular name adopted by academics and officials.	Rural villages where industry has caused pollution (e.g. heavy metals), where lay knowledge of illness is relevant, and subdued protests take place.

References

Agarwal B (1992) The gender and environment debate: lessons from India. Fem Stud 18(1):119–158
Bandyopadhyay J, Shiva V (1988) Political economy of ecology movements. Econ Polit Wkly 23(24):1223–1232
Brown P (1992) Popular epidemiology and toxic waste contamination: lay and professional ways of knowing. J Health Soc Behav 33(3):267–281
Brown P (1997) Popular epidemiology revisited. Curr Sociol 45(3):137–156
Bullard R (1990) Dumping in dixie: race, class, and environmental quality. Westview Press, Boulder
Carrere R, Lohman L (1996) Pulping the South: industrial tree plantations and the world paper economy. Zed, London
Fischer-Kowalski M (1997) Society's metabolism: on the childhood and adolescence of a rising conceptual star. In: Redcliff M, Woodgate G (eds) The international handbook of environmental sociology. Edward Elgar, Cheltenham, pp 119–137
Gerber J-F (2011) Conflicts over industrial tree plantations in the south: who, how and why? Glob Environ Chang 21(1):165–176
Gonzalez-Martinez AC, Schandl H (2008) The biophysical perspective of a middle income economy: material flows in Mexico. Ecol Econ 68(1–2):317–327
Guha R, Martinez-Alier J (1997) Varieties of environmentalism. Essays north and south. Oxford University Press, Delhi
Haberl H et al (2007) Quantifying and mapping the human appropriation of net primary production in earth's terrestrial ecosystems. Proc Natl Acad Sci U S A 104(31):12942–12947
Hornborg A (1998) Towards an ecological theory of unequal exchange: articulating world system theory and ecological economics. Ecol Econ 25:127–136
Hornborg A (2006) Footprints in the cotton fields: the industrial revolution as time–space appropriation and environmental load displacement. Ecol Econ 59(1):74–81
Hornborg A (2009) Zero-sum world: challenges in conceptualizing environmental load displacement and ecologically unequal exchange in the world-system. Int J Comp Sociol 50(3–4):237–262
Kapp KW (1950) The social costs of private enterprise. Harvard University Press, Cambridge, MA
Martinez-Alier J (2005) The environmentalism of the poor. A study of ecological conflicts and valuation. Oxford University Press, Delhi
Martinez-Alier J (2009) Social metabolism, ecological distribution conflicts, and languages of valuation. Capital Nat Social 20(1):58–87
Martinez-Alier J, O'Connor M (1996) Ecological and economic distribution conflicts. In: Costanza R et al (eds) Getting down to earth: practical applications of ecological economics. Island Press, Washington D.C
Martinez-Alier J et al (2011) Between science and activism: learning and teaching ecological economics with environmental justice organisations. Local Environ 16(1):17–36
Martinez-Alier J, Temper L, Demaria F (2014) Social metabolism and environmental conflicts in India. Indialogs 1(1):51–83
Mc Cully P (1996) Silenced rivers: the ecology and politics of large dams. Zed, London
Moore JW (2000) Sugar and the expansion of the early modern world-economy: commodity frontiers, ecological transformation, and industrialization. Review (Fernand Braudel Center) 23(3):409–433
Muradian R, Walter M, Martinez-Alier J (2012) Hegemonic transitions and global shifts in social metabolism: implications for resource-rich countries. Introduction to the special section. Glob Environ Chang 22(3):559–567
Padel F, Das S (2010) Out of this earth: East India adivasis and the aluminium cartel. Orient Black Swan, Hyderabad

Pérez Manrique PL, Brun J, González-Martínez AC, Walter M, Martínez-Alier J (2013) The biophysical performance of Argentina (1970–2009). J Ind Ecol 17(4):590–604
Pérez-Rincón MA (2014) Conflictos Ambientales en Colombia. In: Luis Jorge G (ed) Minería en Colombia: Control público, memoria y justicia socio-ecológica, movimientos sociales y posconflicto. Contraloría General de la República de Colombia, Bogota, pp 253–326
Raikes P, Jensen MF, Ponte S (2000) Global commodity chain analysis and the French Filière approach: comparison and critique. Econ Soc 29(3):390–417
Robbins P (2004) Political ecology: a critical introduction. Blackwell, Oxford
Russi D, Gonzalez AC, Silva-Macher JC, Giljum S, Martinez-Alier J, Vallejo MC (2008) Material flows in Latin America: a comparative analysis of Chile, Ecuador, Mexico, and Peru, 1980–2000. J Ind Ecol 12(5–6):704–720
Singh SJ, Krausmann F, Gingrich S, Haber H, Erb KH, Lanz P, Martinez-Alier J, Temper L (2012) India's biophysical economy, 1961–2008: sustainability in a national and global contex. Ecol Econ 72:60–69
Temper L, Martinez-Alier J (2013) The god of the mountain and Godavarman. Net Present Value, indigenous territorial rights and sacredness in a bauxite mining conflict in India. Ecol Econ 96:79–87
Vallejo MC, Pérez-Rincón MA, Martinez-Alier J (2011) Metabolic profile of the Colombian economy from 1970 to 2007. J Ind Ecol 15(2):245–267
West J, Schandl H (2013) Material use and material efficiency in Latin America and the Caribbean. Ecol Econ 94:19–27

Chapter 15
Human Values and Sustainable Development

Irina A. Shmeleva

Abstract These chapter overviews most prominent theories of human values as seen from the point of view of psychology (S. Schwartz) and sociology (R. Inglehart). Exploring the social values at the macro level, it aims to explain differences in sustainable development performance in various countries by the predominant values in society. It highlights the importance of value differences in various stakeholder groups, paying attention to sustainability professionals, ethnic minorities, civil servants, lawyers, teachers and students. It was shown that intellectual autonomy according to Schwartz and emancipative values according to Inglehart are best predictors for peace index, life expectancy, Yale Environmental Performance index and recycling levels. Among the countries with the highest scores of indices mentioned are Sweden, the Netherlands and Germany; at the opposite end of the spectrum are Yemen, Nigeria and Ghana. The experience of Nordic countries in this light seems to be most relevant and is explored in more detail from the policy point of view.

Keywords Environmental values • Sustainable development • Psychology • Sociology • Universalism • Power • Intellectual autonomy • Self-expression • Emancipative values • Peace index • Life expectancy • Yale Environmental Performance Index • Recycling levels • Behavioral change • Policy • Nordic model

15.1 Introduction

This chapter deals with the changes of values in modern society. There is currently a great interest in environmental values as they could explain the causes of unsustainable lifestyles and could help to find the ways of overcoming the global environmental crisis. According to the Club of Rome "values lie at the heart of the common future" and are considered a priority theme for modern investigation and discourse. Analysis of theoretical insights and empirical research has shown that human values

I.A. Shmeleva (✉)
Institute of Sustainable Development Strategies, St Petersburg, Russia

ITMO University, St Petersburg, Russia
e-mail: director@isdstrategy.ru; irina_shmeleva@hotmail.com

S. Shmelev (ed.), *Green Economy Reader*, Studies in Ecological Economics 6,
DOI 10.1007/978-3-319-38919-6_15

determine the formation of ecological consciousness, pro-environmental behavior, behavioral change and, finally, the implementation of sustainable development policy.

During the last decades, the interest in the study of environmental values, applying methodologies and results by S. Schwartz and R. Inglehart, has grown considerably. A value of *universalism* (Schwartz and Bilsky 1987) has been shown to be a highly significant component of the value system being in opposition to and conflict with the *power* values. Universalism corresponds to understanding, tolerance, and protection of the well-being of all peoples and nature and historically evolved from the need of individuals and groups to survive. This type of value is activated when people become aware of the scarcity of the natural resources they depend on, and when failure to protect the environment can lead to the societal degradation. According to Schwartz (2005a), *universalism* unites two types of relationships: concern for the wellbeing of the whole world and concern for the local environment. It is underpinned by such aspects as tolerance, broad outlook, social justice, equality, world peace, beauty, unity with nature, wisdom, environmental protection, as well as inner harmony and spiritual life. As different actors are involved in the promotion and implementation of environmental and sustainable development policy process, it becomes important to study value structures of different stakeholder groups.

European psychologists initiated several UN research programmes focusing on the applications of psychological knowledge and skills to facilitate the solution of global environmental problems as early as 1997 (Shmeleva 2012). In "Psychology of Sustainable Development" volume (Schmuck and Schultz 2002) some crucial questions were posed: "what enables some people to live their lives in accordance with the vision of sustainability?", "which aspects of different cultures are compatible with sustainable development and which are not?". This volume presents the evidence that psychological science can contribute to the sustainable development. As it was mentioned at the Seventh Annual Psychology Day at the United Nations titled "Psychological Contribution to Sustainable Development: Challenges and Solutions for the Global Agenda"– the role of psychology increases as there is a paradigm shift in conceptualization of the development process from mainly economic growth to human centered sustainable development where psychology needs to play a crucial role (American Psychological Association 2014).

The most influential theories of human values in the cultural context of late twentieth century were introduced in sociology by R. Inglehart and psychology by S. Schwartz. S. Schwartz methodology is used as a basis for studying values (Schwartz 2004) within the European Social Survey (2014), R. Inglehart methodology is used in the World Values Survey (2015), both covering many countries and cultures. We will discuss two of them in the context of sustainable development.

Contemporary research results show interdependence of values, pro-environmental attitudes and pro-environmental behavior (Schultz and Zelezny 1998, 2003; Didz et al. 2002; Raudsepp 2001; Schultz 2001; Niit et al. 2004; Reser and Bentrupperbaummer 2005; Vlek and Steg 2007; Steg et al. 2014).

15.2 Environmental Values from a Psychological Perspective

Psychologists suggest that different types of motivations (motivational goals) in a given situation influence environmental behavior: hedonic goals, gain goals and normative goals (Steg et al. 2014). One of the ways to encourage pro-environmental behavior is to strengthen the normative goals, as normative considerations are important predictors of pro-environmental behavior. Normative goals according to researchers are influenced by such individual factors as biospheric values and situational factors that can activate (or deactivate) different types of values (Steg et al. 2014). Along with affecting the strength of goals, values can activate personal norms that are considered to be the feelings of moral obligations to act pro-environmentally (in different cultures using similar process of norm activation) as different research results had shown according to Steg et al. (2014). Researchers also reveal that values affect behavior via environmental self-identity, strengthening the significance of normative considerations. The model suggested by Steg et al. (2014) attracts attention to the fact that situational factors also play an important role in activating or deactivating particular values and affect the dominance of goals in a particular period of time. The authors argue that the research results they gained have significant environmental policy implications as they demonstrated that normative considerations are important predictors of pro-environmental behavior but they need activation and support from the situational cues that could be implied by policies.

Contemporary psychological research considers values as a motivational construct and an element of cognition (Rokeach 1973; Schwartz and Bilsky 1987; Schlöder 1993; Leontief 1998; Schwartz and Bardi 2001; Schwartz 2005a, b). Personal values are emerging and developing under the influence of external and internal factors. External factors are represented by the elements of the micro-environment (values, shared by membership groups and reference groups) and macro-environment (the system of basic human values, social institutions, mass media). Internal factors include age, sex, temperament features, inherited talents, skills, needs and the level of self-consciousness. Values therefore are ideas, ideals and goals, which are aimed at by both an individual human being and the society as a whole.

Most of the studies of values and environmental attitudes until present have been carried out with student and teacher samples. Their professional orientation and activities, with rare exceptions, have not been taken into account; their ethnic decent has been taken into consideration, however it was not connected with their professional orientation. In our research we have tested a hypothesis on specificity of value structures in groups, professionally orientated towards solving environmental problems, working for sustainable development, involved in education for sustainability, as well as in groups, ethnic descent of which assumes a close connection with the environment.

Values are the subject of interdisciplinary study and lie at the intersection of at least three areas: philosophy and, specifically, its branch of axiology (axios –

'valuable', logos – 'notion', 'science'), sociology and psychology. The problem of values occupies an important place in the structure of sociological and psychological knowledge. The notion of 'value' has been introduced into sociology by Thomas and Znanetski (1918) but has existed in philosophy prior to that. The notion of value orientations has been introduced by Parsons (1951). Sociological understanding of values allows to interpret them as 'commonly held convictions regarding the goals, a human being could strive for'. Value orientations, therefore, act as an important factor, regulating and determining the motivation of a person and are the determinants of the social knowledge acquisition process. An individual explores the world through the prism of values and considers the social world through the prism of the social values. Values and norms form a united normative system, which regulates behavior of people and social groups. Value and norms system provides guidance in choosing a mode of actions, selects and chooses ideals, builds goals and oversees the means of reaching these goals. The notion of values belongs to both sociology and psychology.

Values and norms are a part of consciousness of an individual and social consciousness and a part of culture, therefore they could be studied and measured both at the level of individual distinctions and at the level of culture, reflected in the social norms, customs and traditions of social groups (Schwartz and Bilsky 1987; Schlöder 1993; Leontief 1998; Lebedeva 2000; Schwartz and Bardi 2001; Schwartz 2005a, b).

The concepts of values explicitly or implicitly exist in sociological and psychological theories of G. Allport, S. Freud, E. Fromm, K. Chorny, E. Ericson, K. Dichter, K. Klackhorn, M. Rokich, S. Schwartz, P. Sorokin, V. Frankl, V. Yadov and others. Values according to P. A. Sorokin are meanings, which people invest in the same material objects and spiritual phenomena. He differentiated between four universal values: knowledge; love and a desire of productive creation; family; religious attitude to life (Sorokin 1992). Value in Allports's understanding is some kind of personal meaning. The value scale according to Allport includes: (1) theoretical, (2) social, (3) political, (4) religious; (5) aesthetic, (6) economic; (7) conflicting values. Allport stated that the conflict between values could be the source of their development (Allport 1955).

Understanding the values, according to Frankl, gives them a shared and universal meaning. He understood human values as 'universals of meaning', i.e. meanings inherent to the majority of the members of society and humanity as a whole during the history of its development. Values are ideas, ideals, goals, which a human being and a society strive for. There are commonly held values: (1) universal (love, prestige, respect, safety, knowledge, money, things, nationality, freedom, health); (2) social (inter-group and individual values) – political, religious, defining the 'normal' behavior for society or a group; (3) individual (personal). The values are united in systems, forming a hierarchical structure, which changes with age and life circumstances (Frankl 1959).

One can find a plethora of definitions of values: we have found over 180. Many authors agree that the notion of values is often used as an analytical tool to connect macro-social phenomena with individual attitudes and behavior.

According to Kluckhohn, values are explicit or implicit concepts of the desired, characterizing the individual or a group and defining the choice of types, means and goals of behavior. Values determine orientation in the reality; help differentiate between the necessary and unnecessary, suitable and unsuitable, good and evil. A human being is guided by values in organizing his or her life. The society needs values to determine the most valuable goals and forms of behavior (Kluckhorn and Murray 1953).

Oldemeyer (1978) proposed four groups of values: interpersonal relations values, values in the sphere of human-nature relations, ego-oriented values and transcendent values. Oldenmeyer differentiated normative (or still dominating) values and alternative values, which become more significant over time. He thus postulated that the value structure is changing over time.

In early 1990s Schlöder carried out research using the methodology by Oldemeyer (1978) focused on dominant and alternative values of students belonging to different professional groups (first, psychology and education, and second, finance, law and engineering). The research has shown that both groups exhibited high levels of alternative socio-environmental values and high ranks were observed for classic liberal values (Schlöder 1993). Alternative values were consistently preferred to the dominant values in his sample. Schlöder concluded that Oldemeyer's hypothesis of changing value structures was confirmed. Schlöder assumed that the research results reflected the situation emerged in West Germany in the 1980s. He discovered a higher significance of humanitarian and cooperation values and a skeptical withdrawal from the ego-oriented values of achievement. Such a situation is observed in a country in the period of economic stability and confidence.

15.3 Schwartz Value Theory

The Schwartz Value Theory is the most widely used methodological tool for the analysis of human values today (Schwartz 2005a, b). Schwartz started to develop his concept in mid 1980s and it has finally taken shape by 1992. Since then it is employed in a range of cross-cultural research projects and is a part of European Social Survey, which is upgraded with new data every year. The theory includes two conceptual frameworks: one for the study of individual values and another for the study of cultural values. Schwarz identifies the following features of values (Schwartz 2005a):

1. Values are beliefs, connected directly to emotions, but not to objective 'cold' ideas;
2. Values are motivational constructs, they relate to the goals that people plan to achieve;
3. Values are abstract aims, which make them different from norms and attitudes that are related to specific action, objects or situations;
4. Values direct the choice and evaluation of actions, policies, people and events;

5. Values are ordered by importance in relation to one another and form a system of value priorities that characterize people as individuals.

The values act as criteria used by people to estimate other people, actions, behavior patterns, events to be good or bad, legal or illegal, something that one needs to avoid or pursue. We acknowledge our values only when our acts or judgments require difficult decisions in a situation of a value conflict.

The values identified in Schwartz theory have roots in religious, philosophical and psychological theories. Basic human values can be found in most cultures (Schwartz and Bilsky 1987). According to Schwartz theory, these values are likely to be universal in a sense that they are 'grounded in one or more of three universal requirements of human existence, with which they help to cope': needs of individuals as biological organisms, needs for coordinated social interaction and needs of groups for survival and welfare.

In its classic form, Schwartz differentiates between ten basic human values, which are presented below (Schwartz 2005a, b). Each of the ten values is defined in terms of the broad goal it signifies.

Self-direction is defined by the goal of independent thought and action, expressed through creativity, freedom, choosing own goals, curiosity, independence, self-respect, intelligence and privacy.

Stimulation is defined by the goal of excitement, novelty and challenge in life, manifested in a varied life, an exciting life, daring.

Hedonism is defined by the goal of pleasure or sensuous gratification for oneself, expressed through life enjoyment and self-indulgence.

Achievement is defined by the goal of personal success through demonstrating competence according to social standards, manifested in ambition, successfulness, capability, intelligence, self-respect, influence and social recognition.

Power is defined by the goal of control or dominance over people and resources, highlighting social status and prestige and expressed through authority, wealth, social power, social recognition, preservation of one's public image. According to Schwartz, both power and achievement values focus on social esteem, however achievement values e.g. successfulness and ambition emphasize the active demonstration of successful performance in concrete interaction, whereas power values e.g. authority and wealth emphasize the attainment or preservation of a dominant position within a more dominant social system.

Security is defined by the goal of safety, harmony, stability of relationships and of self. There are two sub-types of security values: individual and group, which can be expressed by social order, family security, national security, reciprocation of favours, health and a sense of belonging.

Conformity is defined by the goal of restraining action, inclinations and impulses likely to upset or harm others and violate social expectations or norms, which can manifest itself in obedience, self-discipline, politeness, honoring parents and elders, being loyal and responsible.

Tradition is defined by the goal of respect, commitment and the acceptance of the customs and ideas that one's culture or religion provides, which often take form of religious rites, beliefs and norms of behavior. This type of value manifests itself in respect for tradition, humbleness, devotion, accepting one's portion, moderation and spirituality.

Benevolence is defined by the goal of preserving and enhancing the welfare of those with whom one is in frequent personal contact. Benevolence values emphasize voluntary concern for other's welfrare and manifest themselves in helpfulness, honesty, forgiveness, responsibility, loyalty, true friendship, mature love and sense of belonging.

Universalism is defined by the goal of understating, appreciation, tolerance and protection for the welfare of all people and for nature. The values of universalism stem from survival needs of groups and individuals, which are not recognized until people become aware of the scarcity of natural resources and realize that failure to protect the natural environment will lead to the destruction of natural life-support systems. Universalism comprises several types of consideration: concern for nature, people and the world and is manifested in broadmindedness, social justice, equality, world at peace, world of beauty, unity with nature, wisdom, protecting the environment, inner harmony and spiritual life.

Universalism historically evolved from the need of individuals and groups to survive. It is underpinned by such aspects as broad outlook, social justice, equality, world peace, and beauty, unity with nature, wisdom, environmental protection, harmony and spiritual life. Universalism has three potential subtypes – tolerance: broadminded/tolerant, wisdom/mature understanding; societal concern: equality for all; social justice; world of peace and protecting nature: protect the environment; unity with nature; world beauty.

Schwartz asserts (Schwartz 2005a) that the actions, which are motivated by each type of values have psychological, practical and social consequences. According to Schwartz, the values can be represented by a circular structure, depicted in Fig. 15.1, which reflects the complimentary and contradictory nature between values. The closer the values are located in a circle, the more similar the motivational goals, lying behind them are. The further away they are from each other, the more antagonistic such motivational goals are. Competing values are located at the opposite sides of the circle, complimentary are adjacent to each other. For example, universalism and power are located opposite each other as they represent the competing values. Schwartz claims that values have similar meaning in different cultures.

On the outer rim of the circle in Fig. 15.1 two bi-polar dimensions can be found. First, we can observe the opposition between openness to change (which includes self-direction and stimulation) and conservation (which includes security, conformity and tradition). The second pair of opposing dimensions is a contrast between self-enhancement and self-transcendence. This dimension highlights the conflict between the concern for nature and well-being of other people (universalism and benevolence values) and an orientation towards personal success and domination over others (power and achievement values). The only value not included in the bipolar structure is hedonism, which shares the elements of openness to change and self-enhancement.

Fig. 15.1 The value structure according to S. Schwartz theory of values (Reference)

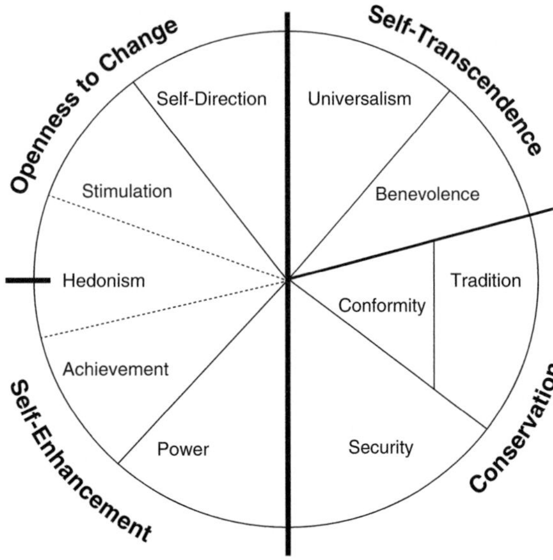

In 2011 based on conceptual definition of original ten values and empirical findings, Schwartz (2011) offered a further detalization of the value structure identifying nineteen values, defined in terms of individual motivational goals (Table 15.1).

During the last 15 years a lot of research showed close correlations between *self-transcendence* values priority (according to S. Schwartz methodology) and environmental attitudes and concerns.

Nearly all research showed a positive tendency: those respondents that estimated the priority of transcendence values higher were more concerned about environment problems than the others. As Schultz and Zelezny (1998, 2003) demonstrated, attitudes that were revealed by New Ecological (Environmental) Paradigm, NEP, (Dunlap et al. 2000; Dunlap 2008) could be definitely predicted by the motivational types (or motivational goals), that were revealed by the Schwartz Value Theory (Schwartz 2005a, b). The regression analysis was used to test the relations between values and *environmental concerns* in cross-cultural comparison analysis of data obtained from different countries.

The results showed that *universalism* is a strong positive predictor of *environmental concerns* and *power* and *tradition* are negative predictors. As M. Raudsepp (2001) indicated, values allow a person to "transfer" the sustainability discourse to the everyday life. Environmental attitudes and pro-environmental behavior can manifest themselves in the everyday activities realized via personal interest in environmental problems, inclusion in nature, beliefs in environmental care and protection, attitudes towards nature preservation and different behavior activities.

Table 15.1 Value conceptual definitions in terms of motivational goals

Value	Description
Self-Direction	Thought Freedom to cultivate one's own ideas and abilities
Self-Direction	Action Freedom to determine one's own actions
Stimulation	Excitement, novelty, and change
Hedonism	Pleasure and sensuous gratification
Achievement	Success according to social standards
Power – Dominance	Power through exercising control over people
Power – Resources	Power through control of material and social resources
Face	Maintaining one's public image and avoiding humiliation
Security – Personal	Safety in one's immediate environment
Security – Societal	Safety and stability in the wider society
Tradition	Maintaining and preserving cultural, family or religious traditions
Conformity Rules	Compliance with rules, laws, and formal obligations
Conformity Interpersonal	Avoidance of upsetting or harming other people
Humility	Recognizing one's insignificance in the larger scheme of things
Benevolence Dependability	Being a reliable and trustworthy member of the in group
Benevolence Caring	Devotion to the welfare of in group members
Universalism Concern	Commitment to equality, justice and protection for all people
Universalism Nature	Preservation of the natural environment
Universalism Tolerance	Acceptance and understanding of those who are different from oneself

15.4 Schwartz Theory of Cultural Values

There are different psychological theories concerning cultural values (Hofstede, Triandis, Schwartz). According to Schwartz (2011) there are seven cultural value orientations that are appropriate for comparisons among cultural groups. These value orientations on a cultural level are related to such societal characteristics as level of corruption, country wealth and democratization level, and they couldn't be used to characterize the values of individual people.

The theory of Cultural Values that S. Schwartz suggested deals with normative value orientations, these orientations underline functioning of societal institutions. The cultural value orientations that form the poles of the conceptual dimensions are ideal-types; the cultures of actual societies are arrayed along the dimensions. These

Table 15.2 Schwartz cultural value theory

Cultural value orientation	Value items
Harmony	A world of beauty, a world at peace, protecting the environment, unity with nature,
Embeddedness	Clean, devout, forgiving, honoring parents and elders, moderate, national security, obedient, politeness, protecting my public image, reciprocation of favors, respect for tradition, self discipline, social order, wisdom
Hierarchy	Authority, humble, social power, wealth
Mastery	Ambitious, capable, choosing own goals, daring, independent, influential, social recognition, successful
Affective Autonomy	Enjoying life, exciting life, pleasure, varied life, self-indulgent
Intellectual Autonomy	Broadminded, creativity, curious, freedom
Egalitarianism	Equality, helpful, honest, loyal, responsible, social justice

orientations are normative responses; they prescribe how institutions should function and how people should behave in order best to deal with the key problems societies face. This theory specifies three bipolar dimensions of culture that represent alternative resolutions to each of three problems that confront all societies.

As Schwartz agues, a societal emphasis on the cultural orientation at one pole of a dimension typically accompanies a de-emphasis on the polar type with which it tends to conflict. Table 15.2 summarizes the culture-value orientations derived from the surveys from 1997 to 2006.

S. Schwartz explains the meanings of each value orientation. As in his Individual Values theoretical framework, values comprise a contradictory set. Different pairs are described below.

Autonomy vs. Embeddedness This value orientation defines the nature of the relations and boundaries between the person and the group. The main question is to what extent are people autonomous vs. embedded in their groups? In autonomy cultures, people are encouraged to cultivate and express their own preferences, feelings, ideas, and abilities, and to find meaning in their own uniqueness.

The empirical research shows that there are two types of autonomy: **intellectual autonomy that** encourages individuals to pursue their own ideas and intellectual directions independently and **affective autonomy** that encourages individuals to pursue affectively positive experience for themselves. **Embeddedness** value orientation indicates how people are viewed as entities embedded in the collectivity in different cultures.

Schwartz indicates that meaning in life is expected to come largely through social relationships, through identifying with the group, participating in its shared way of life, and striving toward its shared goals. «Embedded cultures emphasize maintaining the status quo and restraining actions that might disrupt in-group solidarity or the traditional order».

Egalitarianism vs. Hierarchy The second societal problem is to guarantee that people behave in a responsible manner that preserves the social fabric. That is, people must engage in the productive work necessary to maintain society rather than compete destructively or withhold their efforts. People must be induced to consider the welfare of others, to coordinate with them, and thereby to manage their unavoidable interdependencies. **Egalitarian** cultures seek to induce people to recognize one another as moral equals who share basic interests as human beings. They try to socialize their members to internalize a commitment to cooperate and to feel concern for everyone's welfare. People are expected to act for the benefit of others as a matter of choice.

Hierarchy Hierarchy cultures rely on hierarchical systems of ascribed roles to insure responsible, productive behavior. They define the unequal distribution of power, roles, and resources as legitimate and even desirable. People are socialized to take the hierarchical distribution of roles for granted, to comply with the obligations and rules attached to their roles, to show deference to superiors and expect deference from subordinates.

Harmony vs. Mastery The third societal problem is to regulate people's treatment of human and natural resources. **Harmony** cultures emphasize fitting into the social and natural world, trying to appreciate and accept rather than to change, direct, or exploit. **Mastery** cultures encourage active self-assertion in order to master, direct, and change the natural and social environment to attain group or personal goals.

15.5 Environmental Needs as Post-Materialistic Values

In 1970s Ronald Inglehart proposed a hypothesis that with time 'material values' are replaced with 'post-material values'(Inglehart 2000). Material orientation is an orientation on economic wellbeing and safety and post-material is an orientation on social, aesthetic and ecological aspects of human life. Material values are: (a) a need for safety (strengthening the defense capacity of a country, maintaining order in a country and the crime prevention); (b) a need for material wellbeing, economic growth, countering inflation and economic stability. Post-material values are: (a) human rights and dignity or democratic values (the right to vote at work and in their neighborhood, strengthening the voice of the people in government decision making, attention to every person in society); (b) intellectual, aesthetic and environmental needs.

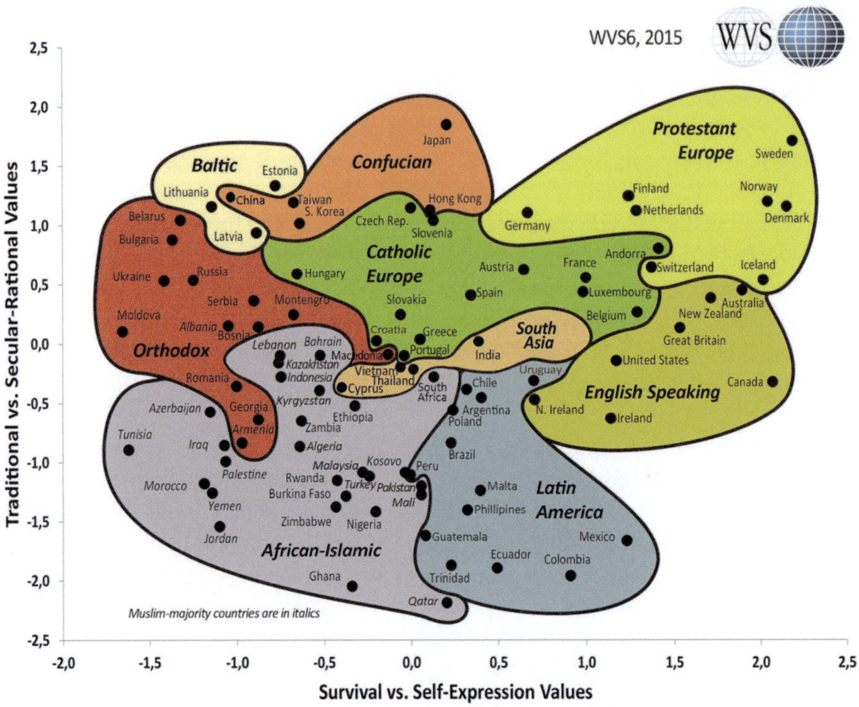

Fig. 15.2 Representation of human values across countries and cultures according to Inglehart (2015)

Since 1981 Inglehart with colleagues have conducted the World Values Survey.[1] The results can be summarized as follows. To a large extent, variation in human values between societies can be explained using two key dimensions: a first dimension of 'traditional vs secular-rational values' and a second dimension of 'survival vs self-expression values' (Fig. 15.2). Traditional values represent religiosity, national pride, and respect for authority, obedience and marriage. Secular-rational values emphasize the opposite on each of these dimensions, people who share them place less emphasis on religion, traditional family values and authority. Survival values relate to the priority of security over liberty, abstinence from political action, distrust in outsiders and a weak sense of happiness. Self-expression axis gives high priority to environmental protection, tolerance, gender equality and rising demands for participation in decision making in economic and political life and represents the opposite to survival values. Inglehart observes that people's priorities shift from traditional to secular-rational values as their sense of existential security increases. This process accelerates with a transition from industrial to knowledge societies.

A subset of self-expression values – emancipative values – expresses a preference for freedom of choice and equality of opportunities. Emancipative values,

[1] http://www.worldvaluessurvey.org/wvs.jsp

therefore, according to Inglehart, involve priorities for lifestyle liberty, gender equality, personal autonomy and the voice of the people. Emancipative values form the key cultural component of human empowerment, a social process, which once set in motion, empowers people to exercise freedoms in their course of actions. According to Inglehart, values are changing over time and are relevant to political changes that countries are experiencing. As Smith et al. (2006) indicate, two dimensions of variation across nations are based on the factor analysis of 47 variables and the results meet the requirements suggested by S. Schwartz that "national averages only be compared for those value items whose meaning is the same across individual persons in all cultural groups" (p.45).

The most recent results of the world values survey presented in Fig. 15.2 depict distribution of countries in a two-dimensional space of Traditional versus Secular-Rational Values and Survival versus Self-Expression values. Inglehart highlights several major cultural groups, based on language (English Speaking), religion (Protestant Europe, Catholic Europe, Orthodox, Confucian and African-Islamic), geographic location (Baltic, South Asia, Latin America). Assignment of countries to cultural groups is not always very precise: for example Poland features on the border between Latin America and African-Islamic, Malta and Philippines are found in the Latin America sector; largely Islamic Albania falls into the Orthodox group, predominantly Orthodox Ethiopia is featured in the African-Islamic and Eastern Orthodox Cyprus appeared in the South Asia group.

According to Davidov et al. (2008), basic human values influence political attitudes and choices through their effect on ideologies. Ideology stands for ideas about means and ends or 'means-ends' philosophy'. It includes views about the present state of the world, in what direction to go and where one wants to be at future points in time (Söderbaum 2008; Söderbaum and Brown 2012).

15.6 Environmental Values in Different Stakeholders Groups

From 2004 to 2010 we carried out research in Russia using S. Schwartz methodology (Schwartz 2005a, b), paying special attention to such contradictory groups of values as *self-transcendence* vs *self-enhancement* values and, especially, *universalism* vs *power*. In our research the method that included two different questionnaires was used: one actualizing individual normative values (57 Questions), the other actualizing individual priority values (40 questions). In the first questionnaire (57) respondents were asked to estimate what were the main principles and motivational goals of their life using the Lickert scale. In the second questionnaire (40 questions) they needed to compare the given behavioral models with behavior that was most typical to them.

As a rule, the values research is done focusing on groups of respondents representing student or teachers samples with homogeneous educational levels (with rare exception as Schlöder 1993), highlighting general sample parameters such as age and gender and never paying attention to professional backgrounds, interests and concerns. In our research, we expected different professional groups dealing with sustainability and environmental problems and students trained in other professional fields to exhibit different value priorities. This expectation was based on the assumption that every professional activity is determined by a complex of direct and indirect motives leading to a future success in a professional activity.

The hypothesis of our research was based on the expectation that a priority of *self-transcendence values* and estimation of the level of *power* (*self enhancement values*) will be different in a sample group involved in the sustainable development process. Our first hypothesis, therefore was that professionals in the field of sustainable development and students who professionally studying ecology will have the highest ranks of *self- transcendence* (*universalism values*). The main aim of the research was therefore to find out if individual values differ in different target groups.

The groups in the research were formed according to their different professional background and orientation, age and status. Groups differed by (1) their orientation in relation to environmental problems – how close or far were they dealing with environmental or sustainable development problems; (2) according to age from 18 till 23, from 23 to 35, from 35 to 65; (3) according to their status: students, young professionals and highly qualified professionals. The sample consisted of more than 400 respondents.

Sub-groups were defined as follows with acronyms found in the figures highlighted in *italics*:

1. Sustainable Development Professionals (university professors, environmental managers, NGO activists) – *Professionals*;
2. Students of Ecology and Resource Management, Geology Department, State University – *Geologists*;
3. Student of Law Department, State University – *Lawyers;*
4. Students of International relations, State University (General group) –*IR Master*;
5. Students of International relations, State University (MA Environment and Development) – *IR ECO*;
6. Students of the Polar Academy representing indigenous community of the Far North, Russia – *Indigenous;*
7. Students of the Polar Academy representing ethnic Russians communities of the Far North – *PO Russians;*
8. Legislative Assembly Members (Saint Petersburg, Russia) – *Legislative Assembly*
9. State Service Academy Excellence training participants – *State Service Academy*;
10. Students of Education (Pedagogy) Department, State University – *Educators.*

Figures 15.3 and 15.4 show how value priorities for 10 main values, discussed above, look like if we present the diversification according to the target groups. The results are shown as rank scores on a scale from 1 to 10. Rank 1 is the highest rank

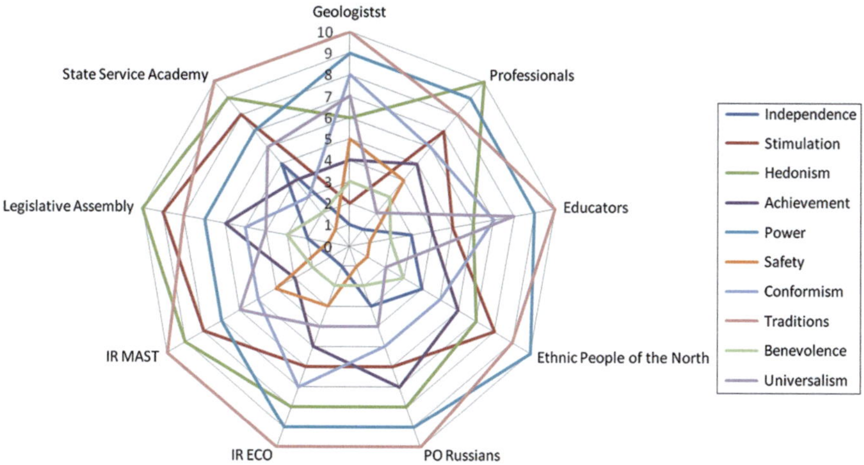

Fig. 15.3 Individual values per groups (normative ideals)

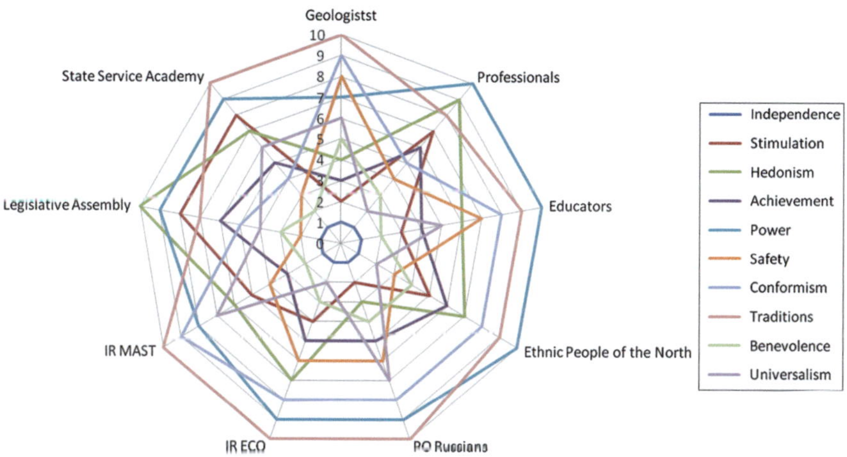

Fig. 15.4 Individual values per groups (individual behavior similarities)

and is placed close to the center; rank 10 is the lowest rank and is depicted at the edge of the circle.

We can see therefore (Fig. 15.3) how different the values priorities are in different groups ranging from 1 to 5 for *safety values* and *independence values*, 2 to 8 for *universalism values*, 10 to 7 for *power values*, 3 to 8 for *conformity values* when respondents were asked about their **normative ideals** values. A big discrepancy in values priorities is also seen in Fig. 15.4 (**individual behavior**), only *independence values* acquiring consensus among different groups.

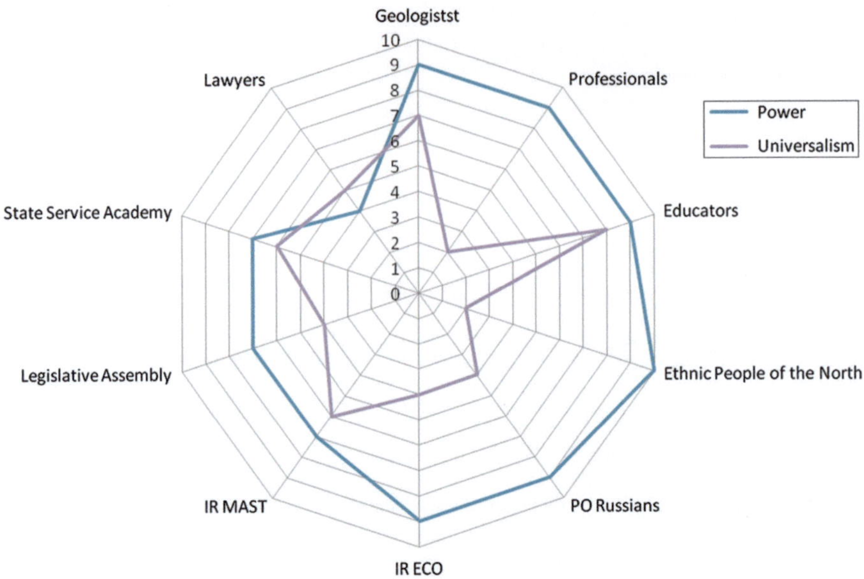

Fig. 15.5 Power and Universalism individual normative values in different stakeholder groups

The presented results give us an understanding of why the average statistical scores of a sample need to be so carefully discussed and interpreted. There are indeed large differences between values structures in various stakeholder groups we need to be aware of.

In this research we were mostly interested in two types of values: *universalism values* that are considered to be predictors of pro-environmental attitudes and behavior change in the direction of sustainable development and *power values* that are according to S. Schwartz theory in contradictory relations to universalism, which could explain the emergence of barriers in promoting sustainable development ideas by the existing interest groups.

Figures 15.5 and 15.6 were designed to show more precisely how *universalism* and *power values* manifest themselves in different target groups in our research results. It is clear from Fig. 15.5 that priority to *universalism* values is given in the group of Sustainability Professionals and Indigenous Ethnic people of the North. The lowest priority to *universalism* values (Protection of Nature, Care about Humans) is given in groups of Geology students and Students of Education. It gives us a chance to think about the lack of education for sustainable development component in the educational systems of the departments mentioned above. We also can notice from the Fig. 15.5 that Power values have the highest priority in the group of Law students (Lawyers) that exceeds the priority of the universalism Values.

A similar situation we can see with the results of PPQ Questionnaire on Fig. 15.6. It's interesting to note the difference showed on Figs. 15.5 and 15.6 between two groups of International Relations students –IR Mast and IR ECO. In the group of IR

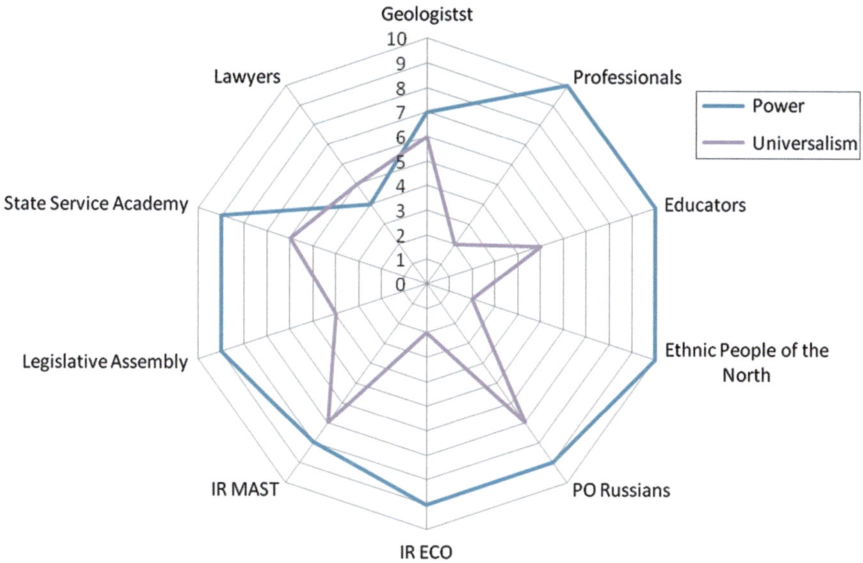

Fig. 15.6 Power and Universalism individual priority values, in different Stakeholder Groups

Eco the priority for *power* values in lower and priority for the *universalism* values is higher than in the group of IR Master, which can be a result of the education courses in Philosophy of Sustainable Development and Regional Environmental Studies that were taught by the author of this chapter to the group of IR Eco students participating in the research as respondents.

Two groups of respondents showing the closest values priority scores especially in such values as *independence*, *universalism* and *power*, require our special attention (Fig. 15.7). These groups differ in age and professional experience. First is the group of Sustainability Professionals aged between 24 till 65, which were educated and professionally involved in the field of environmental management, sustainable development, ecological safety and teaching in this field. The high rank of the *universalism values* and low rank of *power values* can be explained by the high level of knowledge in the field, high level of ecological consciousness, understanding of global environmental problems and personal responsibility. The other group (ethnic people of the North) consisted of student respondents aged between 18 and 24, studying ecology disciplines at the Polar University, but originally coming from the North Polar circle region representing 16 different indigenous communities. Their professional knowledge of the global environmental problems was rather limited, but we can explain the reason for the highest level of the *universalism values* by the local indigenous knowledge they had from their childhood concerning the human relations with nature.

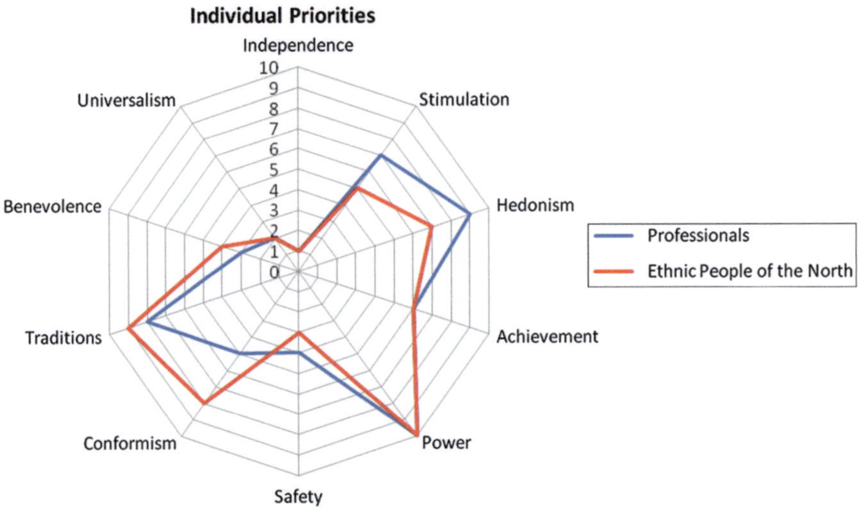

Fig. 15.7 Individual value profiles in Sustainability professionals and ethnic people of the North Stakeholder Group

15.7 Policies, Indicators and Models

Our main interest, however, was in understanding how the self-expression values, which were associated with environmental protection and, especially, the emancipative values, relate to the broad range of sustainability metrics used at the national scale. In other words, what are the psychological drivers of sustainability at a national cultural level? We have established statistically significant correlations between average country *emancipative* value scores (Inglehart 2015) and the Global Peace Index (countries with higher levels of *emancipative* values tend to be more peaceful, Fig. 15.8), Human Development Index (countries with higher levels of emancipative values tend to have higher levels of human development, Fig. 15.9), life expectancy (countries with higher levels of emancipative values tend to have higher life expectancy, Fig. 15.10), Yale Environmental Performance Index, EPI (countries with higher levels of *emancipative* values tend to have higher EPI, Fig. 15.11) and Municipal Solid Waste recycling levels (higher levels of *emancipative* values leading to higher recycling rates, Fig. 15.12).

Figure 15.8 shows that the level of peace in a country is positively correlated with the degree of *emancipative* values. We need to mention that lower levels of the Peace Index scale correspond to a higher degree of peace. On the one hand such countries as Pakistan, Nigeria, Iraq, Yemen, Russia and Egypt have rather low levels of peace and at the same time low level of *emancipative* values. On the other hand, such countries as Sweden, Netherlands, Germany, Australia, New Zealand and Slovenia have the highest levels of peace and, at the same time, the highest *emancipative* values level.

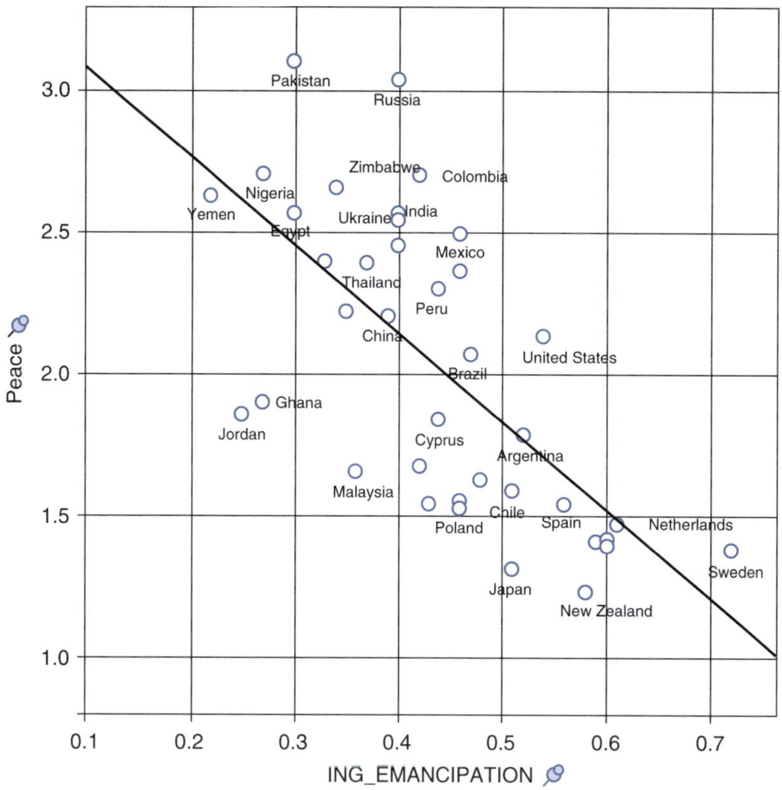

Fig. 15.8 Emancipative values and peace

Figure 15.9 shows positive correlation between the level of *emancipative* values and Human Development Index. The higher the degree to which *emancipative* values are expressed in the country, the higher the Human Development Index. On the one end of the spectrum we see such countries as Sweden, Netherlands, Germany, Australia, Slovenia, Japan and United States with high levels of HDI and high expression of *emancipative* values; on the other side of the spectrum we find such countries as Yemen, Nigeria, Rwanda, Zimbabwe, Pakistan, Ghana with relatively low levels of HDI and low levels of *emancipative* values.

Life Expectancy Index is one of the three constituent parts of the HDI. Figure 15.10 shows the positive correlation between the Life expectancy Index Level and degree of expression of *emancipative* values.

Here we see that the higher the level of expression of *emancipative* values, the higher the Life expectancy index is. Among the countries demonstrating the high level of both indicators are Sweden, Netherlands, Australia, Slovenia, New Zealand and US; on the other end of the spectrum are Yemen, Ghana, Ruanda, Pakistan, India and Iraq.

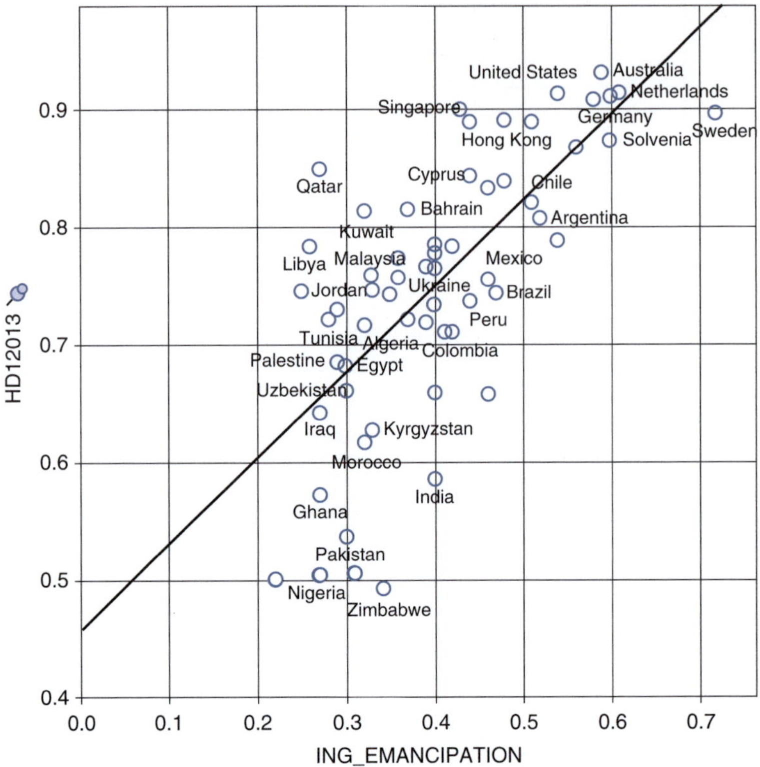

Fig. 15.9 Emancipative values and human development index

Yale Environmental Performance Index (EPI) is an annually calculated composite measure comprising Environmental health dimensions (Health impact, Air quality, Water and Sanitation) and Ecosystems Vitality Dimension (water resource, agriculture, forests, fishery, biodiversity and habitat, climate and energy (Yale, 2014). Fig. 15.11 shows that high levels of EPI and *emancipative* values are observed in Sweden, Australia, Germany, Netherlands, Slovenia and New-Zealand. On the opposite side of the spectrum we observe such countries as Yemen, Pakistan, India, Rwanda, Nigeria, Ghana, Libya.

Concerning recycling rates, of all the countries in the pool, Japan has the highest MSW (Municipal Solid Waste) recycling rate of 77 %, Germany exhibits a rate of 64.5 %, Singapore – 61 %, Australia – 51 %, and Netherlands – 49.8 %. The worst performers on MSW recycling among the countries under consideration are Georgia, Belarus, Azerbaijan, Chile, Egypt, China, Trinidad and Tobago, Palestine, Pakistan and Russia. Recycling rate according to our consideration clearly presents an example of conscious choice at the level of policies, communities and individuals and is positively correlated with the *emancipative* values.

In turn, we found significant correlations between the similar indicators of sustainability and Cultural values according to S. Schwartz methodology (Schwartz

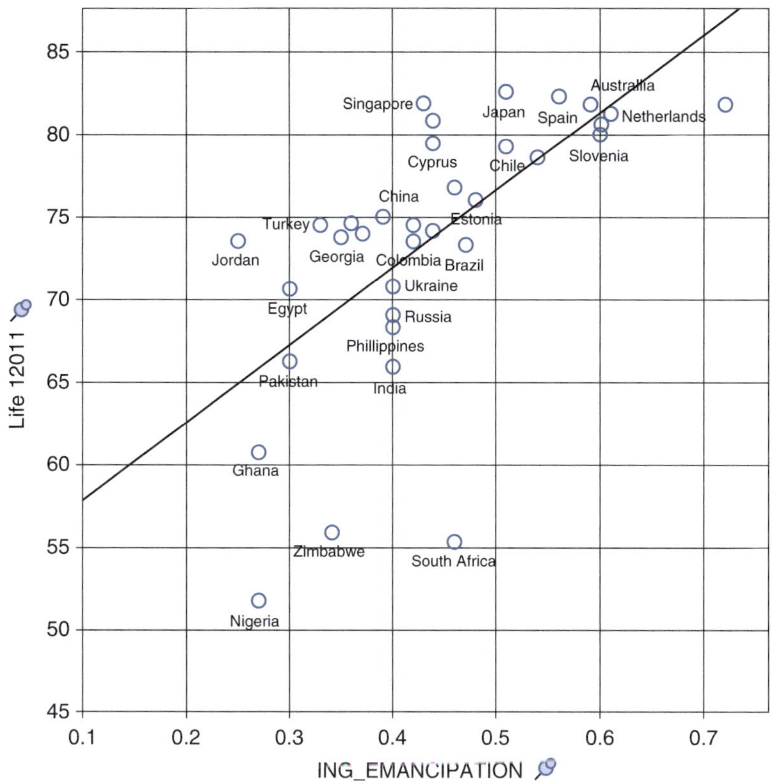

Fig. 15.10 Emancipative values and life expectancy index

2011). Figure 15.13 shows that higher degrees of *intellectual autonomy* correlate with higher level of Peace. On the one hand we observe countries with high level of Peace and high level of *intellectual autonomy* as Sweden, Germany, Netherlands, New Zealand and Spain, on the other hand such countries as Nigeria, Pakistan, Egypt, India, South Africa.

Figure 15.14 demostrates that Yale Environmnetal performance Index and expression of *intellectual autonomy* are both highest in Sweden, Germany, Spain, Netherlands, Slovenia, Japan and New Zealand. On the other hand the lowest EPI scores and lowest *intellectual autonomy* scores are obsrved in Yemen, Pakistan Ghana, India Nigeria, Philippines Zimbabwe, China.

Figure 15.15 presents a positive correlation between the degree of *intellectual autonomy* according to Schwartz and the life expectancy across countries. The highest values of *intellectual autonomy* and life expectancy are observed in Sweden, Germany, Spain, Slovenia, the Netherlands, Japan. On the other hand, the lowest values of *intellectual autonomy* and life expectancy can be seen in Nigeria, South Africa, Zimbabwe, Yemen and Ghana.

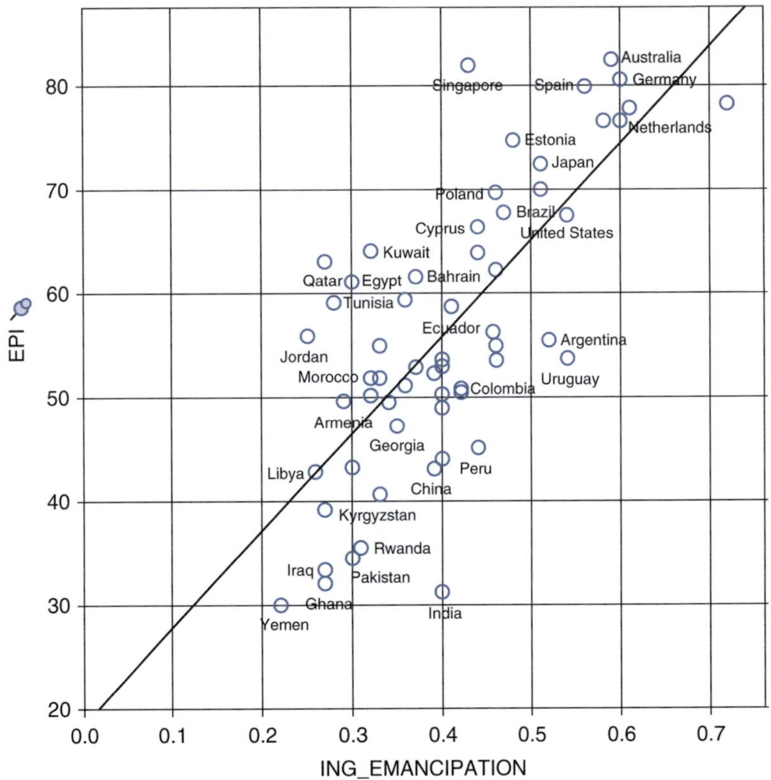

Fig. 15.11 Emancipative values and Yale Environmental Performance Index

As can be seen in Fig. 15.16, there is a certain degree of correlation between the *intellectual autonomy* and a sustainability policy indicator for recycling of municipal solid waste. On the one end of the spectrum we find Germany, Sweden, Netherlands, and Slovenia exhibiting a very high degree of *intellectual autonomy* and high levels of recycling, on the opposite end of the spectrum we observe Pakistan, Yemen, Egypt, Philippines, Thailand, Georgia, and Jordan, where both *intellectual autonomy* and recycling levels are low.

There is a strong negative correlation between the level of *embeddedness* according to S Schwartz and Yale Environmental Performance Index (EPI), which can be seen in Fig. 15.17. Low *embeddedness* and high EPI are observed in Germany, Sweden, Netherlands and Spain. High *embeddedness* and low EPI could be seen in Yemen, Ghana, Pakistan, India and Nigeria.

At the same time, low values of *embededdness* correspond to high degree of peacefulness (low value of the Peace Index). Such situtation can be observed in Germany, Sweden, New Zealand, Netherlands, Spain and Japan (Fig. 15.18). The opposite situation can be found in Pakistan, Yemen, Nigeria, Egypt, Zimbabwe and Philippines.

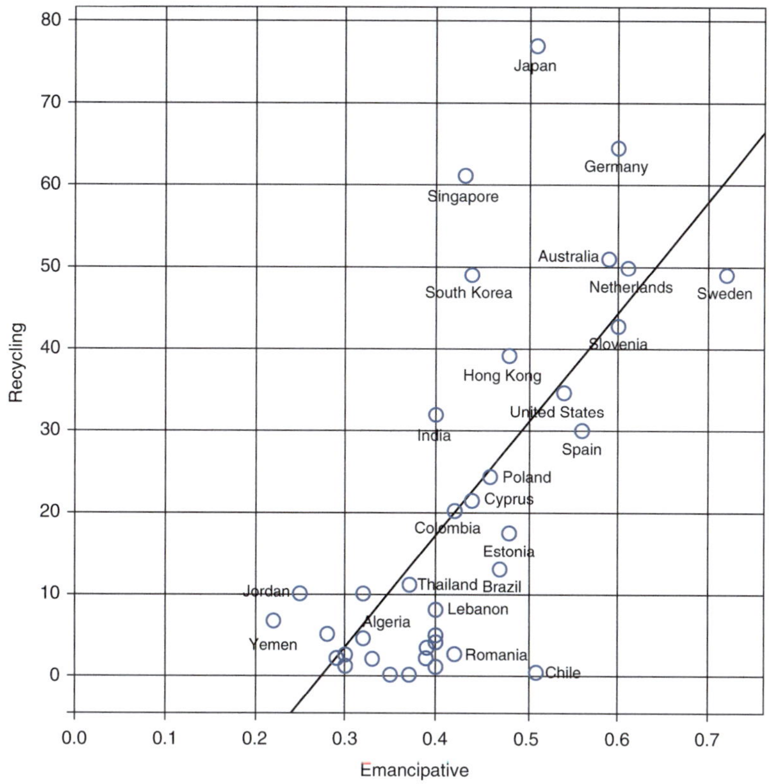

Fig. 15.12 Emancipative values and recycling

Figure 15.19 presents a significant negative correlation between the degree of *embeddedness* and life expectancy in the sample. High degree of *embeddedness* and low life expectancy can be found in Nigeria, Zimbabwe, South Africa, Ghana, Yemen and Pakistan. On the other hand, low levels of *embeddedness* and high levels of life expectancy are observed in Sweden, Germany, Spain, Netherlands and New Zealand.

In Fig. 15.20 we can observe a certain degree of correlation between *harmony* according to Schwartz and Yale Environmental Performance Index. Germany, Spain, Sweden, Slovenia and Estonia exhibit high levels of EPI and *harmony*; at the same time, Yemen, India, Pakistan, Philippines, Colombia and Thailand are characterised with low levels of *harmony* and low levels of EPI.

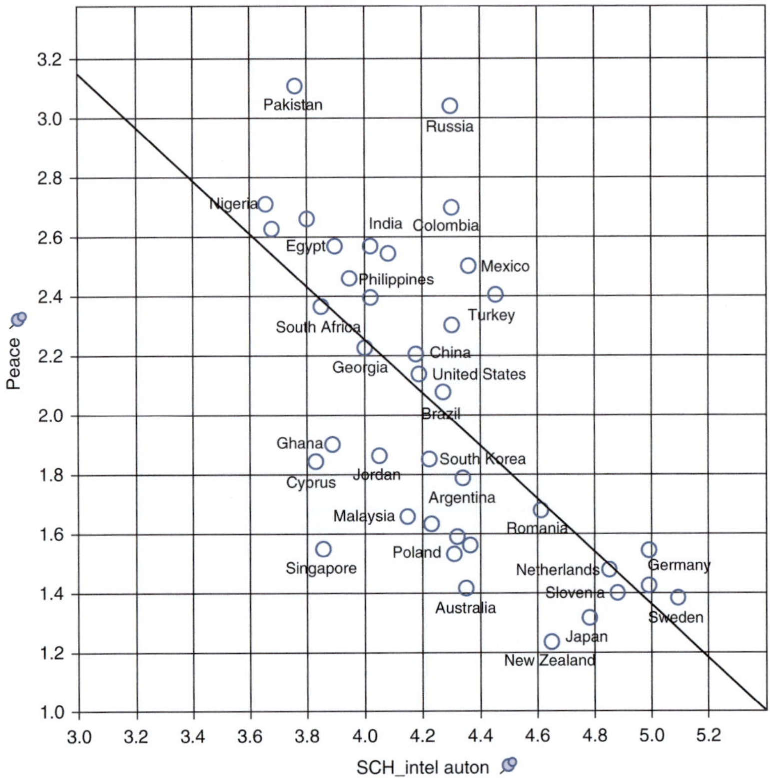

Fig. 15.13 Peace index and Intellectual autonomy

15.8 Discussion: The Nordic Model

Some of the crucial questions one could pose here would be: "which countries managed to achieve the highest levels of environmental, economic and social performance expressed in sustainable development indicators?", "how this achievement was linked to the values these societies hold as important?" and "how institutions and policies shaped such a state of affairs?". In this context, particularly interesting are examples of the Nordic countries: Sweden, Norway, Iceland, Denmark and Finland. As charts presented in this chapter have shown, Sweden is featured consistently very high on sustainabilitty dimensions, such as life expectancy, Yale Environmental performance index, Peace Index and recycling. At the same time, Sweden is characterised with the high levels of *emancipative* values (Ingleheart), high levels of *intellectual autonomy* and low values of *embeddedness* (Schwartz) and low values of *masculinity* (Hofstede). In this light it would be particularly interesting to see what are the socio-political conditions in Sweden and other Nordic countries that allowed such an achievement in sustainability.

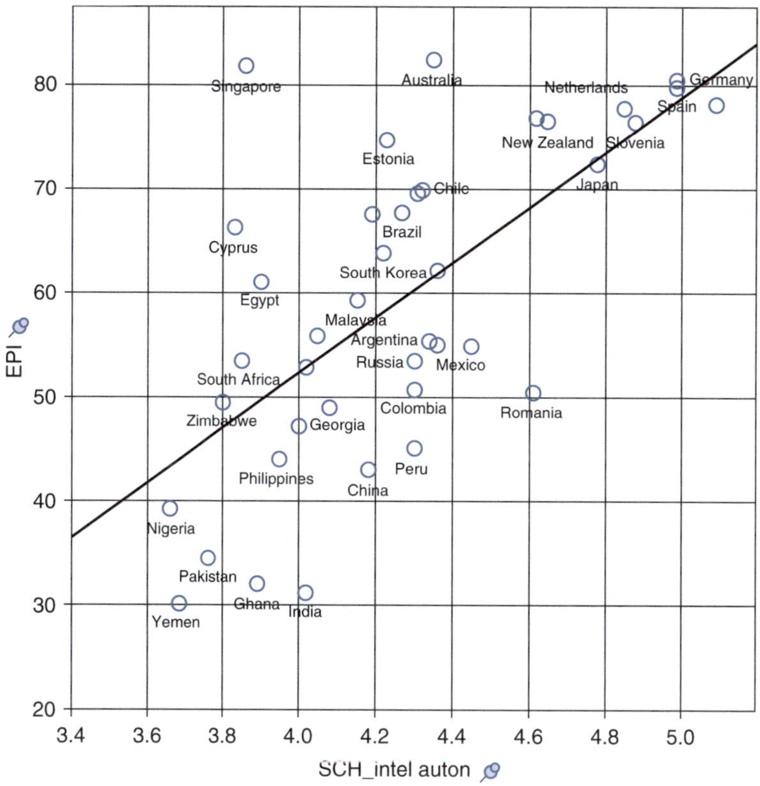

Fig. 15.14 Yale environmnetal performanc index and Intellectual autonomy

Nordic countries form a region in Northern Europe, comprising Denmark, Finland, Iceland, Norway and Sweden and their associated territories, including Greenland. The combined population of the Nordic countries is about 25 million. Nordic countries feature very high in the international competitiveness rankings with Finland occupying 3rd, Sweden - 4th, Denmark 8th and Norway 12th place globally in 2012. In comparison with other countries, the Nordics have low unemployment levels, balanced budgets and low public debts (Hillamo and Kangas 2013).

Principle features of the Nordic model (Andersen et al. 2007) are :

1. a comprehensive welfare state an emphasis on trasfers to housholds & publicity provided social services financed by taxes, with a high notably for wage income and consumption;
2. significant public or private spending on investment in human capital, including child care and education, as well as research and development;

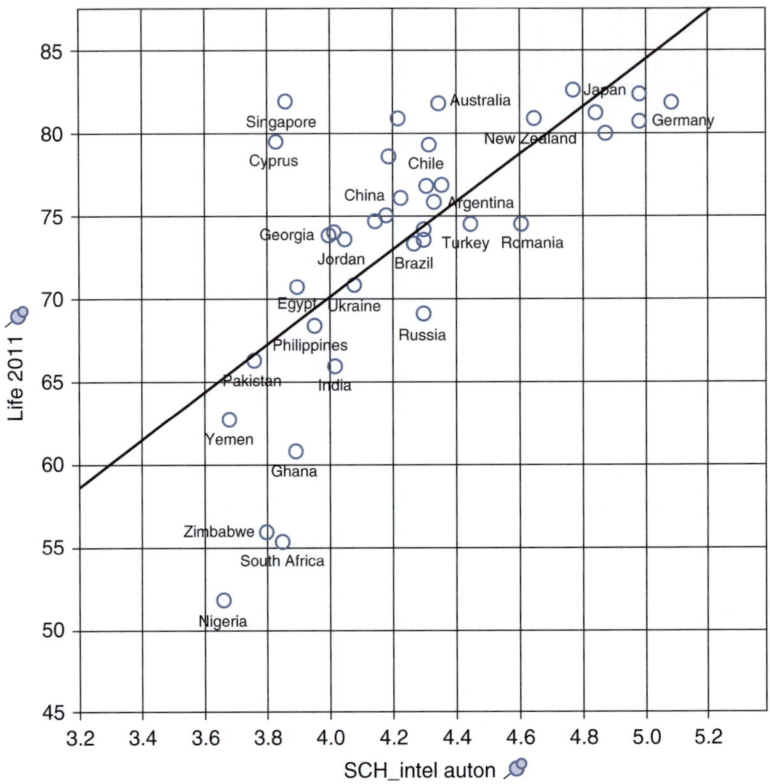

Fig. 15.15 Life expectancy index and Intellectual autonomy

3. a set of labour market institutions that include strong labour units and employer associations. Significant elements of wage coordination relatevely generous benefits a prominent role of active labour market policies.

One of the principle features of the Nordic model is to pursue universal welfare state policies, which means that public programmes, services and transfers are designed to serve everyone in the country. (Hillamo and Kangas 2013). At the same time, the Nordic countries exhibit high degrees of trust in national institutions and their fellow citizens as well as a high degree of satisfaction with life. Nordic countries managed to help reduce gender inequality both in terms of the share of women in top management positions and income gap. The state in Scandinavia developed as a strong and powerful force, but at the same time a transparent and non-corrupt system. Investment in social capital is often seen as a key to success in the Nordic countries, which manifests itself in universal cover of the whole population, the emphasis on equal chances in life, low levels of child poverty, investment in health care and education.

A key feature of the Nordic model is the beneficial and mutually supportive interaction of openness and collective risk sharing. Successful trasformation of their

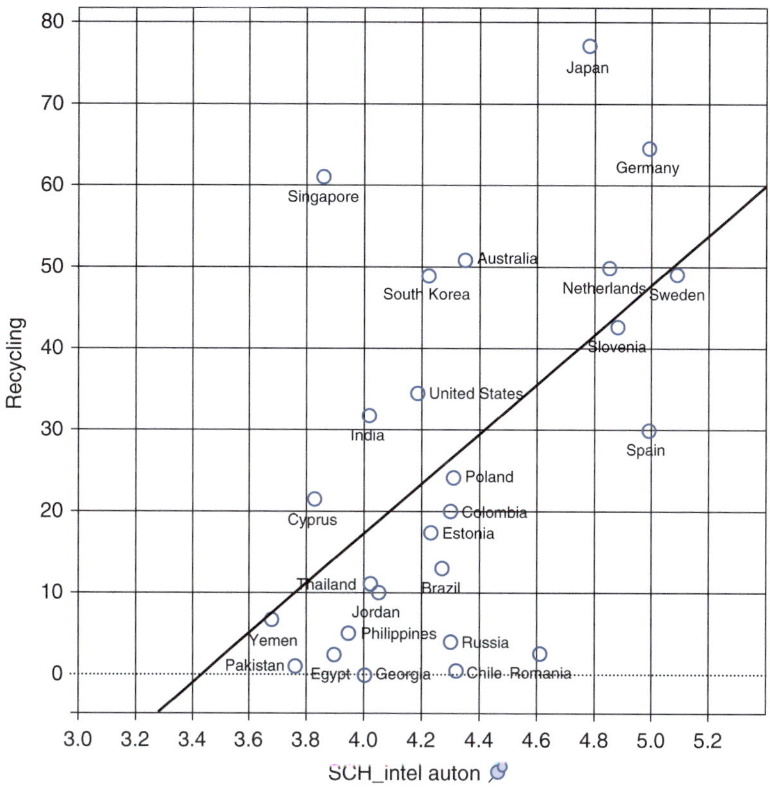

Fig. 15.16 Recycling and Intellectual autonomy

economics in favour of knowledge intensive activities. High quality education for large share of young age cohorts and government support for research and innovation is accompanied by security provided by through collective mechanism for risk sharing. The Nordic countries are egalitarian societies in the sense that income and wealth differenties are smaller than elswere.

In this chapter we have shown that existing psychological and sociological value theories could be used to explain certain sustainability preformance differences across countries. Values indeed play a big role in determining behaviour patterns, and certain types of values, especially intellectual autonomy (Schwartz) and emancipative values (Inglehart) have been shown to be good predictors for sustainability performance at the national level. It is important to pay attention to the value differences between various stakeholder groups, as it was demonstrated how in the context of a particular culture, universalism and power values of sustainability professionals and indigenous people of the north have been found significanly different from the values of other stakeholder groups. Psychological and sociological analysis of individual and cultural values could therefore be very beneficial for designing pathways towards a green economy.

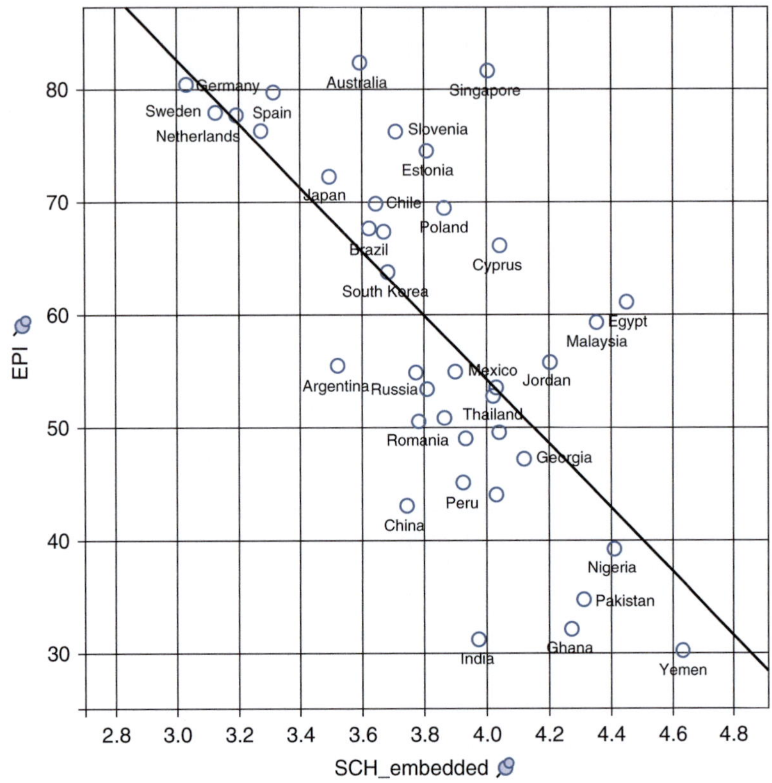

Fig. 15.17 Yale Environmental Performance Index and Embeddedness

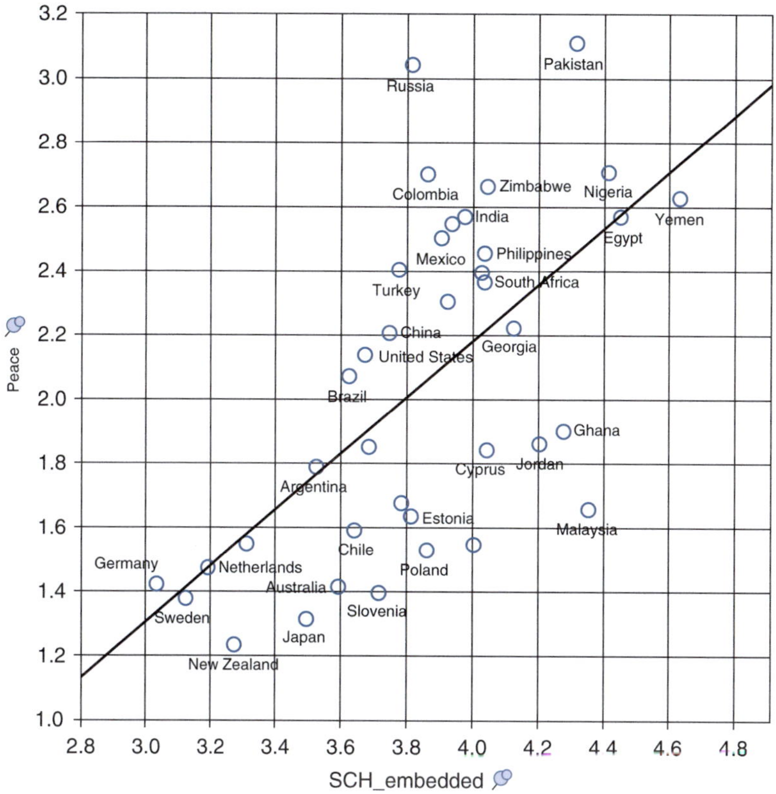

Fig. 15.18 Peace Index and Embeddedness

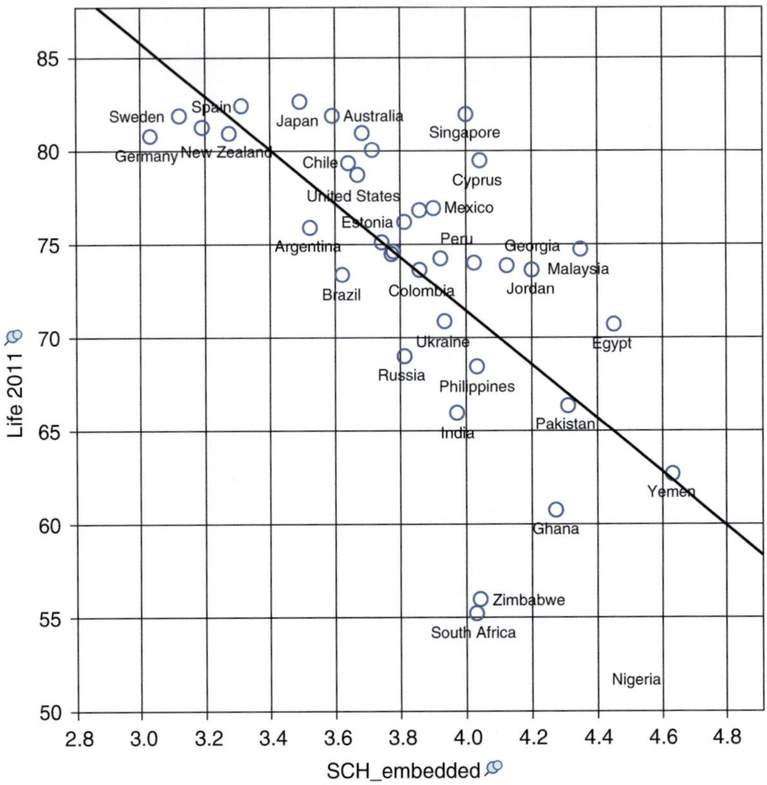

Fig. 15.19 Life expectancy and Embeddedness

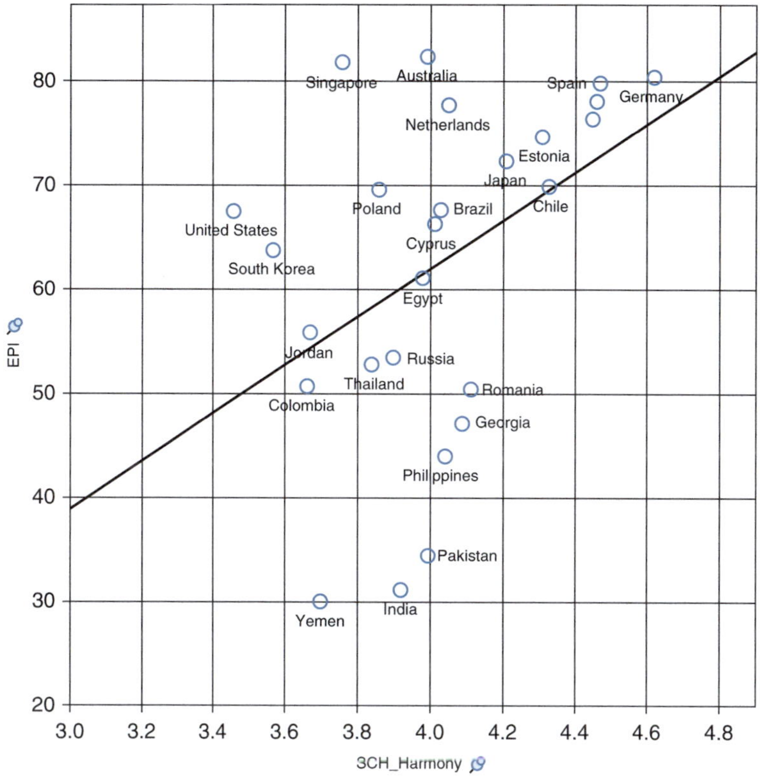

Fig. 15.20 Yale Environmental Performance Index and Harmony

References

Allport G (1955) Becoming: basic considerations for a psychology of personality. Yale University Press, New Haven, p 119
American Psychological Association (2014) Psychology Day at the UN. Retrieved from http://www.apa.org/international/united-nations/psych-day.aspx
Andersen TM, Holmstrom B, Honkapohia S, Korkman S, Soderstrom HT, Vartianen J (2007) The Nordic model. Embracing globalization & sharing risks. Taloustieto, Helsinki, p 165
Davidov E, Schmidt P, Schwartz SH (2008) Bringing values back in the adequacy of the european social survey to measure values in 20 countries. Public Opin Q 72(3):420–445
Didz S, Kalof L, Stern P, Gender C (2002) Values and environmentalism. Soc Sci Q 83(1):351–364
Dunlap RE (2008) Th e new environmental paradigm scale: from marginality to worldwide use. J Environ Educ 40(1):3–18
Dunlap RE, Liere V, Kent D, Mertig AG, Jones RE (2000) Measuring endorsement of the new ecological paradigm: a revised NEP scale. J Soc Issues 56(3):425–442
European Social Survey (2014) http://www.europeansocialsurvey.org/data/themes.html?t=values
Frankl V (1959) Man's search for meaning. Beacon Press, Boston, MA, 2000, 184 pp
Hillamo H, Kangas O (2013) A recipe for a better life: experiences from the Nordic countries. CMI Martti Ahtisaari Centre, Helsinki, p 18
Inglehart R (2000) Globalization and postmodern values. Wash Q 2000(Winter):215–228
Inglehart R (2015) World values survey. http://www.worldvaluessurvey.org/wvs.jsp
Kluckhorn C, Murray HA (1953) Personality in nature, society and culture, 2nd edn. Knopf, New York, p 717
Lebedeva NM (2000) Bazovye zennosti russkich na rubezhe XXI veka [basic values of the Russians at the turn of 21st century]. Psikhologicheskii Zhurnal 21(3):73–87
Leontief DA (1998) Zennostnye Predstavlenija v Individualnom i Gruppovom Soznanii: Vidy, Determinanty i Izmenenija vo Vremeni [Value representations in individual and group consciousness: types, determinants and changes over time]. Psychologicheskoe Obozrenije, N. 1, pp 13–25
Niit T, Kailo H, Link K, Mira R (2004) Relationship between values and environmental attitudes in Estonia and Galicia, XVII International Congress of the IACCP, X'ian, China: the Book of Abstracts, p 110
Oldemeyer E (1978) Zum problem der umwertung von Werten. In Ropohl et al. (1978), pp 11–65
Parsons T. (1951) The social system. Routledge, London, 1991, 404 pp
Raudsepp M (2001) Environmental belief systems: empirical structure and a typology. Trames J Humanit Soc Sci 55/50(3):101–127
Reser JP, Bentrupperbaummer JM (2005) What and where are environmental values? Assessing the impact of current diversity of use of "environmental" and "world heritage" values. J Environ Psychol 25(2):125–146
Rokeach V (1973) The nature of human values. The Free Press, New York, p 438
Schlöder B (1993) Soziale Werte und Werthaltungen : eine sozialpsychologische Untersuchung des Konzepts sozialer Werte und des Wertwandels. Leske u. Budrich, Opladen, p 235
Schmuck P, Schultz WP (eds) (2002) Psychology of sustainable development. Springer, Boston/Dordrecht/London, p 340
Schultz WP (2001) The structure of environmental concern: concern for self, other people and the biosphere. J Environ Psychol 21(4):327–339
Schultz WP, Zelezny L (1998) Values as predictors of environmental attitudes: evidence for consistency across 14 countries. J Environ Psychol 19:255–265
Schultz WP, Zelezny L (2003) Reframing environmental messages to be congruent with american values. Hum Ecol Rev 10(2):26–136

Schwartz SH (2004) A proposal for measuring value orientations across Nations. Chapter 7 https://www.europeansocialsurvey.org/docs/methodology/core_ess_questionnaire/ESS_core_questionnaire_human_values.pdf
Schwartz SH (2005a) Basic human values. Their content and structure across countries. In: Tamayo A, Porto JB, (eds) Valores e comportamento nas organizacoes [Values and behavior in organizations]. Petropolis: Vozes, Brazil, pp 21–64
Schwartz SH (2005b) Robustness and fruitfulness of a theory of universal in individual human values. In: Tamayo A, Porto JB, (eds) Valores e comportamento nas organizacoes [Values and behavior in organizations]. Petropolis: Vozes, Brazil, pp 56–95
Schwartz SH (2011) Values: cultural and individual. In: van de Vijver FJR, Chasiotis A, Breugelmans SM (eds) Fundamental questions in cross-cultural psychology. Cambridge University Press, Cambridge/New York, pp 463–493
Schwartz SH, Bardi A (2001) Values hierarchy across cultures. Taking similarities perspective. J Cross-Cult Psychol 32(3):268–290
Schwartz SH, Bilsky W (1987) Toward a universal psychological structure of human values. J Pers Soc Psychol 53(3):550–562
Shmeleva IA (2012) Systemic psychological description of ecological consciousness in the context of global environmental problems. In: Shmelev SE, Shmeleva IA (eds) Sustainability analysis: an interdisciplinary approach. Palgrave Macmillan, New York, pp 129–148
Smith PB, Bond MH, Kagitcibasi C (2006) Understanding social psychology across cultures. Living and working in a changing world. Sage, London/Thousand Oaks/New Dehli
Söderbaum P (2008) Understanding sustainability economics. Towards pluralism in economics. Earthscan, London/Sterling, p 158
Söderbaum P, Brown J (2012) Making actors, paradigms, and ideologies visible in global governance for sustainability. In: Shmelev S, Shmeleva I (eds) Sustainability analysis: an interdisciplinary approach. Palgrave Macmillan, Basingstoke/New York, p 335
Sorokin P (1992) Chelovek. Civilizacia. Obschestvo. [Man. Civilization. Society], Politizdat, Moscow, 543 pp
Steg L, Bolderdijk JW, Keizer K, Perlaviciute G (2014) An integrated framework for encouraging pro-environmental behaviour: the role of valued situational factors and goals. J Environ Psychol 38:104–115
Thomas W, Znanetski F (1918) The polish peasant in Europe and America. University of Illinois Press, Champaign, IL, 1996, 152 pp
Vlek C, Steg L (2007) Human behavior and environmental sustainability: problems, driving forces and research topics. J Soc Issues 63(1):1–19
Yale Centre for Environmental Law and Policy (2014) Environmental Performance Index Full Report and Analysis

Chapter 16
Building a Sustainable and Desirable Economy-in-Society-in-Nature

Robert Costanza, Gar Alperovitz, Herman Daly, Joshua Farley, Carol Franco, Tim Jackson, Ida Kubiszewski, Juliet Schor, and Peter Victor

Abstract In this chapter we describe what an "ecological economy" could look like and how we could get there. We believe that this future can provide full employment and a high quality of life for everyone into the indefinite future while staying within the safe environmental operating space for humanity on earth. Developed countries have a special responsibility for achieving those goals. To get there, we need to stabilize population; more equitably share resources, income, and work; invest in the natural and social capital commons; reform the financial system to better reflect real assets

R. Costanza (✉) • I. Kubiszewski
Crawford School of Public Policy, Australian National University, Canberra, Australia
e-mail: robert.costanza@anu.edu.au; ida.kubiszewski@anu.edu.au

G. Alperovitz
The Democracy Collaborative and Department of Government and Politics, University of Maryland, College Park, MD, USA
e-mail: press@democracycollaborative.org

H. Daly
Emeritus, School of Public Policy, University of Maryland, College Park, MD, USA
e-mail: hdaly@umd.edu

J. Farley
Department of Community Development and Applied Economics, and Gund Institute for Ecological Economics, University of Vermont, Burlington, VT, USA
e-mail: Joshua.Farley@uvm.edu

C. Franco
Woods Hole Research Centre, Falmouth, MA, USA
e-mail: carolfrancobillini@gmail.com

T. Jackson
Centre for the Understanding of Sustainable Prosperity (CUSP), University of Surrey, Guildford, UK
e-mail: t.jackson@surrey.ac.uk

J. Schor
Department of Sociology, Boston College, Chestnut Hill, MA, USA
e-mail: juliet.schor@bc.edu

P. Victor
Faculty of Environmental Studies, York University, Toronto, Canada

S. Shmelev (ed.), *Green Economy Reader*, Studies in Ecological Economics 6,
DOI 10.1007/978-3-319-38919-6_16

and liabilities; create better measures of progress; reform tax systems to tax "bads" rather than goods; promote technological innovations that support well-being rather than growth; establish "strong democracy," and create a culture of well-being rather than consumption. In other words, a complete makeover. Several lines of evidence show that these policies are mutually supportive and the resulting system is feasible. The substantial challenge is making the transition to this better world in a peaceful and positive way. There is no way to predict the exact path this transition might take, but we hope that painting this picture of a possible end-point and some milestones along the way will help make this choice and this journey a more viable option.

Keywords Genuine progress indicator • Ecological economics model • Natural capital • Social capital • Human well-being

16.1 Rationale and Objectives

The current mainstream model of the economy is based on a number of assumptions about the way the world works, what the economy is, and what the economy is for (Table 16.1). These assumptions arose in an earlier period. In this "empty-world" context, built capital was the limiting factor, while natural capital was abundant. It made sense, in that context, not to worry too much about environmental "externalities," since they could be assumed to be relatively small and ultimately solvable. It made sense to focus on the growth of the market economy, as measured by GDP, as a primary means to improve human welfare. It made sense, in that context, to think of the economy as only marketed goods and services and to think of the goal as increasing the amount of these goods and services produced and consumed.

Key Points

- Growth in material consumption is unsustainable: there are fundamental planetary boundaries.
- Growth in material consumption beyond a threshold already reached by many is undesirable: it has negative effects on social and natural capital and in overdeveloped economies does not increase well-being.
- Viable alternatives exist that are both sustainable and desirable, but they require a fundamental redesign of the entire "regime."

But the world has changed dramatically. We now live in a world relatively full of humans and their built capital infrastructure. In this new context, we have to reconceptualize what the economy is and what it is for. We have to first remember that the goal of the economy should be to sustainably improve human well-being and quality of life. We have to remember that material consumption and GDP are merely means to that end, not ends in themselves. We have to recognize, as both ancient wisdom and new psychological research tell us, that too much of a focus on material

Table 16.1 The basic characteristics of the current economic model, the green economy model, and the ecological economics model

	Current economic model	Green economy model	Ecological economics model
Primary policy goal	**More:** Economic growth in the conventional sense, as measured by GDP. The assumption is that growth will ultimately allow the solution of all other problems. More is always better.	**More but with lower environmental impact:** GDP growth decoupled from carbon and from other material and energy impacts.	**Better:** Focus must shift from merely growth to "development" in the real sense of improvement in sustainable human well-being, recognizing that growth has significant negative by-products. More is not always better.
Primary measure of progress	GDP	Still GDP, but recognizing impacts on natural capital.	Index of Sustainable Economic Welfare (ISEW), Genuine Progress Indicator (GPI), or other improved measures of real welfare.
Scale/carrying capacity/ role of environment	Not an issue, since markets are assumed to be able to overcome any resource limits via new technology, and substitutes for resources are always available.	Recognized, but assumed to be solvable via decoupling.	A primary concern as a determinant of ecological sustainability. Natural capital and ecosystem services are not infinitely substitutable and real limits exist.
Distribution/poverty	Given lip service, but relegated to "politics" and a "trickle-down" policy: a rising tide lifts all boats.	Recognized as important, assumes greening the economy will reduce poverty via enhanced agriculture and employment in green sectors.	A primary concern, since it directly affects quality of life and social capital and is often exacerbated by growth: a too rapidly rising tide only lifts yachts, while swamping small boats.

(continued)

Table 16.1 (continued)

	Current economic model	Green economy model	Ecological economics model
Economic efficiency/allocation	The primary concern, but generally including only marketed goods and services (GDP) and market institutions.	Recognized to include natural capital and the need to incorporate the value of natural capital into market incentives.	A primary concern, but including both market and nonmarket goods and services, and effects. Emphasis on the need to incorporate the value of natural and social capital to achieve true allocative efficiency.
Property rights	Emphasis on private property and conventional markets.	Recognition of the need for instruments beyond the market.	Emphasis on a balance of property rights regimes appropriate to the nature and scale of the system, and a linking of rights with responsibilities. Includes larger role for common-property institutions in addition to private and state property.
Role of government	Government intervention to be minimized and replaced with private and market institutions.	Recognition of the need for government intervention to internalize natural capital.	Government plays a central role, including new functions as referee, facilitator, and broker in a new suite of common-asset institutions.
Principles of governance	Laissez-faire market capitalism.	Recognition of the need for government.	Lisbon principles of sustainable governance.

Costanza et al. (1997)

consumption can actually reduce our well-being (Kasser 2002). We have to better understand what really does contribute to sustainable human well-being (SHW) and recognize the substantial contributions of natural and social capital, which are now the limiting factors to improving SHW in many countries. We have to be able to distinguish between real poverty, in terms of low quality of life, and merely low monetary income. Ultimately we have to create a new vision of what the economy is and what it is for, and a new model of the economy that acknowledges this new "full-world" context and vision.

Some argue that relatively minor adjustments to the current economic model will produce the desired results. For example, they argue that by adequately pricing the depletion of natural capital (e.g., putting a price on carbon emissions) we can address many of the problems of the current economy while still allowing growth to continue. We call this approach the "green economy" (GE) model (Table 16.1). Some of the areas of intervention promoted by GE advocates, such as investing in natural capital are necessary and we should pursue them. However, we do not agree that they are sufficient to achieve sustainable human well-being. We need a more fundamental change, a change of our goals and paradigm as discussed in the remainder of this report.

16.1.1 Some Background

The World Bank (WB) and the International Monetary Fund (IMF), founded at the Bretton Woods conference at the end of World War II, were chartered to speed economic development, stabilize the world economy, and end poverty. These institutions have relied largely on the current economic model as described above and in Table 16.1. The inability of these institutions and the later World Trade Organization (WTO), whose origins can also be traced to the Bretton Woods conference, to fully achieve their original goals of improving lives in the developing world and stabilizing the global economy has given rise to many critics, who are no longer marginalized voices of the displeased. These include former World Bank economists, the Group of 77 (G-77), and, increasingly, the millions of people in developed countries who have taken to the streets in protest. The policies under fire include removing barriers that check corporate access to a country's resources and often involve suspension of social and environmental legislation. Such policies can even over-ride national laws instituted through democratic processes. For example, the WTO once ruled that the United States Clean Air Act was a barrier to free trade. Such policies are antithetical to the goal of developing in a way that is sustainable, democratic, and equitable. They are also by no means agreed-upon in a broad consensus but are rather the dictates of a few powerful countries and their attendant organizations. Lending countries and their economists drove these policies, and borrowing nations have had little say in their implementation. Loans have required cuts in government salaries and privatization of social services. The conditional loans foisted upon many Latin American countries resulted in massive unemployment and devastating economic crises. In short, the execution of this model of the economy has led to unemployment, falling worker wages, biodiversity loss, environmental degradation, and disintegration of the social fabric.

Critics of the current model are many, and a coherent and viable alternative is sorely needed. Our purpose in this report is to lay out a new model of the economy based on the worldview and principles of ecological economics (Costanza 1991; Costanza et al. 1997; Daly and Farley 2004). These include the ideas that:

1. our material economy is embedded in society which is embedded in our ecological life-support system, and that we cannot understand or manage our economy without understanding the whole, interconnected system;
2. growth and development are not always linked and that true development must be defined in terms of the improvement of sustainable human well-being, not merely improvement in material consumption; and
3. a balance of four basic types of assets (capital) are necessary for sustainable human well-being: built, human, social, and natural capital (financial is merely a marker for real capital and must be managed as such).

Before describing this new model, we provide a bit more background on why the current model is both unsustainable and undesirable.

16.1.2 Growth in Material Consumption Is Unsustainable: There Are Fundamental Planetary Boundaries

Historically, human recognition of our impact on the earth has consistently lagged behind the magnitude of the damage we have imposed, thus seriously weakening efforts to control this damage (Costanza et al. 2007a). Even today, technological optimists and others ignore the mounting evidence of global environmental degradation, including climate disruption. Even some serious observers draw comfort from arguments such as the following:

- GDP figures are still increasing throughout much of the world.
- Life expectancies are still increasing in many nations.
- Evidence of human-caused climate disruption is still not absolutely definitive.
- Some claims of environmental damage have been exaggerated.
- Some previous predictions of environmental catastrophe have not been borne out.

Each of these statements is correct. However, not one of them is a reason for complacency, and indeed, taken together they should be viewed as powerful evidence of the need for an innovative approach. GDP and other current measures of national income accounting are notorious for overweighting market transactions, understating resource depletion, omitting pollution damage, and failing to measure real changes in well-being. For example, the Index of Sustainable Economic Welfare (ISEW), and a variation called the Genuine Progress Indicator (GPI), show significantly reduced improvement in real gains despite great increases in resource-depleting throughput. The ISEW and GPI also show increases in life expectancies in many nations, clearly indicating improvements in welfare; but unless accompanied by corresponding decreases in birth rates, such increases are warnings of acceleration in population growth, which will compound all other environmental problems. More details about these and other indicators of well-being are provided in Sect. 16.1.3.

The pervasiveness of uncertainty about the basic nature of our ecological life-support systems and the recognition that complex systems often exhibit rapid, nonlinear changes and threshold effects emphasizes the need for building precautionary minimum safety standards into our policies (Rockström et al. 2009).

Only relatively recently, with advances in environmental sciences, global remote sensing, and other monitoring systems, has a more comprehensive assessment of local and global environmental deterioration become possible. Evidence is accumulating with respect to accelerating loss of vital rain forests, species extinctions, depletion of ocean fisheries, shortages of freshwater in some areas and increased flooding in others, soil erosion, depletion and pollution of underground aquifers, decreases in quantity and quality of irrigation and drinking water, and growing global pollution of the atmosphere and oceans (even in the polar regions), including global climate disruption by carbon dioxide enrichment and other greenhouse gases (Rockström et al. 2009; MEA 2005). Obviously the exponential growth of human populations, recently surpassing 7 billion, is rapidly crowding out other species before we have begun to understand fully our dependence on species diversity.

Even more fundamentally, our planet's ability to provide an accommodating environment for humanity itself is being challenged by our own activities. The environment—our life-support system—is changing rapidly from the stable Holocene state of the last 12,000 years, during which we developed agriculture, villages, cities, and contemporary civilizations, to an unknown future state of significantly different conditions. We have entered what Paul Crutzen (Crutzen 2002) has identified as a whole new geologic era—the Anthropocene.

One way to address this challenge is to determine "safe boundaries" based on fundamental characteristics of our planet and to operate within them. "Boundaries" here mean specific points related to a global-scale environmental process beyond which humanity should not go. Identifying our planet's intrinsic, nonnegotiable limits is not easy, but recently a team of scientists has specified nine areas that are most in need of well-defined planetary boundaries (Rockström et al. 2009). These nine areas are (1) climate change, (2) biodiversity loss, (3) excess nitrogen and phosphorus production, (4) stratospheric ozone depletion, (5) ocean acidification, (6) global consumption of freshwater, (7) change in land use for agriculture, (8) air pollution, and (9) chemical pollution (Fig. 16.1). Johan Rockström and colleagues estimate that humanity has already transgressed three of these boundaries: climate change, biodiversity loss, and nitrogen production, with several others rapidly approaching the safe boundary.

Clearly, remedial policy responses to date have been local, partial, and inadequate. Early policy discussions and the resulting responses tended to focus on symptoms of environmental damage rather than basic causes, and policy instruments tend to be ad hoc rather than carefully designed for efficiency, fairness, and sustainability. For example, in the 1970s emphasis centered on end-of-pipe pollution which, while a serious problem, was actually a symptom of expanding populations and inefficient technologies that fueled exponential growth of material and energy throughput while threatening the recuperative powers of the planet's life-support systems.

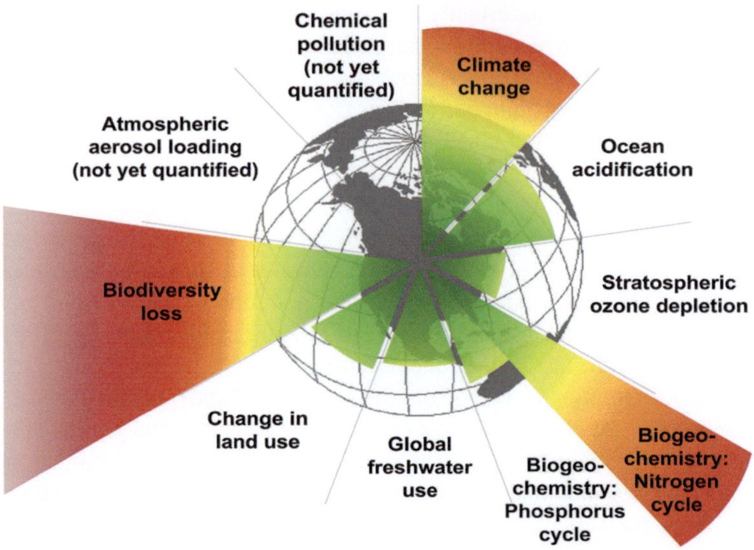

Fig. 16.1 Planetary Boundaries (Rockström et al. 2009; Steffen et al. 2011)

As a result of early perceptions of environmental damage, people learned a lot about policies and instruments for attacking pollution. These insights will help in dealing with the more fundamental and intractable environmental issues identified here.

The basic problems for which we need innovative policies and management instruments include:

- unsustainably large and growing human populations, as well as growing per capita consumption levels that are fast approaching, or already exceed, planetary boundaries;
- highly entropy-increasing technologies that deplete the earth of its resources and whose unassimilated wastes poison the air, water, and land; and
- land conversion that destroys habitat, increases soil erosion, and accelerates loss of species diversity, and which, coupled with resource extraction and waste emissions, decreases the ecosystem services that support humanity.

These problems are all evidence that the material scale of human activity is rapidly approaching, or already exceeds, the safe operating space for humanity on the earth.

We argue throughout this report that in addressing these problems we should adopt courses of action based on:

- recognition of the planetary boundaries the earth places on the type and scale of economic activity;

- fair distribution of resources and opportunities among groups within the present generation, between present and future generations, and between humans and other species; and
- economically efficient[1] allocation of resources that adequately accounts for protecting the stocks of natural and social capital.

Homo sapiens is at another turning point in its relatively long and (so far) inordinately successful history. Our species' activities on the planet have now reached such a scale that they are beginning to affect the ecological life-support system itself. The entire concept of economic growth (defined as increasing material consumption) must be rethought, especially as a solution to the growing host of interrelated social, economic, and environmental problems. What we need now is real economic and social development (qualitative improvement without growth in resource throughput) and an explicit recognition of the interrelatedness and interdependence of all aspects of life on the planet. We need to move from an economics that ignores this interdependence to one that acknowledges and builds on it. We need to develop an economics that is fundamentally "ecological" in the broadest sense and in its basic view of the problems that our species currently faces.

16.1.3 Growth in Material Consumption Beyond a Certain Point Is Undesirable: It Has Negative Effects on Well-Being and on Social and Natural Capital

There is a substantial body of new research on what actually contributes to human well-being and quality of life. While there is still much ongoing debate, this new science clearly demonstrates the limits of conventional economic income and consumption in contributing to well-being. For example, psychologist Tim Kasser, in his 2003 book *The High Price of Materialism* (Kasser 2002), points out that people who focus on material consumption as a path to well-being are actually less satisfied with their lives and even suffer higher rates of both physical and mental illness than those who do not focus so much on material consumption. Material consumption beyond real need is a form of psychological "junk food" that only satisfies for the moment and ultimately leads to depression, Kasser says.

[1] "Economically efficient" simply means that increasing marginal costs and diminishing marginal benefits from an activity are in balance. Marginal costs and benefits should be measured in terms of contributions to the sustainable welfare of humans and other species. Precise measurement of these contributions is not currently possible. Conventional economists emphasize purely monetary costs and benefits, which are determined by willingness to pay, and hence fail to reflect costs and benefits for those with limited purchasing power. Under these conditions, an efficient allocation is one that maximizes monetary value. While measurements may be fairly precise, this narrow goal is inappropriate.

Economist Richard Easterlin has shown that well-being tends to correlate well with health, level of education, and marital status and shows sharply diminishing returns to income beyond a fairly low threshold. He concludes (Easterlin 2003) that

> people make decisions assuming that more income, comfort, and positional goods will make them happier, failing to recognize that hedonic adaptation and social comparison will come into play, raise their aspirations to about the same extent as their actual gains, and leave them feeling no happier than before. As a result, most individuals spend a disproportionate amount of their lives working to make money, and sacrifice family life and health, domains in which aspirations remain fairly constant as actual circumstances change, and where the attainment of one's goals has a more lasting impact on happiness. Hence, a reallocation of time in favor of family life and health would, on average, increase individual happiness.

British economist Richard Layard synthesizes many of these ideas and concludes that current economic policies are not improving well-being and happiness and that "happiness should become the goal of policy, and the progress of national happiness should be measured and analyzed as closely as the growth of GNP (gross national product)" (Layard 2005).

Economist Robert Frank, in his book *Luxury Fever* (Frank 1999), also concludes that some nations would be better off—that is, overall national well-being would be higher—if we actually consumed less and spent more time with family and friends, working for our communities, maintaining our physical and mental health, and enjoying nature.

On this last point, there is substantial and growing evidence that natural systems contribute heavily to human well-being. In a paper published in the journal *Nature* (Costanza et al. 1997), the annual, nonmarket value of the earth's ecosystem services was estimated to be substantially larger than global GDP. This estimate was admittedly a rough first cut, but the goal of this paper was to stimulate interest and research on the topic of natural capital and ecosystem services. It has certainly had that effect. The paper is one of the most highly cited in the ecology/environment area in the last 15 years and it has stimulated a huge amount of discussion, research, and policy follow-up. For example, the UN Millennium Ecosystem Assessment (MEA 2005) was a global update and compendium of ecosystem services and their contributions to human well-being. The Economics of Ecosystems and Biodiversity (TEEB) Synthesis report (Sukhdev and Kumar 2010) is a more recent contribution to this rapidly increasing field of study and policy. The World Bank has recently announced its Wealth Accounting and Valuation of Ecosystem Services (WAVES) project. The new Intergovernmental Platform on Biodiversity and Ecosystem Services (IPBES) is also in the formation stages (www.ipbes.net). Finally, the recently established Ecosystem Services Partnership (ESP) is a global effort to coordinate the thousands of researchers and practitioners around this topic (www.es-partnership.org).

So, if we want to assess the "real" economy—all the things that contribute to real, sustainable, human well-being—as opposed to only the "market" economy, we have to measure and include the non-marketed contributions to human well-being from nature; from family, friends, and other social relationships at many scales; and from health and education. What does such a more comprehensive, integrative definition of well-being and quality of life look like?

16.1.3.1 An Integrative Definition of Quality of Life and Well-Being[2]

When we evaluate the state of human affairs or propose policies to improve it, we typically proceed from assumptions about the characteristics of a good life and strategies for achieving them. We might suppose, for example, that access to particular resources is a part of a good life and, therefore, that increasing economic production per-capita is an appropriate goal. Unfortunately, our underlying assumptions are rarely tested and established. We therefore need a more basic approach to defining well-being or quality of life (QOL) that, in turn, can guide our efforts to improve humans' experience. Examinations of QOL often fall under two headings:

1. So-called "objective" indicators of QOL include, for example, indices of economic production (i.e., GDP), literacy rates, life expectancy, and other data that can be gathered without a subjective evaluation being made by the individual being assessed (although, of course, we must acknowledge that subjective judgments of the researcher are involved in the process of defining and gathering "objective" measures as seen in the case, for example, of selecting a proxy for "literacy"). Objective indicators may be used singly or in combination to form summary indexes, as in the UN's Human Development Index (HDI) (United Nations Development Programme 1998), the Index of Sustainable Economic Welfare, or Genuine Progress Indicator. To the extent that such a measure can be shown to be valid and reliable across assessment contexts (admittedly a difficult task), these relatively objective measures may help us gather standardized data that are less vulnerable to social comparison and local adaptation. For example, a valid measure should minimize the degree to which QOL is largely a function of comparing one's life to others' in one's locale, in the media, or some other narrowly construed group; a person's QOL should not be considered high simply because others in the locale are more miserable.
2. Subjective indicators of QOL gain their impetus, in part, from the observation that many objective indicators merely assess the opportunities that individuals have to improve QOL rather than assessing QOL itself. Thus economic production may best be seen as a *means* to a potentially (but not necessarily) improved QOL rather than an end in itself. In addition, unlike most objective measures of QOL, subjective measures typically rely on survey or interview tools to gather respondents' own assessments of their lived experiences in the form of self-reports of satisfaction, happiness, well-being, or some other near-synonym. Rather than presume the importance of various life domains (e.g., life expec-

[2]Much of this section is taken from reference 19. Costanza et al. (2007a, b, c) Quality of life: An approach integrating opportunities, human needs, and subjective well-being. Ecological Economics 61: 267–276.

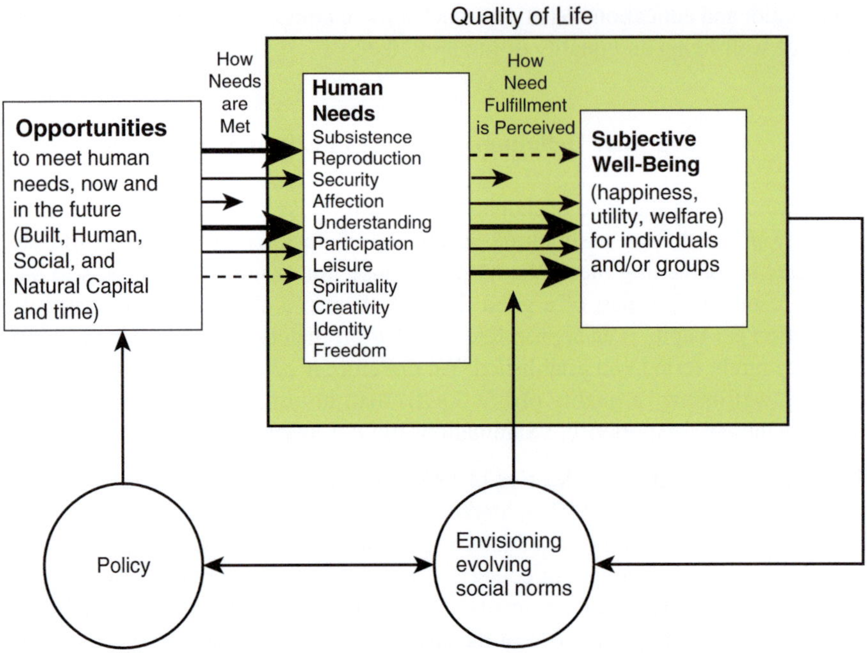

Fig. 16.2 Quality of Life (QOL) as the interaction of human needs and the subjective perception of their fulfillment, as mediated by the opportunities available to meet the needs (Costanza et al. 2007)

tancy or material goods), subjective measures can also tap the perceived significance of the domain (or "need") to the respondent. Diener and Suh provide convincing evidence that subjective indicators are valid measures of what people perceive to be important to their happiness and well-being (Diener and Suh 2003). Nevertheless, there are individuals who cannot provide subjective reports or whose subjective reports may not be as trustworthy in reflecting their true welfare because of the internalization of cultural norms (Nussbaum and Glover 1995), mental illness, lack of information, or other reasons.

What seems best, then, is to attempt an approach to QOL that combines objective and subjective approaches. Our integrative definition of QOL is as follows: QOL is the extent to which objective human needs are fulfilled in relation to personal or group perceptions of subjective well-being (Fig. 16.2). Human needs are basic needs for subsistence, reproduction, security, affection, etc. (see Table 16.1 and below). SWB is assessed by individuals' or groups' responses to questions about happiness, life satisfaction, utility, or welfare. The relation between specific human needs and perceived satisfaction with each of them can be affected by mental capacity, cultural context, information, education, temperament, and the like, often in quite complex ways. Moreover, the relation between the fulfillment of human needs and overall subjective well-being is affected by the (time-varying) weights individuals, groups, and cultures give to fulfilling each of the human needs relative to the others.

With this definition, the role of policy is to create opportunities for human needs to be met, understanding that there exists a diversity of ways to meet any particular need. Built, human, social, and natural capital s represent one way of categorizing those opportunities. Time is also an independent constraint on the achievement of human needs.

Social norms affect both the weights given to various human needs when aggregating them to overall individual or social assessments of SWB, and also policy decisions about social investments in improving opportunities. Social norms evolve over time due to collective population behavior (Azar 2004). The evolution of social norms can be affected by conscious shared envisioning of preferred states of the world (Costanza 2000a).

As we said, one convenient way to summarize the opportunities for meeting human needs is to group them into four basic types of assets or "capital" that are necessary to support the real, human-well-being-producing economy: built capital, human capital, social capital, and natural capital.

We refer to these assets as "capital" in the sense of a stock or accumulation or heritage—a patrimony received from the past and contributing to the welfare of the present and future. Clearly our use of the term "capital" is much broader than that associated with capitalism. These assets, which overlap and interact in complex ways to produce all benefits, are generally defined as follows:

- ***Natural capital:*** The natural environment and its biodiversity. Among other things, natural capital is needed to provide ecosystem goods and services. These goods and services are essential to basic human needs such as survival, climate regulation, habitat for other species, water supply, food, fiber, fuel, recreation, cultural amenities, and the s required for all economic production.
- ***Social and cultural capital:*** The web of interpersonal connections, social networks, cultural heritage, traditional knowledge, and trust, and the institutional arrangements, rules, norms, and values that facilitate human interactions and cooperation between people. These contribute to social cohesion; strong, vibrant, and secure communities; and good governance, and help fulfill basic human needs such as participation, affection, and a sense of belonging.
- ***Human capital:*** Human beings and their attributes, including physical and mental health, knowledge, and other capacities that enable people to be productive members of society. This involves the balanced use of time to fulfill basic human needs such as satisfying employment, spirituality, understanding, skills development, creativity, and freedom.
- ***Built capital***: Buildings, machinery, transportation infrastructure, and all other human artifacts and services that fulfill basic human needs such as shelter, subsistence, mobility, and communications.

We recognise that human, social, and produced assets depend entirely on the natural world, and that natural capital is therefore ultimately non-substitutable.

Sustainability therefore requires that we live off the interest (sustainable yields) generated by natural capital without depleting the capital itself.

To think of nature, the biosphere, the earth as a form of capital is a way of recognizing its importance to the economy, an importance that is often overlooked. Ecological economics understands economies as embedded in cultures and societies, which are embedded in the geobiosphere. This means that economies rely on the geobiosphere to provide materials and energy and accommodate all the wastes that economic activity inevitably produces. Natural capital is similar to built capital (buildings, machines, infrastructure, warehouses) in that it provides goods (e.g., minerals, fossil fuels) and services (e.g., pollination, flood control) without which economies could not function.

In speaking of "natural capital" we are using the term "capital" in its physical, not financial sense, e.g., a carpenter's stock of tools or a factory assembly line. A herd of livestock is a capital stock that yields a flow of new members. The physical herd converts grass, water, etc., into new animals. The net increment is income or sustainable yield. The constant herd is capital, reproducing stock. This is a physical stock-flow relation independent of financial arrangements. Indeed the word "capital" derives from "capitas," the number of heads the herdsman has in his livestock. Similar stock-flow relationships hold for forests, fisheries, and other populations. The problems arise when the physical descriptive term "natural capital" is converted into financial monetary terms, and especially when natural growth rates are converted into monetary yields of different physical stocks, and then compared to the rate of interest on a stock of money in the bank. But reasonable rejection of financialization of nature should not keep us from recognizing the physical importance of natural capital as a stock that yields desired flows.

But natural capital is also very different from built capital. First of all, built capital is made from natural capital. In other words, nature can exist without built capital, but built capital cannot exist without nature. There is an essential hierarchy limiting the extent to which built capital can substitute for natural capital, and they are better thought of complements than substitutes.

Second, built capital represents a"fund" that provides a"service," as, for example, a lathe provides a service when it is used to shape wood. The lathe does not end up embodied in the wood. Natural capital can also be a fund that provides services, such as when a forest provides habitat for forest creatures. But natural capital can also be a stock out of which a supply of material flows. So the forest that provides habitat as a fund-service is also a stock of trees that supplies a flow of wood (the very wood used on the lathe.) Services do not deplete funds. Flows do deplete stocks, which can however be regenerated if renewable. Since materials flowing from natural capital are usually sold through markets, and ecosystem services often are not, there is an ever-present tendency to overuse natural capital for the flows it can provide to the detriment of its capacity to provide services.

A third and more profound reason for differentiating between natural and built capital is that built capital is simply an object for the benefit of humans. That is why it exists. When built capital no long provides a useful service, it is demolished. Nature, of which humans are an integral part, is much more than that. Nature is populated by countless species, many of whom are sentient, experience a range of emotions, learn, and live in communities of their own making. Reverence for all life acknowledges that the rest of nature has rights and that a fair distribution of resources needs to acknowledge those rights. Thus, thinking of built capital and natural capital as substitutes is not appropriate, as a common designation of both of them as forms of capital might otherwise suggest.

With these caveats in mind, we employ the concept of natural capital in this report cognizant of its limitations (Third World Network 2012).

16.1.3.2 Are We Making Progress?

Given this definition of well-being and quality of life, are we really making progress? Is the mainstream economic model really working, even in the developed countries? One way to tell is through surveys of people's life satisfaction, which have been relatively flat in the United States and many other developed countries since about 1975, in spite of a near doubling in per capita income (Fig. 16.3) (Hernández-Murillo and Martinek 2010).

A second approach is an aggregate measure of the real economy that has been developed as an alternative to GDP called the Index of Sustainable Economic Well-Being (ISEW) or a variation called the Genuine Progress Indicator (GPI).

Let's first take a quick look at the problems with GDP as a measure of true human well-being. GDP is not only limited—measuring only marketed economic activity or gross income—it also counts all of this activity as positive. It does not separate desirable, well-being-enhancing activity from undesirable, well-being-reducing activity. For example, an oil spill increases GDP because someone has to clean it up, but it obviously detracts from society's well-being. From the perspective of GDP, more crime, more sickness, more war, more pollution, more fires, storms, and pestilence are all potentially good things, because they can increase marketed activity in the economy.

GDP also leaves out many things that *do* enhance well-being but are outside the market. For example, the unpaid work of parents caring for their own children at home does not show up; but if these same parents decide to work outside the home to pay for childcare, GDP suddenly increases. The nonmarketed work of natural capital in providing clean air and water, food, natural resources, and other ecosystem services does not adequately show up in GDP either; but if those services are damaged and we have to pay to fix or replace them, then GDP suddenly increases. Finally, GDP takes no account of the distribution of income among individuals. But it is well known that an additional dollar of income produces more well-being if one is poor rather than rich. In fact, GDP is maximized by allocating resources to those with the greatest willingness to pay. In a highly unequal society, a rich person may be willing to pay more for drinking water to flush their toilets than a poor family can pay to prevent a child from dying of dysentery. It is also clear that a highly skewed income distribution has negative effects on a society's social capital.

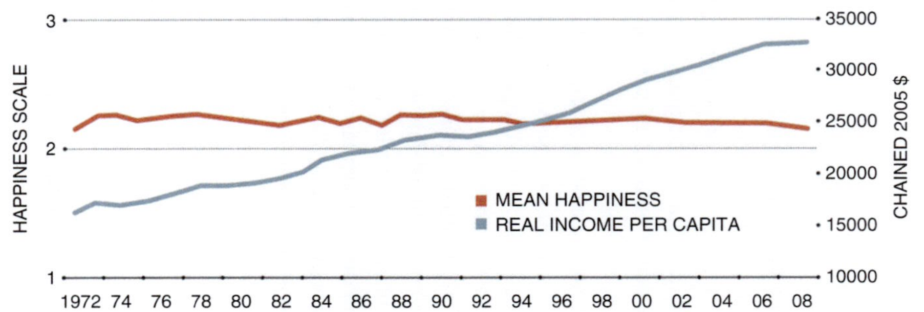

Fig. 16.3 Happiness and Real Income in the United States, 1972–2008. NOTE: Mean happiness (left scale) is the average reply from respondents to the U.S. General Social Survey. The survey question asks: "Taken all together, how would you say things are these days? Would you say that you are not too happy, pretty happy or very happy?" These values were coded as 1, 2 and 3, respectively (Hernández-Murillo and Martinek 2010)

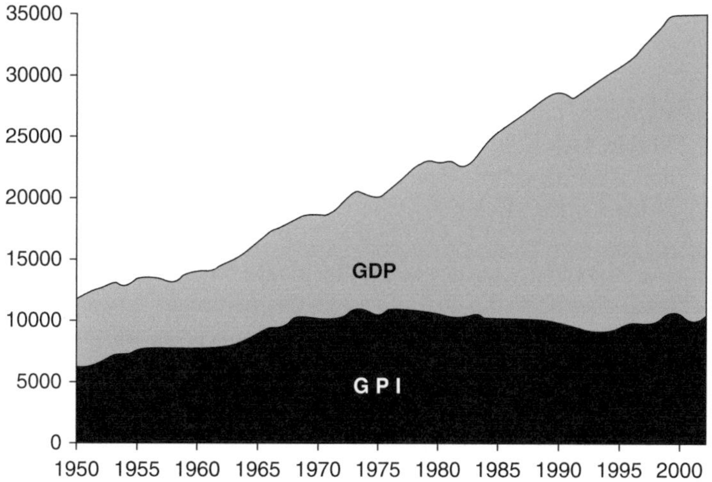

Fig. 16.4 GDP (Gross Domestic Product) and GPI (Genuine Progress Indicator) for the U.S. from 1950 to 2005) (Talberth et al. 2007)

The GPI addresses these problems by separating the positive from the negative components of marketed economic activity, adding in estimates of the value of non-marketed goods and services provided by natural, human, and social capital, and adjusting for income-distribution effects. While the measure is by no means a perfect representation of the real well-being of nations, GPI is a much better approximation than GDP. As many have noted, it is much better to be approximately right in these measures than precisely wrong.

Comparing GDP and GPI for the United States Fig. 16.4 shows that, while GDP has steadily increased since 1950, with the occasional dip or recession, GPI peaked in about 1975 and has been flat or gradually decreasing ever since (Talberth et al. 2007). From the perspective of the real economy, as opposed to just the market economy, the United States has been in recession since 1975. As already mentioned, this picture is also consistent with survey-based research on people's stated life-satisfaction. The United States and several other developed countries are now in a period of what Herman Daly has called "uneconomic growth," where further growth in marketed economic activity (GDP) is actually reducing well-being, on balance, rather than enhancing it. In terms of the four capital s, while built and some aspects of human capital have grown, social and natural capital have declined or remained constant, more than canceling out the gains in built and human capital.

GPI is certainly not the perfect indicator of well-being or quality of life (QOL) and there are several other alternatives under active discussion (Costanza et al. 2009; Stiglitz et al. 2010). As we discussed earlier, QOL is a complex interaction of objective and subjective factors and the relationships among them, and sustainable human well-being is an active area of research. Nevertheless, GPI is certainly a better approximation to the objective elements of well-being than GDP, a function for which GDP was never designed. In addition, GPI data for the United States and other countries seem to match subjective well-being surveys much better than income or GDP data.

16.1.3.3 Viable Alternatives Exist That Are Both Sustainable and Desirable, but They Require a Fundamental Redesign of the Entire "Regime"

A new model of the economy consistent with our new full-world context (Table 16.1) would be based clearly on the goal of sustainable human well-being. It would use measures of progress that clearly acknowledge this goal (e.g., GPI instead of GDP). It would acknowledge the importance of ecological sustainability, social fairness, and real economic efficiency.

Ecological sustainability implies recognizing that natural and social capital s are not infinitely substitutable by built and human capital and that real biophysical limits and planetary boundaries exist to the expansion of the market economy. Climate change is perhaps the most obvious and compelling of these limits.

Social fairness implies recognizing that the distribution of wealth is an important determinant of social capital and quality of life. The conventional economic model, while explicitly aimed at reducing poverty, has bought into the assumption that the best way to do this is through growth in GDP. This has not proved to be the case, and explicit attention to distribution issues is sorely needed. As Robert Frank has argued (Frank 2007), economic growth beyond a certain point sets up a "positional arms race" that changes the consumption context and forces everyone to consume too much of positional goods (like houses and cars) at the expense of nonmarketed, nonpositional goods and services from natural and social capital. Increasing inequality of income actually reduces overall societal well-being, not just for the poor but across the income spectrum. Wilkinson and Pickett (2009) have produced empirical data that show a strong correlation between income inequality in OECD countries and a whole range of health and social problems. Large income inequality is as detrimental to the well-being of the rich as to the poor.

Real economic efficiency implies including all resources that affect sustainable human well-being in the allocation and management system. Our current market-focused allocation system excludes most non-marketed natural and social capital assets and services that are huge contributors to human well-being. The current economic model ignores this and therefore does not achieve real economic efficiency. A new, sustainable model would measure and include the contributions of natural and social capital in ways that go well beyond the market. This would better approximate real economic efficiency.

The new model would also acknowledge that a complex set of property rights regimes is necessary to adequately manage the full range of resources that contribute to human well-being. For example, most natural and social capital assets are part of the commons. Making them private property does not work well. When a resource is non-rival (meaning that use by one person does not leave less for others to use), then market prices will ration access to those who can afford to pay, even though additional use incurs no additional costs. The clearest example of this is information. In fact, for information that protects the environment or provides other social benefits—for example, an inexpensive, carbon-free energy technology—additional use actually reduces social costs. The value of such resources is paradoxically maximized at a price of zero (or less). Since the private sector will not provide products for free, the public sector must be responsible for their protection and provision. On the other hand, when resources are rival, meaning that use by one person leaves less for others, leaving them as open-access resources (with no property rights) does not work well either. What is needed is a third way to *propertize* these resources without privatizing them. Several new (and old) common-property-rights systems have been proposed to achieve this goal, including various forms of common-property trusts. These are described in detail later in this report.

The role of government also needs to be reinvented. In addition to government's role in regulating and policing the private market economy, it has a significant role to play in expanding the commons sector, which can propertize and manage non-marketed natural and social capital assets. It can also help develop new common-ownership models at various levels of scale that are not driven by growth principles, and can play a planning and coordinating role to help manage a reduced-growth regime (Alperovitz 2011). Government also has a major role to play in facilitating societal development of a shared vision of what a sustainable and desirable future would look like. As Tom Prugh and colleagues (Prugh et al. 2000) have argued, a strong democracy, based on developing a shared vision, is an essential prerequisite to building a sustainable and desirable future.

One way to look at our goals for the new economy is shown in (Fig. 16.5). This figure combines planetary boundaries (Fig. 16.1) as the "environmental ceiling" with basic human needs as the "social foundation" (Raworth 2012). This creates an environmentally sustainable and socially desirable and just "doughnut" as the space within which humanity can thrive.

In the remainder of this report we more fully develop these ideas, beginning with a vision of what such a sustainable and desirable society living within the doughnut could look like.

16.2 What Would a Sustainable and Desirable Economy-in-Society-in-Nature Look Like?

The most critical task facing humanity today is the creation of a shared vision of a sustainable and desirable society, one that can provide permanent prosperity within the biophysical constraints of the real world in a way that is fair and equitable to all of humanity, to other species, and to future generations. Recent work with businesses and communities indicates that creating a shared vision is the most effective engine for change in the desired direction (Costanza 2000b).

Key Points

- To better articulate and communicate the goal, we need to envision the resulting society and how the pieces might fit together.

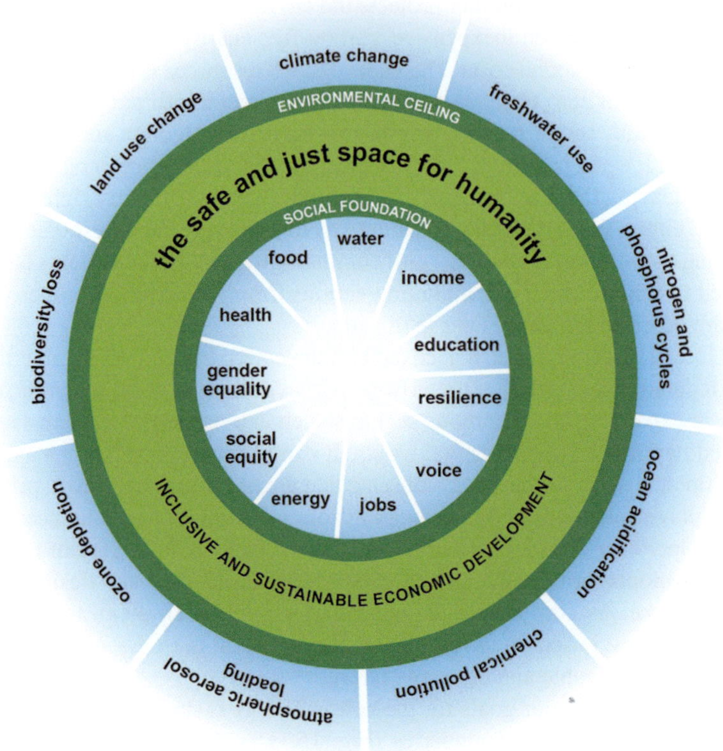

Fig. 16.5 A safe and just space for humanity—the sustainable and desirable doughnut (Raworth 2012)

In the previous sections we have sketched out the general characteristics of this world and how it differs from our current society: it is ecologically sustainable, fair, efficient, and secure. Here we put all the policies together and develop the implications for the whole system. We need to fill in the details in a coherent vision that is tangible enough to motivate all kinds of people to work toward achieving it. Without a coherent, relatively detailed, shared vision of what a sustainable society could look like, there will be no political will nor united effort to take us from here to there. The default vision of continued, unlimited increases in material consumption is inherently unsustainable and undesirable, as we have pointed out, but we cannot break away from this vision until a credible and widely shared alternative is created.

Below we sketch out one version of such a vision as a starting point.[3] There are several other visioning exercises that have created similar descriptions, including the Great Transition Initiative (www.gtinitiative.org) and the Future We Want

[3]This vision is adapted from one created at a workshop held at Oberlin College in January 2001.

(www.futurewewant.org). Ultimately, this vision must be shared and further developed through participatory democratic processes.

If humanity is to achieve a sustainable and desirable future, we must create a shared vision detailing what we as a society want to sustain and incorporating the central shared values that express our hopes for the future. This vision must incorporate a diversity of perspectives and be based on principles of fairness, respect, and sustainability.

This draft vision is divided into five parts: (1) worldviews, (2) built capital, (3) human capital, (4) social capital, and (5) natural capital, encompassing the basic elements of the ecological economics framework. This vision is written from the perspective of the year 2050, describing the world we have achieved by implementing the policies outlined in previous sections.

16.2.1 Worldview

Our worldview no longer divides the planet into "humans vs. nature." People now recognize that humans are a part of nature, one species among many, and must obey the laws and constraints imposed on all of nature. Nevertheless, humans bear responsibility that other creatures do not—we don't blame deer for overgrazing—yet we expect humans to recognize they're "overgrazing" and stop it. We recognize that nature is not something to be subjugated, but instead is something we depend upon absolutely to meet physical, psychological, cultural, and spiritual needs. We recognize that natural resources are scarce and must be invested in. Our goal is to create conditions conducive to life in the broadest sense.

For centuries the worldview of mechanistic physics dominated Western society. Within this worldview, each action has an equal and opposite reaction, and only by studying systems at smaller and smaller scales can we come to fully understand these reactions. As more and more people have come to understand the inherent complexity of ecosystems and human systems, we have come to realize that results cannot always be predicted and that irreducible uncertainty dominates the provision of life-support services by healthy ecosystems.

An ecological worldview of complexity and indeterminacy, inspired by nature as mentor—holistic, integrated, and flexible—has replaced the worldview of mechanistic physics. Unfettered individualism is appropriate and even necessary in a world of vast frontiers and unlimited elbowroom. Individualism is still extremely important in 2050, but is far more tempered by a concern for the common good. This has led to a system where communities promote individual liberty as long as individual actions do not have a negative impact on the community. Individuals in return accept that they are a part of society, and it is unfair and illegal (even uneconomic) to impose costs on society for private gain. This attitude was necessary to wean ourselves of our dependence on heavily polluting single-occupancy vehicles, for example.

Further, ever-increasing consumption is no longer considered an integral component of human needs as it was in the early part of the century. People pay attention to their other needs and desires, such as joy, beauty, affection, participation, creativity, freedom, and understanding. Building strong community helps us meet these needs, while working ever harder to pay for more consumption deprives us of the time and energy required to fulfill them. Thus, status is not conferred by high incomes and high consumption (individual ends), but rather by contribution to civil society and community ends. With the recognition that consumption beyond limit is not only physically unsustainable but also does little to improve our quality of life, we now understand that a "steady-state" economy—prosperous but within planetary boundaries—is our goal. A steady-state economy does not mean an end to development; it simply means that we limit the input of raw materials into our economic system and their inevitable return to the ecosystem as waste to a level compatible with the ecological constraints imposed by a finite planet with finite resources. We now live happily and well within the safe operating space of our planet. We do not know the precise location of these planetary boundaries, and they are subject to change. Therefore, "adaptive management" has become the guiding principle.

The economy is now powered by our incoming solar energy—direct sunlight captured by solar panels—as well as wind, hydro, and the traditional forms of solar energy capture (agriculture, forestry, and fisheries). Economic production now focuses on quality, not quantity, on everyone having enough, and on fulfilling employment. Rather than the earlier focus on the production of goods, we now focus on the production of the services provided by goods and how those services are distributed. We do not need cars, we need transportation. We do not need televisions, we need entertainment and information. Goods are only a means to an end—the larger end of sustainable human well-being—and by recognizing this our economy has developed as never before without growing in physical terms.

16.2.2 Built Capital

Built capital is the human-made infrastructure used to meet human needs. Technological advance over the last century has had a large impact on the type of built capital we find in 2050. Different priorities have had as much or even greater impact.

Housing Communities have been dramatically redesigned to integrate living space, community space, and workspace with recreational needs and nature. Workspace includes the stores that supply our everyday needs as well as production facilities for most of the goods those stores supply. People now live very close to where they work, where they shop, and where they play. The huge cities of the early twenty-first century did not disappear, but they have been dramatically reorganized. Cities are now aggregations of smaller communities in close physical proximity but where each community meets the housing, employment, social, recreation, and shopping needs of those who live there. The "20-min neighborhood" idea—that all basic services should be no more than a 20-min walk away—has taken hold as an urban design principle. Natural areas have also made a big comeback in cities. The specifics of community size and design are, of course, determined by local physical and cultural conditions, and there is enormous diversity.

In addition to these very practical aspects, communities have been designed as soul-satisfying spaces that resonate with our evolutionary history. Most communities include natural areas and incorporate parks and other green spaces (though "green" is a misnomer in drier parts of the world, where xeriscaping is the norm), and such spaces also serve as common space for community members. They also foster social interaction and community. Rather than something new, this is simply a resurgence of a millennial tradition of settlement patterns.

Because community space is abundant and well designed, private homes are generally smaller (hence cheaper and easier to care for) and are much more energy efficient. Private lawns have virtually disappeared, though lawn-like community green spaces still exist, and private gardens abound. Private gardens in fact meet a substantial portion of community food needs. Walking and bicycle riding have effectively become the dominant forms of transportation, except in the worst weather. Rapidly increasing energy costs provided the initial incentive, but people then discovered the enormous benefits of such pedestrian communities.

One of the biggest impacts was simply getting people out of their cars. Walking to work, to the store, to community meeting places, or to nature preserves brings people into direct contact with the other members of the community. People walking together in the same direction naturally converse, establishing friendships, informing each other of current events, and discussing issues of relevance to the community. In fact, developing community and social capital has become one of many explicit goals for designing built capital. Modern communities are very healthy places for humans and other species. Invigorating exercise and nurturing social interaction have replaced the stress of hour-long commutes, road rage, and the pollution of vehicle exhaust, improving both physical and mental health. Air quality is very high. Many roads and parking lots have become redundant, and in their spaces stand parks, streams, and greenways, providing clean air, clean water, and healthy recreation, among numerous other vital ecosystem services. The dramatic reduction in impervious areas has reduced flooding and allowed the land and the ecosystems it sustains to filter water, restoring waterways to health.

With scarcer resources, the practice of destroying still useful buildings to build others on the same site has diminished, and stable populations have further decreased the need for new construction. But from time to time new buildings are still required. Ecologically designed "living buildings" have become the norm for new construction.

Transportation As already mentioned in the description of communities, single-occupancy vehicles are now rare. The dominant modes of transportation within communities are walking and bicycling; between communities people use high-speed rail. Public transportation is important within communities and is designed to transport goods as well as passengers, making it convenient for grocery shopping and the like. Because so many people use public transportation, it is abundant and extremely convenient. Rail is common, but so are electric buses and taxis. "Traffic" is a thing of the past, and public transportation gets people around much more quickly than private vehicles used to, at a fraction of the cost. Dramatically fewer vehicles on the roads has also cut maintenance costs to a fraction of what they were, and new roads are unnecessary. Some people still own private vehicles, but these vehicles are expensive and their owners pay a higher share of costs of road-maintenance coasts. Most communities have electric cars, such as ZipCars, available for rent when private transportation is absolutely required. When not being driven, these cars provide electric energy storage.

Energy Renewable resources now meet virtually all of the world's energy needs. The conversion from hydrocarbons was facilitated by continuous increases in efficiency of energy use, combined with appropriate full-cost pricing of all energy sources, including environmental and health costs and risks of the full fuel cycle. Photovoltaic tiles are ubiquitous roofing materials, and roofs alone meet over half the world's energy needs. Large-scale hydropower has decreased in importance as more and more rivers are restored to their natural states, but low-impact mini-turbines are increasingly common. In spite of the abundance of nonrenewable, non-polluting forms of energy, energy-efficiency research is still very important and advances are still being made in both renewable-energy supply and demand management. The "smart grid" has done much to help this transition. In many places municipalities and/or cooperatives now locally manage the generation, supply, and distribution of renewable energy resources, keeping prices affordable and ownership democratically controlled.

Industry Industry has changed dramatically. Industrial design is now based on closed-loop systems in imitation of nature, where the waste product from one industry becomes the feedstock of the next. Wasted heat from industrial processes is used to heat nearby homes and workspaces. When possible, industrial production uses local materials to meet local needs, and wastes (the few that are not put to use) are processed locally. Most smaller-scale industries consist of a mix of locally owned proprietary firms and smaller corporations on the one hand, and cooperatives and new community-based commons institutions on the other (Alperovitz 2011). While these characteristics do not always maximize productive efficiency, the benefits outweigh the costs.

First, local production dramatically reduces transportation costs, helping to compensate for sometimes-higher production costs. Second, it makes communities directly aware of the environmental impacts of production and consumption. Costs of waste disposal are not shifted elsewhere. Third, industries are more a part of their communities. Most of them are locally owned by the workers they employ, by new cooperative and municipal institutions, and by the people whose needs they meet.

Rather than simply trying to maximize returns to shareholders, industries strive to provide healthy, safe, secure, and fulfilling working conditions for workers.

Those who produce goods and those who consume them know each other, so workers take particular pride in the quality of what they produce.

Fourth, the decentralization of the economy means that the economy as a whole is much less susceptible to business cycles, increasing job and community stability—a central requirement of local sustainability planning in general. Fifth, an emphasis on local ownership and production for local markets has reduced the importance of trade secrets and patents; competition has been replaced to some extent by cooperation.

Sixth, a significant number of larger firms are structured as public and quasi-public enterprises jointly owned with the workers involved. They are designed, on the one hand, to help target and anchor jobs to help achieve local stability, thereby also supporting sustainability planning, and on the other, to be less dependent on very short-term profit considerations necessary to meet stock market expectations that foster excessive growth.

Finally, decreased competition has led to a dramatic decrease in the size of the advertising industry. This means that money once spent on convincing people to buy one brand over another is now spent on making those products better—or simply not spent, making those products more affordable.

Markets and competition, of course, still play an important role. Industries are free to sell to distant communities, though having to pay the full cost of transportation provides a natural barrier. Still, this threat of competition means that communities need not rely solely on the good will of local industries to keep prices low. Trade secrets play less of a role in competition than in the past due to the resurgence of sharing information. The development of open-source software shows that freely sharing knowledge can lead to more rapid technological innovation than the profit motive provided by privatizing knowledge through patents. The problems with patents have became more obvious with the tremendous growth in green technologies, which have proven themselves capable of slowing climate change, reducing pollution, and decreasing demands on scarce ecosystem resources, but only by being used on a large scale. Patents on these technologies (and the accompanying monopoly profits) would mean that much of the world would be unable to afford them. The global community has come to realize that it cannot afford the price of people not using these technologies.

Fortunately, the free flow of information has led to impressive new innovations, often making patents obsolete. Some industries retain substantial economies of scale, using fewer resources per unit when producing in enormous factories. This is still the case for solar cells, for example. Large corporations still exist to produce such goods, but many are structured in ways that broaden representation on boards and in certain cases entail public ownership or joint public/worker ownership. Corporate charters have largely changed to the "benefit corporation" model that explicitly acknowledges a firm's responsibility to produce a social benefit rather than merely a private profit.

16.2.3 Human Capital

Human capital was defined in the early part of the century as the practical knowledge, acquired skills, and learned abilities of an individual that make him or her potentially productive and thus equip him or her to earn income in exchange for labor.

The definition of human capital itself has changed—no longer emphasizing solely productivity in terms of income exchanged for labor. The primary emphasis instead is now on knowledge, skills, and abilities that make people productive members of society. The goals of society are far more than simply earning income. Education is now integrated into everyday life, not simply something we do for a few hours a day before we grow up. And it is not always confined to classrooms—schools are an institution, not a physical place. Nature offers us an amazing laboratory every time we step outside, and is valued every bit as much in urban settings as in rural. This is even truer in 2050, when our communities are designed to maximize exposure to healthy ecosystems. Education about civic responsibilities and roles is heavily stressed, and such topics are taught by direct exposure to the decision-making process or hands-on participation in activities that benefit the community. Youth are schooled in civic responsibility by actively participating in the community. And what better place to learn skills required for economic production than at the workplace? Apprenticeships are now an integral part of the learning process. Technology also plays an important role in education. Online learning environments are used where appropriate but by no means replace direct interaction. Education is now an interactive balance between online tools and content acquisition, and on-the-ground problem solving in the community.

Education and science no longer focus solely on the reductionist approach, in which students are only taught to analyze problems by breaking them down into their component parts. While the reductionist approach and analysis still play an important role in education, the emphasis is now on synthesis—how to rebuild the analyzed components of a problem into a holistic picture to solve problems. Synthesis is critical for understanding system processes, and system processes dominate our lives.

Beyond analysis and synthesis, learning also now emphasizes communication. Researchers skilled at communication are able to more readily share ideas, and ideas grow through sharing. Workers skilled at communication are able to work together to solve production problems. Citizens skilled at communication are able to contribute to the ever-evolving vision of a sustainable and desirable future that is the motivating force behind policy and governance. Citizens are also able to communicate their knowledge with each other, so that education, livelihood, family, and community become a seamless whole of lifelong learning and teaching, everyone simultaneously a student and teacher.

Education also now emphasizes much more than just scientific understanding of the material world. Critical thinking and research are important, but so are creative expression and curiosity. Knowledge and science are not portrayed as value-neutral endeavors; students now learn that the very decision of what to study is a moral choice with broad implications for society. The goal of education is to cultivate wisdom and discernment, to cultivate the emotional maturity to allow responsible decision-making in every type of human endeavor.

The whole notion of work has also changed, and the word itself has lost the connotation of an unpleasant chore. Work hours have been reduced through work sharing and more generous leave policies to allow for a more reasonable balance of family and work life. Moreover, people now recognize the absurdity of applying technology to the problem of producing more goods to be consumed during leisure time regardless of the drudgery involved in the production process itself. Instead, to recruit the needed workers, industry is now forced to redirect some of its technological prowess toward making work itself a pleasurable part of our days that engages both mental and physical skills. A typical job now involves far more variety, not only to make work more exciting and interesting, but also to take advantage of the full range of a person's skills. There is less distinction between what would have earlier been considered gainful employment and volunteer work.

Everyone participates in civil society, both in decision-making and in maintaining the public space. This is not an onerous chore, but a pleasurable time for socializing with neighbors and community. Nor does it take time away from private lives, since the typical work week in traditional jobs now averages only 15 h. Education deemphasizes the old "more is better" mindset and promotes a greater understanding of the linkages between economic production, nature, human development, and society. This has made people more aware of the true costs of excessive consumption.

With years of technological advance and diminished "needs," society is now able to provide a satisfactory living wage to all who work and to meet the basic needs of those who do not. Participation in the various types of work is expected and supported, but not forced. Because work is now more a fulfilling experience than an onerous necessity, there is little resentment of those who do not work but rather a feeling of concern that these people are not developing their potential as humans. Living in more tightly knit communities where social goals are actively discussed, people now better understand the importance of their work and feel greater obligation to contribute to the common good. Remuneration for work has been restructured to provide the greatest awards to those who provide the greatest amount of service to the community, such as teachers, childcare providers, and so on.

Human capital is also directly related to human populations. The population has stabilized at a level compatible with the safe operating space of our planet.

16.2.4 Social Capital

Social capital refers to the institutions, relationships, and norms that shape the quality and quantity of a society's social interactions. Social capital is not just the sum of a society's institutions, which underpin that society; it is the glue that holds them together.

The dominant form of social capital in the employment and economic sphere in the early part of the century was the market. The interaction between employer and employee was that of buying and selling labor. In this model, employer loyalty exists only as long as the continued employment of the employee increases profits. Employee loyalty exists only as long as no other job offers a greater salary or better fringe benefits (which may include location, working conditions, etc.). The interaction between producer and consumer is even more market-based in this model. People buy a product only as long as it is perceived to provide the greatest value in monetary terms, though admittedly advertising may play as large a role in shaping perceptions as the actual price and quality of the product.

In 2050, worker and worker/community ownership of many industries and local production for local markets has changed these relationships. Such enterprises logically pay more attention to worker and community well-being than enterprises driven by the need to generate shareholder profit. Well-being, of course, includes profit-shares but is increased by working conditions that are healthy, that stimulate creativity, and that create feelings of participation, community, and identity. While not all enterprises are owned in these ways, when a significant percentage of enterprises began to offer these conditions, they put pressure on the others to do so as well. In the absence of strong social capital, local production for local markets can be a disaster. In many cases, it might be inefficient to have a number of firms providing similar products for a small community. This could lead to monopoly provision of certain goods. If the market had remained the dominant form of social capital driving interactions between producers and consumers, high profits and poor quality would have resulted. However, when worker-owners also live in the local community, they have to answer to their neighbors for both the price and quality of what they produce. High-quality production is a source of pride, while low quality and high prices are perceived as incompetence and laziness, decreasing the individual's social standing in the community.

Local currencies also now contribute significantly to locally based production and consumption. Such systems existed in many communities in the early part of the century, such as in Ithaca, New York (www.ithacahours.org) and the Berkshires in western Massachusetts (www.berkshares.org). These currencies are backed only by trust that other members of the community will accept them in exchange for goods and services, and therefore require strong social capital to function. They also build social capital every time a community member accepts the currency. They are virtually immune to national and global economic instability and provide communities with greater autonomy.

For local markets to work, social capital must be strong. As discussed in the section on built capital, the very physical structure of communities now works to create that social capital. Abundant community spaces, parks, and recreation areas stimulate social interaction, build friendships, and generate a sense of responsibility toward neighbors and community. With single-occupancy vehicles almost gone and people living in smaller communities, just getting from place to place brings people in close contact with their neighbors.

At the beginning of the century, public transportation was primarily found only in large cities, and fellow passengers were strangers, not neighbors. Under these circumstances, public transportation did little to build social capital. But this is no longer the case in 2050. Some neighborhoods coalesced around different ethnicities and cultures, and these too served as sources of social capital. However, the world has rid itself of the racism, sexism, regionalism, and other prejudices that were all too prevalent earlier. People have more time for family, and family life is characterized by more balanced gender roles.

The process of government itself now creates social capital. Many countries are no longer weak representative democracies, but strong participatory ones. In a participatory democracy, the people must discuss at length the issues that affect them to decide together how the issues should be resolved. In the old world—of high-pressure jobs, little free time, and large communities of anonymous strangers—this approach to government seemed impractical, unwieldy, and too demanding. Now, with smaller communities of neighbors, a far shorter workweek, and engaged, active citizens, participatory democracy is a privilege of citizenship and not an onerous chore. Of course, this required that civic education form an essential part of education and development of human capital from childhood on. This approach to government is particularly effective at the local level. As citizens come together in regular meetings to discuss the issues and work together to resolve them (even when substantial conflict exists), it creates strong bonds of social capital and plays an essential role in forging a sense of community.

Government, of course, implies action, and action implies purpose. The purpose must be defined by the people, who in these civic meetings also forge a shared vision of the to guide their actions. This vision is not static but must adapt to new information and new conditions as they emerge. Of course, not all issues can be decided on the local level. Institutions are required at the scale of the problems they address. It is at the local level where people will feel the consequences of ecosystem change, for example, but causes may be distant, perhaps in other countries. On the national level it is not feasible to bring together millions of people to discuss the issues and decide on actions, so some form of representation is required. But representatives are now chosen through direct participation by people to whom they have strong social ties and obligations, so these representatives are far more likely to truly represent their communities and not some large corporation that funds their rise to power. Additionally, new intermediary representative institutions on the regional scale exist to bridge the gap between local and national governance.

Social capital, the glue that holds society together, also include basic moral values and ethics such as honesty, fair dealing, care for the disabled, and a common set of cultural practices and expectations that for the majority do not have to be enforced by law. Both markets and government bureaucracies fail without these common values. These values are rooted in community and nurtured by the religions of the world and other systems of thought and practice. Social capital has deep roots, and has been depleted in many areas.

16.2.5 Natural Capital

Natural capital consists of all the world's ecosystems—their structure and processes that contribute to the well-being of humans and every other species on the planet. This includes both mineral and biological raw materials, renewable (solar, wind and tidal) energy and fossil fuels, waste-assimilation capacity, and vital life-support functions (such as global climate regulation) provided by well-functioning ecosystems.

The absolute essentiality of natural capital is now so completely accepted that it is taken for granted that we must protect it if we are to survive and thrive as a species. Any schoolchild is able to tell you that you cannot make something from nothing, so all economic production must ultimately depend on raw material inputs. Economic production is a process of transformation, and all transformation requires energy inputs. It is equally impossible to make nothing from something, so every time we use raw materials to make something, when that product eventually wears out, it returns to nature as waste. It is therefore incumbent upon us to make sure that those wastes can be processed by the planet's ecosystems. Waste-absorption capacity is only one of many critical but still scarcely understood services provided by intact ecosystems. These ecosystem services include regulation of atmospheric gases, regulation of water cycles and the provision of clean water, stabilization of the global climate, protection from ultraviolet radiation, and the sustenance of global biodiversity, among many others. Without these services, human life itself would be impossible.

While by 2050, we have made substantial efforts to protect ecosystem services, uncontrolled human economic activity still has the capacity to damage them sufficiently to threaten our civilization. Obviously, well-functioning ecosystems are composed of the same plants and animals that serve as raw-material inputs to the economy; and, all else being equal, increasing raw-material inputs means diminished ecosystem services. Extraction of renewable raw materials directly diminishes ecosystem services, while the extraction of mineral resources unavoidably causes collateral damage to ecosystems. Ecosystem services are also threatened by waste outputs. While waste outputs from renewable resources are, in general, fairly readily assimilated and broken down by healthy ecosystems, ecosystems have not evolved a similar capacity to break down waste products from mining and industry, concentrated heavy metals, fossil fuels, and synthesized chemicals. In 2050, we have dramatically decreased our reliance on these slow-to-assimilate materials.

Natural capital is also economically important because it provides so many insights into the production process. The more we have learned about how nature produces, the more we have realized the inefficiency, toxicity, and wastefulness of former production techniques. It has now become a standard approach when seeking to solve a production problem to examine healthy ecosystems and strive to understand how they "solve" similar problems.

A recognition and high level of awareness of the importance of natural capital have led to dramatic changes in the way it is treated. The negative environmental impacts of nonrenewable resource use, even more than such materials' growing scarcity, have forced us to substitute renewable resources for nonrenewables, reversing the trend that began with the Industrial Revolution and making renewables more valuable than ever. Passive investment in natural capital stocks—that is, simply letting systems grow through their own reproductive capacity—is insufficient to meet our needs. Active investment is required. We are actively engaged in restoring and rebuilding our natural capital stocks by planting forests, restoring wetlands, and increasing soil fertility. The former philosophy of natural capital as free goods provided by nature has disappeared. This change has required and inspired significant institutional changes. For example, notions of property rights to natural capital have changed. Most forms of natural capital are now recognized as intergenerational assets. For example, legislation in many countries now explicitly prohibits the extraction of renewable resources beyond the rate at which they can replenish themselves, which would leave future populations dependent for survival on nonrenewable resources in danger of exhaustion and for which no substitutes exist.

Property rights to land are explicitly extended to future generations, and there are steep fines or even criminal penalties for leaving land in worse condition than when it was purchased. While ecological factors determine the total amount of natural capital that we can safely deplete, market forces still determine how that natural capital should be allocated. In addition to these fixed limits on resource use, green taxes now force both consumers and producers to pay for the damage caused by resource depletion and waste emission. When these costs are unknown, those undertaking potentially harmful activities are forced to purchase bonds or insurance that guarantee reimbursement to society for whatever damages do occur. These policies have dramatically increased the costs of degrading natural capital. As a result, most countries are rapidly weaning themselves from dependence on nonrenewable resources, having developed renewable substitutes for most of them. Many countries are competing to become global leaders in green technology. While we once relied on hydrocarbons as a feedstock for many industrial processes, we now rely heavily on carbohydrates produced by plants. This allows us to build nontoxic, biodegradable carbon polymers from CO_2 extracted directly from the atmosphere. As this technology came into its own, it helped to stabilize and even reduce atmospheric CO_2. Whether we will be able to continue to reduce global warming is still an open question, but one with growing cause for optimism.

Our understanding of ecosystem function has progressed dramatically and we continue to discover new ecosystem services. Yet for every puzzle we solve, we uncover three others. And we remain unable to accurately predict impacts of human activities on specific ecosystems, in part because of ongoing changes induced by continued global change. While the rate of warming has slowed, ecosystems are still slowly adapting to the impacts of that warming. The precautionary principle therefore now plays a critical role in deciding how we treat the environment when there is doubt over the potential impact of resource extraction or waste emissions on ecosystem goods and services. We choose to err on the side of caution. Continuing ecological-restoration efforts have begun to reverse the massive degradation that took place from 1950 through 2020, but continued global warming still threatens dangerous disruptions in ecosystem services. In keeping with the precautionary principle, we now consider it an imperative to develop extensive ecological buffers and to take the idea of planetary boundaries seriously.

16.3 A Redesign Of "The Economy" Recognizing Its Embeddedness In Society And Nature

To achieve the vision outlined in the previous section will require some fundamental changes. As Meadows has pointed out, there is a spectrum of ways we can intervene in systems (Meadows 2010). She lists 12 leverage points (shown on the right) for changing systems, ranging from changing parameters all the way to changing basic worldviews. We believe that the transition to a sustainable and desirable society will require a fundamental redesign of our system utilizing all of the leverage points. But most fundamentally, it will require changing worldviews, as outlined in the vision section above. Below, we outline some of the policy, governance, and institutional design implications of that change in worldview.

Leverage Points For Changing Complex Systems

12. **Numbers:** Constants and parameters such as subsidies, taxes, and standards
11. **Buffers:** The sizes of stabilizing stocks relative to their flows
10. **Stock-and-Flow Structures:** Physical systems and their nodes of intersection
9. **Delays:** The lengths of time relative to the rates of system changes
8. **Balancing Feedback Loops:** The strength of the feedbacks relative to the impacts they are trying to correct
7. **Reinforcing Feedback Loops:** The strength of the gain of driving loops
6. **Information Flows:** The structure of who does and does not have access to information
5. **Rules:** Incentives, punishments, constraints
4. **Self-Organization:** The power to add, change, or evolve system structure
3. **Goals:** The purpose or function of the system
2. **Paradigms:** The mindset out of which the system—its goals, structure, rules, delays, parameters—arises.
1. **Transcending Paradigms**

The problems we face—overconsumption, overpopulation, fossil fuel use, and destruction of species—are not mainly technical problems. If they were, we'd be able to solve them within a few years. The systems involved are complex and interconnected in ways that make their behavior inherently unpredictable. "As a result, the politics of communities' and nations' efforts to address their sustainability problems is much more important than any technical expertise they can muster" (Prugh et al. 2000).

There are experts aplenty, but we cannot simply consult them for the "best" solutions, because nobody can know what those solutions are in any complete or final sense. The solutions must be explored and tested through a process of continuous adaptive learning. Deciding which options to try means making political choices that affect everyone and require wide support and engagement. A generation after its coinage, the slogan "Power to the People" takes on a new meaning.

Because there can be no permanent solutions in a world that is ecologically and culturally dynamic, these choices will have to be made again and again as circumstances evolve. Therefore, moving toward a sustainable and desirable future will require a radically broadened base of participants and a political process that continuously keeps them engaged. The process must encourage the perpetual hearing, testing, working through, and modification of visions at multiple scales, from local to global.

The key seems to be structuring political systems so that people's decisions matter. What does all this mean? It means the most important issue we all face is democratic control of our lives. In a very real sense, all the issues of poverty, environment, justice, and community boil down to failures of democratic participation. When we complain about corporate power and the destructive effects of "globalization," we are complaining about the absence of democratic decision-making (decision-making by those who are affected by the decisions). We all want democracy. But how much time do we devote to studying how to make democracy really work? How much effort do we spend trying to re-arrange our local communities so that we make decisions by talking together? These are good questions. In sum, how can we turn our vision of a sustainable and desirable world into reality? We can start by learning how to make democracy work—really work—in workplaces, in local communities, in cities, in states, in nations, and globally (Alperovitz 2011). How can that begin to happen? How can we shift our society from "thin democracy" to "strong democracy" (Barber 1984; Barber 1998)?

The key to achieving sustainable governance in the new, full-world context is an integrated (across disciplines, stakeholder groups, and generations) approach based on the paradigm of "adaptive management," whereby policy-making is an iterative experiment acknowledging uncertainty, rather than a static "answer." Within this paradigm, six core principles (the Lisbon principles) that embody the essential criteria for sustainable governance have been identified (Costanza et al. 1998). The six principles together form an indivisible collection of basic guidelines governing the use of common natural and social capital assets.

- ***Principle 1: Responsibility.*** Access to common asset resources carries attendant responsibilities to use them in an ecologically sustainable, economically efficient, and socially fair manner. Individual and corporate responsibilities and incentives should be aligned with each other and with broad social and ecological goals.
- ***Principle 2: Scale-matching.*** Problems of managing natural and social capital assets are rarely confined to a single scale. Decision-making should (1) be assigned to institutional levels that maximize ecological input, (2) ensure the flow of information between institutional levels, (3) take ownership and actors into account, and (4) internalize social costs and benefits. Appropriate scales of governance will be those that have the most relevant information, can respond quickly and efficiently, and are able to integrate across scale boundaries.
- ***Principle 3: Precaution.*** In the face of uncertainty about potentially irreversible impacts to natural and social capital assets, decisions concerning their use should err on the side of caution. The burden of proof should shift to those whose activities potentially damage natural and social capital.
- ***Principle 4: Adaptive management.*** Given that some level of uncertainty always exists in common asset management, decision-makers should continuously gather and integrate appropriate ecological, social, and economic information with the goal of adaptive improvement.
- ***Principle 5: Full cost allocation.*** All of the internal and external costs and benefits, including social and ecological, of alternative decisions concerning the use of natural and social capital should be identified and allocated, to the extent possible. When appropriate, markets should be adjusted to reflect full costs.
- ***Principle 6: Participation.*** All stakeholders should be engaged in the formulation and implementation of decisions concerning natural and social capital assets. Full stakeholder awareness and participation contributes to credible, accepted rules that identify and assign the corresponding responsibilities appropriately.

Below are examples of worldviews, institutions, and technologies that can help move us toward the new economic paradigm. In this case technologies are broadly defined as the applied information that we use to create human artifacts (printing press) as well as the institutional instruments used to help us meet our goals (taxes) (Beddoe et al. 2009). The list is separated into three primary sections: respecting ecological limits, protecting capabilities for flourishing, and building a sustainable macro-economy. These are further elaborated below.

16.3.1 Respecting Ecological Limits

Once society has accepted the worldview that the economic system is sustained and contained by our finite global ecosystem, it becomes obvious that we must respect ecological limits. This requires that we understand precisely what these limits entail, and where economic activity currently stands in relation to these limits.

16.3.1.1 Waste Emission Stocks and Flows

There are several categories of dangerous waste emissions, including nuclear waste, particulates, toxic chemicals, heavy metals, greenhouse gases, and excess nutrients. Here, we focus on just two as examples. One of the most serious problems the planet currently faces is global climate disruption, caused by excessive stocks of greenhouse gases in the atmosphere. Another is the potentially catastrophic effect of excessive nitrogen and phosphorous emissions into aquatic ecosystems. These two categories of waste emissions serve to illustrate the general problem of waste emissions.

Climate change is an example of excessive stocks of waste; flows of the predominant greenhouse gas, carbon dioxide, are harmless if the atmospheric stock is at an acceptable level. Since energy is required to do work, and 86 % of the energy currently used for economic production comes from fossil fuels, economic activity inevitably generates flows of greenhouse gases into the atmosphere with current technologies. Various ecosystem processes, such as plant growth, soil formation, and dissolution of CO_2 into the ocean, are capable of sequestering CO_2 from the atmosphere. However, if flows into the atmosphere exceed flows out of the atmosphere, then atmospheric stocks will accumulate. This represents a critical ecological threshold for flows, and exceeding it, risks runaway climate change with disastrous consequences. At a minimum then, for any type of waste where accumulated stocks are the main problem, emissions must be reduced below absorption capacity. The Intergovernmental Panel on Climate Change (IPCC) estimates that global ecosystems currently absorb about 20 % of anthropogenic emissions. Achieving stable atmospheric stocks of CO_2 requires emissions reductions of 80 %, or else some means to increase the rate at which ecosystems can sequester CO_2.

However, it is also essential to target a sustainable atmospheric stock of CO_2. There is currently considerable debate about what such a stock would be, with two separate levels of uncertainty: first, what level of climate change is tolerable, and second, what level of atmospheric stocks will lead to that level of change. What determines tolerable climate change also has two components. First are the issues of impacts on agriculture, sea level rise, biodiversity loss, and so on. Second is that the threat that warming climate will create positive feedback loops leading to an even warmer climate, causing runaway climate change. There is widespread agreement that 2° C is the maximum acceptable level of change. The *Stern Review on the Economics of Climate Change* argued that we should ideally target 440 ppm (ppm) CO_2e,[4] which the report estimated would impose a 6 % chance of exceeding 2° change, but that 550 ppm was a more feasible target even though it would impose a 29 % risk of exceeding 2° (Stern 2007). More recently, Stern has concluded that 440-ppm is the maximum acceptable limit. NASA climatologist James Hansen, in contrast, argues that 350 ppm is the maximum acceptable level, though he is vague about whether this is CO_2 itself or CO_2e (Hansen et al. 2008). These are all different

[4] CO_2e is short for CO_2 equivalent. It is measured by converting all greenhouse gases into their CO_2 equivalent in terms of greenhouse effect.

estimates of the critical ecological thresholds for stocks. Current stocks are in the vicinity of 390 ppm CO_2, and 435 CO_2e.

There is growing evidence that current stocks are indeed already too high. There is clear evidence of global climate change in current weather patterns, and scientists predict that, even if society currently reduced emissions to zero, the climate would continue to warm for another 30 years. Furthermore, the oceans are beginning to acidify as they sequester more CO_2. Acidification threatens the numerous forms of oceanic life that form carbon based shells or skeletons, such as mollusks, corals, and diatoms.

The weight of evidence suggests that we have already exceeded the critical ecological threshold for atmospheric stocks. This means that we must reduce flows by more than 80 % or increase sequestration until atmospheric stocks are reduced to acceptable levels. At this point flows could be set equal to absorption capacity, with the caveat that it does not lead to excessive acidification of the ocean. If we accept that all individuals are entitled to an equal share of CO_2 absorption capacity, then the wealthy nations would need to reduce net emissions by 95 % or more. If we believe that wealthy nations should be held accountable for accumulated stocks, they would essentially need to reduce net emissions to zero or less.

Nitrogen and phosphorous emissions are somewhat different. As emission levels increase, they cause excessive growth of plant life, which rapidly sequesters the pollutants. In other words, sequestration rates increase in response to increasing emissions. However, the excessive growth of plant life can seriously disrupt aquatic ecosystems. As the plants die, the bacteria that consume them utilize much of the available oxygen, causing massive dead zones. In this case, the target of emissions reductions is primarily the flow, not the stock.

The rule for limiting waste emissions is that flows cannot be allowed to exceed absorption capacities nor disrupt critical ecological processes. If accumulated stocks already disrupt critical ecological processes, then flows must be reduced below absorption capacity until stocks are reduced to acceptable levels. Quantitative restrictions are preferable to price signals, since the latter are ineffective in the presence of growing demand.

16.3.1.2 Renewable Resource Stocks, Flows, Funds, and Services

All economic production requires the transformation of raw materials provided by nature. To a large extent, society can choose the rate at which it harvests these raw materials. Whenever extraction rates of renewable resources exceed their regeneration rates, stocks will decline. Extraction typically becomes more expensive as stocks decline, reducing economic benefits. At some point, the regeneration capacity of declining stocks will decline as well. Eventually, the stocks will reach a point at which they are no longer capable of regenerating. The first rule for renewable resource stocks is that extraction rates must not exceed regeneration rates, thus maintaining the stocks to provide appropriate levels of raw materials at an acceptable cost.

However, this simple result ignores the fact that if renewable resources are not used for economic production, they otherwise serve as the structural building blocks of ecosystems. A particular configuration of ecosystem structure generates critical ecosystem services, including both life-support services (without which no species can survive) and the capacity of ecosystems to reproduce themselves. These services are diminished when the structure is depleted or its configuration changed. We cannot simply treat ecosystem structure as a stock that yields a flow of raw materials. We must also treat it as a fund that yields a flux of services over time. The generation of this flux of services does not require the physical transformation of ecosystem structure, and flux occurs at a rate over which we have little control.

The second rule for resource extraction and land use conversion is that they must not threaten the capacity of the ecosystem fund to provide essential services. Furthermore, the marginal economic gains from conversion cannot exceed the marginal ecological costs. In short, we face a macro-allocation problem: determining how much ecosystem structure can be converted to economic production and how much must be conserved in order to supply ecosystem services. If we proceed rationally, the first units of economic production satisfy our most pressing needs. As economic output increases, it goes to satisfy less pressing needs and wants. Furthermore, if we strive to minimize the ecological costs of conversion, we sacrifice the least important components of our ecosystem funds first. As we convert more and more, we most sacrifice increasingly important components, and hence pay increasingly higher ecological costs. When the rising marginal costs of conversion exceed the diminishing marginal benefits, then continued conversion to economic production becomes uneconomic. Our limited understanding of ecosystem structure and function, and the dynamic nature of ecological and economic systems, mean that we cannot pinpoint some precise optimum. However, it is increasingly obvious that economic growth has already become uneconomic. Rates of resource extraction must therefore be reduced to below regeneration rates in order to restore ecosystem funds to desirable levels.

16.3.1.3 Unacceptable Tradeoffs: Ecological and Economic Thresholds

The necessity for imposing ecological limits on resource extraction and waste emission is straightforward. Failure to respect these limits means ecological catastrophe. However, respecting ecological limits in the short run is likely to impose unacceptable economic costs. Take, for example, the case of CO_2 emissions from fossil fuels. The marginal costs of continued emission rates are unacceptably high. However, our economy is deeply dependent on fossil fuels. Very few of us can own or consume anything that did not require fossil fuels, including food. The economic costs of reducing emissions by over 80 % in the short run would be unacceptably high.

Food systems are even more important than fossil fuels. Almost 1 billion people are currently malnourished. The global population is expected to increase by another 2 billion by 2050, and rising incomes will likely increase the demand for animal protein, which requires far more land and resources to produce than plant foods.

The UN Food and Agriculture Organization therefore estimates that we must increase global food production by 70 % by 2050, or face malnutrition and even starvation for the world's poor (Food and Agriculture Organization of the United Nations 2009). Clearly, the benefits of agriculture are extremely high. At the same time, of the nine planetary boundaries discussed by Rockström and colleagues, agriculture is the leading threat to five of them (loss, nitrogen and phosphorous loading, land use change, and freshwater use) and a major contributor to several others (Rockström et al. 2009). The last significant source of wild food, oceanic fisheries, is also serious depleted, posing significant threats to marine ecosystem services (Worm et al. 2006). Even current levels of food production may have unacceptably high ecological marginal costs, and increasing output by 70 % certainly would. Goodland and Anhang have determined that the lifecycle and supply-chain impacts of livestock production account for at least half of anthropogenic greenhouse gases in the form of methane (Goodland and Anhang 2009). Since methane is a more potent greenhouse gas than CO_2 and has a shorter half-life in the atmosphere, a reduction of flows of methane now will have a larger and quicker effect on global warming than CO_2 reductions. As a result, a 25 % reduction in meat production would almost fully achieve the goals of the recent (failed) international climate conferences. Replacing livestock products with alternatives can also decrease forest burning and allow for substantial regeneration of forest (Goodland and Anhang 2009). So it is the only available strategy for both reducing emissions and increasing carbon capture on a large scale in the timeframe during which it is widely agreed that climate change must be addressed.

16.3.1.4 Redirecting Technology Toward Sustainable Solutions

Conventional economists have long assumed that technological progress would overcome any resource constraints and allow endless economic growth (Simon 1981). A far less challenging, but still formidable, goal for technological progress would be to help stave off the looming crises already caused by endless growth described above. To do this, we would need to make rapid progress on alternative energy technologies and develop alternative approaches to agriculture. Given the urgency of the problem, we must assess various types of institutions and disseminate these technologies as quickly as possible.

Today, much research and development is performed by corporations driven by economic incentives. But, there are a number of serious problems inherent to market driven research. First, it can be difficult and expensive to make information excludable (i.e. to prevent people from benefiting from information unless they pay). The private sector is unlikely to produce non-excludable information, since other firms can simply copy it at low cost, giving them a competitive edge over the firm that actually invested in it. Patents can make information relatively excludable, but then anyone who uses that information in subsequent inventions must pay for the right to do so. Unfortunately technologies that generate public goods (such as climate stability) or that meet the needs of the poor (such as affordable food) produce no revenue to pay patent royalties. Such royalties are therefore an added deterrent to generating these technologies. For example, some scientists developed golden rice, a genetically modified strain that produces vitamin A and improves quality of life for the malnourished poor. However, after developing this technology, the scientists discovered that they had potentially infringed on 70 separate patents, which have proved a serious obstacle to distributing the rice to poor farmers (Kowalski 2002).

The solution to the conflict between food production and ecosystem services would appear to be agro-ecology—projects that increase the provision of ecosystem services from agricultural land and also increase food production and farmer income from ecological restoration (De Schutter 2010). However, the private sector generally fails to invest in agro-ecology (Vanloqueren and Baret 2009), favoring instead technologies that increase market production at the expense of ecosystems.

Alternative energy supplies are also critical. However, the energy sector is among the least innovative of all industries, investing only about 6 % as much in research and development as the manufacturing sector (Avato and Coony 2008). Private sector investment in energy technology (research development and employment) has in fact fallen steadily since the 1980s, and accounts for only 0.03 % of sales revenue in the United States (Coy 2012).

Cooperative, public-sector investment efforts, in contrast, would address these problems. The public sector by definition is interested in the provision of public goods. Research financed by the public sector can be made freely available for all to use, eliminating the costs of protecting intellectual property rights. A meta-study of returns to research and development typically conducted by the public sector found average annual rates of return of 80 % (Alston et al. 2000).

Markets are simply ill-suited for producing information at lowest possible cost. The most important input into new technologies is existing knowledge; information is like grass that grows longer the more it is grazed. When patents raise the price of accessing this knowledge, it raises the price of developing new information.

Furthermore, markets reduce the value of information once it has been developed. If a firm develops a clean, decentralized, inexpensive, and safe alternative to fossil fuels, it would be able to sell the technology at a very high cost, potentially too high for firms in developing countries to afford. These firms would then continue to burn coal and other fossil fuels, leading to continued global climate change. Paradoxically, the value of information is maximized at a price of zero, but at this price there is zero incentive for markets to provide the technology. The solution is not to create private property rights that reduce the value of information, but rather the cooperative, public provision of green technologies that are freely available for all to use.

Since many of the most serious threats to global ecosystems were caused by the excessive consumption of the wealthiest nations, those same nations should provide the bulk of the funding required for R&D in the green technologies that solve those problems. Ideally, all nations would contribute to such an effort to the best of their abilities. Many economists are worried that some nations would free-ride on investments by others. However, free-riding on certain technologies would help protect the environment and also provide benefits to those countries that made the initial investments.

16.3.1.5 Stabilization of Population

One potential solution to these apparently irreconcilable goals is to stabilize or even reduce global populations. With a world population that is surpassing 7 billion, increasing in food and energy prices due to lack of resources (Brown 2011), slowing of development in already underdeveloped countries due to overpopulation (Bloom and Canning 2004; Birdsall et al. 2003), and a lack of jobs (Cincotta et al. 2003), there has been a refocusing on population stability, often in the form of family-planning policies. Family-planning has been proven to be very cost effective (Singh et al. 2010): for every dollar spent on family planning, the United Nations has found that two to six dollars can be saved in the future on other development goals (Department of Economic and Social Affairs UN 2009). Recently the United States and the United Kingdom once again increased their foreign aid funding towards international family planning (Bongaarts and Sinding 2011).

An estimated one-third of global births is the result of unintended pregnancy (Bongaarts 2009). More than 200 million women in developing countries would prefer to delay their next pregnancy or not have any more children at all (Singh et al. 2003). However, several barriers prevent many of these women from making a conscious choice: lack of access to contraceptives, risk of side effects, cultural values, or opposition from family members (Carr and Khan 2004; Sedgh et al. 2007).

One of the major impacts of such population growth is the negative impact it is having on the earth's life-supporting ecosystem services (Speidel et al. 2009; Ehrlich and Ehrlich 1991; Wilson 2003). It has been estimated that about half of the productivity of the earth's biosystems has been diverted to human use (United Nations Development Programme UNEP, World Bank, and World Resources Institute 2003;Brown 2004). As population continues to increase, competition for these increasingly scarce resources will intensify globally. The disconnect between the "haves" and the "have nots" will also become more visible as living standards drop below survival level (Brown and Institute EP 2008).

However, if we do succeed in stabilizing, or even decreasing, the global population, other problems become apparent. With a non-growing population, the average age of the population increases, creating a situation where more retirees exist relative to workers. Addressing this problem may require higher taxes, extensions of retirement age, and/or pension reductions (Jackson 2009).

16.3.2 Protecting Capabilities for Flourishing

16.3.2.1 Sharing the Work

In a zero-growth or contracting economy, working-time policies are essential for two main reasons: to achieve macro-economic stability and to protect people's jobs and livelihoods. In addition, reduced working hours can increase flourishing by improving the work/life balance. Specific policies should include: reductions in working hours; greater choice for employees about working time; measures to combat discrimination against part-time work as regards grading, promotion, training, security of employment, rate of pay, health insurance, etc.; and better incentives to employees (and flexibility for employers) for family time, parental leave, and sabbatical breaks (Jackson 2009).[5]

However, achieving hourly reductions will require structural changes in the operation of labor markets. Indeed, even the proximate causes of rising hours are complex. In the United States, factors include the movement of women into full-time career jobs, an upward shift in work norms made possible by the growing power of employers relative to employees, and the collapse of hourly wages at the bottom of the wage distribution (which necessitates longer hours to avoid costly declines in household income) (Schor 2005a). Higher levels of income inequality have also led workers to prefer longer hours (Bell and Freeman 2001; Bowles and Park 2005).

Workers' preferences for income and consumer goods affect the determination of hours but are mainly endogenous, i.e., they adjust to the level of hours, income, and consumption that the market delivers, rather than exogenous preferences that drive the market. The phenomenon of preference endogeneity, preferences that adapt to market outcomes, rather than being fixed, may be more important than has heretofore been recognized (Schor 2005a). This endogenous preference view is the reverse of the conventional wisdom, which is that workers' exogenous preferences determine the level of hours. It is also quite different from historical accounts that emphasize consumer desires and unionizing strategy as the leading variable in determining hours, and hence the level of output and growth (Cross 1993).

To date, no detailed empirical studies linking environmental degradation and hours of work exist. Yet, in the simplest models, in which hours are correlated with income and hence consumption, a reduction in hours *ceteris paribus* (other factors being held equal) would reduce impact (Schor 2005a). The increased presence of Western media and advertising, the expansion of transnational corporations into domestic markets in the global South, and the development in the South of large middle classes with disposable income are part of a process of rapid growth in branded consumer goods worldwide. In addition to cultural products these include apparel, vehicles, consumer electronics, fast food, travel and tourism, and a range of household durables. In general, this shift is associated with much higher levels of environmental impact (Durning 1992).

[5] Much of this section was take from reference 70. Jackson (2009) Prosperity without growth: Economics for a finite planet: Earthscan/James & James.

However, many of the productivity gains of the past 200 years were driven by a shift from human labor to fossil fuels. There is therefore a distinct possibility that a dramatic reduction in fossil fuel use will lead to a shift from capital to labor. It takes approximately 5000 h of human labor to generate the work in a barrel of oil (Farley 2010). At US$100 a barrel, labor can only compete with oil at $0.02/h.

16.3.2.2 Tackling Systemic Inequality

Social inequality can express itself in many forms besides income inequality, such as life expectancy, poverty, malnourishment, and infant morality (Acemoglu and Robinson 2009). Inequality can be seen between countries but also within countries and small communities. Inequality can drive other social problems (such as over-consumption), increase anxiety, undermine social capital, and expose lower income households to higher morbidity and lower life satisfaction (Jackson 2009).

In the United States civil service, military, and universities, income inequality ranges within a factor of 15 or 20. Corporate America has a range of 500 or more. Many industrial nations are below 25 (Daly 2010). One solution to such inequity is to have people who have reached their weekly or monthly working wage limit either work for nothing at the margin, if they enjoy their work, or devote their extra time to hobbies, public service, or their family. The demand left unmet by those at the top will be filled by those who are below the maximum.

A sense of community, necessary for democracy, is hard to maintain across the vast income differences found in the United States. The main justification for such differences has been that they stimulate growth, which will 1 day filter down, making everyone rich. This may have had plausibility in an empty world, but in our full world, it is unrealistic.

Without aggregate growth, poverty reduction requires redistribution. Complete equality is unfair; unlimited inequality is unfair. Fair limits to the range of inequality need to be determined, i.e., a minimum income and a maximum income (Daly 2010). Studies have also shown that the majority of adults would be willing to give up personal gain in return for reducing inequality they see as unfair (Almås et al. 2010; Fehr and Falk 2002).

Other redistributive mechanisms and policies have also been well-established and could include revised income tax structures as discussed above, improved access to high-quality education, anti-discrimination legislation, implementing anti-crime measures and improving the local environment in deprived areas, and addressing the impact of immigration on urban and rural poverty (Jackson 2009). New forms of cooperative ownership (as in the Mondragón model), or of public ownership, as is common in many European nations, can also help constrain internal pay ratios.

16.3.2.3 Strengthening Human and Social Capital

Satisfaction of basic human needs requires a balance between social, built, human, and natural capital (and time). Policy and culture help to allocate the four types of capital defined earlier as a means for providing these opportunities.

One institution that helps build social capital is a strong democracy. A strong democracy is most easily understood at the level of community governance, where all citizens are free (and expected) to participate in all political decisions affecting the community. Interactive discussion plays an important role. Broad participation requires the removal of distorting influences like special interest lobbying and funding of political campaigns (Farley and Costanza 2002). In fact, the process itself helps to satisfy myriad human needs, such as enhancing the citizenry's understanding of relevant issues, affirming their sense of belonging and commitment to the community, offering opportunity for expression and cooperation, strengthening the sense of rights and responsibilities, and so on. Historical examples include the town meetings of New England or the system of the ancient Athenians (with the exception that all citizens must be represented, not simply the elite) (Prugh et al 2000; Farley and Costanza 2002).

Participating in society demands that attention be paid to the underlying human and social resources required for this task. Creating resilient social communities is particularly important in the face of economic shocks. Specific policies are needed to create and protect shared public spaces; strengthen community-based sustainability initiatives; reduce geographical labor mobility; provide training for jobs in sustainability; offer better access to lifelong learning and skills; place more responsibility for planning in the hands of local communities; and protect public service broadcasting, museum funding, public libraries, parks and green spaces (Jackson 2009).

16.3.2.4 Expanding the "Commons Sector"

Most resource allocation done today is through markets, which are based on private property rights. Private property rights are established when resources can be made "excludable," i.e., one person or group can use a resource while denying access to others. However, many resources essential to human welfare are "non-excludable," meaning that they are difficult or impossible to exclude others from benefiting from these resources. Examples include oceanic fisheries (particularly those beyond the economic exclusion zone), timber from unprotected forests, and numerous ecosystem services, including the waste absorption capacity for unregulated pollutants.

In the absence of property rights, open access to resources exists—anyone who wants to may use them, whether or not they pay. However, individual property rights owners are likely to overexploit or under-provide the resource, imposing costs on others, which is unsustainable, unjust, and inefficient. Private property rights also favor the conversion of ecosystem structure into market products regardless of the difference in contributions that ecosystems and market products have on human welfare. Hence, the incentives are to privatize benefits and socialize costs.

All *scarce* resources are *rival*, meaning that use by one person leaves less of the resource (in quality or quantity) for others to use. Many resources, however, are non-rival, which means that use by one person does not leave less for others to use. When this is true there is no competition for use and the resource is not scarce in an economic sense, even if total supply is inadequate. Examples include streetlights, many different ecosystem services (e.g., climate stability, flood regulation, scenic beauty), and information. Price rationing in this case reduces use and hence value to society without affecting quantity, which is inefficient. For example, if someone develops a cheap, clean solar energy technology and then patents it (which makes it excludable), it can be sold at a price. A positive price will reduce use, leading to less substitution away from competing energy sources, such as coal, and society as a whole suffers. Markets will only provide non-rival resources if they are made excludable and can be sold at a price, but this creates artificial scarcity. Paradoxically, the value of non-rival resources to society is maximized at a price of zero, but at that price markets will not provide it (Kubiszewski et al. 2010).

The solution to these problems lies with common or public ownership. Public ownership can be problematic due to the influence of money in government, which frequently results in the government rewarding the private sector with property rights to natural and social assets. An alternative is to create a commons sector, separate from the public or private sector, with common property rights to resources created by nature or society as whole, and a legally binding mandate to manage them for the equal benefit of all citizens, present and future. The misleadingly labeled “tragedy of the commons” (Hardin 1968) results from no ownership or open access to resources, not common ownership. Abundant research shows that resources owned in common can be effectively managed through collective institutions that assure cooperative compliance with established rules (Ostrom 1990; Pell 1989; Feeny et al. 1990).

Resources that are rival but non-excludable would need to be “propertized” (made excludable) to prevent over-use (Barnes 2006). Governments—or in the case of global resources such as atmospheric waste absorption capacity or oceanic fisheries, a global coalition of governments—are generally required to create and enforce property rights, but could turn these rights over to the commons sector as a common assets trust (CAT) (Barnes 2006). The trust would cap resource use at rates less than or equal to renewal rates, which is compatible with inalienable property rights for future generations. Since the resources under discussion were created by nature, and enforcement of property rights requires the cooperative efforts of society as a whole, rights to the resource should also belong to society as a whole. Individuals who wish to use the resource for private gain must compensate society for the right to do so. This could be achieved through a cap-and-auction scheme, in which the revenue is shared equally among all members of society, or else invested for the common good (Barnes et al. 2008). Preventing the re-sale of the temporary use-rights would reduce the potential for speculation and private capture of rent. Under common ownership, both costs and benefits accrue to society as whole, and the two are likely to be

Table 16.2 Rivalry, excludability, and suitable institutions for allocation

	Excludable *(rationing is possible)*	Non-excludable *(rationing is not possible)*
Rival and scarce (rationing is desirable)	*Potential market resources*:	*Open access resources*:
	Price rationing may be appropriate, rent should be captured for commons sector by taxes or royalties.	"Propertization" via collective action is required. Private use rights can be auctioned off by commons sector.
	Examples: land, timber, oil, absorption capacity for regulated wastes, use of airwaves	Examples: many aquifers, oceanic fisheries, absorption capacity for unregulated wastes
Rival and abundant (rationing is not desirable, except to prevent scarcity)	*Club or toll good:*	*Public good:*
	Price rationing may be appropriate to prevent scarcity; rent should be captured by commons sector.	Economic growth and ecological degradation are likely to increase scarcity over time. Common sector management is appropriate to prevent scarcity.
	Examples: toll roads, golf courses, ski resorts, private beaches, parks with entrance fees, etc.	Examples: oxygen, public beaches
Non-rival (rationing is not desirable; value maximized at a price of zero)	*Inefficient market good:*	*Public good:*
	Price rationing causes artificial scarcity. Common sector provision and ownership would be more efficient.	Commons sector must ensure adequate provision by preventing degradation or investing in provision.
	Example: patented information	Examples: open source information, many ecosystem services.

brought into balance. Taxes on waste emissions and resource extraction can serve the same purpose as a cap-and-auction system.

When a resource is non-rival, excludable property rights are inappropriate, but lack of property rights eliminates private sector incentives to provide the resource. The solution is common investment and common use. The commons sector must invest in the provision of non-rival ecosystem services and in green technologies that help provide and protect such services. Everyone would be free to use the non-rival ecosystem services, but not to degrade the ecosystem structure that sustains them. The means to invest in non-rival resources can be obtained from auctioning off access to rival resources. For example, the CAT could auction off the right to greenhouse gas absorption capacity, and then invest the revenue in carbon-free energy technologies.

When a resource is privately owned but generates economic rent, or is used in a manner that socializes costs and privatizes benefits, taxation can achieve the same goals as common ownership, as discussed in Sect. 16.4. Table 16.2 summarizes appropriate property rights for different categories of resources.

If the public sector shirks its duties to manage our shared social and natural inheritance for the common good, we require a commons sector to ensure sustainability and a just distribution of resources. Once these two goals have been achieved, the market will be far more effective in its role of allocating scarce resources towards the products of highest value, then allocating those products towards the individuals that value them the most.

16.3.2.5 Removing Communication Barriers and Improving Democracy

With the invention of television, political advertisements became a critical outlet for candidates to broadcast their message and to sway voters. However, the decentralized nature of the Internet "allows citizens to gain knowledge about what is done in their name, just as politicians can find out more about those they claim to represent" (Street 2001). As a means of two-way communication, the Internet provides voters the ability to speak out about their government's behavior without leaving their homes. For the Internet to transform the idea of electronic democracy, universal access is critical, but technological, financial, and social barriers currently prevent such universal accessibility (Street 2001). Removal of these and other barriers to engagement and deliberation thus becomes a major goal for replacement of the current plutocracy with real democracy.

Unlike television, very low technological and financial barriers exist to establishing a presence on the Internet. This has the effect of decentralizing information production, and returns control of the distribution of information to the audience, providing a venue for dialogue instead of monologue (Gore 2007). Opinions and services previously controlled by small groups or corporations are now shaped by the entire population. Television news networks, sitcoms, and Hollywood productions are being replaced by e-mail, Wikipedia, YouTube, and millions of blogs and forums, all created by the same billions of people who are the audience for the content.

In 2008, the United States presidential election marked the first election year in which more than half of the nation's adult population became involved in the political process by using the Internet as a source of news and information. Rather than simply receiving uni-directional news, approximately one-fifth of the people using the Internet used websites, blogs, social networking sites, and other forums to discuss, comment, and question issues related to the election (Smith 2009).

16.3.3 Building a Sustainable Macro-Economy

The central focus of macroeconomic policies is typically to maximize economic growth. This is evident in the definition of a recession as two consecutive quarters with no economic growth. Lesser goals include price stabilization and ensuring full employment. Meadows argues that changing goals is the second most powerful lever for changing complex systems (Meadows 2010). If society instead adopts the central economic goal of sustainable human well-being, macroeconomic policy will change radically. The goal will be to create an economy that offers meaningful employment to all, that balances investments across the four types of capital to maximize well-being. Recession would be redefined as unacceptable or increasing rates of poverty, misery, inequality, and unemployment, or unsustainable levels of throughput. Such goals would lead to fundamentally different macroeconomic policies and rules. Changing the rules is the third most effective of Meadow's places to intervene in a system.

16.3.3.1 Changing the Institutions: Monetary Reform for Sustainability and Justice

The current monetary system is inherently unsustainable. The base of the money supply in almost all countries is coins and bills printed by governments, and money that governments create when they provide credit to banks during purchases. Government money spent into existence is then destroyed by taxes. Taxes in fact are what give the government the power to create money: everyone accepts government currency because they require it to pay taxes. In the modern era, national currencies are backed by the taxation power of the government. However, this government money (also known as vertical money) is now only a small fraction of the money supply in most economies.

Most of our money supply is now a result of fractional reserve banking. Banks are required by law to retain a percentage of every deposit they receive; the rest they loan at interest. However, loans are then deposited in other banks, which in turn can lend out all but the reserve requirement. The net result is that the new money issued by banks, plus the initial deposit, will be equal to the initial deposit divided by the fractional reserve. For example, if a government credits $1 million to a bank and the fractional reserve requirement is 10 %, banks can create $9 million in new money, for a total money supply of $10 million. Fractional reserve requirements may not even limit the amount of money created. Banks will typically loan money to any investor who they believe offers a high probability of repayment. If the amount they lend exceeds their reserves, they can borrow money from other banks or the Federal Reserve Bank to make up the deficit. If there is too much borrowing of this type, it threatens to drive up the interest rate. If the Federal Reserve Bank is trying to target interest rates, it will be forced to buy securities from banks to increase bank reserves and the money supply. Regardless of whether the fractional reserve or investor demand determines total money supply, most money is today created as interest-bearing debt. Total debt in the United States, adding together consumers, businesses, and the government, is about $50 trillion dollars. This is the source of the national money supply.

When the loans are repaid, the new money is destroyed. However, the borrowers must repay the loans plus interest and the banks initially loaned out enough to repay only the principal. Either new government expenditures or new loans are required to pay back the interest.

There are several serious problems with this system. First, it is highly destabilizing. When the economy is booming, banks will be eager to loan money and investors will be eager to borrow, which leads to a rapid increase in money supply. This stimulates further growth, encouraging more lending and borrowing, in a positive feedback loop. A booming economy will stimulate firms and households to take on more debt relative to the income flows they use to repay the loans. This means that any slowdown in the economy will make it very difficult for borrowers to meet their debt obligations. Borrowers can sell assets to meet their obligations, but this will drive down the price of assets, for example, home values. Eventually some borrowers will be forced to default. Banks are likely to lose the confidence of other borrowers and will be unwilling to make new loans, which the borrowers require to pay back interest, leading to more defaults. Repayment of loans will exceed creation of loans, leading to a shrinking money supply. Outstanding loans will continue to grow exponentially, even as output diminishes as a result of less money available for investment. Widespread default on the debt becomes inevitable. The result is a self-reinforcing downward economic spiral, leading to recession or worse. The poor usually bear the brunt of the resulting suffering.

Second, the current system systematically transfers resources to the financial sector. Borrowers must always pay back more than they borrowed. At 5.5 % interest, homeowners will be forced to pay back twice what they borrowed on a 30-year mortgage. Conservatively speaking, interest on the $50 trillion total debt of the United States must be at least $2.5 trillion a year, one-sixth of our national output. Currently, banks can borrow money from the Federal Reserve Bank at almost zero percent, then charge 20 % or more on credit card debt.

Third, the banking system will only create money to finance market activities that can generate the revenue required to repay the debt plus interest. Since the banking system currently creates far more money than the government, this system prioritizes investments in market goods over public goods, regardless of the relative rates of return to human well-being. Studies find that government investments in public goods regularly generate 25–60 % non-diminishing annual rates of return, in monetary measures (López and Galinato 2007). There is no reason to believe that returns would be any less when the investments are targeted towards the new macroeconomic goals.

Fourth, and most important, the system is ecologically unsustainable. Debt is a lien on future production. Debt grows exponentially, obeying the abstract laws of mathematics. Future production, in contrast, confronts ecological limits and cannot possibly keep pace. Interest rates exceed economic growth rates even in good times. Eventually, the exponentially increasing debt must exceed the value of current real wealth and potential future wealth, and the system collapses. However, in the effort to stave off an economic crisis and the unacceptable misery, poverty, and unemployment it will cause, policy makers will pursue endless economic growth, unsustainable on a finite planet. The system forces us to choose between unsustainable growth and misery.

In order to address this problem, the public sector must reclaim the power to create money, a constitutional right in the United States and most other countries, and take away from the banks the right to do so by gradually moving towards 100 % fractional reserve requirements. This would allow banks only to loan money on time deposits, in which case the owner of the money forgoes the right to use it while it is loaned to someone else. Banks would be restricted to the role that most people believe they play anyway—serving as an intermediary between those who want to save their money and those who want to borrow it. The current recession is an ideal time to implement this change, since banks are currently loaning far less than allowed by fractional reserves. Reserve deposits in the United States are currently about $1.4 trillion greater than required by law.

The public sector could create money in several different ways. First, the government could simply spend money into existence to provide the public goods that the private sector will not supply, to invest in social and human capital, to create jobs, to rebuild the national infrastructure, and to restore the natural systems that sustain us all. Such spending would end the recession (as previously defined) without increasing the national debt and without systematically transferring interest to the already wealthy. Second, the government could loan money into existence interest-free. Money could be loaned directly to the private sector to finance critical economic activities, such as food production and alternative energy, or it could be loaned to state and local governments (SLGs) to meet their needs. SLGs would also have the option of loaning money interest free or spending it on public goods.

Third, in order to minimize disruption as we change from the current system, the government could make time deposits in banks that serve the common good, allowing them to carry on with business as usual. The public, however, would have control over the money supply.

Ironically, many economists argue that the public sector cannot be trusted to print and spend money—that it will create too much and spend it irresponsibly. The United States government, however, printed $1.6 trillion in government bonds in a single year to finance its deficit, which must be paid back with interest. Issuing interest-free currency is much less risky; it would be difficult for the government to under-perform the private sector when measured by the new goals for macroeconomic policy. At the very least, voters have some control over governments, and none over the banking sector.

There is, however, no free lunch. The government cannot and should not endlessly spend money into existence. The goal must be to achieve a steady state with sustainable levels of throughput, which will likely require a significant reduction in market activity in the wealthy nations, and thus a reduction in the total money supply required to support the economy. When money is loaned into existence, it will be destroyed when it is repaid. State and municipal governments would need to use tax revenue to repay the federal government, but would not need to pay fees to investment banks to issue municipal bonds, nor interest to bond holders. When money is spent into existence, it can be destroyed through taxes, which would play a critical role in regulating the money supply. To ensure that too much money does not flood the economy, any new expenditure could be matched by future taxes, imposed at the same time the expenditure takes place. Rather than a tax, borrow, and spend policy, the government would explicitly pursue a policy of spend, then tax (which, many argue, is actually the way the system currently works anyway). There will no doubt be errors as we shift towards a steady state economy, resulting in occasional recessions or booms. The government however could spend extra money into existence to alleviate misery, poverty, and unemployment during times of recession, and raise taxes if throughput becomes excessive. The monetary system would be counter-cyclical, not pro-cyclical. Government would never need to borrow money and pay it back with interest. There would be no debt. With no exponentially growing debt and no interest payments, there would be no pressure to choose between unacceptable misery or endless growth. The feedback signal of a rising price index would government when to stop creating money.

Fiscal reform is also required to meet the goals of macroeconomic policy. This section is limited to a discussion of taxes, which are a powerful tool for changing economic behavior. The other half of fiscal policy is expenditure, which would be subsumed under monetary policy as described above.

Conventional economists generally look at taxes as a drag on the economy, albeit necessary to finance government expenditures. The reasoning is that taxes increase costs, leading to a reduction in output, and disequilibrium between marginal costs and marginal benefits, resulting in a deadweight loss of economic surplus. They are seen as a significant drag on economic growth. From a more holistic perspective, however, taxes are an effective tool for internalizing negative externalities into market prices, therefore reducing deadweight loss, and for improving income distribution.

16.3.3.2 Tax Bads, Not Goods

A perennial conflict in tax policy is taxing to raise revenue versus taxing to change behavior. Induced behavioral change aims at avoiding the tax, and this naturally reduces revenue. The policy of shifting the tax base from value added to throughput (that to which value is added) encounters this conflict in a different way. Taxing value added (labor and capital) tends to reduce incentives to enterprise and work, and to use untaxed resources lavishly. Taxing the resource flow would lead to emphasizing resource efficiency, and using less resources (more untaxed recycled resources and more labor and capital) to the extent possible, which is a desired behavioral change, but would reduce revenue. Yet depletion and pollution remain "bads" even if reduced, so there is a good case for further raising the tax on them if revenue needs require it, while value added remains something we want to increase, so we would still want to avoid taxing it.

A shift in the burden of taxation from value added (economic goods, such as income earned by labor and capital) to throughput flow (ecological bads, such as resource extraction and pollution), is critical in shifting towards sustainability (Daly 2010). Such a reform would internalize external costs, thus increasing efficiency (Daly 2010). It is possible to impose throughput taxes on resource depletion or on waste emissions. Taxing the origin and narrowest point in the throughput flow induces more efficient resource use in production as well as consumption, and facilitates monitoring and collection. For example, there are far fewer oil wells than there are sources of CO_2 emissions. In either case, taxes will increase prices and induce efficiency in resource use. One disadvantage of green taxes is that the level of pollution is determined by price, rather than the ecosystem's capacity to absorb waste. Prices can adjust to ecological constraints more rapidly than ecosystems can respond to the price signals (Daly and Cobb 1994). We discuss below quantitative limits as an alternative.

Many people call for a gradual revenue-neutral tax shift, rather than a set of new taxes. This approach would begin by forgoing a certain dollar amount of revenue from the most regressive taxes, for example, payroll or sales taxes, which currently take a larger percentage of income from the poor than from the rich, while simultaneously collecting the same amount from the best resource severance tax. Then, as the next step, get rid of the second worst tax and substitute the second best resource tax, and so on. As discussed below, however, increasing tax revenue may be desirable.

The logic of ecological tax reform has been broadly accepted for at least a decade and has been implemented in varying degrees across Europe. But progress towards this goal has been painfully slow. In the United Kingdom, the proportion of taxation from green taxes is now lower than it was in 1997. There's an urgent need to achieve an order of magnitude step-change in the structure of taxation. A sustained effort by government is now required to design appropriate mechanisms for shifting the burden of taxation from incomes onto resources and emissions (Jackson 2009).

16.3.3.3 Tax What We Take, Not What We Make

Taxes should also be used to capture unearned income, or rent, in economic parlance. Green taxes are a form of rent capture, since they charge for the private use of resources created by nature. However, there are many other sources of unearned income in society.

Most obviously, the word "rent" is associated with land. Land is available in a fixed supply which cannot respond to market signals, and is an essential input into all economic activities—even the least tangible economic activities must take place on some physical substrate. The value of land is created by nature and society as a whole, not by individual effort. For example, if a government builds a light rail or subway system—more sustainable alternatives to private cars—adjacent land values typically skyrocket, providing a windfall profit for landowners. New technologies also increase the value of land, due to its role as an essential input into all production (Gaffney 2009). Because the supply of land is fixed, any increase in demand results in an increase in price. Landowners therefore automatically grow wealthier independent from any investments in the land. Furthermore, speculative demand creates a positive feedback loop, in which rising prices increase demand, leading to bubbles and busts in land markets, which can trigger national and even global recessions. High taxes on land values (but not on improvements to land, such as buildings) allow the public sector to capture this unearned income. Similarly, public ownership through land trusts and other means, as is increasingly common, allows for public capture of the unearned income. This removes any reward from land speculation, thus stabilizing the economy. It also drives down land prices. Mortgage payments will be replaced by tax payments, so there will be no negative impact on new landowners. If land values fall, so do payments, dramatically decreasing the likelihood of default and foreclosure. Fixed stocks of land means that it exhibits perfectly inelastic supply, so landowners cannot pass tax increases on to renters.

Growing demand and increasing scarcity of natural resources also drive up their price, generating windfall profits for resource owners. The depletion taxes discussed above should increase in tandem with price increases, capturing the rent for the public sector.

16.3.3.4 Taxation to Reduce Inequality

Income inequality can have very pernicious effects on human well-being. Figure 16.6 below shows the relationship between inequality and an index of health and social problems across OECD countries.

Inequality is also closely related to taxation policies. Figure 16.7 shows the highest marginal income tax bracket in the United States, along with the share of income captured by the wealthiest 0.1 %. However, taxes on capital gains, which account for a significant share of the income of the top 0.1 %, are not included in this figure. The top capital gains tax dropped from 28 % to 20 % in 1997, which accounts for the dramatic increase in income inequality beginning that year.

There is also a strong correlation between tax rates and social justice, as evident from Fig. 16.8. High tax rates that contribute to income equality appear to be closely related to human well-being. This suggests that tax rates should be highly progressive, perhaps asymptotically approaching 100 % on marginal income. The measure of tax justice should not be how much is taxed away, but rather how much income remains after taxes. For example, hedge fund manager John Paulson earned $4.9 billion in 2010 (Goldstein 2011). If Paulson had to pay a flat tax of 99 %, he would still retain nearly $1 million per week in income. Presumably, most of this income was taxed at the current capital gains tax rate of 15 %, which also applies to a large share of hedge fund manager income. Increasing his tax rate to 99 % (which might entail a marginal tax rate of 99.99 %, depending on the tax schedule) would allow the government to hire 84,000 teachers at $49,000 per year.

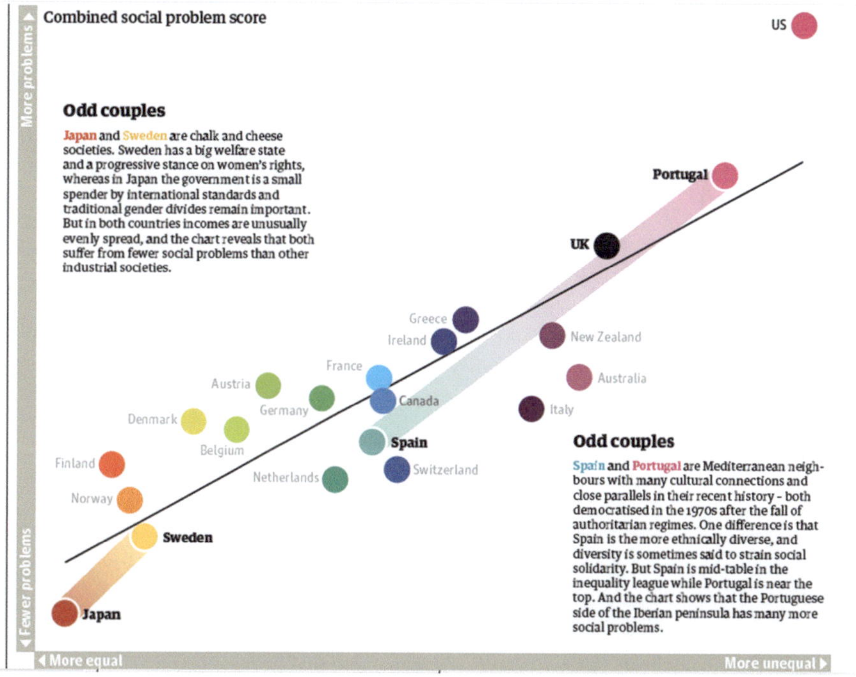

Fig. 16.6 Relationship between income inequality and social problems score in OECD countries (Wilkinson and Pickett 2009)

16.3.3.5 Increasing Financial and Fiscal Prudence

The monetary reform proposed above requires significant political will, which may be slow in coming. Other policies for achieving financial and fiscal prudence may be required in the meantime.

For over the past decade, debt-driven consumption has pushed economic growth globally. However, our relentless pursuit of that growth as the end goal has contributed to the global economic crisis. A new era of financial and fiscal prudence needs to: increase the regulation of national and international financial markets; incentivize domestic savings, for example through secure (green) national or community-based bonds; outlaw unscrupulous and destabilizing market practices (such as short selling); and provide greater protection against consumer debt (Jackson 2009). Governments must pass laws that restrict the size of financial sector institutions, eliminating any that impose systemic risks for the economy. "Too big to fail" is "too big to exist."

Certain governmental policies have promoted the financial turmoil of the past few years. Reforming these policies would reduce the distortions within the financial markets, eliminate the too-big-to-fail problem, and prevent the government from manipulating housing credit. These reforms would include: (1) smarter micro-prudential regulation of banks, (2) macro-prudential regulation of bank capital and liquidity standards, (3) creation of credible plans for reforming large, complex

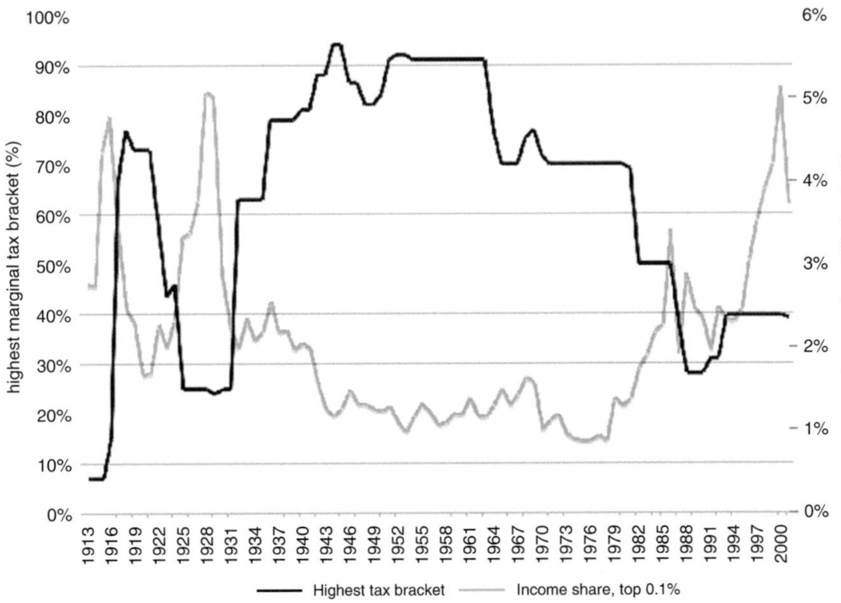

Fig. 16.7 Time series of income in the highest tax bracket in the U.S. (black) and income share in the top 0.1 % of households (grey) from 1913 to 2002 (Daly and Farley 2004)

banks, (4) elimination of leverage subsidies as a means of promoting homeownership, (5) removal of barriers to stockholder discipline of bank management, (6) policies that promote improvement in counter-party risk management (Calomiris 2010), and (7) encouraging sustainable local development through new and existing community, municipal, and state development banking institutions.

16.3.3.6 Improving Macro-Economic Accounting

Unlimited economic growth is not only impossible, it is undesirable. GDP measures costs, not benefits, as illustrated by recent declines in energy and food supply, increasing both their prices and share in GDP even as the benefits they generate decline. An indicator of welfare should measure years of satisfying life, encompassing both quality and quantity.

A large body of literature exists critiquing the value of GDP as a wellbeing measure (Anderson 1991). Its primary limitations include the following:

1. Failure to account for externalities, both positive (household labor, volunteering, ecosystem services) and negative (pollution, crime, or cancer) (Costanza et al 2009).
2. Counting the depletion of natural capital as income.
3. Ignoring thresholds beyond which increasing GDP no longer contributes to quality of life. As GDP increases, overall quality of life often increases up to a point.

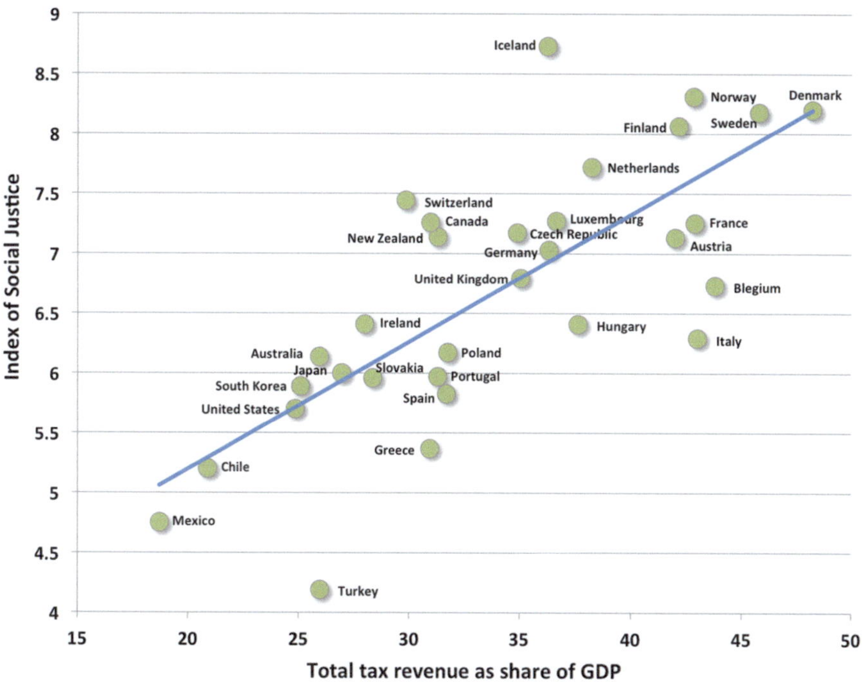

Fig. 16.8 Relationship between tax revenue as a percent of GDP and index of social justice in OECD countries (Goldstein 2011)

Beyond this point, increases in GDP are offset by the costs associated with increasing income inequality, loss of leisure time, and natural capital depletion (Talberth et al 2007; MaxNeef 1995).

4. Failure to account for inequality.
5. Failure to account properly for changes in the asset base, which affect our future consumption possibilities (Jackson 2009).
6. Concentration on flows, when capital stocks may be a better measure of quality of life. Society should seek to minimize the flows required to sustain these stocks (Boulding 1968).

GDP does, however, belong as an indicator of economic efficiency. The more efficient we are, the less economic activity, raw materials, energy, and work it requires to provide satisfying lives. Real efficiency reduces environmental impacts and increases leisure time. As a major cost of providing satisfying lives, GDP does frequently move in parallel with welfare. In the same way, countries that spend more on medical care tend to have better indicators of health. However, concluding that we should therefore maximize medical expenditures, a cost, is absurd. When GDP rises faster than life satisfaction, efficiency declines. Our goal should be to minimize GDP, subject to maintaining a high and sustainable quality of life. The real problem with recession is not that it decreases GDP but that it undermines quality of life by increasing unemployment, poverty, and suffering (Beddoe et al 2009).

In 1969, the United States came to the end of a four-decade decline in income inequality and poverty. People then consumed about half as much per capita as they do today. The genuine progress indicator (GPI), a measure of welfare designed to adjust for the inadequacies of GDP, reached a plateau around this time, and has since declined (Talberth et al 2007). Subjective measures of well-being, such as the percentage of people who consider themselves "very happy," have steadily declined since then as well (Layard 2005). Empirical evidence therefore suggests that a return to 1969 per-capita consumption levels would not make us worse off. On the contrary, returning to 1969 consumption levels would presumably lower our resource depletion, energy use, and ecological impacts by half, so there is every reason to believe that dramatically lowering our per-capita consumption could actually make us better off (Farley et al. 2010).

A number of ways of measuring national-level progress has been proposed, developed, and used to address this growing realization that GDP is a measure of economic quantity, not economic quality or welfare, let alone social or environmental well-being. The measures also address the concern that GDP's emphasis on quantity encourages depletion of social and natural capital and other policies that undermine quality of life for future generations.

In general, these new measures can be categorized as (1) indexes that address the issues described above by making "corrections" to existing GDP accounts, (2) indexes that measure aspects of well-being directly, (3) composite indexes that combine approaches, and (4) indicator suites. Like GDP, all these measures are abstracted indicators, not comprehensive reports on the heart and soul of individual communities. However, some can and are being used to inform local and regional decisions. This is an improvement on the misuse of GDP and economic growth as a proxy for well-being (Costanza et al. 2009).

National accounts should focus on well-being and societal progress as we defined above. Such accounts will provide policy-makers a better chance to react appropriately to financial crises, climate change, and oil price shocks (New Economics Foundation 2008). By utilizing national accounts focused on well-being, a well-being screen will be applied to every policy proposal, allowing a shift away from narrow, income-driven costs/benefits analysis to a wider range of potential impacts on personal and social well-being (New Economics Foundation 2009).

16.3.3.7 Improving Macro-Economic and Regional Coordination

Unless planned with care, moving towards a reduced-growth and reduced-time economy could cause many disruptions at the level of firms, communities, and individuals. Current coordination and planning strategies are limited in general, and are focused largely on growth in particular. A new infrastructure capable of generating specific sectoral, geographic, and time allocating alternatives will be required so that choices between alternative paths can become policies rather than scenarios. Developing ways in which larger-order coordination and planning choices can be presented to publics for democratic consideration and decision-making is an essential requirement of the new direction proposed (Alperovitz and Faux 1984).

16.4 Example Policy Reforms

16.4.1 Reversing Consumerism

Economic policy has focused almost entirely on promoting continuous growth in GDP. Economic growth often translates into more, instead of better consumption, excessive material and fossil fuel use, and increased waste. The culture of consumerism has developed, in part at least, as a means of enhancing consumption-driven economic growth. But it has had damaging psychological and social impacts on people's well-being. There is a need to systematically dismantle incentives for excessive material consumption and unproductive status competition (Kasser 2002; Frank 1999).

Excess consumption is driven in part by artificially low prices that fail to reflect full social and environmental costs. Natural resource prices fail to reflect demand by future generations or the degradation of ecosystem services caused by resource extraction. Export-oriented economies often fail to impose or enforce labor and environmental regulations in order to keep prices down. Wages, particularly in poor developing countries, are frequently inadequate to meet basic needs, and working conditions are often dangerous, debilitating, and degrading (Arrow et al. 2004), contributing to a decline in workers' well-being (Schor 2005b). We need to have effective labor and environmental policies in place that prevent the exploitation of foreign workers and internalize environmental costs. When we account for the real costs of labor, resource use, and externalities, then import prices will increase and the demand and consumption for these goods/services in rich countries will decrease. Also, the increase in labor wages will benefit the poor in developing countries, raising their purchasing power and improving their livelihoods (Schor 2005b). High levels of consumption in rich countries may promote excessive resource degradation in poor countries, which jeopardizes well-being in the poorer countries.

Income inequality also drives excessive consumption. Once basic needs are met, relative income and status may be more important than total income. Consumption decisions are driven by comparisons with a reference group and the pursuit of status (Schor 2005b, Kallis 2011). Status, however, requires consuming more status goods than one's peers and creates a never-ending treadmill. When the extremely wealthy spend more, less wealthy individuals on the fringes of their social circles also feel compelled to do so, followed by the even less well on the fringes of their circles, in what economist Robert Frank describes as an "expenditure cascade" (Frank 2007).

In the presence of growing income inequality, this leads to a cycle of excessive work and indebtedness that can dramatically decrease quality of life. Partly as a result of the status treadmill, increases in labor productivity, education, skills, etc., have led to increases in production and consumption of goods and services, instead of more leisure time, earlier retirements, more holidays, etc.

Status, however, requires consuming more status goods than one's peers and creates a never-ending treadmill. When the extremely wealthy spend more, less wealthy individuals on the fringes of their social circles also feel compelled to do so, followed by the even less well on the fringes of their circles, in what economist Robert Frank describes as an "expenditure cascade" (Frank 2007). In the presence of growing income inequality, this leads to a cycle of excessive work and indebtedness that can dramatically decrease quality of life. Partly as a result of the status treadmill, increases in labor productivity, education, skills, etc., have led to increases in production and consumption of goods and services, instead of more leisure time, earlier retirements, more holidays, etc.

Decreases in consumption in some goods and services can have rebound effects, leading to increases in consumption elsewhere (van den Bergh 2011). For example, when people save money by driving a more fuel-efficient car or by increasing the energy efficiency of their homes, they may spend their savings on a holiday flight, resulting in a net increase in energy use (Sorrell 2007). Similar results can occur on larger scale, when increases in the efficiency of resource use lead to greater marginal benefits and an increase in total use (Polimeni et al. 2008). In order to decrease consumption, all prices need to reflect real costs (environmental, social, and climate externalities). This will help achieve changes in consumption behavior and will limit, or even decrease, rebound effects. Policies should also target the composition of production and consumption to ensure that rebound effects are minimized. We can also decrease consumption through decreases in work time, which will translate into less purchasing power and thus less consumption and environmental degradation. By decreasing income and spending (income caps), it will also limit rebound effects (van den Bergh 2011; Kallis 2011). However it does not guarantee a shift to cleaner consumption (van den Bergh 2011). A cap-auction-trade scheme, rather than a tax, avoids the rebound effect by simply limiting quantity; any demand rebound just bids up price.

Improvements in technological efficiency are necessary, but not sufficient. They are more appealing to all because of their apolitical nature and mostly because they do not challenge production and consumption. However, there is an extensive literature showing how improvements in technological efficiency have led to increases in production and consumption due to a decrease in relative prices of products/services (Jackson 2009; Schor 2005b; Kallis 2011; Knight et al 1970; Victor 2010). Some benefits of improvements in energy efficiency are offset by an increase in the demand for the product or service due to a decrease in price (Knight et al 1970).

The increase in overall productivity through technological innovation has led to an increase in consumption and use of high quality energy and material resources, while avoiding the real social and environmental costs. Technological innovation also means a decrease in labor; the more efficient it becomes, the fewer workers are needed to produce the same level of outputs. This would work as long as the economy continues to grow and offsets labor productivity, but if there is a slowdown in the economy, then increasing productivity may also lead to increasing unemployment (Jackson and Victor 2011).

For many politicians, growth (increases in production and consumption) equals more jobs, thus attempts to decrease productivity growth are seen to reduce welfare (Jackson and Victor 2011). However, decreases in productivity growth can be achieved by shifting from a product-based economy to a more service-based economy, since services are usually considered less material- and energy-intensive (Jackson and Victor 2011). But it all depends on the type of services that are pursued; activities in the service sector can heavily depend on high levels of material and energy consumption (i.e., tourism and retail distribution). A focus on activities that promote social interaction and community engagement (farmers markets, crafts, community green projects, among others) will reduce labor productivity growth. The green service sector (less material and energy intensive) will also contribute to a reduction of GHG emissions (Jackson and Victor 2011).

We should also look at productivity growth as an opportunity for increasing leisure instead of consumption (Schor 2005b). One approach to decreasing material and energy consumption is to reduce the time spent working. Less hours of work will limit production and consumption. Working less typically leads to reduced spending and also a shift to lower-impact forms of consumption: taking the bike instead of the car; cooking at home instead of buying fast food (Schor 2011).

In addition, other regulations or policies that have been identified to decrease and/or reverse consumerism are:

- Taxing luxury consumption (Frank 1999; Kallis 2011; Howarth 1996): progressive taxes are necessary to disincentivize over-consumption, which has been pursued at the expense of increases in free time and environmental quality. For example, the book *Luxury Fever* has proposed a shift in the United States tax code to exempt savings and tax only consumption at very progressive rates (Frank 1999). Similarly, Howarth has proposed taxing status goods that increase energy and resource consumption (Howarth 1996). Such policies could even benefit the rich by decreasing the level of consumption required to exhibit status, while leading to environmental benefits.
- Redirecting consumption from private status goods to public goods (investing in the commons), which will increase welfare (Victor and Rosenbluth 2007)). Government can offer tax reductions or preferential investment conditions for activities that generate or protect public goods, such as green services to disincentivize energy and material intensive production and consumption. The rich could even benefit from higher taxes to fund these public goods: their status will be unaffected by across-the-board income reductions, while they will benefit from more public goods (Frank 2012).
- Increasing employment in specific service sectors (health, green projects, community based projects, etc.) (Kallis 2011; Nørgård 2010).
- Shifting the traditional focus of investment towards renewable energy, public goods, green (resource-efficient) technology, climate adaptation and mitigation, etc.
- Redistributing surpluses from private consumption to communal activities—urban food gardens, recycling, car-pooling—since communal activities tend to reduce conspicuous consumption.
- Incentivizing voluntary self-restrictions (Kallis 2011; Martínez-Alier et al. 2010).

- Cap-and-auction policies for waste emissions that would internalize externalities and promote a shift towards cleaner consumption (van den Bergh 2011).
- Promoting and improving communication and the diffusion of information to reduce consumption, which would incentivize voluntary reductions in consumption and more socially desirable decisions; peer pressure plays a key role in consumption. This could be achieved by restoring the requirement for public service messages in exchange for private sector use of the airwaves.
- Directly controlling commercial advertising and media. The advertisement of status goods increases consumption since it encourages people to seek more income and to pursue wants that did not exist before. Regulation of advertising can lead to a change in individual/societal preferences (van den Bergh 2011; Kallis 2011). Commercial advertising represents a social cost and the regulation of advertising will likely affect compositional consumption, increase well being, and decrease environmental degradation. Other measures might include banning advertising to children and in public spaces, establishing commercial-free zones and times, taxing advertising, and funding the right of reply to advertisers' claims (Jackson 2007; Gannon and Lawson 2009):

  Banning advertising in public spaces: The Clean City Laws of São Paulo, Brazil. This law, introduced in 2007, completely bans outdoor advertising in the city and fines those who break it. The state of Vermont similarly bans billboards.

  Banning advertising for children: Stockholm decided in 1991 to prohibit ads targeting children under 12 years. Greece does not allow war toy advertisements at all and any toy advertisements are prohibited between 7:00 AM and 10:00 PM. The U.K. does not allow the advertisement of alcohol to youths and requires ads to convey the size of the toys and what the toys can really do.

  Tax advertising: Advertising is currently considered a business expense, exempt from taxation. This exemption should be removed, and an additional tax imposed on companies that spend more than a certain amount on advertising based on the rationale that advertising could be viewed as market externality that increases consumerism.

16.4.2 Expanding the Commons

To realize the transition to the new economic system we envision, it is necessary to greatly expand the commons sector of the economy, the sector responsible for managing existing common assets and creating new ones. Some assets, such as resources created by nature or by society as a whole, should be held in common because this is more just. Other assets, such as information or ecosystem structures (for example, forests), should be held in common because this is more efficient. Still other assets, such as essential common-pool resources and public goods, should be held in common because this is more sustainable.

One option for expanding and managing the commons sector is to create "common asset trusts" at various scales. Trusts, such as the Alaska Permanent Fund and regional land trusts, can propertize the commons without privatizing them (Costanza et al 2010). Barnes (2006) provides more specific examples of existing or proposed local, regional, national, and global initiatives for expanding the commons sector:

16.4.2.1 Local Initiatives

1. **Land trusts:** There are various types of land trusts. One type is meant to protect land from development and degradation, which can be achieved via direct ownership of the land or by ownership of easements that restricts its use (e.g., the Marin Agricultural Land Trust, the Pacific Forest Trust, the Vermont Land Trust). Another type is meant to keep housing affordable. Land is held in a trust, while houses on the land are sold on the condition that the owner cannot profit from rising land values when the land is resold (e.g., the Champlain Housing Trust) (Swann 1972; Davis and Jacobus 2008).
2. **Conservation trusts:** Conservation funds for the protection of biodiversity that have been created since the 1990s through debt-swap funding or grants. These trusts were created with an endowment that allowed them to cover their short- and long-term needs (e.g., Bhutan Conservation Trust, The Mgahinga and Bwindy Impenetrable Forest Conservation Trust, and Colombian National Protected Areas Conservation Trust) (Adams and Victurine 2011).
3. **Terrestrial and marine protected areas:** Established for the protection and maintenance of biodiversity (marine sanctuaries, wildlife refugees, national parks, etc.).
4. **Surface water trusts:** Acquisition of water rights to protect fish, other species, or aquatic ecosystems. This has also led to changes in agricultural practices like switching crops and changing irrigation patterns. A good example is the Oregon Water Trust.
5. **Groundwater trusts:** Permit issuance to limit the amount of water withdrawn from the aquifers, e.g., Edward Aquifer Authority in Texas.
6. **Community gardens:** Food production for neighborhoods and communities and promote community engagement.
7. **Farmers markets:** Commercial commons that provide fresh and local food, social interaction and engagement, awareness and importance of local produce, and other functions.
8. **Public spaces:** Spaces for social interaction that can be created by governments or reclaimed from urban spaces by neighbors or communities. Studies have shown that green public spaces can increase social inclusion for immigrant youth (Seeland et al 2009), protect against negative health impacts of stressful life events (van den Berg et al 2010), and improve health overall and reduce income related health inequalities (Mitchell and Popham 2008).
9. **Internet:** Using the Internet to remove communication barriers and improve democracy. Unlike television and other broadcast media, the Internet has very

low technological and financial barriers for individuals seeking a presence there. This has the effect of decentralizing the production and distribution of information by returning control to the audience, providing a venue for dialogue instead of monologue. Opinions and services that were previously controlled by small groups or corporations are now shaped by the entire population. Television news networks, sitcoms, and Hollywood productions are being replaced by e-mail, Wikipedia, YouTube, and millions of blogs and forums—all created by the same millions of people who are the audience for the content (Costanza et al 2010).

16.4.2.2 Regional Initiatives

1. **Air trusts:** An example of a regional air trust is the Regional Greenhouse Gas Initiative (RGGI), a cap-and-auction program in the U.S. Northeast, in which most revenues are dedicated to energy efficiency measures. This not only helps mitigate the distributional impacts by generating cost savings for households (RGGI Inc. 2011), but also helps to reduce GHG emissions far more than the caps themselves (Cowart 2008). The European Union Emission Trading System is a cap-and-trade program that puts a cap on GHG emissions from businesses and creates a market for carbon allowances (UE Climate Action). However, most emission allowances are awarded directly to polluters, creating enormous windfall profits for firms. The goal, however, is to auction off half of emissions by 2013, which should help address this problem (Capoor and Ambrosi 2009), and move towards the creation of common property rights to GHG absorption capacity. The United States cap-and-trade program for SO_2 emissions was successful at reducing pollution, but since it awarded emissions rights to polluters (Burtraw and Mansur 1999), it is really an example of the public sector transferring common assets to the private sector (which nonetheless may be superior than leaving them as open access resources).
2. **Watershed trusts:** To protect waterways, fish, and wildlife from agricultural run-off through the promotion of best management practices and sustainable agriculture. An example is the Southeastern Wisconsin Watersheds Trust for the Greater Milwaukee Watersheds.
3. **Land value tax:** These taxes capture some of the value of land for society as a whole, while providing numerous additional benefits. Harrisburg, Pennsylvania, for example, introduced a split tax on real estate, in which the tax on land far exceeded the tax on buildings. This made it necessary for owners of abandoned or degraded buildings to restore or replace them, in order to generate the income required to pay the tax, or sell the land to someone who would. The result was a revitalization of the urban center and an increase in its value as a public space.
4. **Buffalo Commons:** First proposed in 1987 for the social and ecological restoration of the Great Plains, the main purpose of the Commons is to re-establish a corridor between now-fragmented prairie lands for the bison and other wildlife to move freely along as well as to promote the health and sustainability of the land.

5. **Regional planning authorities:** These would begin to develop sustainable economic plans for regional implementation, building upon the lessons (positive and negative) of the Tennessee Valley Authority, the Appalachian Regional Commission, and numerous other modern regional efforts, including those in Canada, Australia, and within and between European Union member states such as in Torino, Ireland, and elsewhere (Bradshaw 1988; Collits 2007; Glasson and Marshall 2007; Hodge and Robinson 2002).

16.4.2.3 National Initiatives

1. **An American Permanent Fund:** The rationale for this fund would be similar to that of the Alaska Permanent Fund, i.e., to distribute common-property income equally to every citizen of the United States. Most of the income of the American Permanent Fund would originate from pollution permits (especially for CO_2), but also from the commons' share of corporate profit. The Fund would contribute to decreasing carbon emissions and improving overall well-being.
2. **Common tax credits:** The rationale behind this tax is that the wealthier segment of American society owes more to the commons than what they pay to the federal government in taxes. So government would increase taxes on the wealthier while giving them the option to either pay those taxes or contribute to a commons trust. An incentive to do the latter would be a 100 % tax credit (Barnes 2006).
3. **National planning**: To help achieve local economic stability, to help distribute work and time in appropriate ways, and to manage potential dislocations caused by reduced growth.

16.4.2.4 Global Initiatives

1. At a larger scale, a proposed Earth Atmospheric Trust could help to massively reduce global carbon emissions while also reducing poverty. This system would comprise a global cap and trade system for all greenhouse gas emissions (preferable to a tax, because it would set the quantity and allow price to vary); the auctioning of all emission permits before allowing trading among permit holders (to send the right price signals to emitters); and a reduction of the cap over time to stabilize atmospheric greenhouse gas concentrations at a level equivalent to 350 ppm of carbon dioxide. The revenues resulting from these efforts would be deposited into the Earth Atmospheric Trust, administered transparently by trustees who serve long terms and have a clear mandate to protect earth's climate system and atmosphere for the benefit of current and future generations. A designated fraction of the revenues derived from auctioning the permits could then be returned to people throughout the world in the form of a per-capita payment. The remainder of the revenues could be used to enhance and restore the

atmosphere, invest in social and technological innovations, assist developing countries, and administer the Trust (Costanza and Farley 2010).
2. International agreements are critical for the success of national climate policies and strategies. Through an international agreement, countries will not suffer for having strict national policies in place; they won't lose their comparative position. This will work in favor of the acceptability of the policies. As a result, there will be a shift toward clean, instead of dirty, production and consumption. It will also incentivize technological change (van den Bergh 2011).
3. A third possible global initiative is the "green paper gold" introduced by Joseph Stiglitz to promote investment in green infrastructure (UNEP 2009; Smith et al 2009). According to Stiglitz, green paper gold, also known as special drawing rights, are "a kind of global money, issued by the International Monetary Fund, which countries agree to exchange for dollars or other hard currencies." Stiglitz has argued that SDRs could be used to promote investment in the developing world and expanding the global commons or "global public goods" (UNEP 2009).

Government has a role to play in protecting and expanding the commons. When government is responsible for a common, it should act as its trustee and should be accountable for it. Government should also increase the allocation of property rights to commons trusts and contribute with the purchasing of former pieces of the commons, now privatized (e.g., through long-term tax-exempt bonds). Common asset trusts of the kind we have described are a mechanism for governments to fulfill these duties.

16.4.3 Implications of Systematic Caps on Natural Resources

A lasting prosperity requires much closer attention to the ecological limits of economic activity. Identifying and imposing strict resource and emission caps is vital for a sustainable economy. The contraction and convergence model developed for climate-related emissions should be applied more generally. Declining caps on throughput should be established for all non-renewable resources. Sustainable yields should be identified for renewable resources. Limits should be established for per-capita emissions and wastes. Effective mechanisms for imposing caps on these material flows should be set in place. Once established, these limits need to be built into the macro-economic frameworks.

Cap and Trade Ownership of the quotas is initially public; the government auctions them to individuals and firms. The revenues go to the treasury and could be used to replace regressive taxes, such as the payroll tax, and to reduce income tax on the lowest incomes, or else to increase investments in public goods or energy efficiency measures that benefit the poor. Once purchased at auction, the quotas can be freely bought and sold by third parties, just as can the resources whose rate of depletion they limit. The trading allows efficient allocation, the auction serves just distribution, and the cap serves the goal of sustainable scale. However, free trading threatens speculative investments and other forms of gaming the market to capture rent. More frequent auctions of permits that could not subsequently be traded could avoid this risk. The same logic can be applied to limiting the off-take from fisheries and forests. With renewables, the quota should be set to approximate sustainable yield. For nonrenewables, sustainable rates of absorption of resulting pollution or the rate of development of renewable substitutes may provide a criterion (Daly 2010). It's worth noting that in a survey conducted in Vermont, only 5.8 % of respondents favored distributed revenue equally among households; 64.2 % favored investing it in natural resources, 14.2 % favored investing it public goods such as education and healthcare, and the remainder favored some mix of dividends and public investments (Kirk 2010).

The idea of a carbon tax and other pollution taxes as a replacement for payroll taxes has gotten political support. It has been recognized that it makes more sense to tax what we burn instead of what we earn (Barnes and McKibben 2010). A very popular method, the Alaskan Permanent Fund, pays a dividend to the citizens of Alaska from the fossil fuel revenue the state collects (Barnes and McKibben 2010). This model is known as "cap and dividend," "where some fraction of the revenues of an auction on emissions allowance is returned to citizens on an equal per capita basis" (Kunkel and Kammen 2011). However, in the case of fossil fuel use, where prices are determined at the global level, and not influenced by extraction rates in any single state, this leads to citizen pressure to "drill, baby, drill," increasing outputs and revenue. In the case of cap and auctions on emissions, local caps would determine prices. Given the highly inelastic demand for fossil fuels (and hence for the waste absorption capacity for CO_2), the tighter the cap, the greater the total revenue, since every 1 % restriction in quantity would lead to a greater than 1 % increase in price.

Cap and dividend is considered by some to be a fair and transparent model, since it is based on the amount of carbon-based energy a person consumes. The more a person consumes, the more he/she would have to pay. It would also have a progressive distributional effect; poor people usually consume less energy than the middle class and the rich (Kunkel and Kammen 2011). For cap and dividend to work, there would have to be a cap on fossil fuel supplies. It is much easier and more cost-effective to have an economy-wide cap on suppliers than emitters. Companies that sell fossil fuel would have to buy permits equal to the carbon content of the fuels they sell. Then, once a year there would be an auditing to make sure the companies have enough permits; if they don't, they would have to pay a high penalty. The number of permits would be reduced every year, decreasing the amount of carbon that enters the economy. As the carbon cap declined, prices would increase and private capital would shift to cleaner alternative technologies and cleaner production and consumption.

Another important element of this model is the dividend, which would be paid equally to every American once a month. As carbon prices increase, so would the dividend, and this in turn would increase the livelihoods of the poor (Barnes and McKibben 2010; Kunkel and Kammen 2011).

However, from a global perspective, a cap and dividend regime in the United States or other wealthy country may be unfair. Both Europe's existing cap and any of the proposed caps in the United States far exceed a fair share of global absorption capacity, and completely fail to account for past contributions to the carbon stock. As discussed previously, reducing flows to ecologically sustainable levels in the short run would likely cause economic collapse, with the worst impacts likely to be borne by the poor. Perhaps the most sustainable, fair, and efficient approach would be for rich countries to invest revenue in making existing infrastructure more energy efficient, and in investing in new, open-source technologies for alternative energy and energy efficiency. This would be more sustainable since it would accelerate the rate at which we develop new technologies and reduce emissions; it would be more fair because it would put the burden of developing new technologies on the wealthy countries, and because the poor would likely benefit most from more energy efficient housing and infrastructure; and it would be more efficient because information is non-rival and should therefore be open access to all, which requires public sector investment, as explained above. Currently, the United States energy sector invests only 0.03 % of sales in R&D, which is clearly inadequate given the importance of developing low carbon energy (Coy 2010).

A variation on the cap-auction-trade mechanism is the commons asset trust, for example, the Earth Atmospheric Trust described above (Barnes et al 2008). In this mechanism, as in the cap-auction-trade, caps are established around a resource. However, in this case a trust manages the sale of permits and the revenue from the auction. It can adjust the availability of permits, depending on need, though ultimately resource use cannot exceed planetary boundaries. The trust would provide equal dividends to the citizens (in a national system) or to countries for distribution to their populations (in an international system), or else invest revenues in public goods. The benefit of providing dividends directly to the population is that it provides some mitigation to the inevitable price increases passed down to consumers (Barnes and McKibben 2010). However, households and businesses frequently fail to adopt energy efficiency measures with high rates of return (Nauclér and Enkvist 2009). This may be especially true for poor households that lack the resources, knowledge, and initiative required to undertake such investments. Recycling revenue into energy efficiency investments with high rates of return would effectively increase total benefits, and could therefore benefit poor households even more than dividends.

An alternative and intermediate option is also available by returning some fraction of the annual revenues as dividends to the population, but using the remainder for other purposes related to preserving and enhancing the common assets, such as atmosphere and climate. This would allow for rewarding people that have a lower carbon footprint to be rewarded as well as for providing funds for related projects like researching and developing renewable energy, deploying renewable energy technologies in developing countries, paying for ecosystem services like carbon sequestration, etc. (Costanza and Farley 2010).

National environmental policies nearly all result in internalizing previously uncounted ecological and social costs. This naturally increases prices relative to those in countries that do not internalize these costs, putting domestic firms at a competitive disadvantage in international trade if the country's international policy is free trade. In this case national and international policies are inconsistent. An international policy consistent with national cost internalization would require moving away from free trade by imposing cost-equalizing tariffs on imports produced under conditions that do not internalize these costs. This is protection, to be sure—but it is protection of an efficient national policy of cost internalization, not protection of an inefficient national firm. Without such protection, or international agreement on cost-internalizing measures, there would be a competitive, cost-externalizing race to the bottom. Globalization (free trade coupled with free capital mobility) seeks to substitute the transnational corporation for the nation as the controlling economic power. Existing traditional community at the national level is sacrificed to the abstraction of a very tenuous "global community."

16.4.4 Sharing Work Time

We need labor policies that allow and encourage shorter work time. Reductions in work time are one of the most cited policies to sustain full employment (or at least decrease unemployment) without increasing output, and to protect workers' livelihoods (Jackson 2009; van den Bergh 2011; Schor 2011).

Work-share programs are considered one of the best ways to respond to a short-term decrease in economic activity. Sharing work time can help reduce, and even prevent, layoffs and also serve as a stabilizer when the economy is slow or the country is facing an imminent recession. Work-share programs help avoid re-hiring and re-training costs and would work best if implemented during the early months of the economic downturn (Schor 2011). In the United States, work sharing has helped save jobs. In 2009, work sharing saved 166,000 jobs, three times more than in 2008. Jack Reid, the Democratic senator from Rhode Island, has introduced work-share bills in Congress (in 2009 and 2010) in an effort to encourage more states to implement such programs. Currently 20 states across the United States operate work-share programs (Schor 2011).

Shorter working hours will improve the work-life balance. Having more time to spend with family and engaging in social interactions has been found to increase subjective well-being, which could lead to decreases in consumption (Golden and Wiens-Tuers 2006; Kasser and Brown 2003; Kasser and Sheldon 2009). Some of the benefits of shorter work hours are less stress and work pressure as well as more time for activities like gardening, child care, meals, volunteer work, social interactions, and so on (Golden and Wiens-Tuers 2006). Kasser and Brown found that people with more leisure time have a smaller ecological footprint (Kasser and Brown 2003). Schor also found similar results: there is a significant positive correlation between work hours and the ecological footprint (Schor 2005a).

There are different types of hours reduction that can be used: reduced average hours per job, reduced average annual hours per person, shorter total hours per working life, etc. The different types of hour reduction will have diverse welfare and economic impacts, which is why it is important to have a just distribution of hours to ensure political feasibility in the long run. Ultimately, environmental degradation will depend on total number of hours worked per capita, which is a function of average hours per job per person and the employment-to-population ratio (Schor 2005a).

Increases in productivity of capital and labor can be accomplished through increases in production and consumption, increases in leisure, or a combination of the two. Thus a greater proportion of any future gains in productivity being taken as an increase in leisure will decrease the rate of unemployment and reduce environmental degradation (Victor and Rosenbluth 2007). The shift to policies that channel productivity growth into increases in free time instead of increases in income will impact the product mix and/or the composition of consumption and can increase environmental degradation because of time-use rebound effects. According to a study on the household production function, timesaving innovations in the production of a service result in an increase in the demand for that service. If the service is energy intensive (i.e., transportation), then the energy demand will increase (Schor 2005a; Binswanger 2001). Thus, the time-use rebound effect will depend on the type of activity that increases as work hours are reduced and there is more free time available. At the household level, families with more purchasing power and less time will invest in time-saving activities, products such as faster transportation and fast food, which are both more energy intensive and require less time (Jalas 2002).

From the production side, if the economy is slowing down (decreases in GDP) or going into recession, it would be necessary to reduce work hours in order to decrease or even avoid unemployment (assuming increases in population). From the consumption side, keeping or increasing work hours will lead to increases in productivity growth (GDP growth), which is translated into increased income and consumption (Knight et al 1970). Working hours affect income and fuel the spending culture, which Knight and colleagues have called the "work and spend" cycle (Knight et al 1970). When a society is in a "work and spend" cycle, advertising and marketing are more effective in promoting consumption. Furthermore, the increases in productivity growth, translated into increases in production and consumption, lead to increases in environmental degradation.

Society has been focusing on green and more efficient technology to decrease energy consumption and GHG emissions, however technological efficiency is necessary but not sufficient. Consumption, energy use, and GHG emissions are closely interconnected and depend on how increasing productivity is achieved, through increases in income or through decreases in work hours. Nässén and colleagues analyzed the income effect of shorter working hours and how consumption and energy use is affected, and found a strong relationship between income and energy use (Nässén et al 2009). Thus a decrease in work time/income of 1 % leads to a decrease in energy use of 0.89 %. However, when analyzing the time effect of shorter work hours—how changes in work hours affect time use off work and, in turn, energy use—the results show that a decrease in work hours by 1 % leads to an increase in energy use of 0.06 % and a respective increase in CO_2 of 0.02 %. If we calculate the net effect of both, the sum of income and time effects, shorter work hours will lead to decreases in energy use of 83 % and decreases in CO_2 of 0.85 % (Nässén et al 2009). Rosnick and Weisbrot found the same positive significant relationship between work hours and energy use (Rosnick and Weisbrot 2007). They showed that a 1 % increase in work hours per worker increased energy use by 1.32 % (controlling for GDP/h, worker/population, and temperature). They estimated that if European Union workers worked as many hours as U.S. workers, there would be an 18 % increase in energy consumption in the European Union.

Juliet Schor argues that there are four main barriers/challenges related to labor costs that disincentivize firms to support decreases in work hours (Schor 2005a):

1. Firms increase wages above market clearing levels to raise the cost of job loss. Thus longer working hours lead to increases in the cost of job loss.
2. Employment related costs (hiring costs, training costs, fringe benefits, etc.) are structured based on the worker and not on hours worked.
3. Workers paid annual salaries instead of per-hour wages tend to work more. Schor found that working for an annual salary instead of a per-hour salary increases the number of work hours up to 100–150 per year (Schor 2005a).
4. An upward-sloping labor supply function will cause the firm to prefer longer hours to avoid salary increases or decreases in worker quality.

Many firms also do not take into consideration workers' preferences for shorter hours. Thus, ion contrast to what the dominant paradigm of neoclassical economics states, workers do not prefer to work more to increase future income and hence consumption. On the contrary, according to several studies (Knight et al 1970; Otterbach 2010), workers are willing to forgo future increases in income in exchange for a reduction in work hours, since future income is less valued. For example, using International Social Survey Programme survey data for 21 developed countries, Otterbach and Sanne showed evidence indicating that, in countries with higher GDP, people prefer to work less even if this means earning less income (Otterbach 2010; Sanne 1992). However, it is important to note that workers are averse to decreases in present income because of habit formation (preferences adapt to current income and consumption levels). Furthermore, firms that do allow shorter work hours can, and many times do, penalize workers for choosing them by denying medical insurance, pensions, opportunity for career trajectory jobs or promotions, and so on (Schor 2005a).

Surveys done before the 2008 crash indicate that 30–50 % of Americans expressed a preference for fewer work hours, even for less pay (Schor 2010). Germany responded to the 2008 crash primarily through the adjustment of hours, and as a result unemployment rates barely increased. This was achieved through the combination of a federal scheme to replace lost wages (which accounted for about 20 % of the reduction in hours), private bargains between employers and unions, canceled overtime, and flexible use of vacation and other time off (Schor 2010). There has also been an increase in leisure time in various OECD countries (Jackson and Victor 2011).

General policies that would help achieve shorter working hours include:

1. Compensation for reducing working time: a package deal to receive compensation for reducing or sharing work hours (New Economics Foundation 2009).
2. Limiting overtime through disincentives to employees and/or raising the overtime premium to make it more expensive for firms to use overtime (New Economics Foundation 2009; Golden and Wiens-Tuers 2006). High levels and increases in income inequality have been identified as one of the reasons workers prefer to work longer hours (Bowles and Park 2005).
3. Standardizing working hours and building flexibility for workers into the labor economy (van den Bergh 2011; Schor 2011). Examples of the latter might include:

   (a) A federal law that allows shorter hours of work to be compensated through at least partial unemployment insurance, to offset the forgone income. States now have the option under federal law to apply for this but many have not done so.
   (b) Government hiring on an 80 % schedule. Government is a big employer and this would have a ripple effect. Policymakers could also structure tax credits to give incentives to employers who hire on 80 % schedules, which would enable more people to be brought back into the labor force than if hiring were done on the full-time schedule.

4. Promoting self-employment and considering adopting the Danish example of "flexicuity" (a combination of flexibility in the labor market, protection for the self-employed, and labor market policy) (New Economics Foundation 2009).
5. Structurally restricting the flow of increased future income in order to reduce consumption. People are more willing to forgo future increases in income and consumption than cuts in current income and consumption (Schor 2005a).
6. As for firms, some incentives that would encourage the firm to accept shorter work time include (Schor 2005a):

   (a) Removing the firms' upper-limit payments to social welfare funds.
   (b) Shifting the responsibility for social welfare to outside entities, like unions, the state, etc. In some cases it may help to create a market for hours, so unions can bargain for workers.
   (c) Ensuring cost-neutral work time reductions through the provision of state subsidies to compensate the firm or through the structure of the deals that are struck with the workers.

7. Transforming a percentage of future productivity gains into shorter work time, but for a large part of the population and not just for a some workers (Nässén et al. 2009).
8. Ensuring basic citizens' income to help equalize wages/income disparities and ensure that workers would be more willing to reduce work hours (Jackson and Victor 2011).

9. Increasing diversity in labor contracts to allow for shorter work time, early retirement, regular sabbaticals, etc., and at the same time ensuring pension systems as safety nets for workers.

16.5 Are These Policies Consistent And Feasible?

We have so far presented a brief vision of what a sustainable and desirable "ecological economy" would look like, and a summary list of some of the policies we think would be required in order to get there. This begs the important question of whether these policies taken together are consistent and whether they are sufficient to achieve the goals we have articulated. Can we have a global economy that is not growing in material terms but that is sustainable and provides a high quality of life for most (if not all) people? While we can never really know the answer to this question until we actually try it out in practice, we can provide a few lines of evidence to help anticipate whether such an economy-in-society-in-nature can work. These include lessons from history, modern day small-scale examples, and modeling studies. We will briefly discuss each of these lines of evidence in turn.

16.5.1 Lessons from History[6]

Human history has traditionally been cast in terms of the rise and fall of great civilizations, wars, specific human achievements, and extreme natural disasters (e.g., earthquakes, floods, plagues). This history tends to leave out, however, the important ecological and climatic context and the less obvious interactions which shaped and mediated these events. The capability to integrate human history with new data about the natural history of the earth at global scales and over centuries to millennia has only recently become possible. It is a critical missing link that is needed in order to provide a much richer picture of how (and why) the planet has changed in historical times, and how (and why) past human societies have either been able to sustain themselves or have collapsed.

Socio-ecological systems are intimately linked in ways that we are only beginning to appreciate (Redman 1999; Schellnhuber et al. 2004; Steffen et al. 2004; Diamond 2005). One major challenge in linking human and environmental change is the development of a new integrated analytical modeling paradigm that reveals the complex web of causation across multiple spatial and temporal scales, while allowing important emergent properties and generalities to rise above the details. Only with such a paradigm can we survey the past and test alternate explanations rigorously. To develop this integrated understanding, a project of the global change

[6]This section relies heavily on Costanza et al. 2007a.

research community has been initiated, titled Integrated History and Future of People on Earth (IHOPE) (Hibbard et al. 2010).

The big, general questions that the IHOPE activity is aimed at addressing can be summarized as the following:

1. What are the complex and interacting mechanisms and processes resulting in the emergence, sustainability, or collapse of socio-ecological systems?
2. What are the pathways to developing and evaluating alternative explanatory frameworks, specific explanations, and models (including complex systems models) by using observations of highly variable quality and coverage?
3. How do we use knowledge of the integrated history of the earth for understanding and creating the future?

It has often been said that if one fails to understand the past, one is doomed to repeat it. IHOPE takes a much more "hopeful" and positive attitude. If we can really understand the past, we can create a better, more sustainable, and desirable future.

Getting back to the original intention of this section, we can ask: Have there ever been non-growing economies that have been sustainable? Actually, this question needs to be turned around, since for the vast majority of human history, economies have grown at very low to zero rates. If anything, from an historical perspective, it is the phenomenal rate of growth of recent economies that is the anomaly. However, we also know that many historical societies have collapsed (Costanza et al. 2007a; Diamond 2005) and many of them were not what we would call "desirable." On the other hand, there were a few successful historical cases in which decline did not occur, including the following (Diamond 2005; Weiss H, Bradley 2001):

- Tikopia Islanders have maintained a sustainable food supply and non-increasing population with a bottom-up social organization.
- New Guinea features a silviculture system more than 7000 years old with an extremely democratic, bottom-up decision-making structure.
- Japan's top-down forest and population policies in the Tokugawa-era arose as a response to an environmental and population crisis, bringing an era of stable population, peace, and prosperity.

Understanding the history of how humans have interacted with the rest of nature can help clarify the options for managing our increasingly interconnected global system. However, we know from history that non-growing societies are feasible. We also know that sustainable societies are possible. As we learn more about the details of historical societies' interaction with the rest of nature, we can use that knowledge to help design a better, more sustainable, and desirable future.

16.5.2 Small-Scale Examples

There are many small-scale examples of sustainable communities that can serve as models. Many groups and communities around the world are involved in building a new economic vision and testing solutions. There are far too many to list all, but here are a few examples:

- Transition town movement (www.transitionnetwork.org)
- Global EcoVillage Network (gen.ecovillage.org)
- Co-Housing Network (www.cohousing.org/)
- Wiser Earth (www.wiserearth.org)
- Sustainable Cities International (www.sustainablecities.net)
- Center for a New American Dream (www.newdream.org)
- Democracy Collaborative (www.community-wealth.org)
- Portland, Oregon, Bureau of Planning and Sustainability (www.portlandonline.com/bps/)

All of these examples embody the vision, worldview, and policies we have elaborated to some extent. Their experiments collectively provide evidence that the policies are feasible at a smaller scale. The challenge is to scale up some of these models to society as a whole.

The problem is that we live in a globalized world and it is difficult to generate larger scale examples that are independent enough from the world to actually try something significantly different. In a sense, we need a total "regime shift" to a new system (Beddoe et al. 2009) and that often requires at least a partial collapse of the existing order.

Nevertheless, even though the world is still largely enmeshed in the conventional economic paradigm, several cities, states, regions, and countries are further along the path we outline than others. Examples include Portland, Oregon; Stockholm and Malmö, Sweden; London, U.K.; the states of Vermont, Washington, and Oregon in the U.S.; Germany, Sweden, Iceland, Denmark, Costa Rica, Bhutan; and many others.

One way to look at this transition is shown in Fig. 16.9, which plots the percent change in ecological footprint by country (an indicator of change in material and energy throughput) against per-capita fair share of the ecological footprint relative to global bio-capacity (an indicator of the scale of the economy, with 1 indicating "optimal" scale) (O'Neill DW in press). This divides the graph into four quadrants, with the center of the graph representing countries that are closest to steady state. In the upper right quadrant are countries whose ecological footprint is increasing and is above their optimal scale. This is "undesirable growth." In the upper left quadrant are countries that are still above their optimal scale but whose ecological footprint is decreasing. This is "desirable degrowth." Likewise, countries that are below their optimal scale are either experiencing "undesirable degrowth" if their ecological footprint is decreasing or "desirable growth" if their footprint is increasing.

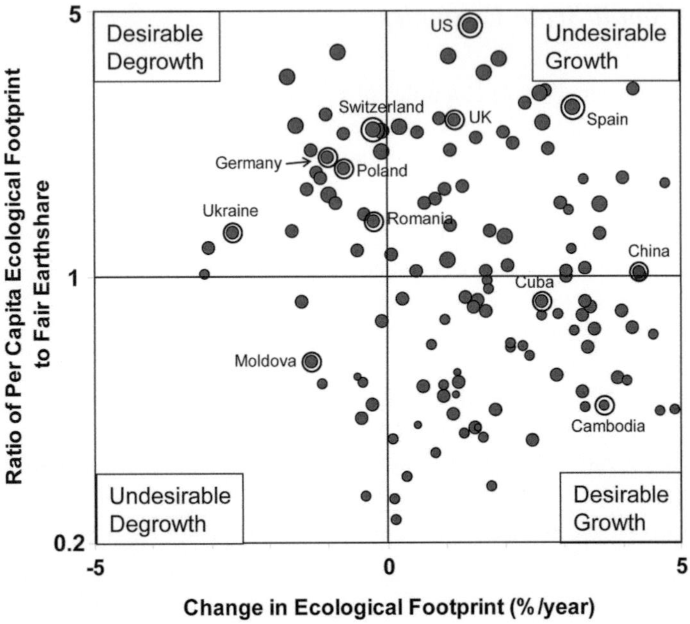

Fig. 16.9 Relationship between change in ecological footprint and distance from per-capita biocapacity by country (O'Neill in press)

The policies we have recommended in this report would drive countries toward the center of this graph. Depending on the country, this could involve either growth or degrowth of material and energy throughput and the scale of the economy, accompanied by an improvement in human well-being broadly defined.

The transition to the world we envision will be a process of directed cultural evolution (Beddoe et al. 2009). To direct this process, we need to generate, communicate, and broadly discuss more smaller-scale experiments that embody the vision and policies we have articulated.

However, a third line of evidence for the feasibility of our vision is based on *simulating* how these societies might work.

16.5.3 Modeling Studies

There are several integrated modeling studies that provide evidence that a sustainable, non-growing economy is both feasible and desirable. Below we briefly describe three of them.

16.5.3.1 World3[7]

The World3 model has been the subject of three influential books, beginning with *The Limits to Growth* (Meadows et al. 1972), continuing with *Beyond the Limits* (Meadows et al. 1992) and ending with the recent, 30-year update (Meadows et al. 2004). World3 is a globally aggregated systems dynamics model broken into five sectors: population, capital, agriculture, nonrenewable resources, and persistent pollution, and containing 16 state variables (i.e., population, capital, pollution, and arable land), 100 variables total, and 80 fixed parameters (Meadows et al. 1972).

Because of the influence of the original book (several million copies were sold), this model has been the topic of intense scrutiny, debate, misunderstanding, and, one could argue, willful misinformation over the years. One interesting bit of misinformation that has been persistently circulating is the idea that the model's "predictions" have been proven totally wrong by subsequent events (Economist 1997). In fact, the model's standard run scenario, made in 1972, fits the data so far very well (Turner 2008). The model's forecasts of collapse under certain scenarios did not start to occur until well past the year 2000. The true tests of this model's forecasts will arrive in the coming decades.

World3 has been criticized on methodological grounds (Cole et al. 1973). The most often cited difficulties are that it does not include prices explicitly, that it assumes resources are ultimately limited, and that it does not present estimates of the statistical uncertainty of its parameters. In fact, World3 is a viable and effective method to reveal the implications of the primary assumptions about the nature of the world that went into it. That is all that can be claimed for any model. These assumptions, or "pre-analytic visions," need to be made clear and placed in direct comparison with the corresponding assumptions of the alternatives, in this case the "unlimited growth model." As Meadows and colleagues have repeatedly pointed out, the essential difference in pre-analytic visions centers around the existence and role of limits: thermodynamic limits, natural resource limits, pollution absorption limits, population carrying capacity limits, and most importantly, the limits of our understanding about where these limits are and how they influence the system (Meadows et al. 1992; Meadows et al. 2004). The alternative unlimited growth model assumes there are no limits that cannot be overcome by continued technological progress, while the limited growth model assumes that there are limits, based on thermodynamic first principles, observations of natural ecosystems, and understanding of basic planetary boundaries (Rockström et al. 2009). Ultimately, we do not know which pre-analytic vision is correct (they are, after all, assumptions), so we have to consider the relative costs of being wrong in each case (Costanza 2000b; Costanza et al. 2000).

[7] This and the following section are adapted from [164. Costanza, Leemans, Boumans, and Gaddis (2007c) Integrated global models. In: Costanza R, Graumlich L, Steffen W, editors. Sustainability or collapse? An integrated history and future of people on earth. Cambridge, MA: MIT Press. pp. 417–446.

Finally, while the discussions of World3 often point to the limited vs. unlimited growth assumptions as a key difference from conventional economic models, they do not take the opportunity to look at the relative costs and benefits of being right or wrong in those assumptions. If one does this, one can easily see that the cost of assuming no limits and being wrong is the collapse scenarios shown by World3, while the cost of assuming limits and being wrong is only mildly constrained growth (Boumans et al. 2002, #485).

16.5.3.2 Gumbo

The Global Unified Metamodel of the BiOsphere (GUMBO) (Boumans et al. 2002) was developed by a working group at the National Center for Ecological Analysis and Synthesis (NCEAS) in Santa Barbara, California. Its goal was to simulate the integrated earth system and assess the dynamics and values of ecosystem services. It is a "metamodel" in that it represents a synthesis and a simplification of several existing dynamic global models in both the natural and social sciences at an intermediate level of complexity. GUMBO is the first global model to include the dynamic feedbacks among human technology, economic production and welfare, and ecosystem goods and services within the dynamic earth system. GUMBO includes five distinct modules or "spheres": the atmosphere, lithosphere, hydrosphere, biosphere, and anthroposphere. The earth's surface is further divided into 11 biomes or ecosystem types, which encompass the entire surface area of the planet: open ocean, coastal ocean, forests, grasslands, wetlands, lakes/rivers, deserts, tundra, ice/rock, croplands, and urban. The relative areas of each biome change in response to urban and rural population growth, gross world product (GWP), and changes in global temperature. Among the spheres and biomes, there are exchanges of energy, carbon, nutrients, water, and mineral matter. In GUMBO, ecosystem services are aggregated to seven major types, while ecosystem goods are aggregated into four major types. Ecosystem services, in contrast to ecosystem goods, cannot accumulate or be used at a specified rate of depletion. Ecosystem services include soil formation, gas regulation, climate regulation, nutrient cycling, disturbance regulation, recreation and culture, and waste assimilation. Ecosystem goods include water, harvested organic matter, mined ores, and extracted fossil fuel. These 11 goods and services represent the output from natural capital, which combines with built capital, human capital, and social capital to produce economic goods and services and social welfare. The model calculates the marginal product of ecosystem services in both the production and welfare functions as estimates of the shadow prices of each service.

Historical calibrations from 1900 to 2000 for 14 key variables for which quantitative time series data were available produced an average R^2 of 0.922. A range of future scenarios to the year 2100 representing different assumptions about future technological change, investment strategies, and other factors have been simulated. The scenarios include a base case (using the "best fit" values of the model parameters over the historical period) and four initial alternative scenarios. These four alternatives are the result of two variations (a technologically optimistic set and a skeptical set) concerning assumptions about key parameters in the model, arrayed against two variations (a technologically optimistic and a skeptical set) of policy settings concerning the rates of investment in the four types of capital (natural, social, human, and built). They correspond to the four scenarios laid out by Costanza (2000b) and are very similar to the four scenarios used in the Millennium Ecosystem Assessment (MEA 2005).

Like World3, GUMBO can produce scenarios of global steady state or overshoot and decline. Achieving a steady state is possible with investment and population priorities similar to the ones outlined in the previous sections of this report, indicating that the policies are internally consistent.

16.5.3.3 LowGrow[8]

More recently, the "LowGrow" model of the Canadian economy has been used to describe the possibility of constructing an economy that is not growing in GDP terms but that is stable, with high employment, low carbon emissions, and high quality of life (Victor and Rosenbluth 2007; Victor 2008). LowGrow was explicitly constructed as a fairly conventional macroeconomic model calibrated for the Canadian economy, with added features to simulate the effects on natural and social capital. shows the simplified structure of LowGrow. Aggregate (macro) demand is determined in the normal way as the sum of consumption expenditure (C), investment expenditure (I), government expenditure (G), and the difference between exports (X) and imports (I.) Their sum total is GDP measured as expenditure. There are separate equations for each of these components in the model, estimated with Canadian data from about 1981 to 2005, depending on the variable. Production in the economy is estimated by a Cobb-Douglas production function in which macro supply is a function of employed labor (L) and employed capital (K). The time variable (t) represents changes in productivity from improvements in technology, labor skills, and organization. The production function is shown as macro supply at the bottom of. It estimates the labor (L) and employed capital (K) required to produce GDP allowing for changes in productivity over time.

There is a second important link between aggregate demand and the production function. Investment expenditures (net of depreciation), which are part of aggregate demand, add to the economy's stock of capital, increasing its productive capacity. Also, capital and labor become more productive over time. It follows that, other things equal, without an increase in aggregate demand these increases in capital and productivity reduce employment. Economic growth (i.e., increases in GDP) is needed to prevent unemployment rising as capacity and productivity increase.

Population is determined exogenously in LowGrow, which offers a choice of three projections from Statistics Canada. Population is also one of the variables that

[8] Adapted from 173. Victor (2008) Managing without growth: Slower by design, not disaster. Cheltenham, UK: Edward Elgar Publishing.

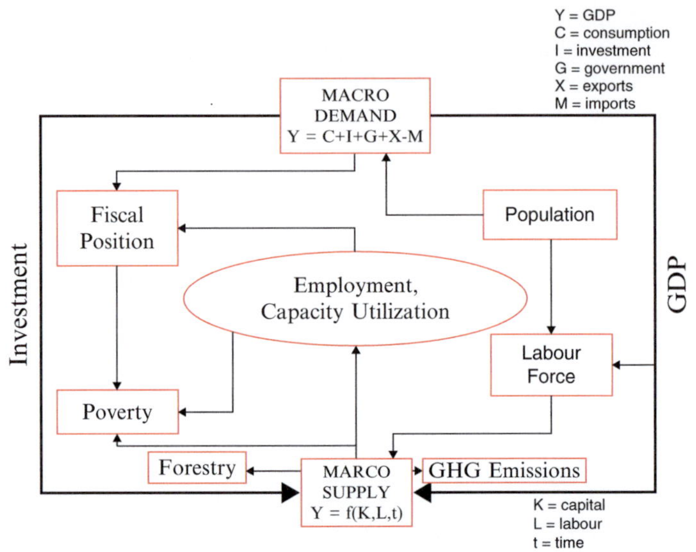

Fig. 16.10 The high level structure of LowGrow (Victor 2008)

determines consumption expenditures in the economy. The labor force is estimated in LowGrow as a function of GDP and population.

There is no monetary sector in LowGrow. For simplicity it is assumed that the Bank of Canada, Canada's central bank, regulates the money supply to keep inflation at or near the target level of 2 % per year. LowGrow includes an exogenously set rate of interest that remains unchanged throughout each run of the model. A higher cost of borrowing discourages investment, which reduces aggregate demand. It also raises the cost to the government of servicing its debt (Fig. 16.10).

The price level is not included as a variable in LowGrow, although the model warns of inflationary pressures when the rate of unemployment falls below 4 % (effectively full employment in Canada).

LowGrow includes features that are particularly relevant for exploring a low/no-growth economy. LowGrow includes emissions of carbon dioxide and other greenhouse gases, a carbon tax, a forestry sub-model, and provision for redistributing incomes. It measures poverty using the UN's Human Poverty Index (i.e., HPI-2 for selected OECD countries). LowGrow allows additional funds to be spent on health care and on programs for reducing adult illiteracy (both included in HPI-2) and estimates their impacts on longevity and adult literacy with equations from the literature.

Implications of changes in the level of government expenditures can be simulated in LowGrow through a variety of fiscal policies, including an annual percentage change in government expenditure that can vary over time, and a balanced budget. LowGrow keeps track of the overall fiscal position of all three levels of government combined (federal, provincial, and municipal) by calculating total revenues and expenditures and estimating debt repayment based on the historical

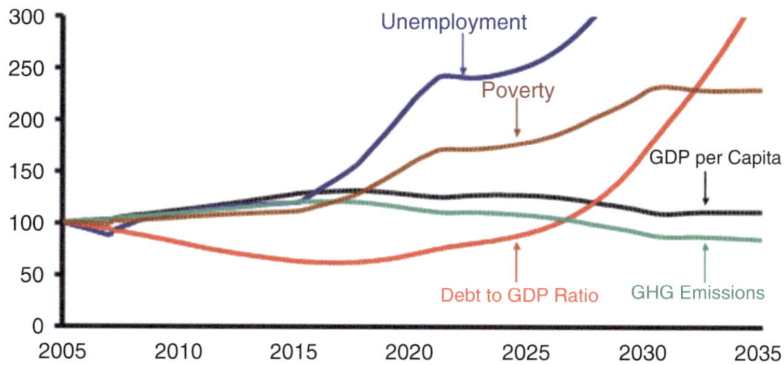

Fig. 16.11 A no-growth disaster (Victor 2008)

record. As the level of government indebtedness declines, the rates of taxes on personal incomes and profits in LowGrow are reduced endogenously, broadly consistent with government policy in Canada.

In LowGrow, as in the economy that it represents, economic growth is driven by net investment (which adds to productive assets), growth in the labor force, increases in productivity, growth in the net trade balance, growth in government expenditures, and growth in population. Low- and no-growth scenarios can be examined by reducing the rates of increase in each of these factors singly or in combination.

Economic growth is desired not only for what it offers in terms of increased living standards but also out of fear of what might happen if a modern economy deliberately tried to wean itself off growth. Such fears are well-founded. Modern economies and their public, private, and not-for-profit institutions, as well as individual citizens, have come to rely on growth. They expect it, they plan for it, they believe in it.

Several scenarios have been run with LowGrow to look at the feasibility of a low- or no-growth economy. Adjusting to life without economic growth could be a wrenching experience and a lot could go wrong, as shown in Fig. 16.11. In this scenario, zero growth in GDP and GDP per capita is achieved around 2030 by eliminating growth in government expenditures, productivity, and population, and achieving zero net investment and net trade balance over a period of years starting in 2010. GDP per capita rises slightly until all the factors contributing to growth are extinguished and then drops back to the same level as at the start of 2005. Meanwhile, the unemployment rate literally goes off the chart, causing a dramatic rise in poverty. The debt-to-GDP ratio also rises to untenable heights, largely because of the massive increase in income support paid to the rising number of unemployed. Certainly, the human misery entailed in such a scenario is to be avoided if at all possible (Fig. 16.11).

However, a wide range of low- and no-growth scenarios can be examined with LowGrow. Some are not much better than the no-growth disaster just described, but others offer more promise. One such promising scenario is shown in Fig. 16.12.

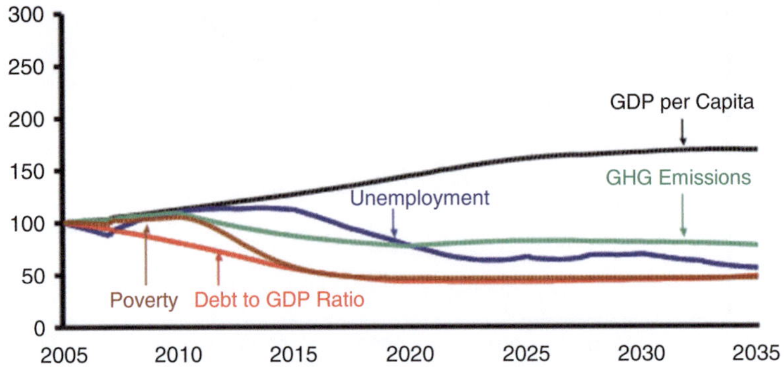

Fig. 16.12 A better low/no growth scenario (Victor 2008)

Compared with the business as usual scenario, GDP per capita grows more slowly, leveling off around 2028, at which time the rate of unemployment is 5.7 %. The unemployment rate continues to decline to 4.0 % by 2035. By 2020 the poverty index declines from 10.7 to an internationally unprecedented level of 4.9, where it remains, and the debt-to-GDP ratio declines to about 30 % and is maintained at that level to 2035. Greenhouse gas emissions are 31 % lower at the start of 2035 than 2005 and 41 % lower than their high point in 2010. These results are obtained by slower growth in government expenditures, net investment, and productivity; a positive net trade balance; cessation of growth in population; a reduced workweek; a revenue-neutral carbon tax; and increased government expenditure on anti-poverty programs, adult literacy programs, and health care.

The contrast between the no-growth disaster (Fig. 16.11) and the sustainable and desirable no-growth scenario (Fig. 16.12) is striking and naturally raises questions about what makes the difference. The no-growth disaster scenario is based on a systematic elimination of all of the factors represented in LowGrow that contribute to growth without any compensating adjustments. The better no/low-growth scenario results from a wide range of policy measures, some more controversial than others, that would be required to transform the business as usual scenario into the kind of scenario illustrated in Fig. 16.12. In summary, these policy measures include:

- **Investment:** reduced net investment, a shift from investment in private to public goods through changes in taxation and expenditures.
- **Labor force:** stabilization through changing age structure of the population and population stabilization.
- **Population:** stabilization through changes to immigration policy.
- **Poverty:** trickle down replaced with focused anti-poverty programs that address the social determinants of illness and provide more direct income support.
- **Technological change:** slower, more discriminating, and preventative rather than end-of-pipe, through technology assessment and changes in the education of scientists and engineers.
- **Government expenditures:** a declining rate of increase.

- **Trade**: a stable, positive net trade balance (and diversification of markets).
- **Workweek:** shorter and with more leisure, through changes in compensation, work organization and standard working hours, and active market labor policies.
- **Greenhouse gases:** a revenue neutral carbon tax.

To complement these policies:

- **Consumption:** more public goods and fewer positional (status) goods, through changes in taxation and marketing.
- **Environment and resources:** limits on throughput and use of space through better land use planning and habitat protection and ecological fiscal reform.
- **Localization:** fiscal and trade policies to strengthen local economies.

These are precisely the policies that we have elaborated in the previous sections of this report. No model results can be taken as definitive, since models are only as good as the assumptions that go into them. But what World3, GUMBO, and LowGrow have provided is some evidence for the *consistency* and *feasibility* of these policies, taken together, to produce an economy that is not growing in GDP terms, but that is sustainable and desirable.

16.6 Conclusions

The world is at a critical turning point. This turning will not come overnight, however. In fact we are probably already in the middle of it. It will take decades. But it is a time of real choices: (1) we can attempt to continue business as usual, pursuing the conventional economic growth paradigm that has dominated economic policy since the end of World War II; (2) we can pursue an environmentally sensitive version of this model and attempt to achieve "green growth"; or (3) we can pursue a more radical departure from the mainstream that does not consider growth to be the real goal at all, but rather sustainable human well-being, acknowledging uncertainty and the complexity of understanding, creating, and sustaining well-being (Table 16.1). This report has described option 3, which entails a change in worldview, vision, and goals that would have far-reaching implications and will demand a substantial departure from business as usual. However, we believe it is the only option that is both sustainable and desirable on our finite planet.

In this report we have sketched a vision of what this "ecological economics" option might look like and how we could get there. We believe that this option can provide full employment and a high quality of life for everyone into the indefinite future while staying within the safe environmental operating space for humanity on earth. Developed countries have a special responsibility for achieving those goals. To get there, we need to stabilize population; more equitably share resources, income, and work; invest in the natural and social capital commons; reform the financial system to better reflect real assets and liabilities; create better measures of progress; reform tax systems to tax "bads" rather than goods; promote technological innovations that support well-being rather than growth; establish "strong democracy," and create a culture of well-being rather than consumption. In other words, a complete makeover.

These policies are mutually supportive and the resulting system is feasible. It is not merely a utopian fantasy. In fact, it is as usual that is the utopian fantasy. We will have to create something different and better or risk collapse into something far worse.

The substantial challenge is making the transition to a better world in a peaceful and positive way. There is no way to predict the exact path this transition might take, but we hope that painting this picture of a possible end-point and some milestones along the way will help make this choice and this journey a more viable option.

Acknowledgements This study is part of the Sustainable Development in the twenty-first Century (SD21) project. The project is implemented by the Division for Sustainable Development of the United Nations Department of Economic and Social Affairs, contract number 20465. Support from the European Commission is gratefully acknowledged.

This publication has been produced with the assistance of the European Union. The contents of this publication are the sole responsibility of the United Nations Department of Economic and Social Affairs and can in no way be taken to reflect the views of the European Union.

We thank David Le Blanc, project officer, for guidance and extremely useful feedback on earlier versions of this report. We also thank Ron Coleman and Thomas Prugh for editorial assistance. The ideas in this report have been published elsewhere by the authors and several sections are taken almost verbatim from these prior publications. The co-authors all agree on the general concepts and ideas but do not necessarily completely endorse every specific argument. This report is a synthesis and compendium.

References

Acemoglu D, Robinson J (2009) Foundations of societal inequality. Science 326:678–679

Adams JS, Victurine R (2011) Permanent conservation trusts: a study of the long-term benefits of conservation endowments. Conservation Trust Funds

Almås I, Cappelen AW, Sørensen EØ, Tungodden B (2010) Fairness and the development of inequality acceptance. Science 328:1176–1178

Alperovitz G (2011) America beyond capitalism: reclaiming our wealth, our liberty, and our democracy. Democracy Collaborative Press, New York

Alperovitz G, Faux J (1984) Rebuilding America: a blueprint for the new economy. PsychoBabel Books, Oxford

Alston JM, Marra MC, Pardey PG, Wyatt TJ (2000) Research returns redux: a meta-analysis of the returns to agricultural R&D. Aust J Agric Resour Econ 44:185–215

Anderson V (1991) Alternative economic indicators. Routledge, London

Arrow K, Dasgupta P, Goulder L, Daily G, Ehrlich P et al (2004) Are we consuming too much? J Econ Perspect 18:147–172

Avato P, Coony J (2008) Accelerating clean energy technology research, development, and deployment. World Bank, Washington, DC

Azar OH (2004) What sustains social norms and how they evolve? The case of tipping. J Econ Behav Organ 54:49–64

Barber BR (1984) Strong democracy: participatory politics for a new age. University of California Press, London
Barber BR (1998) A place for us: how to make society civil and democracy strong. Hill and Wang, New York
Barnes P (2006) Capitalism 3.0: a guide to reclaiming the commons. Berrett-Koehler Publishers, San Francisco
Barnes P, McKibben W (2010) A simple market mechanism to clean up our economy. Solutions 1:30–40
Barnes P, Costanza R, Hawken P, Orr D, Ostrom E et al (2008) Creating an earth atmospheric trust. Science 319:724
Beddoe R, Costanza R, Farley J, Garza E, Kent J et al (2009) Overcoming systemic roadblocks to sustainability: the evolutionary redesign of worldviews, institutions, and technologies. Proc Natl Acad Sci 106:2483–2489
Bell LA, Freeman RB (2001) The incentive for working hard: explaining hours worked: differences in the US and Germany. Labour Econ 8:181–202
Binswanger M (2001) Technological progress and sustainable development: what about the rebound effect? Ecol Econ 36:119–132
Birdsall N, Kelley AC, Sinding SW (2003) Population matters: demographic change, economic growth, and poverty in the developing world. Oxford University Press, Oxford
Bloom DE, Canning D (2004) Global demographic change: dimensions and economic significance. Popul Dev Rev 34:17–51
Bongaarts J (2009) Human population growth and the demographic transition. Philos Trans R Soc B: Biol Sci 364:2985–2990
Bongaarts J, Sinding S (2011) Population policy in transition in the developing world. Science 333:574–576
Boulding K (1968) Beyond economics: essays on society, religion, and ethics. University of Michigan Press, Ann Arbor
Boumans R, Costanza R, Farley J, Wilson MA, Portela R, Rotmans J, Villa F, Grasso M (2002) Modeling the dynamics of the integrated earth system and the value of global ecosystem services using the GUMBO model. Ecol Econ 41:529–560
Bowles S, Park Y (2005) Emulation, inequality, and work hours: was Thorsten Veblen right? Econ J 115:F397–F412
Bradshaw MJ (1988) Regions and regionalism in the United States. University Press of Mississippi, Jackson
Brown LR (2004) Outgrowing the earth: the food security challenge in an age of falling water tables and rising temperatures. Earth Policy Institute, London
Brown LR (2011) The new geopolitics of food. Foreign policy. The FP Group, Washington, DC
Brown LR, Institute EP (2008) Plan B 3.0: mobilizing to save civilization. WW Norton, New York
Burtraw D, Mansur E (1999) The environmental effects of SO2 trading and banking. Environ Sci Technol 33:3489–3494
Calomiris CW (2010) Financial innovation, regulation, and reform. In: Spence M, Leipziger D (eds) Globalization and growth: implications for a post-crisis world. The World Bank, Washington DC
Capoor K, Ambrosi P (2009) State and trends of the carbon market 2009. World Bank, Washington DC
Carr D, Khan M (2004) The unfinished agenda: meeting the need for family planning in less developed countries. Population Reference Bureau, Washington, DC
Cincotta RP, Engelman R, Anastasion D (2003) The security demographic: population and civil conflict after the Cold War. DTIC Document
Cole HSD, Freeman C, Jahoda M, Pavitt KLR (1973) Models of doom: a critique of the limits to growth. Universe Publishing, New York
Collits P (2007) Planning for regions in Australia. In: Thompson S (ed) Planning Australia: an overview of urban and regional planning. Cambridge University Press, Melbourne

Costanza R (1991) Ecological economics: the science and management of sustainability. Columbia University Press, New York
Costanza R (2000a) Social goals and the valuation of ecosystem services. Ecosystems 3:4–10
Costanza R (2000b) Visions of alternative (unpredictable) futures and their use in policy analysis. Conserv Ecol 4:5
Costanza R, Farley J (2010) What should be done with the revenues from a carbon cap and auction systems? Solutions 1:33
Costanza R, dArge R, de Groot R, Farber S, Grasso M et al (1997) The value of the world's ecosystem services and natural capital. Nature 387:253–260
Costanza R, Andrade F, Antunes P, van den Belt M, Boersma D et al (1998) Principles for sustainable governance of the oceans. Science 281:198–199
Costanza R, Daly M, Folke C, Hawken P, Holling CS et al (2000) Managing our environmental portfolio. Bioscience 50:149–155
Costanza R, Graumlich L, Steffen W, Crumley C, Dearing J et al (2007a) Sustainability or collapse: what can we learn from integrating the history of humans and the rest of nature? Ambio 36:522–527
Costanza R, Fisher B, Ali S, Beer C, Bond L et al (2007b) Quality of life: an approach integrating opportunities, human needs, and subjective well-being. Ecol Econ 61:267–276
Costanza R, Leemans R, Boumans R, Gaddis E (2007c) Integrated global models. In: Costanza R, Graumlich L, Steffen W (eds) Sustainability or collapse? An integrated history and future of people on earth. MIT Press, Cambridge, MA, pp 417–446
Costanza R, Hart M, Posner S, Talberth J (2009) Beyond GDP: the need for new measures of progress. Frederick S. Pardee Center for the Study of the Longer-Range Future, Boston
Costanza R, Farley J, Kubiszewski I (2010) Adapting institutions for life in a full world. In: Assadourian E (ed) 2010 state of the world. The Worldwatch Institute, Washington, DC
Cowart R (2008) Carbon caps and efficiency resources: how climate legislation can mobilize efficiency and lower the cost of greenhouse gas emission reduction. Vermont Law Rev 33:201–223
Coy P (2010) The other U.S. energy crisis: lack of R&D: R&D neglect is holding back innovative energy technologies. Bloomberg Business Week
Coy P (2012) The other U.S. energy crisis: lack of R&D: R&D neglect is holding back innovative energy technologies. Bloomberg Businessweek. New York: Bloomberg L.P
Cross G (1993) Of time and money: the making of consumer culture. Routledge, London
Crutzen PJ (2002) The effects of industrial and agricultural practices on atmospheric chemistry and climate during the Anthropocene. J Environ Sci Health A 37:423–424
Daly HE (2008) Ecological economics and sustainable development, selected essays of Herman Daly. Edward Elgar Publishing, Cornwall
Daly HE (2010) From a failed-growth economy to a steady-state economy. Solutions 1:37–43
Daly HE, Cobb JB Jr (1994) For the common good: redirecting the economy toward community, the environment, and a sustainable future, 2nd edn. Beacon, Boston
Daly HE, Farley J (2004) Ecological economics: principles and applications. Island Press, Washington
Davis JE, Jacobus R (2008) The city-CLT partnership: municipal support for community land trusts. Lincoln Institute of Land Policy, Cambridge, MA
De Schutter O (2010) Report submitted by the special rapporteur on the right to food. United Nations Human Right Council, New Yock
Department of Economic and Social Affairs UN (2009) What would it take to accelerate fertility decline in the least developed countries? United Nations, Dept. of Economic and Social Affairs, New York
Diamond J (2005) Guns, germs, and steel: the fates of human societies. WW Norton, New York
Diener E, Suh EM (2003) National differences in subjective weil-being. In: Kahneman D, Diener E, Schwarz N (eds) Well-being: the foundations of hedonic psychology. Russell Sage, New York

Durning A (1992) How much is enough? The consumer society and the future of the Earth. W. W. Norton, New York
Easterlin RA (2003) Explaining happiness. Proc Natl Acad Sci 100:11176–11183
Economist (1997) Plenty of gloom. Econ pp. 19–20
Ehrlich PR, Ehrlich AH (1991) The population explosion. Simon & Schuster, New York
Farley J (2010) Ecological Economics. In: Lerch D, Heinberg R (eds) The Post Carbon Reader – Managing the 21st Century's Sustainability Crises, Watershed Media, Healdsburg
Farley J, Costanza R (2002) Envisioning shared goals for humanity: a detailed, shared vision of a sustainable and desirable USA in 2100. Ecol Econ 43:245–259
Farley J, Aquino A, Daniels A, Moulaert A, Lee D et al (2010) Global mechanisms for sustaining and enhancing PES schemes. Ecol Econ 69:2075–2084
Feeny D, Berkes F, McCay BJ, Acheson JM (1990) The tragedy of the commons: twenty-two years later. Hum Ecol 18:1–19
Fehr E, Falk A (2002) Psychological foundations of incentives. Eur Econ Rev 46:687–724
Food and Agriculture Organization of the United Nations (2009) How to feed the world in 2050. UN FAO, Rome
Foundation NE (2009) The cuts won't work: the second report of the green new deal group. New Economics Foundation, London
Frank RH (1999) Luxury fever: why money fails to satisfy in an era of excess. Simon and Schuster, New York
Frank RH (2007) Falling behind: how rising inequality harms the middle class. University of California Press, Berkeley
Frank R (2012) Higher taxes help the richest, too. New York Times, New York
Gaffney M (2009) The hidden taxable capacity of land: enough and to spare. Int J Soc Econ 36:328–411
Gannon Z, Lawson N (2009) The advertising effect: how do we get the balance of advertising right. Compass, London
Glasson J, Marshall T (2007) Regional planning. Routledge, Milton Park
Golden L, Wiens-Tuers B (2006) To your happiness? Extra hours of labor supply and worker well-being. J Socio-Econ 35:382–397
Goldstein M (2011) Paulson, at $4.9 billion, tops hedge fund earner list. Reuters: Thomson Reuters
Goodland R, Anhang J (2009) Livestock and climate change: what if the key actors in climate change are cows, pigs and chickens? World Watch 22:10–19
Gore A (2007) The assault on reason. Penguin Pr, New York
Hansen J, Sato M, Kharecha P, Beerling D, Berner R et al (2008) Target atmospheric CO2: where should humanity aim? Open Atmos Sci J 2:217–231
Hardin G (1968) The tragedy of the commons. American Association for the Advancement of Science, Washington, D.C, pp 1243–1248
Hernandez-Murillo R, Martinek CJ (2010) The dismal science tackles happiness data. Reg Econ pp. 14–15
Hibbard KA, Costanza R, Crumley C, van der Leeuw S, Aulenbach S et al (2010) Developing an integrated history and future of people on earth (IHOPE): research plan. IGBP Secretariat, Stockholm, p 40
Hodge G, Robinson IM (2002) Planning Canadian regions. University of British Columbia Press, Vancouver
Howarth RB (1996) Status effects and environmental externalities. Ecol Econ 16:25–34
Jackson T (2007) Sustainable consumption. In: Atkinson G, Dietz S, Neumayer E (eds) Handbook of sustainable development. Edward Elgar Publishing, Cornwall, p 254
Jackson T (2009) Prosperity without growth: economics for a finite planet. Earthscan/James & James, New York
Jackson T, Victor P (2011) Productivity and work in the 'green economy': some theoretical reflections and empirical tests. Environ Innov Soc Transit 1:101–108

Jalas M (2002) A time use perspective on the materials intensity of consumption. Ecol Econ 41:109–123
Kallis G (2011) In defence of degrowth. Ecol Econ 70:873–880
Kasser T (2002) The high price of materialism. The MIT Press, Cambridge
Kasser T, Brown KW (2003) On time, happiness, and ecological footprints. In: DeGraaf J (ed) Take back your time!: fighting overwork and time poverty in America. Berrett-Koehler Publishers, San Francisco, pp 107–112
Kasser T, Sheldon K (2009) Time affluence as a path toward personal happiness and ethical business practice: empirical evidence from four studies. J Bus Ethics 84:243–255
Kirk DJ (2010) Allocating Vermont's trust: dividends or public investment from carbon cap and auction revenues. University of Vermont, Burlington
Knight R, Schor JB, Rose E (In Progress) Work hours, consumption, and climate change: a cross-national analysis of OECD countries, 1970–2007
Kowalski SP (2002) Golden rice: a case study in intellectual property management and International capacity building. Pierce Law Faculty Scholarship Series Paper 7
Kubiszewski I, Farley J, Costanza R (2010) The production and allocation of information as a good that is enhanced with increased use. Ecol Econ 69:1344–1354
Kunkel CM, Kammen DM (2011) Design and implementation of carbon cap and dividend policies. Energ Policy 39:477–486
Layard R (2005) Happiness: Lessons from a new science. The Penguin Press, New York
López R, Galinato GI (2007) Should governments stop subsidies to private goods? Evidence from rural Latin America. J Public Econ 91:1071–1094
Martínez-Alier J, Pascual U, Vivien FD, Zaccai E (2010) Sustainable de-growth: mapping the context, criticisms and future prospects of an emergent paradigm. Ecol Econ 69:1741–1747
MaxNeef M (1995) Economic growth and quality of life: a threshold hypothesis. Ecol Econ 15:115–118
Meadows DH, Randers J, Meadows DL (2004) Limits to growth: the 30-year update. Chelsea Green Publishing Company, White River Junction, 368 p
MEA (2005) Ecosystems and human well-being: synthesis. Island Press, Washington, DC
Meadows D (2010) Leverage points: places to intervene in a system. Solutions 1:41–49
Meadows DH, Meadows DL, Randers J, Behrens WW (1972) The limits to growth. Club of Rome, Rome
Meadows DH, Meadows DL, Randers J (1992) Beyond the limits: confronting global collapse, envisioning a sustainable future. Chelsea Green Publishing, White River Junction
Mitchell R, Popham F (2008) Effect of exposure to natural environment on health inequalities: an observational population study. Lancet 372:1655–1660
Nässén J, Larsson J, Holmberg J (2009) The effect of work hours on energy use: A micro-analysis of time and income effects. La Colle sur Loup. pp. 1–6
Nauclér T, Enkvist PA (2009) Pathways to a low-carbon economy: Version 2 of the global greenhouse gas abatement cost curve. McKinsey & Company, 26–31 p
New Economics Foundation (2008) Triple crunch: Joined-up solutions to financial chaos, oil decline and climate change to transform the economy. New Economics Foundation, London
Nørgård JS (2010) Sustainable degrowth through more amateur economy, Barcelona
Nussbaum M, Glover J (1995) Women, culture, and development: a study of human capabilities. Oxford University Press, Oxford
O'Neill DW (in press) Measuring progress in the degrowth transition to a steady state economy. Ecological Economics
Ostrom E (1990) Governing the commons: the evolution of institutions for collective action. Cambridge University Press, Cambridge/New York
Otterbach S (2010) Mismatches between actual and preferred work time: empirical evidence of hours constraints in 21 countries. J Consum Policy 33:143–161
Pell D (1989) In: Berkes F (ed) Common property resources: ecology and community-based sustainable development. Belhaven, London

Polimeni JM, Mayumi K, Giampietro M, Alcott B (2008) The Jevons paradoxnext and the myth of resource efficiency improvements. Earthscan, Sterling
Prugh T, Costanza R, Daly HE (2000) The local politics of global sustainability. Island Press, Washington, DC
Raworth K (2012) A safe and just space for humanity: can we live within the doughnut? Oxfam International, London
Redman CL (1999) Human impact on ancient environments. University of Arizona Press, Tucson
RGGI Inc. (2011) Investment of proceeds from RGGI CO2 allowances. Regional greenhouse gas initiative. Online: http://www.rggi.org/docs/Investment_of_RGGI_Allowance_Proceeds.pdf
Rockström J, Steffen W, Noone K, Persson Å, Chapin FS et al (2009) A safe operating space for humanity. Nature 461:472–475
Rosnick D, Weisbrot M (2007) Are shorter work hours good for the environment? A comparison of U.S. and European energy consumption. Int J Health Serv 37:405–417
Sanne C (1992) How much work? Futures 24:23–36
Schellnhuber HJ, Crutzen PJ, Clark WC, Claussen M, Held H (2004) Earth system analysis for sustainability. The MIT Press and Freie Universität Berlin, Berlin
Schor JB (2005a) Sustainable consumption and worktime reduction. J Ind Ecol 9:37–50
Schor JB (2005b) Prices and quantities: unsustainable consumption and the global economy. Ecol Econ 55:309–320
Schor JB (2010) The work-sharing boom: exit ramp to a new economy? Yes! Magazine. Positive Futures Network, Bainbridge Island
Schor JB (2011) Combating consumerism and capitalism: a decade of no logo. WSQ: Women's Stud Q 38:299–301
Sedgh G, Hussain R, Bankole A, Singh S (2007) Women with an unmet need for contraception in developing countries and their reasons for not using a method. Guttmacher Institute, New York, pp 5–40
Seeland K, Dubendorfer S, Hansmann R (2009) Making friends in Zurich's urban forests and parks: the role of public green space for social inclusion of youths from different cultures. Forest Policy Econ 11:10–17
Simon HA (1981) Man and his tools: technology and the human condition. Intermediair bibliotheek, Amsterdam
Singh S, Darroch JE, Vlassoff M, Nadeau J (2003) Adding it up: the benefits of investing in sexual and reproductive health care. Guttmacher Institute and United Nations Population Fund (UNFPA), New York
Singh S, Darroch J, Ashford L, Vlassoff M (2010) Adding it up: the costs and benefits of investing in family planning and maternal and newborn health. Guttmacher Institute, New York
Smith A (2009) The Internet's role in campaign 2008. Pew Internet Am Life Project 15
Smith B, Costello T, Brecher J (2009) Bailing out the planet. Transnational Institute
Sorrell S (2007) The rebound effect: an assessment of the evidence for economy-wide energy savings from improved energy efficiency. Energy Group for the Technology and Policy Assessment function of The New Economics Foundation UK Energy Research Centre, Sussex
Speidel JJ, Weiss DC, Ethelston SA, Gilbert SM (2009) Population policies, programmes and the environment. Philos Trans R Soc B: Biol Sci 364:3049–3065
Steffen W, Sanderson A, Tyson PD, Jager J, Matson PM et al (2004) Global change and the earth system: a planet under pressure. Springer, Heidelberg
Steffen W, Rockström J, Costanza R (2011) How defining planetary boundaries can transform our approach to growth. Solutions 2:59–65
Stern N (2007) The economics of climate change: the stern review. Cambridge Press, Cambridge
Stiglitz JE, Sen A, Fitoussi JP (2010) Mismeasuring our lives: why GDP doesn't add up. The New Press, New York
Street J (2001) Electronic democracy. Int Encycl Soc Behav Sci 7:43–97
Sukhdev P, Kumar P (2010) The Economics of Ecosystems and Biodiversity (TEEB). European Communities Brussels

Swann RS (1972) The community land trust: a guide to a new model for land tenure in America. Center for Community Economic Development, Cambridge, MA
Talberth J, Cobb C, Slattery N (2007) The genuine progress indicator 2006: a tool for sustainable develoment. Redefining Progress, Oakland
Third World Network (2012) Statement of the plurinational state of Bolivia at 12th special session of the governing council/global environment ministers forum of the UNEP. Third World Network, Nairobi
Turner GM (2008) A comparison of the limits to growth with 30 years of reality. Glob Environ Chang—Hum Policy Dimens 18:397–411
UNEP (2009) Global green new deal: a policy brief. United Nations Environment Programme
United Nations Development Programme (1998) Human development report. Oxford University Press, New York
United Nations Development Programme UNEP, World Bank, and World Resources Institute (2003) World resources 2002–2004: decisions for the Earth: Balance, voice, and power. Washington, DC
van den Berg AE, Maas J, Verheij RA, Groenewegen PP (2010) Green space as a buffer between stressful life events and health. Soc Sci Med 70:1203–1210
van den Bergh JCJM (2011) Environment versus growth — a criticism of "degrowth" and a plea for "a-growth". Ecol Econ 70:881–890
Vanloqueren G, Baret PV (2009) How agricultural research systems shape a technological regime that develops genetic engineering but locks out agroecological innovations. Res Policy 38:971–983
Victor PA (2008) Managing without growth: slower by design, not disaster. Edward Elgar Publishing, Cheltenham
Victor P (2010) Questioning economic growth. Nature 468:370–371
Victor PA, Rosenbluth G (2007) Managing without growth. Ecol Econ 61:492–504
Weiss H, Bradley RS (2001) What drives societal collapse? Science 291:609–610
Wilkinson RG, Pickett K (2009) The spirit level: why greater equality makes societies stronger. Bloomsbury Press, New York
Wilson EO (2003) The future of life. Vintage, New York
Worm B, Barbier EB, Beaumont N, Duffy JE, Folke C et al (2006) Impacts of biodiversity loss on ocean ecosystem services. Science 314:787–790

Index

S. Shmelev (ed.), *Green Economy Reader*, Studies in Ecological Economics 6,
DOI 10.1007/978-3-319-38919-6

D

E

O

P

R

MIX
Papier aus verantwortungsvollen Quellen
Paper from responsible sources
FSC® C105338

If you have any concerns about our products,
you can contact us on
ProductSafety@springernature.com

In case Publisher is established outside the EU,
the EU authorized representative is:
Springer Nature Customer Service Center GmbH
Europaplatz 3, 69115 Heidelberg, Germany

Printed by Libri Plureos GmbH
in Hamburg, Germany